IET TELECOMMUNICATIONS SERIES 58

Digital Communications

Other volumes in this series:

Volume 9 **Phase noise in signal sources** W.P. Robins
Volume 12 **Spread spectrum in communications** R. Skaug and J.F. Hjelmstad
Volume 13 **Advanced signal processing** D.J. Creasey (Editor)
Volume 19 **Telecommunications traffic, tariffs and costs** R.E. Farr
Volume 20 **An introduction to satellite communications** D.I. Dalgleish
Volume 26 **Common-channel signalling** R.J. Manterfield
Volume 28 **Very small aperture terminals (VSATs)** J.L. Everett (Editor)
Volume 29 **ATM: the broadband telecommunications solution** L.G. Cuthbert and J.C. Sapanel
Volume 31 **Data communications and networks, 3rd edition** R.L. Brewster (Editor)
Volume 32 **Analogue optical fibre communications** B. Wilson, Z. Ghassemlooy and I.Z. Darwazeh (Editors)
Volume 33 **Modern personal radio systems** R.C.V. Macario (Editor)
Volume 34 **Digital broadcasting** P. Dambacher
Volume 35 **Principles of performance engineering for telecommunication and information systems** M. Ghanbari, C.J. Hughes, M.C. Sinclair and J.P. Eade
Volume 36 **Telecommunication networks, 2nd edition** J.E. Flood (Editor)
Volume 37 **Optical communication receiver design** S.B. Alexander
Volume 38 **Satellite communication systems, 3rd edition** B.G. Evans (Editor)
Volume 40 **Spread spectrum in mobile communication** O. Berg, T. Berg, J.F. Hjelmstad, S. Haavik and R. Skaug
Volume 41 **World telecommunications economics** J.J. Wheatley
Volume 43 **Telecommunications signalling** R.J. Manterfield
Volume 44 **Digital signal filtering, analysis and restoration** J. Jan
Volume 45 **Radio spectrum management, 2nd edition** D.J. Withers
Volume 46 **Intelligent networks: principles and applications** J.R. Anderson
Volume 47 **Local access network technologies** P. France
Volume 48 **Telecommunications quality of service management** A.P. Oodan (Editor)
Volume 49 **Standard codecs: image compression to advanced video coding** M. Ghanbari
Volume 50 **Telecommunications regulation** J. Buckley
Volume 51 **Security for mobility** C. Mitchell (Editor)
Volume 52 **Understanding telecommunications networks** A. Valdar
Volume 53 **Video compression systems: from first principles to concatenated codecs** A. Bock
Volume 54 **Standard codecs: image compression to advanced video coding, 3rd edition** M. Ghanbari
Volume 905 **ISDN applications in education and training** R. Mason and P.D. Bacsich
Volume 59 **Dynamic ad hoc networks** H. Rashvand and H. Chao (Editors)

Digital Communications
Principles and systems

Ifiok Otung

The Institution of Engineering and Technology

Published by The Institution of Engineering and Technology, London, United Kingdom

The Institution of Engineering and Technology is registered as a Charity in England & Wales (no. 211014) and Scotland (no. SC038698).

First published 2014

The Institution of Engineering and Technology
Michael Faraday House
Six Hills Way, Stevenage
Herts, SG1 2AY, United Kingdom

www.theiet.org

British Library Cataloguing in Publication Data
A catalogue record for this product is available from the British Library

ISBN 978-1-84919-611-6 (hardback)
ISBN 978-1-84919-612-3 (PDF)

Typeset in India by MPS Limited
Printed in the UK by CPI Group (UK) Ltd, Croydon

Go back in time until you can go no further,
For trillions of years if you have to.
What do you see at this beginning?
You see a particle, or you see God.
Neither is more scientific, neither more religious.
Both involve a leap of faith beyond reason.

This book is dedicated to God, Who
In the beginning created the heavens and the earth.

Contents

Preface

Complexity is always a combination of simplicity.
The art of solving a complex problem is to break it into its simple parts.

No efforts were spared in pursuing the primary aim of this book, which is to break the complex concepts and principles of Digital Communications into their simple parts. The approach is first to use such means as lucid graphs and diagrams, intuitive reasoning, analogies, computer simulations, etc. to give the reader an unclouded insight into the engineering concept and underlying physical considerations, and a clear appreciation of the parameters involved as well as a good feel for the interplay of these parameters. Only then is a carefully measured mathematical discourse brought in to better encapsulate the concepts and to consolidate and extend learning. It is hoped that the multiple and often fresh perspectives so imparted will help students and practising engineers gain a sound mastery of this important subject.

A worldwide digital and wireless communication revolution has taken place in the last 20 years, which has created a high demand in industry for graduates with in-depth expertise in digital transmission techniques and a sound and complete understanding of their core principles. This book aims to help fill this need by giving advanced undergraduate, postgraduate and research students as well as practising engineers a thorough grounding in the core principles and design issues of digital communications. The book recognises that although Digital Communications is developing at a fast pace, the core principles remain the same. It therefore concentrates on giving the reader a thorough understanding of core principles and extensive coaching in the solution of practical problems drawn from various application areas. The intention is that after studying the material presented, the student will have a solid foundation free of knowledge gaps, and will be fully equipped to undertake digital communication systems analysis, design and computer simulations, and to deal with specialised applications and follow advances in the technology.

Chapter 1 is aimed at stimulating the reader's interest, easing them into the subject and erecting crucial knowledge pegs on which a more detailed discussion is hung in subsequent chapters. This is done through a non-mathematical *overview of digital communication*, covering building blocks, signal processing tasks, general features and design considerations. This sets the stage for the rest of the book, which treats a range of topics in a carefully structured manner aimed at giving the reader a full comprehension of complex concepts, fresh insights into underlying physical considerations and a complete understanding of the parameters involved and the interplay of these parameters.

Chapter 2 on *linear channels and systems* covers the concepts of impulse response and transfer function of linear time invariant systems and uses both graphical and mathematical approaches in dealing with the convolution operation.

A careful classification of systems is given and the ideas of causality, memory, stability, memorylessness, delayed response and instantaneous response are fully clarified. Signal distortions and multipath propagation are also treated and signal strength calculation in free space and mobile radio channels is presented in some detail.

Chapter 3 discusses the characterisation, analysis and simulation of *non-linear systems*. The concept of quadrature and in-phase components of a band pass signal is introduced for application in determining the response of a non-linear system to a narrow band input signal. Intermodulation distortion analysis is treated in some detail, leading to the calculation of carrier-to-interference ratios in non-linear amplifiers.

Chapter 4 on *sampling of base band and band pass signals* presents a detailed treatment of the subject. Alias distortion and aperture effects are quantified, and the design of anti-alias filters is discussed. The common pitfalls and confusion associated with band pass signal sampling are cleared up completely through a careful treatment of uniform and quadrature sampling approaches.

Chapter 5 deals with *quantisation and PCM*, giving an in-depth study of scalar quantisation, logarithmic quantiser design, A-law and μ-law PCM and lossy data compression through differential quantisation. Performance measures, design parameters, trade-offs and drawbacks and detailed analysis of signal-to-quantisation noise ratio (SQNR) are presented.

Chapter 6 is devoted to *source coding and lossless data compression*. Source entropy and measure of information content are discussed, followed by a detailed treatment and comparison of various entropy coding methods, including Huffman, Lempel-Ziv and Arithmetic coding.

Chapter 7 covers *line codes* in detail, including a detailed classification as well as autocorrelation and spectral analysis, and the design of basic and advanced line codes.

Chapter 8 on *transmission through band limited AWGN channels* deals with inter-symbol interference (ISI) and bandwidth constraints on symbol rate, and with noise constraint on bit rate. Information theory is delved into for the evaluation of channel capacity and the specification of conditions for reliable transmission through noisy channels. Shannon's theorems are introduced and the full implications of the Shannon-Hartley information capacity law are explored. The design of filters that eliminate ISI (including Nyquist, raised cosine and duobinary signalling) is treated in detail, as is the design of a matched filter that optimises symbol detection in the presence of noise.

Chapter 9 covers *transmitted digital signals*, including a detailed discussion of signal orthogonality and the geometric and complex representation of signals, digital modulation, and coherent detection and correlation receivers.

Chapter 10 on *noise impact in digital transmission* covers the characterisation of random signals and the quantification of noise in communication systems as well as detailed bit error rate analysis of various binary and M-ary transmission systems.

Chapter 11 on *error control coding* presents a detailed discussion of the general principles, applications and limitations of error detection and correction and the manner in which they facilitate an efficient trade-off between signal power and bandwidth. ARQ techniques are discussed, including link utilisation analysis. Linear block codes are also presented using graphical, matrix, polynomial and shift register circuit implementation approaches.

Chapter 12 briefly outlines the steps involved in the analysis and design of communications links, and provides link power budget templates for use in satellite communication downlink and uplink design.

Beyond Chapter 1 it is assumed that the reader has a good background in calculus, although in the early part of the book a lot of explanation of mathematical manipulations is provided wherever deemed necessary. Competence in Fourier transform methods is also assumed, but Appendix C gives a comprehensive summary of the relations and properties used in the book. Bit error rate analysis employs the complementary error function, an extensive listing of values of which is given in Appendix D. Tables of character codes (Morse, EBCDIC, ASCII, etc.) are included in Appendix A and a list of useful trigonometric identities is provided in Appendix B.

Acknowledgements

This book grew out of my lectures to university students and practising engineers over the years. In preparing those lectures I benefited immensely from the work of the MATLAB software designers as well as authors such as Bernard Sklar, Simon Haykin, Mischa Schwartz, Timothy Pratt, Barry Evans, Zhili Sun, Fuqin Xiong, Paul Young, William Stallings, Simon Saunders and others too numerous to mention. To these pillars of the subject I extend my most sincere thanks. They may recognise their influence in some of the pages that follow. However, this book brings a unique approach and fresh insight into the subject, motivated by a passion for simplicity and shaped by years of collaboration with industry and training of engineers and students from around the world.

I have been privileged to work with wonderful colleagues and to teach, supervise and learn from an international and multicultural blend of undergraduate, postgraduate and research students over many years. Because of them I have sought to understand it more completely and express it more simply. They have helped me become a better person and teacher in the process and I owe them all a depth of gratitude. My heartfelt appreciation goes particularly to Dr Sahena Begum, Mr Leshan Uggalla, Mr Abdulkareem Karasuwa, Mr Kufre Ekerete, Professor Khalid Al-Begain, Mrs Jana Bauer and all members of CEMAS and the Mobile and Satellite Communications Group at USW for their generous support and contribution in various ways to our joint efforts in research and scholarship.

This book could not have materialised without Nicki Dennis the IET Commissioning Editor. I thank her for her faith in the project and her professional support and patience during the many months it took to juggle full university responsibilities and writing.

I am forever grateful to my family for their unstinting and selfless support over the years and in particular during the months of August and September 2013 when I devoted every waking moment to the book's completion. My wife Buchi helped me out every now and again when I was stuck for just the right word to express an idea and was a constant source of encouragement. Our children Ifiok Junior, Andikan, Yama, Kima and Sara were all very thoughtful fans of the project and supported me at every step of the way. So to my family, thanks for the sacrifice, thanks for believing and thanks for everything.

As I write these words I can see it through the door of my study swinging in the air, dangling from the palatial home it built. For the past six weeks I have watched it grow, inspired by its sense of mission and full of admiration of its hunting skills. The unnamed spider that spun a web in my garden and worked daily on its edifice while I wrote this book, that spider, part of a wonderful world created by an Indescribable God, gets a mention too.

Ifiok Otung
South Wales, United Kingdom
September 2013

Chapter 1

Overview of digital communication

In Chapter One

- ✓ Building blocks of a digital communication system: A clear discussion of the major components of a communication system and their functions, as well as an introduction to different classes of digital communication systems.
- ✓ Signal processing: A lucid and non-mathematical explanation of the signal processing tasks performed in a digital communication system.
- ✓ Features of digital communications and why it is preferred to analogue communications in most modern applications.

1.1 Introduction

Digital communication is basically concerned with the delivery or transmission of a sequence of numbers from one point or node to another. The sequence of numbers usually represents textual, audio, graphical or video information, or a combination of these, and the source and destination points may be separated by distances ranging from less than a metre to many thousands of kilometres. Applications of digital communication are now both ubiquitous and indispensable in society. The exchange of text messages between two people, the instant broadcast of a tweet to a group of followers, the remote control of a device nearby in a room or far away in outer space, the collection and distribution of physical data by various types of sensor devices, a mobile or fixed telephone conversation, Internet access, TV reception; these are just a few of the countless facets and applications of digital communication today.

This chapter presents an overview of digital communications and lays a *crucial foundation* on which we will build further knowledge of the subject in later chapters. It deals with the building blocks, signal processing and features of digital communications, and is a non-mathematical introduction to a modern digital communication system. Design considerations are emphasised, including performance objectives and system resource utilisation.

We discuss the building blocks of a communication system and the signal processing tasks usually carried out at the sending and receiving ends of a digital communication system. We also justify the digital revolution that has taken place in telecommunications by outlining the advantages and disadvantages of modern digital communication compared to analogue communication.

In this chapter we will set the stage for the rest of the book. You are encouraged to study it carefully in order to gain a good appreciation of the overall picture of digital communications and to become familiar with the terminologies and concepts that will receive a more detailed treatment in subsequent chapters.

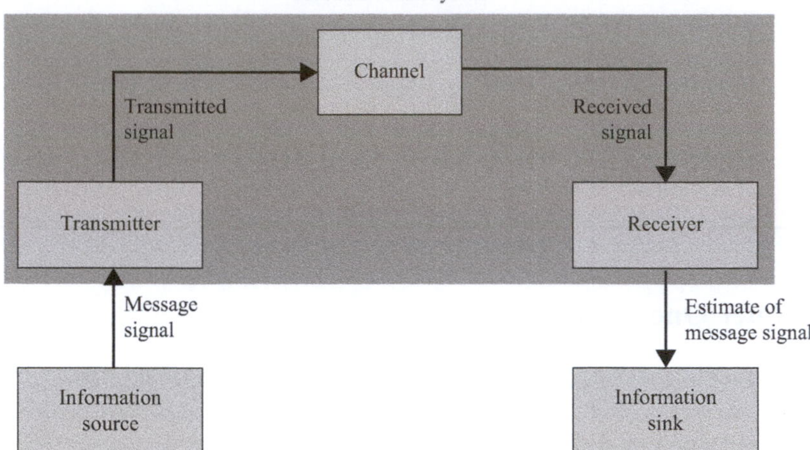

Fig. 1.1 Generic block diagram of a communication system

1.2 Building blocks of a digital communication system

Fig. 1.1 shows a generic block diagram of a communication system. There are three main building blocks, namely a *channel* that provides a connection or transmission path between a *transmitter* at the sending end and a *receiver* at the receiving end or destination. The channel may include one or more transmission media, namely metallic line, radio and optical fibre. The transmitter and receiver are physically separated by distances ranging from a few metres (e.g. the remote control of a TV set) to thousands of kilometres (e.g. a telephone call from Lagos Nigeria to London UK via satellite). The input to the communication system is a *message signal* provided by an *information source* at the sending end, whereas the output is an *estimate* of the message signal delivered to an *information sink* at the destination.

1.2.1 Transmitter and receiver

The transmitter transforms the message signal into a *transmitted signal*, which is placed into the channel. It is designed to

Produce an output signal — the transmitted signal — that is in a form suitable for passage through the particular channel with (i) minimum *distortion* and *attenuation* (i.e. reduction in strength), and (ii) minimum interference to other users of adjacent channels; and

Achieve the above while making the smallest possible demand on system resources such as *bandwidth* and *power*.

In transforming the message signal the transmitter performs a series of *signal processing* tasks that we may view simply as a packaging process. Except where *lossy compression* is involved, the transformation is reversible, meaning that we can obtain the original message signal by 'unpacking' the transmitted signal.

The *received signal* at the output of the channel serves as input to the receiver and is in general a reduced and distorted version of the transmitted signal. The main causes of the distortion are occasional temporary unavailability of the channel (e.g. due to overloading, path blockage, clipping, mistiming, etc.), and the addition to the transmitted signal of (i) *noise* in the channel and front end of the receiver, (ii) other unwanted (i.e. *interference*) signals and (iii) delayed versions of the transmitted signal due to the existence of multiple transmission paths (called *multipath*) in the channel. Reduction in signal strength arises in the transmission medium as part of the transmitted signal is *absorbed* (e.g. heat dissipation in metallic lines) and *diverted* away from the receiver in question due to *scattering* and (in the case of a radio wave front) *spreading*.

The receiver has the task of extracting an estimate of the original message signal from the received signal and delivering this to a user or information sink. In the absence of the *transmission impairments* identified above, the transmitted and received signals would be identical and the receiver would have a straightforward un-packaging task in which it simply undoes in reverse order each operation performed at the transmitter. However, in practice the recovery of the original message signal from a weak and highly impaired received signal is a very challenging task that continues to exercise the minds of communication system researchers. For now it will be sufficient to state that the receiver is designed to

Match the characteristics of the channel,
Provide maximum distinction between contributions from the wanted signal and those from sources of distortion, and
Undo each of the signal processing functions of the transmitter.

1.2.2 Analogue and digital communication systems

The primary distinction between analogue and digital communication systems is in the type of message signal: an *analogue signal* in the former and a *digital signal* in the latter. This has ramifications on the signal processing carried out at the transmitter (and receiver), which is different for both systems and depends on the type of channel involved.

In particular, the transmitted signal in a digital communication system consists of a set of distinct short-duration waveforms (called *pulses* or *symbols*) one of which is transmitted in each interval. A binary digital communication system employs two symbols. Fig. 1.2 shows one simple example. In general, M symbols are used in an *M-ary* digital communication system, with each symbol representing $\log_2 M$ bits. For example, each symbol in an 8-ary system such as 8-PSK conveys 3 bits of information. The receiver does not need to reproduce the transmitted waveform, but merely to determine which symbol was transmitted during each interval and then to *regenerate* the digital signal or bits associated with that symbol. The receiver's crucial task is therefore that of *symbol detection*. And so long as there is no misidentification of symbols at the receiver, the (digital) message signal delivered by a digital communication system to an information sink will be an *exact* copy of the original (digital) message signal received from the information source. One of the great attractions of digital communications is that with advances in semiconductor technology it is possible to implement highly sophisticated signal processing, which greatly enhances the chances of error-free transmission while minimising the amount of system resources consumed.

The transmitted signal in an analogue communication system is on the other hand a continuous waveform lasting the duration of the message signal. A parameter of this waveform is continuously varied in synchrony with the variations of

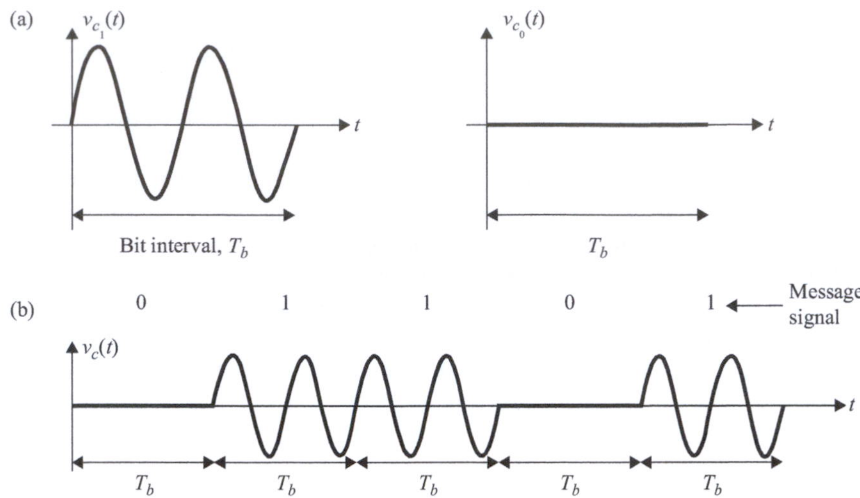

Fig. 1.2 Transmitted signal in a binary amplitude shift keying (ASK) digital communication system: (a) symbols $v_{c_1}(t)$ and $v_{c_0}(t)$ that represent bits 1 *and* 0*, respectively; (b) transmitted waveform $v_c(t)$ for a specified bit stream (the message signal)*

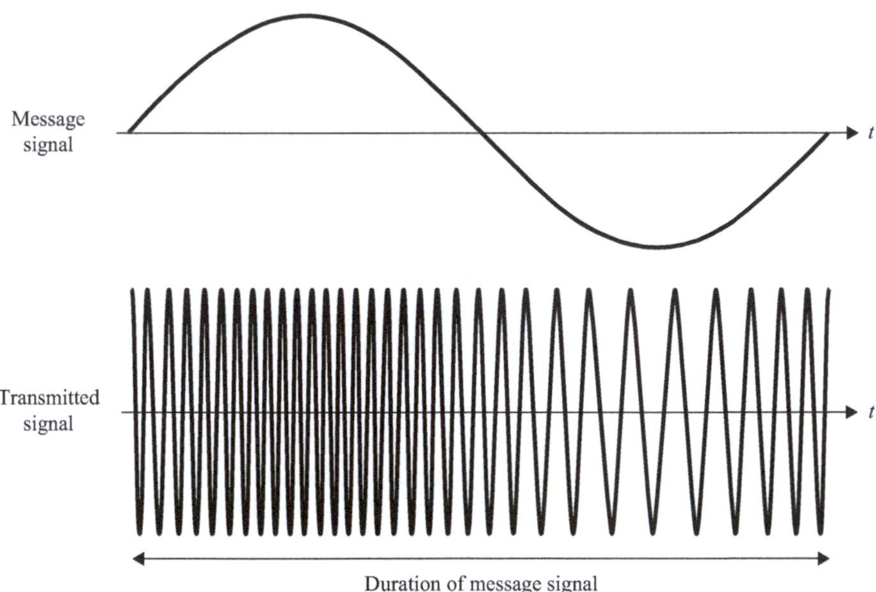

Fig. 1.3 Example of a transmitted signal in an analogue communication system. The transmitted signal is a frequency modulated carrier

the analogue message signal — a process known as *modulation*. An example is shown in Fig. 1.3 where the varied parameter is the rate of oscillation of the transmitted waveform or *carrier*. The task of the receiver is to somehow track variations of this parameter in the received waveform and hence to reconstruct an estimate of the original message signal — a process known as *demodulation*. But contributions by noise and other distortions to variations in this parameter cannot be fully distinguished and separated from the legitimate variations caused by the original message signal. The reconstructed signal is therefore always at best only a *close copy* of the original message signal.

One of the measures of quality of service (QoS) in a digital communication system is the bit error rate (BER) at the input of the information sink, and in an analogue system the signal-to-noise ratio (SNR) at the same point. These quantities are defined as follows:

$$\begin{aligned} \text{BER} &= \frac{\text{Number of bits in error}}{\text{Total number of bits}} \\ \text{SNR} &= \frac{\text{Average signal power}}{\text{Average noise power}} = \frac{P_s}{P_n} \\ &= [(P_s \text{ in dBm}) - (P_n \text{ in dBm})] \text{ dB} \end{aligned} \tag{1.1}$$

The design target is usually a BER of 10^{-4} for voice and 10^{-7} for data traffic in digital communication systems, and an SNR of 30 dB for voice and 45 dB for video in analogue systems.

If the lowest positive frequency component of the transmitted signal is much larger than zero we have a *band pass* or *modulated* communication system, whereas if it is equal or very close to zero (i.e. DC) we have a *base band* communication system. In the former, the message signal is applied to modulate a suitable high frequency sinusoidal waveform or *carrier* at the transmitter, whereas in the latter the transmitted signal is a base band waveform. It is worth emphasising that when the message signal is digital, the modulation process yields step changes in the state of the sinusoidal carrier resulting in the transmission of *sinusoidal* or band pass pulses or symbols as illustrated in Fig. 1.2. Thus modulation in digital communication simplifies to a process of *symbol generation*. In the case of digital base band systems, this symbol generation process is called *line coding* in which suitable rectangular voltage pulses are selected to represent the bit stream.

These sinusoidal pulses (in band pass systems) and rectangular pulses (in base band systems) are usually passed through an appropriate filter for *pulse shaping* before transmission in order to curtail their frequency content. This reduces the amount of spreading experienced by the pulses when they pass through a bandwidth-limited channel, and hence the incidence of *inter-symbol interference* (ISI). Pulse spreading is a type of distortion in which pulses arriving at the detection point of a receiver have been stretched beyond their intervals and hence contribute some energy to pulses in adjacent intervals. Clearly, base band communication systems can only be implemented using metallic line transmission media such as wire pairs and coax, which will pass the low frequency components of the transmitted waveform. The use of optical fibre and radio requires modulated transmission.

To round up our comments on analogue and digital communication systems, we consider Fig. 1.4 that gives an illustration of *analogue, sampled, quantised* and *digital* signals:

An analogue signal is a *continuous-value* and *continuous-time* physical quantity. This means that an analogue signal $g(t)$ is defined at each of the infinite number (or *continuum*) of time instants within its duration ($t = 0 \rightarrow 8.5$ seconds in this example); and can take on any of the *continuum* of values within its range ($g = 1.5$ V$\rightarrow$8.0 V in this case). Most naturally occurring signals are analogue. One example is the acoustic pressure variation generated by a

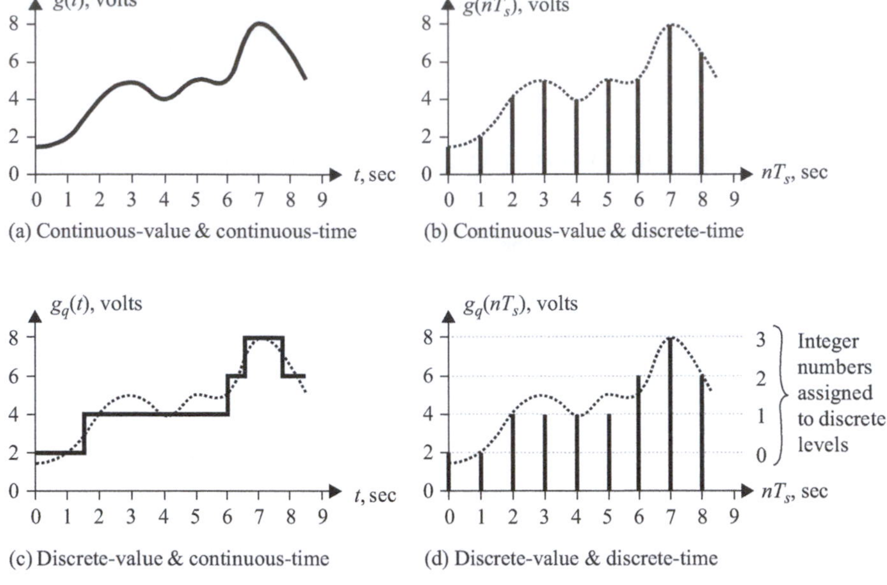

Fig. 1.4 (a) Analogue; (b) sampled; (c) quantised; and (d) digital signals

source of sound in a compressible medium (e.g. air), which a microphone converts into an analogue *audio* signal. Another is the variation in intensity and colour of light reflected by a scene, which a video camera converts into an analogue *video* signal. Ambient temperature and atmospheric pressure are also analogue quantities.

The sampled signal $g(nT_s)$ is a *continuous-value* and *discrete-time* signal obtained by taking samples of $g(t)$ at intervals of T_s. It can take on any of the *continuum of values* in the range of $g(t)$, but is defined only at *discrete time instants* $t = nT_s$, where $n = 0, 1, 2, 3, \ldots$ Thus the sampled signal $g(nT_s)$ is simply a sequence of numbers $g(0)$, $g(T_s)$, $g(2T_s)$, $g(3T_s), \ldots$, where T_s is called the *sampling period* or *sampling interval*, and its reciprocal $f_s = 1/T_s$ is the *sampling frequency*. It may come as a surprise, but if we transmit only this sequence of numbers then the original signal $g(t)$ can be reconstructed from $g(nT_s)$ at the destination, provided the chosen sampling interval T_s was small enough to capture the variations of $g(t)$. We will later have more to say on this matter, a fundamental concept in digital communications, but a little thought will show for example that if $g(t)$ is constant then only one sample is sufficient for reconstructing $g(t)$; and if $g(t)$ varies linearly with time then two samples are adequate. In general however $g(t)$ is neither constant nor linear and the rule for correct sampling (known as the *sampling theorem* — which we will derive in Chapter 4) is that f_s should be at least twice the bandwidth of $g(t)$. The benefit of transmitting $g(nT_s)$ rather than $g(t)$ is that the communication system resource can be allocated to other user signals during the unused intervals, thus allowing simultaneous accommodation of multiple users on the same transmission link — a process known as *multiplexing*, which increases communication system capacity and reduces service cost to each user.

The quantised signal $g_q(t)$ is a *discrete-value* and *continuous-time* signal obtained by approximating $g(t)$ at every time instant to the nearest of a set of

N allowed discrete levels in the range of $g(t)$. Note that sampling is not involved; rather $g_q(t)$ is defined at every point in the continuum of time instants within the duration of $g(t)$ but it only has discrete values in the range of $g(t)$. This is a quantisation process that introduces irreversible errors called *quantisation distortion*. That is, $g(t)$ cannot be reconstructed from $g_q(t)$ and a destination that receives $g_q(t)$ is stuck with a distorted signal and has absolutely no means of identifying and excluding the distortion. However, we can make the distortion as small as we wish by increasing N, the number of *quantisation levels*, thereby reducing the approximation error between $g_q(t)$ and $g(t)$ at each time instant t. In the limit $N\rightarrow\infty$, $g_q(t)$ and $g(t)$ become identical, and there is no distortion. But of course in this limit there is no quantisation. N is usually chosen as an integer power of 2 — i.e. $N=2^k$, where k is an integer — to allow k-bit binary words to be fully utilised to represent the quantisation levels numbered from 0 to $N-1$. In the example shown in Fig. 1.4c, $N=4$ and the discrete levels are numbered (see right of Fig. 1.4d) from 0 to 3 starting at the lowest level.

The digital signal $g_q(nT_s)$ is a *discrete-value* and *discrete-time* signal obtained in practice by quantising the sampled signal $g(nT_s)$, although sampling the quantised signal $g_q(t)$ would yield exactly the same result. Noting that the value of $g_q(nT_s)$ at each sampling instant corresponds to one of the quantisation levels that are numbered from 0 to $N-1$, we see that a digital signal is simply a sequence of *integer numbers*. In the example shown in Fig. 1.4d, the numbered quantisation levels assumed by $g_q(nT_s)$ are the sequence of integers $\{0, 0, 1, 1, 1, 1, 2, 3, 2\}$, and this fully represents the signal $g_q(nT_s)$ from which the original signal $g(t)$ can be reconstructed at a destination subject to the small distortion introduced by quantisation.

We emphasise therefore that the message signal of a digital communication system is always a sequence of integer numbers (usually expressed in binary) received from an information source in one location and delivered to an information sink in another location. Thus an information source converts information and intelligence into a bit stream, and a digital communication system delivers this bit stream to an information sink, which then has the role of interpreting the received bit stream and extracting information and intelligence from them for the benefit of an end user.

1.2.3 Information source

The information source acts as an interface between the communication system and the outside world and provides the message signal that is processed by the transmitter. Fig. 1.5 shows the four main classes of information source, namely data, audio and video input devices and sensors.

The data input device directly generates a digital message signal or data. There is a wide variety many of which work in conjunction with a computer system or other suitable data terminal equipment (DTE). Examples include the *keyboard* that generates character codes (e.g. ASCII and Unicode) in the form of integer numbers that represent textual information. The *mouse, trackball, joystick* and *touch screen* all respond to finger and hand movement and may generate data under the control of a computer program. Other devices read stored data and present them to the communication system. These include *barcode reader, laser scanner, magnetic ink character reader, magnetic strip reader, optical character reader, optical mark*

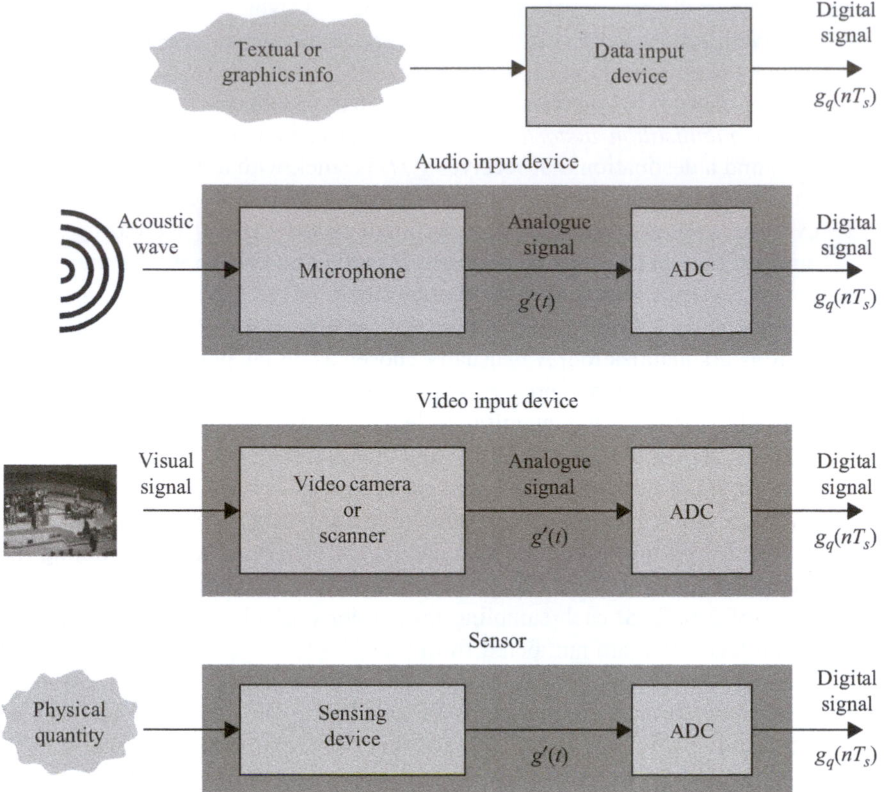

Fig. 1.5 Types of information source

reader, punched card reader, optical disc drive, magnetic disk drive, tape reader, etc. The *digitising tablet* converts graphics information in the form of sketches on the tablet into data.

Audio input devices include the music keyboard, musical instruments digital interface (MIDI), cassette deck, CD player and a host of other devices that read stored audio information and/or generate audio signals in the form of time-varying voltage waveforms. But Fig. 1.5 shows only the most basic audio input device, the microphone, which converts acoustic pressure variation generated by a source of sound into an analogue *audio* signal. This is then converted into a digital signal by an analogue-to-digital converter (ADC). *The most commonly transmitted audio signals are speech* (or *voice*) and *music*, which have different bandwidth requirements.

The most basic video input devices are the *video camera* for movable three-dimensional images and the *scanner* for still two-dimensional images. These convert a visual signal — comprising light intensity (or luminance) and colour (or chrominance) information — into an analogue video signal, which is then digitised in an ADC. Other video input devices include those that read pre-recorded video signals and/or generate time-varying video voltage waveforms.

Sensors measure physical quantities such as temperature, pressure, mass, and convert the measurement into an electrical signal that serves as the message signal input for the communication system. Sensors are used in telemetry systems to

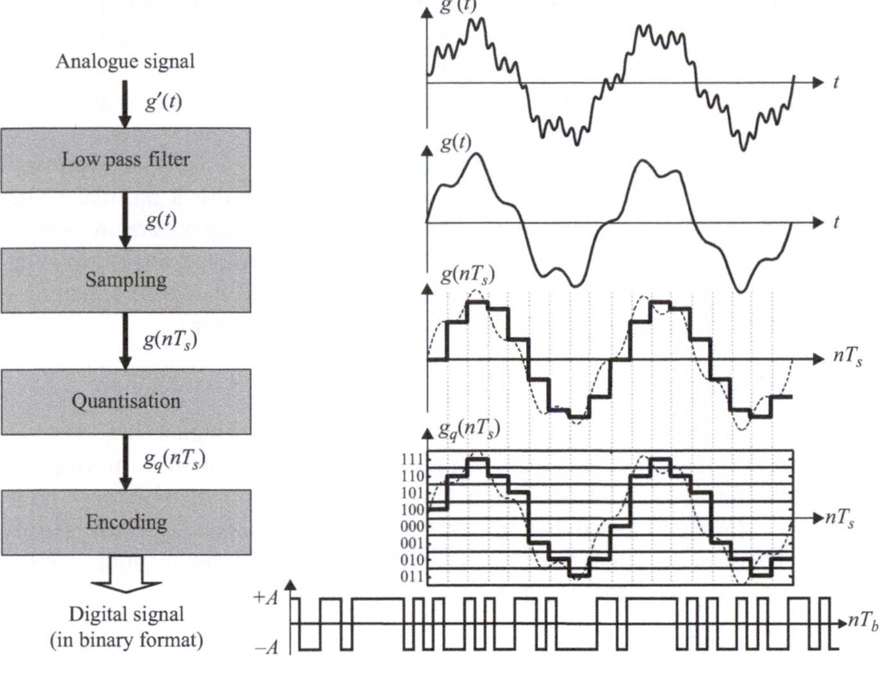

Fig. 1.6 ADC steps

obtain information from remote or inaccessible locations, e.g., monitoring the status of equipment on satellites, detecting the accumulation of ice on the wings of an aircraft. They are also used in numerous systems including automatic data-logging systems, security systems, safety systems, traffic control systems, manufacturing systems and process control systems. Fig. 1.5 shows an arrangement in which the sensing device produces an analogue output that is fed through an ADC to generate the required digital message signal.

The process of analogue-to-digital conversion (ADC) allows message signals such as voice and video that are originally in analogue form to be transmitted in a digital communication system. Fig. 1.6 shows the basic steps involved along with waveforms to illustrate the signal transformation of each stage:

The low pass filter (LPF) removes frequency components in the analogue signal above a certain value f_{max} whose contributions are considered insignificant to the acceptable fidelity of transmission. In speech telephony for example, $f_{max} = 3.4$ kHz. This step yields a band limited signal, which can then be correctly sampled at a rate that satisfies the sampling theorem. In practice, components above f_{max} cannot be completely removed but are sufficiently attenuated so that what remains of them is below the level of root-mean-square (rms) quantisation error (or noise).

The band limited signal $g(t)$ is sampled at a rate $f_s > 2f_{max}$, which (i) satisfies the sampling theorem and thereby avoids *alias* distortion, and (ii) allows a realisable LPF to be used at the destination to reconstruct $g(t)$.

The entire signal range is divided into N quantisation intervals, an equal number ($N/2$) of intervals for the positive and negative values. The number of intervals chosen is a compromise between a large value to reduce quantisation noise and a small value to minimise bandwidth utilisation, but usually $N=2^k$, where k is the number of bits per sample, and the system is referred to as a k-bit ADC. The illustration in Fig. 1.6 is for $N=8$ and a uniform quantisation scheme whereby the intervals are equally spaced.

The final step involves numbering the quantisation levels from 0 to $N/2-1$ for the positive range and similarly for the negative. These numbers are expressed in binary and an extra bit (MSB) is added to distinguish a positive value (MSB = 1) from a negative value (MSB = 0). The resulting bit stream — with bit duration $T_b = T_s/k$, shown in Fig. 1.6 as a bipolar voltage waveform — is the binary digital signal representation of the original analogue signal and is the message signal for our digital communication system.

1.2.4 Information sink

The information sink is the final destination of the transmitted signal. It may serve as an interface between human users and the communication system — making the transmitted information understandable through sight or sound, or it may be a repository — storing the transmitted information for later processing, re-transmission or output. There are therefore three types of information sink, namely audio output devices, visual display devices and storage devices.

The audio output device (Fig. 1.7a) consists of a digital-to-analogue converter (DAC) — discussed shortly — followed by a loudspeaker. The DAC output is converted by the loudspeaker to proportionate mechanical vibrations or acoustic waves at the same frequencies, which are perceived as sound.

Visual display devices include the cathode-ray tube (CRT) and flat-panel displays (used for example in computer screens, TV monitors, LED and liquid crystal displays) that present what is usually called a *soft copy* of the received message signal. The other type of visual display device is the printer that produces a printout or *hard copy* of the message. The display device (Fig. 1.7b) essentially converts an electrical voltage signal (the message signal) into a visual signal that is either painted on screen or on paper.

Many communication services are designed to store the received message signal for visual or audio display at a later time. In electronic mail (email) the text

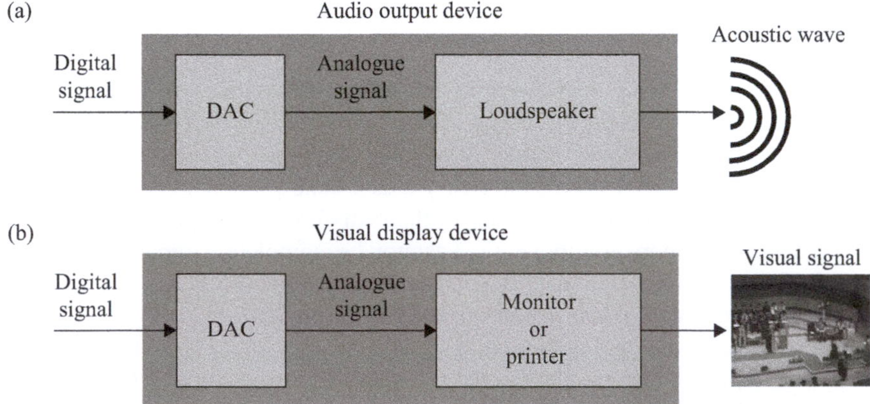

Fig. 1.7 Information sink: (a) audio output device converts electrical signal into sound; (b) visual display device converts electrical signal into a soft or hard visual copy

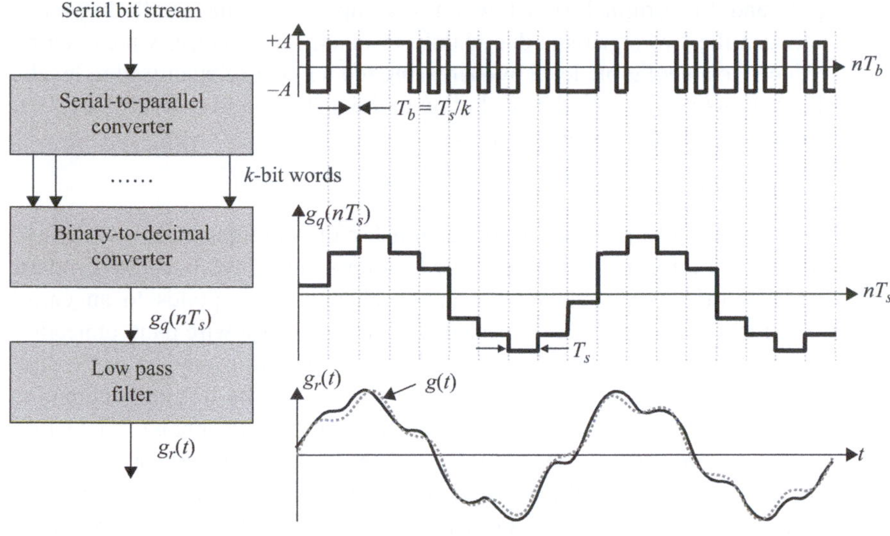

Fig. 1.8 DAC steps

message signal is stored at the recipient's email server machine or 'post office' until the recipient accesses the server to obtain a soft copy (screen display) or hard copy (printout) of the message. Voicemail service provides for the storage of digitised speech message signal, which is played back later to the authorised recipient at their request. In such cases, a suitable storage medium serves as the information sink. The desirable characteristics of a storage medium include large storage capacity, high access and data-reading speeds, portability, durability, low cost, data integrity and reusability. The four types of storage media most commonly used as information sinks in communication systems include magnetic tapes, magnetic discs, optical discs and flash memory.

Digital-to-analogue conversion (DAC) reverses the ADC process and restores a digital signal to its original analogue format. Fig. 1.8 shows the steps involved in the DAC process. Using the ADC output bit stream of Fig. 1.6 as input to the DAC, waveforms are included in Fig. 1.8 to illustrate the signal transformation that takes place at each step:

A serial-to-parallel converter, implemented using a shift register, converts the serial input bit stream into a parallel block of k bits, where k is the number of bits per sample used at the information source.

Each block of k bits is converted in a binary-to-decimal converter to a voltage pulse of duration equal to the sampling interval T_s at source, and amplitude equal to (or, more correctly, proportional to) the quantisation level represented by the code word. For example, assuming a normalised range ± 8 V, and recalling that the MSB is a sign bit and that $k = 3$, we see that the code word 111 is converted to a positive pulse of amplitude 7 V, 010 to a negative pulse of amplitude 5 V, 100 to a positive pulse of amplitude 1 V and so on. The result is a sequence of variable amplitude pulses – referred to as *pulse amplitude modulated* (PAM) signal, denoted $g_q(nT_s)$, which you may recognise is identical to the quantiser output in Fig. 1.6.

The PAM signal is passed through a suitable LPF, known as a *reconstruction filter*, which smooths it to yield the analogue signal $g_r(t)$. Fig. 1.8 shows both

$g_r(t)$ and the original signal $g(t)$ for comparison. The small difference between the two signals is due mainly to quantisation noise, which can be reduced to a negligible level by increasing the number of quantisation levels. There is also a very small contribution to the distortion from *aperture effect,* which we will discuss in Chapter 4.

1.2.5 Channel

The channel essentially provides a connection for the communication system to effect a transfer of information between two locations, say A and B. In the simplest case, where multiplexing is not involved, the channel corresponds to an entire communication link (i.e. physical transmission path) such as a wire pair connecting A to B. However, in most cases the channel corresponds to a portion of the link resource such as a frequency band or time slot. Note that the link may comprise several transmission media. For example, an international telephone connection from London to Lagos may involve a wire pair from the local subscriber to the local exchange, optical fibre links for connection of local exchange through switching centres to the international gateway exchange, and a radio link for the international connection via satellite.

Fig. 1.9 illustrates that the flow of information may be in only one direction from A to B, in which case the communication system is referred to as *simplex* (SX); or it may be in both directions on the *same* communication link, in which case the system is *duplex*. In a simplex system, information typically originates from one transmitter and has one or more receivers as its destination. Communication equipment at destination does not have the capability of responding through the *same* communication link.

Examples of simplex systems include audio broadcast (AM radio, FM radio, music services and digital radio), television broadcast (satellite, cable and terrestrial), paging services, telemetry, remote control units and radar. Note that interactive digital television broadcast is a simplex communication system since the customer responds via a telephone link. Information transmission from the

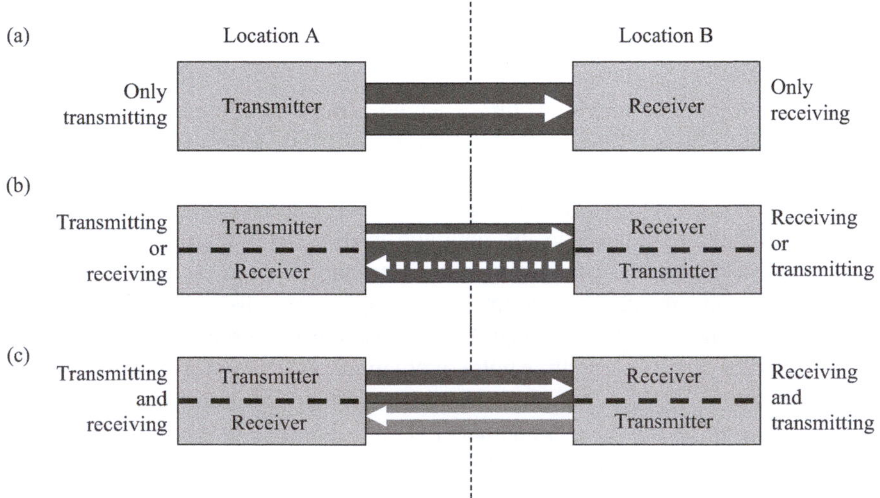

Fig. 1.9 (a) Simplex; (b) half duplex; and (c) full duplex systems

television service provider to the customer and from the customer to the provider is carried out on two different links. Also, *radar* — a system that measures electromagnetic signals reflected off objects (such as ship, aircraft, missile, storm, hydrometeor) in order to determine their range and velocity — is a simplex system. In this case, although the signal may be reflected by the object back towards the transmitter, giving an appearance of bi-directional information flow, there is strictly speaking only one transmitter that is co-located with one receiver. The reflecting object is neither a transmitter nor a receiver.

When information can flow in both directions on the same link, the communication system is referred to as *duplex*. Both locations A and B are equipped with *trans*mitter and re*ceiver* units — collectively referred to as a *transceiver*. The system may be designed in such a way that simultaneous communication in both directions is possible. This is a *full duplex* (FDX) system and requires a separate channel being allocated for each direction of communication. If on the other hand information can only flow in one direction at a time, then the system is referred to as *half duplex* (HDX). A full duplex system may be further classified as either *symmetric* in which the data rate capability is the same in either direction, or *asymmetric* in which case there is a higher data rate capability in one direction (usually the downlink from a so-called server to the user).

An example of a full duplex system is public telephony (both fixed and mobile). Internet access via asymmetric digital subscriber line (ADSL) in the public switched telephone network (PSTN) is also full duplex, as is computer interconnection in local area networks (LAN). An example of a half duplex system is the *walkie-talkie* used for wireless voice communication between two locations. Transmission in either direction uses the same radio frequency band. The handset at each location can be switched between transmit and receive modes, so that at any given time one location transmits while the other receives.

In both simplex and duplex systems, if communication takes place between only two transceivers, then the system may be further described as a *point-to-point* communication system. If there is one transmitter or transceiver communicating with several receivers or transceivers, we have a *point-to-multipoint* system. If there are many intercommunicating transceivers (as in a LAN, or a video conference system linking more than two locations), then we have what is called a *multipoint-to-multipoint* communication system. In this last case, information flow between two transceivers or *nodes* is essentially bi-directional and therefore a simplex multipoint-to-multipoint system is not possible. The Internet is a multipoint-to-multipoint communication system whereas a radio or television broadcast system is a good example of a point-to-multipoint simplex system. Fig. 1.10 shows a few illustrative examples.

The connection may be *direct* or *switched*. Direct linking involves the existence of a permanent link between a pair of nodes. For point-to-point communication this is the natural choice. Direct linking is also the preferred option for point-to-multipoint radio systems by virtue of the natural tendency of the radio wave front to spread over a wide area known as the *coverage zone* thereby providing a permanent link to all nodes in this zone. But this method is prohibitively both expensive and cumbersome for line or non-broadcast radio connections between multiple nodes. Fig. 1.11 illustrates how the required number of links increases rapidly with the number of nodes: 1 link for two nodes; 3 links for 3 nodes; 6 links for 4 nodes; 10 links for 5 nodes; and in general $n(n-1)/2$ links for

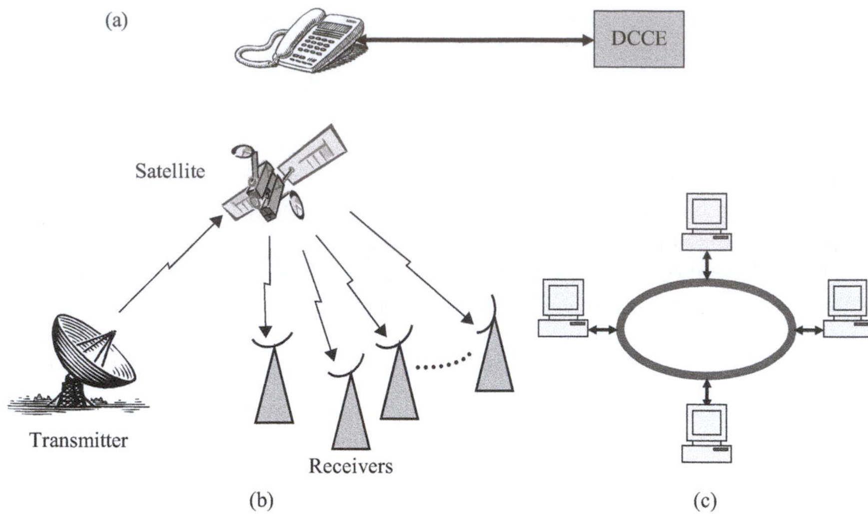

Fig. 1.10 (a) Point-to-point link between a telephone handset and a digital cell centre exchange (DCCE); (b) point-to-multipoint broadcast satellite service; (c) multipoint-to-multipoint token ring LAN

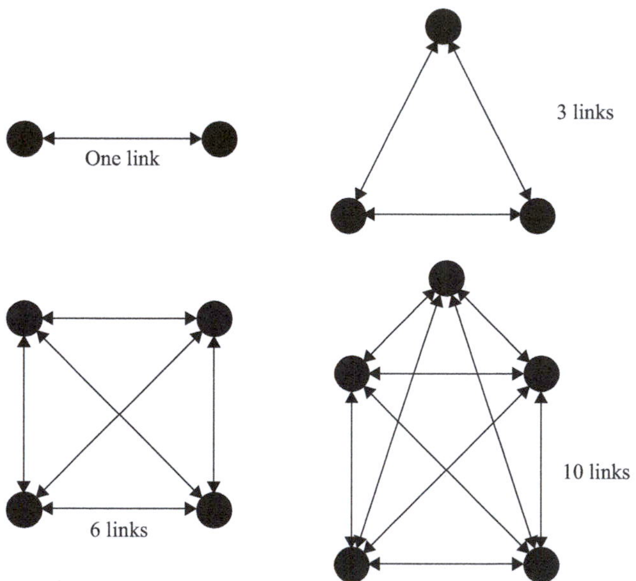

Fig. 1.11 Direct linking of nodes

n nodes. Clearly direct linking cannot be used to build even a modest-sized telephone network in which every handset is connected to every other handset.

Some form of switching is always required in a communication network whereby a non-permanent connection is created on demand between any pair of nodes. Switched connections may be achieved through *circuit switching* or *packet switching*. In circuit switching a channel is found during *call setup* and is set aside exclusively for the transmission between a pair of nodes. A channel thus dedicated is not used during idle intervals in the transmission (or *call*) and is released for other use only when the call is over. Packet switching allows the nodes on the network to share the channels and indeed the entire communication system resource in a dynamic manner. Messages are broken into properly labelled segments called packets before transmission and are reassembled at the destination. Packets are transmitted using non-dedicated channels that remain available for use by other nodes during any idle moments. Circuit switching tends to be used for voice transmission (as in the telephone network) whereas packet switching is well suited to computer data transmission, which is usually of a *bursty* nature.

Over short time intervals, the transmission characteristic of the channel may be *stable* (i.e. *time-invariant*) — as in optical fibre, or *time varying* — as in a mobile radio link. A *distortionless* channel is highly desirable in which the received signal is a *scaled* and *propagation-delayed* but otherwise exact copy of the transmitted signal. In practice however the channel causes significant *signal distortion*. If the channel is both *linear* and time-invariant then it is fully characterised by its transfer function $H(f)$ or impulse response $h(t)$. In the frequency domain the transmitted signal (or channel input) consists of a band of sinusoidal signal components, and $H(f)$ describes how the channel changes the amplitude and shifts the phase of each sinusoid. Importantly, the channel output (i.e. received signal) consists only of sinusoids contained in the input. In the time domain the input signal can be treated as a train of impulses (i.e. samples of infinitesimally small duration), and $h(t)$ specifies the output corresponding to each impulse input. The situation is more complicated in a *non-linear* channel such as one that involves non-linear amplification at some point along the link. Here, in addition to a modification of the amplitudes and phases of the input sinusoidal components, the channel also introduces entirely new components at integer multiples of the input frequencies — known as *harmonic products*, and at sums and differences of the harmonic frequencies — known as *intermodulation products*.

The main task for the transmitter and receiver is to protect the information conveyed by the communication system against the effects of channel distortions, and to do this with a minimum expense of transmission bandwidth and signal power. This feat is more easily achieved in digital communications where a range of sophisticated signal processing can be implemented to ensure *error-rare* transmission in the face of severe channel distortions.

1.3 Signal processing tasks

A digital communication system performs a sequence of signal processing tasks, the selection of which depends on the type of system and communication service. Figs. 1.12 and 1.13 show typical block diagrams of digital base band and modulated communication systems, respectively. The tasks featured in these diagrams are discussed below. Additional functions are often performed in addition to those shown. For example, *encryption* may be carried out whereby the transmitter processes the message — in this context referred to as *plaintext* — and converts it into a *cipher text,* which is disguised in some way to ensure security. The process may be accomplished by scrambling the bits of the plaintext using an *encryption key*. Only authorised receivers will have the correct *decryption key* with which to *decipher* or *decrypt* the cipher text back to plaintext.

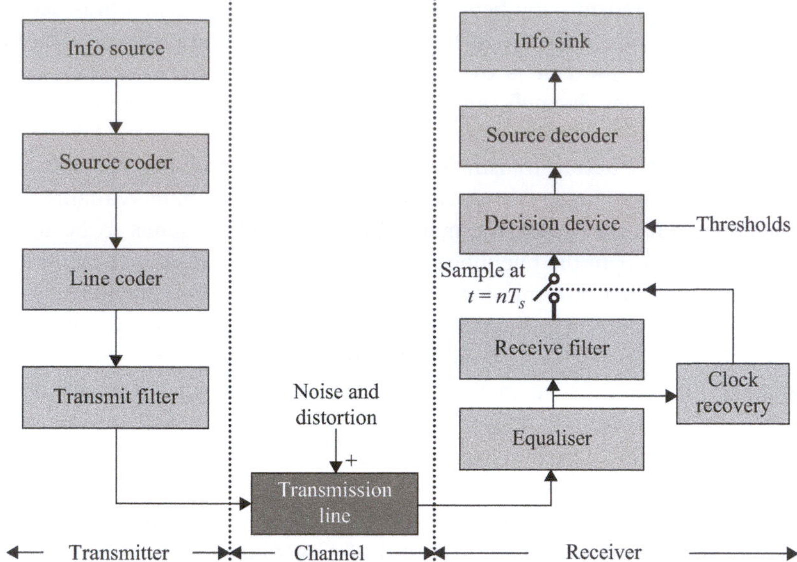

Fig. 1.12 Digital base band communication system

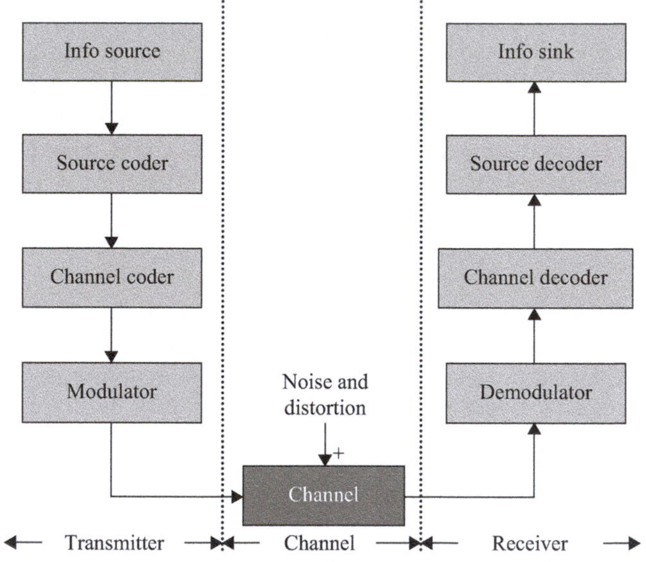

Fig. 1.13 Digital modulated communication system

If the modulated system (Fig. 1.13) is a wireless communication system, then the modulator output must be converted from electrical voltage form into an electromagnetic waveform, which is radiated in the desired direction towards a receiver. The receiver performs a reverse transformation from electromagnetic waves back to electrical voltage. These signal transformations are performed using suitable *transducers* as illustrated in Fig. 1.14. In general the transducers used in radio systems are *antennas* of various designs, whereas infrared systems employ

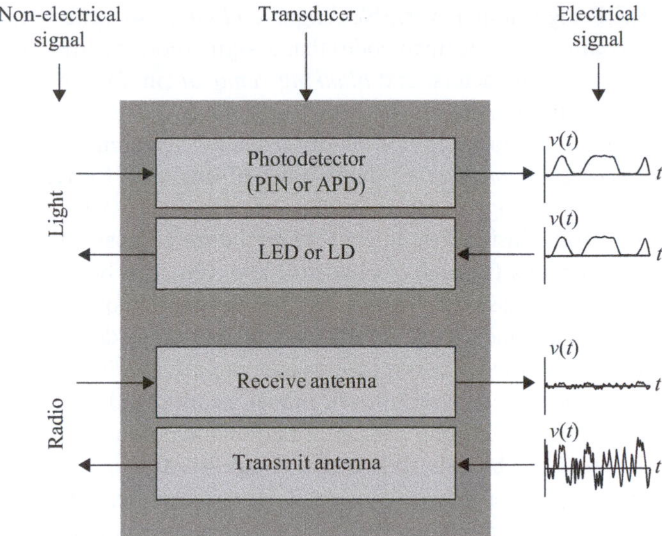

Fig. 1.14 Signal conversions in wireless systems

light emitting diodes (LED) or *laser diodes* (LD) at the transmitter and *photo-detectors* at the receiver. Note that a single antenna usually fulfils the transmit and receive roles.

1.3.1 Source coding

Source coding (or source encoding) deals with the efficient representation of the message signal at the sending end of a communication system. A reverse process of *source decoding* is performed at the receiving end. A number of processes may be involved depending on the type of communication system, but in all cases the output of the source coder is a bit stream. The following are examples of source coding and decoding:

Message formatting is carried out in an information source to present the message in digital format. This includes (i) character coding of text and graphics information using ASCII, Unicode or other suitable character code; and (ii) digitisation of an analogue message signal such as voice and video in an ADC — a process referred to as *pulse code modulation* (PCM) back in the 1930s even though no carrier is involved nor is there any frequency translation from base band as the term modulation would suggest. PCM is extensively used for voice transmission and Chapter 5 deals with the subject. The reverse processes of character decoding and digital-to-analogue conversion (DAC) are performed at the receiving end to convert the recovered message bit stream into user-consumable information.

Data compaction minimises the number of bits used for representing the (digital) message signal *without* incurring any loss of information or detail. This is a lossless data compression process that works by identifying repeating integers or patterns (groups of integers) within the message signal and representing them using fewer integers. It is used mostly on text and on pixel values of display screens to achieve a compression ratio typically <2. Examples of data compaction include *run-length encoding, pattern matching* (e.g. Lempel-Ziv and PKZIP algorithms), *variable-length coding* (also called *statistical* or *entropy coding*, e.g. Huffman code) that assigns shorter code words to more frequent source characters, and *blanking removal* (in TV application only). Barring transmission errors, the receiver *expands* the recovered compacted data to obtain an exact copy of the original message signal.

Lossy data compression reduces the number of bits used for representing the (digital) message signal using a range of techniques, which includes the removal of some insignificant details from the message signal. Much higher compression ratios (typically between 2 and 100) can be achieved than is possible with lossless compression, but the method cannot be used on textual message signals since these contain no insignificant details that may be discarded or modified. Lossy compression exploits the features of the message signal and the characteristics of human hearing and vision to eliminate redundant as well as insignificant information and produce a modified signal of greatly reduced bit rate. When this compressed signal is decompressed (i.e. expanded) or otherwise processed at the receiver, a human observer finds it acceptably similar to the original. Various methods of performing lossy data compression have been devised. These include *predictive coding* (e.g. differential pulse code modulation and delta modulation), a wide range of *low bit rate speech coding* techniques, *transform coding, motion compensation* (applicable to video signals only), *subsampling* (taking fewer samples per second than is required for perfect reproduction — the signal being first filtered to remove high frequency components), *colour table* (applicable in display screens) and *truncation* (using fewer bits per sample).

Multiplexing, in particular TDM, is often a part of source coding. Several message signals are combined for simultaneous transmission on one link as further discussed below.

1.3.2 Multiplexing

The process of multiplexing combines several independent message signals for simultaneous transmission on a single communication link. This is not to be confused with the related process of *multiple access* that deals with how one transceiver (e.g. a satellite transponder) is shared by various geographically dispersed transceivers (e.g. earth stations). There are a number of reasons why a communication system must be able to simultaneously handle signals from multiple and independent sources without mutual interference:

To satisfy the communication needs of a larger number of people. Modern lifestyle has become very dependent on telecommunication so that at any given time in an average city there will be a large number of people needing to make a phone call, send a text message, access the internet, hold a teleconference, etc. If the communication system could handle only one signal at a time, and each user occupied the system continuously for an average duration of three minutes, then only 480 users per day could be serviced, assuming inconvenient times (such as 2.00 a.m.) are not rejected. If such a communication system served a city of one million people, then at this rate it would take nearly 6 years for every person to have just one 3-minute access. Clearly, you couldn't rely on such a system to call an ambulance in a health emergency. By the time it reached your turn on the service queue, you would either have fully recovered or been dead and buried.

- To reduce the cost of the service to each user. This important consideration can be demonstrated by assuming a satellite communication system built exclusively for telephony at a total cost of £300 m, which includes design, construction, launching and maintenance costs over a projected satellite lifetime of 10 years. Allowing a 16% profit margin, the operator must earn (by charging users of the service) a total sum of £348 m during a period of 10 years or 5.22 million minutes. Excluding system idle time of 8 hours per day — you would not normally like to make or receive a phone call during sleeping hours — leaves us with 3.48 million income-yielding minutes over which to recover £348 m. It is easy to see that if the system could handle only one call at a time, then the charge for each call would have to be £100 per minute. However, if we can somehow design the system to handle only up to 24,000 simultaneous calls, then assuming on average 20,000 users every minute, the operator's required earning could be spread out over this number of users, bringing down the charge per user to a mere half pence per minute.
- To allow the coexistence of a multiplicity of telecommunication services in a given geographical area or city. Audio broadcast, television broadcast and mobile communication, to name but a few radio services, must operate simultaneously and independently without mutual interference.
- To improve the exploitation of the available bandwidth of a transmission medium. For example, if a coaxial cable of bandwidth 10 MHz is used to carry one voice signal (of bandwidth 4 kHz), only 0.04% of the cable capacity is being utilised. As the communication distance and hence link cost increases, it becomes more and more important to dramatically increase the utilisation of the cable capacity by somehow packing many voice signals onto the cable medium.
- To allow the use of identical radio systems for the provision of localised broadcast and communication services in different geographical regions. For example, FM radio broadcast can be provided in two different cities using exactly the same carrier frequency of say 98.5 MHz.

To realise the above benefits, there are four multiplexing strategies that may be used separately, but frequently in combination, to simultaneously accommodate multiple users and services in a common transmission medium. Fig. 1.15 provides an illustration of four of these resource-sharing techniques for N users. Three axes are used, namely frequency, which represents the available bandwidth of the transmission medium; time, which represents the instants of usage of the medium; and space, which represents the physical location of the medium.

- In *time division multiplexing* (TDM) the duration of usage of the transmission medium is divided into time slots each of which is allocated to a single user. Thus, each of the N signals has exclusive use of the entire transmission medium during the time slot allocated to it. A useful analogy to TDM is the sharing of the use of a lecture room by four different groups of students, each needing the room for a total period of one hour. We may draw up a schedule dividing the room usage into one-hour time slots so that each group occupies the entire room once in turn; or we may allocate 20-minute time slots so that it takes three slots for each group to complete their business; and so on. There is however an important difference between this analogy and the way TDM is implemented in communication systems. Here there is a

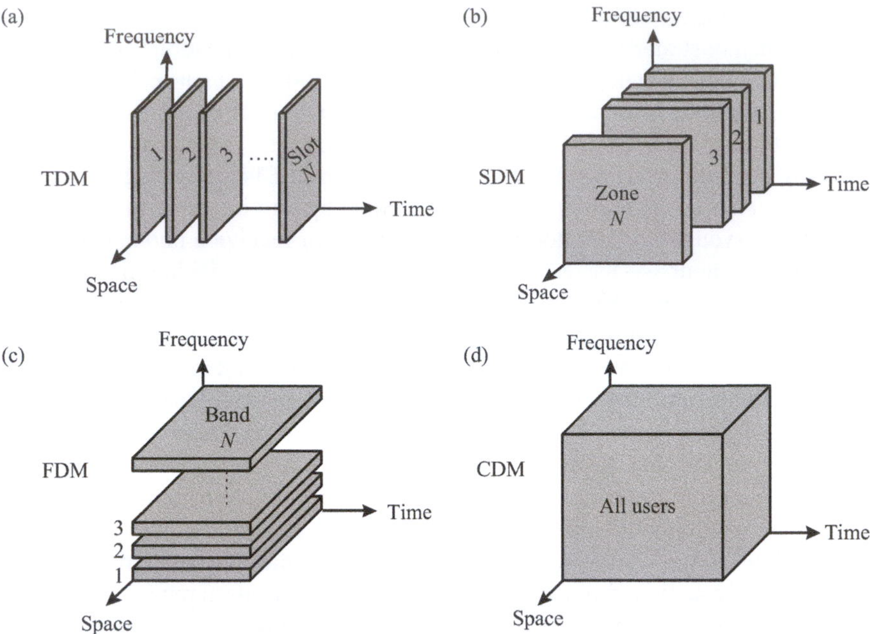

Fig. 1.15 Multiplexing strategies. (a) Time division multiplexing; (b) space division multiplexing; (c) frequency division multiplexing; (d) code division multiplexing

noticeable sense of queuing and waiting for one's turn, whereas in real TDM the time slots are extremely short and are used to convey samples of each signal taken at regular and sufficiently short intervals. Thus, users of the TDM system are totally oblivious of the time-sharing roster and the receiver is able to reconstruct (without distortion) each of the original signals from their samples.

Frequency division multiplexing (FDM) gives each user exclusive use of a separate frequency band (often referred to as a *channel*) for all time. Ideally then, with an average channel bandwidth B_c, and total available bandwidth B_t in the transmission medium, the maximum number of users that can be accommodated is

$$N = B_t/B_c \qquad (1.2)$$

Space division multiplexing (SDM) allocates the same frequency band or all the available bandwidth to more than one user for all time, but user signals of the same frequency are confined to physically separate regions or zones. In closed transmission media it means that each user has exclusive use of a separate line-pair, whereas in open media it requires that the radiated strength of each signal be negligible outside the signal's coverage region. More realistically, the requirement is for the radiated signal strength of a given region to be negligible in neighbouring co-channel regions that reuse the same frequency. In our lecture room analogy, we may apply SDM by allowing all four groups simultaneous use of the room, but with each group seated sufficiently far apart at different corners of the room. As long as the students follow a simple SDM rule of speaking softly (i.e. whispers only, and strictly no shouting), all groups can coexist with little mutual disturbance.

An important area of application of SDM is in cellular mobile communications where the same frequency bands are reused many times. In this way a limited radio spectrum allocation is very efficiently utilised to meet a huge communications demand in a given *serving area* such as a city. For example, in the North American advanced mobile phone system (AMPS) only 25 MHz in the UHF band is available to one operator in a serving area. Of this, 12.5 MHz is for transmission in the forward direction from base station to mobile, and a further 12.5 MHz for transmission in the reverse direction. With 30 kHz per channel and making provision for control channels, it follows that only about 400 users can be accommodated simultaneously in the available bandwidth. This is grossly inadequate to meet the demand for mobile communication services. The use of SDM dramatically increases capacity, enabling the operator to handle tens of thousands of simultaneous calls. A typical SDM or *frequency reuse* plan is shown in Fig. 1.16. The serving area is divided into small zones called *cells*, each of which has one base station for communication with mobile units. A group of cells (enclosed in bold lines in the diagram) across which the entire bandwidth allocation is used up is called a *cluster*. Fig. 1.16 shows a cluster size of 7, but it can also be 3, 4, 9, 12 or multiples of these. The available channels are shared amongst the cells in each cluster. We identify the sets of channels as f_1, f_2, f_3, etc. A mobile unit wanting to make a call is assigned an available channel from the set allocated to its cell. Notice how the frequencies are reused in cells separated by a distance D, meaning for example that calls can be made at the same time in each of the shaded cells using exactly the same set of frequency bands. Obviously, radiated power in each cell must be limited to minimise *co-channel interference*, i.e. interference between cells that use the same frequency. The choice of cell diameter and cluster size is influenced by many factors such as required capacity, acceptable carrier-to-interference ratio (typically $\geq$18 dB). A smaller cell size allows a particular frequency band to be reused more times

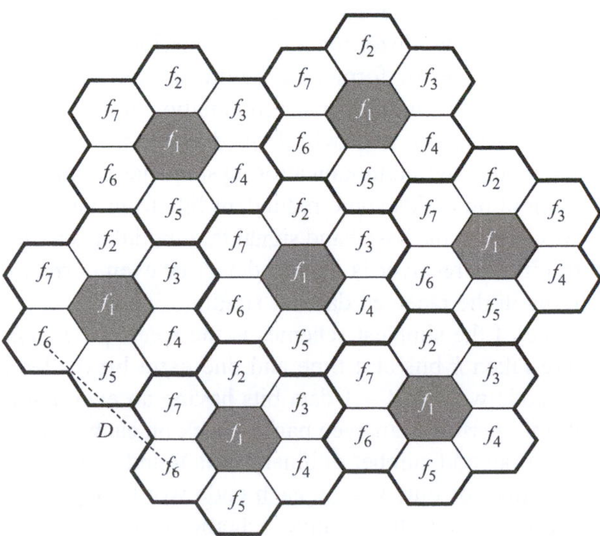

Fig. 1.16 SDM in cellular mobile communications

in the serving area, thus increasing capacity, but *handover* — the process of a mobile unit's transmission being changed from one channel to another as the mobile crosses a cell boundary — occurs more frequently.

Code division multiplexing (CDM) is a kind of free-for-all sharing strategy in which multiple users transmit in the *same frequency* band at the *same time* and in the *same physical medium*. The secret is that each user is assigned a unique pseudo-random code sequence with which their signal is spread over a wide bandwidth giving it a noise-like appearance. A target receiver equipped with exactly the same code sequence is able to extract the wanted signal from the received mix of signals, and to effectively block out the unwanted signals from other users. Returning to our lecture room analogy, the four groups of students may simultaneously share the entire room in CDM fashion, with one group speaking say in *German*, another in *Swahili*, another in *Igbo* and the remaining in *Chinese*. So long as the students understand only the language of their group, then secure and effective communication can take place, with only a slight inconvenience of background noise in each group.

It should be noted that these multiplexing strategies are rarely used in isolation. In fact by taking a broad interpretation of FDM we see that it is inherent in all radio communication systems to allow a multiplicity of services in a given locality. Similarly, to allow the reuse of the same radio band in different regions (or localities in some cases) of the world, SDM is inherent in nearly all radio systems, except, for example, international broadcasting at HF. Thus, if TDM is used on a satellite link at say 6 GHz, then we could describe the system as employing SDM/FDM/TDM. However, we will define multiplexing more restrictively in terms of how multiple signals are combined for transmission on a *common link*. Therefore the satellite system in this example would be regarded simply as a TDM system.

1.3.3 Channel coding

Channel coding (also called channel encoding) is aimed primarily at protecting the message signal against channel distortion and transmission errors. This aim would embrace most of the processes performed in the transmitter after the source coder (and encryptor if present), including carrier modulation for example. In practice however channel coding is more narrowly defined as the process of error control coding. In a digital modulated communication system the channel coder accepts message bits at its input and adds some redundant bits to produce (encoded) data bits at its output, which form the base band signal that modulates a carrier. With the help of these extra bits the receiver is able to detect or even correct certain transmission error patterns in the received data bit stream.

For example, one of the simplest schemes is the *single parity check* in which the message bits are taken k bits at a time and one extra bit (called *parity bit*) is added to produce a code word of $k+1$ data bits having an even number of binary 1s. The receiver simply performs an even parity check on each received code word and declares an error if an odd number of binary 1s is found. An odd parity scheme, which has an odd number of binary 1s in each code word, may also be employed and is equally effective. Thus this simple scheme allows *error detection*, the receiver requesting a retransmission of any code word that is in error. But note that here, as in all error control schemes, only certain error patterns can be detected.

A little thought will show that in this case an even number of bit errors in a code word will pass undetected. However, errors are rare occurrences in a well-designed transmission system. A single bit error in a code word is rare, and two bit errors significantly more rare. The single parity check scheme has been successfully used in operational systems such as the asynchronous transmission of ASCII coded text, where $k=7$. To have an idea of the improvement achieved, consider that if a system incurs on average one character error per transmitted page of text, then with the use single parity check the number of undetected errors will be reduced to one character in about 2500 pages of text.

There is a price to pay for the benefits of channel coding. First, system resources (notably bandwidth) must be devoted to carrying redundant bits. In the single parity check example given above, one in eight or 12.5% of the bits are parity or redundant bits. In other error control schemes, where the receiver is required not only to detect but also to correct the errors, as many as 50% or more of the data bits may be redundant. But channel coding is like an insurance policy: a waste of resources when things go well and a life-saver when they don't. So we are happy to spend valuable system resources or sacrifice system capacity in order to insure against the rare tragedy of transmission errors.

The other price is that of increased system *complexity* especially with the channel decoder at the receiver. Related to this is the issue of extra *processing delay*. How many message bits does the channel coder have to accumulate before producing a code word, and how many data bits does the decoder have to inspect before reaching a decision? Having accumulated the required number of bits, how long does it take to perform the calculations that yield data bits at the coder and message bits at the decoder? Quality of service will be degraded for delay-sensitive traffic such as voice and video if processing time introduces enough extra delay to make the overall transmission delay perceptible.

A wide range of error-control strategies have been devised. Most of the schemes perform the error-control function separately from the process of modulation. These include *linear block codes, cyclic codes, convolutional codes*, and compound codes such as *turbo codes*. Here the input and output of the channel coders and decoders are bit streams. Other schemes, such as *trellis-coded modulation* (TCM), combine channel coding and modulation into one process. In this case we have a bit stream input to the processing block, and an output of transmitted symbols that are fed into the channel.

Most channel coding schemes have the capability of both error detection and error correction. In general a particular scheme can detect more errors than it can correct. For example, a single parity check can detect one error (per code word) but correct none; and a *Hamming code* can detect up to two errors but correct only one error per code word. Furthermore, error detection with *automatic repeat request* (ARQ) is a more reliable error-control strategy than *forward error correction* (FEC). Although it is possible to occasionally *wrongly correct* an error, it is not possible to wrongly detect errors. Also, any error patterns that pass undetected will pass uncorrected but not vice versa. Simplex communication systems however must either ignore transmission errors or rely solely on FEC, since there is no return channel to request a retransmission. Duplex systems on the other hand can deploy either FEC or ARQ schemes, and the latter has been widely used in computer communications. But ARQ schemes suffer from reduced data throughput on paths with large propagation delays such as satellite communication links. We will explore these issues further in Chapter 11.

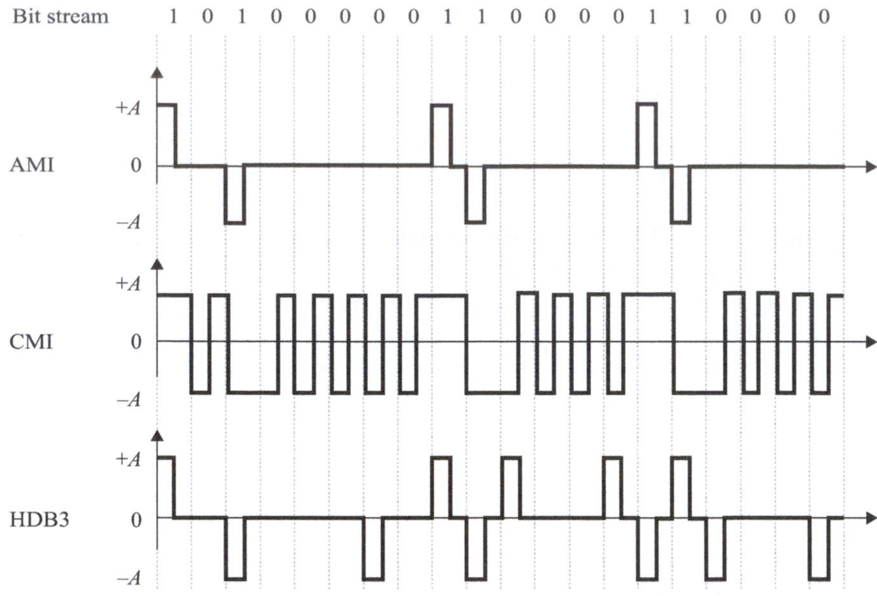

Fig. 1.17 Line code examples: alternate mark inversion (AMI), coded mark inversion (CMI) and high density bipolar with 3 zero maximum (HDB3)

1.3.4 Line coding

Line coding is the process of symbol generation in digital base band communication systems. It fulfils the role played by the combined functions of channel coding and modulation in digital modulated communication systems. The transmitter chooses suitable base band symbols to represent the bit stream output of the source coder and feeds these symbols directly into a transmission line channel. We will show the symbols as rectangular pulses, although in practice they are filtered into non-rectangular shapes before transmission in order to minimise *inter-symbol interference* (ISI). Fig. 1.17 shows the line codes used in Europe for connections between equipment (often within one exchange building) in the interfaces of the digital transmission hierarchy. Line codes are designed to have certain desirable characteristics and to fulfil a number of important functions. The choice of a line code for a particular application is dictated by nine main considerations, namely (1) spectral content; (2) DC content; (3) timing content; (4) bandwidth; (5) transparency; (6) error detection and performance monitoring; (7) code efficiency; (8) error probability; and (9) codec complexity.

We will elaborate briefly on these issues and line code features.

The spectral characteristics or content of the transmitted waveform or symbols — referred to in this context as *coded data* or *line code* — must be matched to the characteristics or frequency response of the transmission medium. A mismatch may result in significant distortion of the transmitted voltage pulses. For example, line transmission systems are easier to design when different parts of the system are ac-coupled using capacitors or transformers to protect their DC bias voltage levels and keep them separate. These coupling elements pass higher frequency (AC) voltages but block zero frequency (DC) voltages. The line code should therefore have no DC offset.

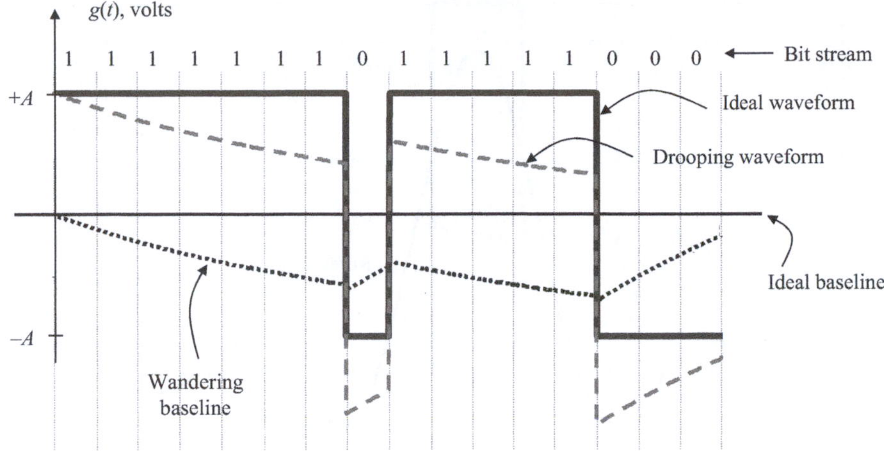

Fig. 1.18 Droop and baseline wander in a binary code

If the coded waveform has significant DC content there will be a *droop* in the constant level as well as *baseline wander*, whereby the received waveform drifts significantly relative to the decision threshold, which is 0 V for the case of a bipolar binary code. See Fig. 1.18.

The line code must combine data and timing information in one signal. It would be very expensive if a separate wire pair or coaxial cable had to be employed to carry the timing information needed at the receiver for synchronising with the transmitter and setting decision or sampling instants. Furthermore, the line code must have a reasonable amount of clock content: The *timing content* of a code is the maximum number of symbols that can occur together without a level transition — a small number indicating high timing content. Ideally, there should be at least one transition in every symbol, but this has a penalty of increased bandwidth requirement.

The bandwidth of the coded waveform must be minimised in order to reduce the vulnerability of the data to noise and inter-symbol interference (ISI). Sudden changes in a signal imply high frequencies in its spectrum. A rectangular pulse (with sharp transitions) transmitted through a low pass transmission medium will spread out as illustrated in Fig. 1.19, with a potential for ISI. Thus, pulse shaping is frequently employed to reduce high frequency components, which also reduces cross-talk since higher frequencies are more readily radiated. By reducing the bandwidth necessary to correctly transmit the coded waveforms pulse shaping cuts the amount of noise power that the receiver inevitably 'admits' in the process of receiving these waveforms.

The line code should have transparency or bit-sequence independence whereby its performance, and in particular its clock content, is acceptable for all patterns of 1s and 0s from the source. If a certain bit pattern would yield unsatisfactory performance then that pattern must be mapped to an acceptable alternative sequence in a way that is recognisable at the receiver.

The line code should allow some amount of error detection and hence straightforward performance monitoring of active transmission lines. This usually involves the use of redundancy in which some code words or symbol

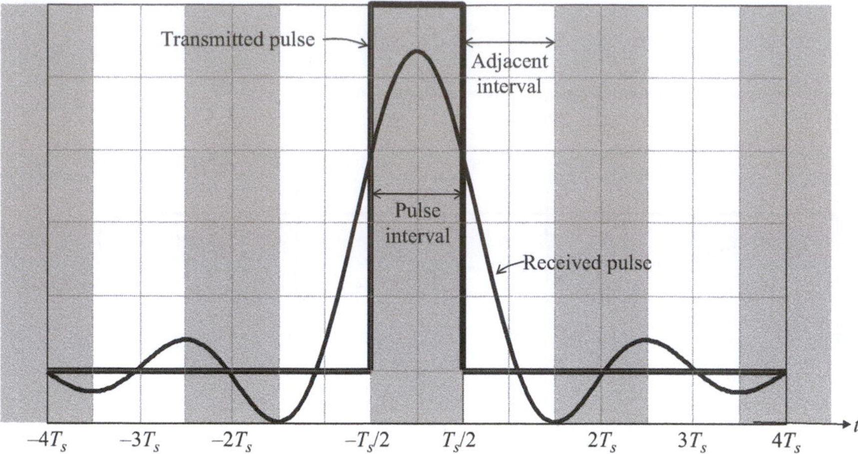

Fig. 1.19 Transmission at rate R_s (= $1/T_s$) baud through an ideal band limited channel of bandwidth $R_s/2$ Hz. Transmitted pulse is rectangular and of duration T_s. Received pulse spreads into adjacent symbol intervals

patterns or sequences are forbidden. A received code word that violates the coding rule in force would then indicate some error.

The line code should maximise code efficiency to allow a lower symbol rate to be used for a given bit rate. In long-distance cable systems, a lower symbol rate reduces the bandwidth of the transmitted waveform. This lessens distortion and hence allows increased repeater spacing that lowers overall system cost. It turns out, however, that codes of high efficiency may lack certain other desirable characteristics. The code selected in practice will involve some compromise and will depend on the priorities of the particular system. Code efficiency is the ratio of actual information content (or bits) per code symbol to potential information content per code symbol; potential information content per code symbol being given by $\log_2$(Code radix), where *code radix* is the number of signalling levels or voltage levels used by the code symbols. For example, the potential information content per code symbol of a binary code (radix = 2) is $\log_2(2) = 1$ bit; that of a ternary code (radix = 3) is $\log_2(3) = 1.585$ bits; and that of a quaternary code (radix = 4) is $\log_2(4) = 2$ bits. Codes with higher radix can therefore convey more information per symbol, but there is increased codec complexity and a higher probability of error. Although multilevel codes (of radix ≥ 4) are very common in modulated communication systems to cope with restricted bandwidth, only codes with radix ≤ 4 are employed in base band systems.

One example of a quaternary code is the 2B1Q line code, which was adopted by ANSI (American National Standards Institute) in 1986 for use on basic ISDN lines. It is also the line code used on DSL local loops. As the name suggests, the 2B1Q code represents two binary digits using one quaternary symbol, i.e. one of four voltage levels. More specifically, the *dibits* 00, 01, 11 and 10 are represented by the voltage levels $-3A, -A, +A$ and $+3A$, respectively, where A is a constant. A related concept to code efficiency is *code redundancy*, which is the ratio between unused and used information per symbol. A code with 100% efficiency has zero redundancy, a code with 50% efficiency has 100% redundancy, and a code with 0% efficiency has infinite redundancy.

For a given channel condition specified by the ratio between signal power and noise power or the ratio between energy per bit (E_b) and noise power per unit bandwidth (N_o), the probability of bit error, called bit error ratio or bit error rate (BER), should be small. Line codes that deliver reliable performance and low BER are preferred.

Finally, the complexity of the encoder and decoder circuits (codec) should be kept to a minimum in order to reduce costs. In general, line codes that can be implemented by simple codecs are used for short links, whereas more efficient but complex and costly codecs are used for long-distance links because they can work with fewer repeaters thereby reducing the overall system cost.

1.3.5 Carrier modulation

Modulation is the process of imposing the variations in a lower-frequency electrical signal (called the *modulating* or *base band* signal) onto a higher frequency signal (called the *carrier*). The carrier signal is usually a sinusoidal signal of frequency f_c that is well above 0 Hz. The end result of the process is that the base band signal is translated to a frequency band centred at f_c. The carrier effectively gives the base band signal a 'ride' through the channel. Modulation serves a number of important roles in modern communication systems:

Modulation is used to obtain a more efficient exploitation of the transmission medium by accommodating more than one user in the same medium at the same time. In most cases, the bandwidth that is available in the medium is much larger than what is required by one user or message signal. For example, the bandwidth available on a coaxial cable is more than 10,000 times the bandwidth of one telephone speech signal. The bandwidth of an optical fibre medium exceeds that of an analogue TV signal by a factor of up to one million; and the radio spectrum is much wider than the bandwidth required by one radio station for its broadcast. Modulation allows the implementation of FDM in which each user's signal is placed in a separate frequency band by modulating an appropriate carrier. As long as the carrier frequencies are sufficiently far apart, the different signals do not interfere with each other. A particular signal can be recovered at the receiver by filtering to exclude unwanted channels, followed by *demodulation* to recover the base band signal conveyed by the carrier. Providers of radio services are able to transmit and receive within the bands allocated to them by using a suitable modulation technique.

Modulation allows us to select a frequency that is high enough to be efficiently radiated by an antenna in radio systems. The power radiated by an antenna may be expressed as $P = I^2R_r$, where I is the rms value of the current signal fed into the antenna and R_r is the antenna's radiation resistance. It turns out that R_r depends on the size of the antenna measured in wavelength units. In general, the size of the antenna must be at least one-tenth of the signal wavelength if the antenna is to radiate an appreciable amount of power. Consider the smallest antennas required to radiate signals at three different frequencies, 3 kHz, 3 MHz and 3 GHz. The wavelengths of these signals, given by the formula 'wavelength = (speed of light)/(radio frequency)', are 100 km, 100 m and 10 cm, respectively. Thus, if we attempted to radiate a 3 kHz speech signal, we would need an antenna that is at least 10 km long.

Such an antenna is prohibitively bulky and expensive, and certainly not suitable for portable applications such as in hand-held mobile telephone units. If on the other hand we use our 3 kHz speech signal to modulate a 3 GHz carrier signal, then it can be efficiently radiated using very small and hence affordable antennas of size $> \approx 1$ cm.

Modulation enables transmission at higher frequencies, thereby exploiting the higher bandwidths available at the top end of the radio spectrum in order to accommodate more users or to transmit signals of large bandwidth. The radio spectrum extends from 3 kHz to 300 GHz and is divided into major frequency bands at decade intervals, namely 3–30 kHz, 30–300 kHz, 300–3000 kHz, ... We see for example that the bands below 3 MHz do not have sufficient bandwidth to accommodate PAL television transmission, which requires a bandwidth of 8 MHz. But the higher bands can accommodate an increasing number of such TV channels, with a maximum of three channels in the high frequency (HF) band (3–30 MHz), rising to 33,750 channels in the extra high frequency (EHF) band (30–300 GHz).

Modulation also allows us to transmit at a frequency that is best suited to the transmission medium. The behaviour of all practical transmission media is frequency dependent. Some frequency bands are passed with minimum distortion, some are heavily distorted and some others are blocked altogether. Modulation provides us with the means of placing the signal within a band of frequencies where noise, signal distortion and attenuation are at an acceptable level within the transmission medium. Satellite communication was pioneered in C-band (4–8 GHz) where both noise (celestial and atmospheric) and propagation impairments are minimum. Ionospheric reflection and absorption become increasingly significant the lower you go below this band, until at about 12 MHz when the signal is completely blocked by the ionosphere. Furthermore, attenuation by tropospheric constituents such as rain, atmospheric gases, fog, cloud water droplets becomes significant and eventually very severe, at higher frequency bands. For this reason, modulation must be used in satellite communication to translate the base band signal to a congenial higher frequency band.

Another example is the optical fibre medium that blocks signals at radio wave frequencies but passes signals in the near-infrared band, particularly the frequencies around 194 and 231 THz. Thus, to use this valuable medium for information transmission we must modulate an optical carrier signal with the base band signal.

There are three basic methods of modulation depending on which parameter of the carrier signal is varied (or modulated) by the message signal. Consider the general expression for a sinusoidal carrier signal:

$$v_c(t) = A_c \cos(2\pi f_c t + \phi) \tag{1.3}$$

This (unmodulated) carrier signal $v_c(t)$ has a *constant amplitude* A_c, a *constant frequency* f_c, an initial phase ϕ and an angle $\theta = 2\pi f_c t + \phi$ that increases *linearly* with time. Varying only the amplitude A_c according to the variations of the modulating signal gives what is known as *amplitude modulation* (AM). *Frequency modulation* (FM) is obtained by varying only the frequency f_c in direct proportion to the value of the modulating signal; and *phase modulation* (PM) is the result of keeping A_c and f_c constant and varying ϕ in direct proportion to the modulating signal value.

In digital communications the base band signal is a bit stream and the three modulation methods are respectively called *amplitude shift keying* (ASK), *frequency shift keying* (FSK) and *phase shift keying* (PSK) to emphasise that the parameters are varied in discrete steps. The number of steps of the parameter generally determines the complexity of the digital modulation scheme. If we modulate the carrier by taking one bit at a time from the base band bit stream, the result is *binary modulation* (or binary shift keying) where the varied carrier parameter can take on one of two values corresponding to binary 1 or 0. This is the simplest scheme and the one most robust to noise.

To implement binary ASK, we take one bit at each clock instant and transmit a sinusoidal carrier of frequency f_c for the duration of the bit. The carrier frequency f_c is chosen to suit the transmission medium. The carrier amplitude is set to A_1 for bit '1' and to A_0 for bit '0'. If either A_1 or A_0 is zero, we have a special type of binary ASK known as *On-Off Keying* (OOK), which is the digital modulation scheme used in optical fibre communication. In binary FSK, same-amplitude and same-phase carriers of frequencies f_1 and f_0 are transmitted for bit 1 and bit 0, respectively. In binary PSK, the carrier amplitude and frequency are fixed, but the carrier is transmitted with a phase ϕ_1 for bit 1 and ϕ_0 for bit 0. Usually $|\phi_1 - \phi_0| = 180°$ in order to use the maximum difference between the two carrier phases.

Fig. 1.20a shows typical ASK, FSK and PSK waveforms for a representative modulating bit stream. As discussed earlier, digital modulation simplifies to a process of symbol generation. The symbols S_{OOK1} and S_{OOK0}, S_{FSK1} and S_{FSK0}, and S_{PSK1} and S_{PSK0} respectively involved in OOK, FSK and PSK are shown in Fig. 1.20b. The symbol rate (or *signalling rate*) is limited by the bandwidth available on the transmission medium. The theoretical maximum symbol rate is twice

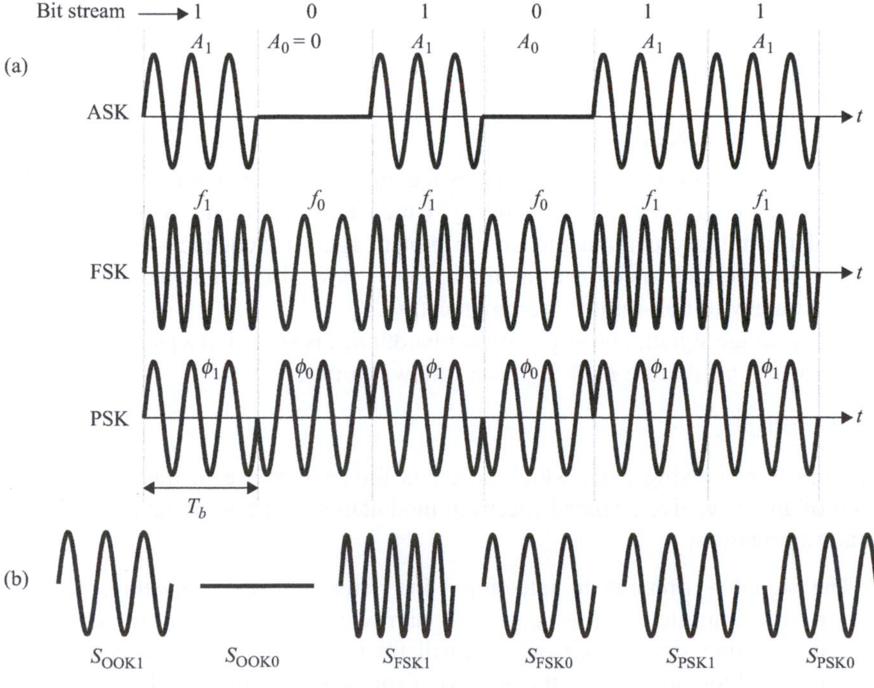

Fig. 1.20 (a) Binary modulation waveforms; (b) symbols used

the transmission bandwidth. Thus in a telephone channel with a bandwidth of about 3.1 kHz, the maximum possible symbol rate is 6200 baud (i.e. symbols per second).

The main drawback of binary shift keying is that bit rate and symbol rate are equal, since each symbol represents one bit. In many applications, a much higher bit rate is desired than can be obtained by increasing the symbol rate, which is limited by the available bandwidth. A higher bit rate can be achieved by using multi-level (also called M-ary) digital modulation, in which M distinct carrier states or symbols are transmitted. Binary modulation is the special case $M=2$. To implement M-ary modulation, we take a group of k bits at a time from the bit stream and represent this group with a unique symbol, where $k=\log_2 M$, or $M=2^k$. For example, taking $k=3$ bits at a time, we have $M=2^3=8$ possible states, namely 000, 001, 010, 011, 100, 101, 110 and 111, which must each be represented by a unique symbol. Each symbol now carries three bits of information and the bit rate is therefore three times the symbol rate. In general, M-ary modulation increases the bit rate according to the relation:

$$\text{Bit Rate} = \log_2 M \times (\text{Symbol Rate}) \tag{1.4}$$

However, the carrier states are now closer together than in binary modulation, making it easier for noise effects to shift one state sufficiently close to an adjacent state to cause misidentification or symbol error at the receiver. For example, the phase difference between adjacent states in an M-ary PSK is $(360/M)°$, which for $M=16$ is only 22.5°. Bearing in mind that these PSK states all have the same energy, carrier amplitude being constant, we see that there is great potential for error if the transmission medium is prone to phase distortion. A combination of *amplitude and phase keying* (APK), also called *quadrature amplitude modulation* (QAM) or *amplitude and phase shift keying* (APSK), is often used, which increases the phase difference between symbol states. For example, the minimum phase difference between states of equal energy is 90° in 16-APK, compared to only 22.5° in 16-PSK.

1.3.6 Spread spectrum

Spread spectrum (SS) modulation involves the transmitted signal being deliberately spread out over a very wide frequency band. This technique was previously designed for military communications to protect against frequency-selective fading, interference and intentional jamming; but is now also employed as a multiplexing strategy in non-military communications.

The message signal, of (unspread) bandwidth B_m, is spread in a pseudo-random manner over a bandwidth $B_c \gg B_m$. The bandwidth ratio

$$G = B_c/B_m \tag{1.5}$$

represents a processing gain, which accounts for an increase in the SNR at the output of an SS receiver. Spread spectrum modulation offers important benefits in signal transmission:

The signal is immune to intentional interference, called *jamming*. A high power jamming signal is necessarily narrow band and will fail to drown the information signal since only a small fraction of the signal energy is corrupted. More accurately, the process of spread spectrum demodulation at the receiver involves the use of a pseudo-random code, which de-spreads the wanted signal back into a narrow band B_m. Interestingly, the effect of this process on the jamming signal is to spread it over a wide band B_c. In this way, the jamming signal energy is rendered insignificant within the narrow band occupied by the recovered wanted signal.

By a similar consideration, SS signals are immune to frequency-selective fading arising from multipath propagation.

An unauthorised receiver cannot recover the information signal from the transmitted SS signal. Simply put, you must have knowledge of the carrier frequency in order to tune into a transmission. And if, as in an equivalent view of SS, the carrier frequency is not fixed but changes pseudo-randomly, then the oscillator frequency at the receiver must change exactly in step for demodulation to be possible. Only authorised receivers will know precisely the pseudo-random sequence of carrier frequencies used at the transmitter.

SS signals have a noise-like appearance to other (unauthorised) receivers. Thus, multiple user transmissions can simultaneously occupy the same frequency band with guaranteed message privacy, provided each user's signal has been spread using a unique pseudo-random code, also referred to as *pseudo-noise* (PN) sequence. This is CDM, which is finding increased non-military applications especially in satellite and mobile cellular communications. Clearly, as the number of users increases a point is reached where the 'background noise' at each receiver becomes excessive, leading to unacceptable bit error rates. However, through a careful selection of PN sequences to minimise their cross-correlation, more users may be flexibly accommodated in a given frequency band by CDM than is possible with FDM and TDM.

Spectrum spreading can be accomplished by various means. In *time-hopping* (TH) SS, the message signal is transmitted in bursts during pseudo-randomly selected time slots. In *frequency-hopping* (FH) SS, the message signal is conveyed on a carrier whose frequency hops pseudo-randomly in value. And in *direct sequence* (DS) SS, the message bit stream of bit duration T_m is multiplied by a PN bit stream of bit duration $T_c \ll T_m$. Hybrid SS techniques are also possible that combine TH, FH and DS; the most common being DS/FH.

1.3.7 Synchronisation

An important requirement in digital communication systems is that identical timing must be maintained between corresponding events at the transmitter and receiver. This *synchronisation* is absolutely crucial for the successful recovery of individual message bits from the composite signal arriving at the receiver input port, and takes place at various stages of the signal recovery process:

Phase synchronisation: The modulation process at the transmitter translates the signal from base band to pass band using a carrier signal. At the receiver, a reverse process of demodulation translates the signal back to base band. *Coherent demodulation* yields minimum susceptibility to noise, but requires a carrier signal to be locally generated at the receiver that is matched in *frequency* and *phase* to the incoming carrier. That is, the zero crossings of the locally generated (receiver) carrier occur at the same time instants as those of the incoming (transmitter) carrier. This type of carrier synchronisation is called *phase synchronisation* and involves using a *phase-locked*

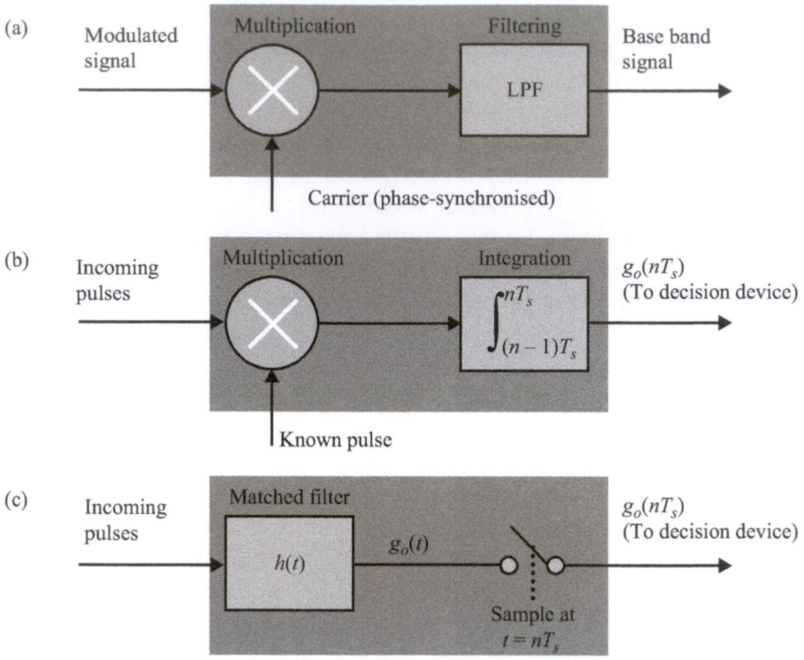

(a) Modulated signal → Multiplication → Filtering (LPF) → Base band signal; Carrier (phase-synchronised)

(b) Incoming pulses → Multiplication → Integration $\int_{(n-1)T_s}^{nT_s}$ → $g_o(nT_s)$ (To decision device); Known pulse

(c) Incoming pulses → Matched filter $h(t)$ → $g_o(t)$ → Sample at $t = nT_s$ → $g_o(nT_s)$ (To decision device)

Fig. 1.21 (a) Coherent demodulation; (b) coherent detection using a correlation receiver; (c) coherent detection using a matched filter

loop (PLL) to vary the phase of a locally generated carrier in step with the channel-induced phase variations in the incoming carrier. Strictly speaking, coherent demodulation (Fig. 1.21a) occurs in analogue communication where we seek to reconstruct the base band waveform shape. The digital equivalent is *coherent detection*, which may be implemented using a *correlation receiver* (Fig. 1.21b) or a *matched filter* (Fig. 1.21c). Here our aim is simply to identify the pulse or symbol that was sent during a given interval of duration T_s. But note that phase synchronisation is involved in all three cases and is implied by the descriptions 'known' and 'matched' in Figs. 1.21b and 1.21c respectively.

Frequency synchronisation: Receiver design may be greatly simplified by employing *noncoherent demodulation* or detection where knowledge of only the incoming carrier (or pulse) frequency is required at the receiver. Carrier phase information is not used. That is, we need to match only the rate but not the instants of zero crossings of the incoming carrier. This type of carrier synchronisation is called *frequency synchronisation*. The circuit simplification afforded by noncoherent detection however comes at a price, namely a reduced robustness to noise when compared to coherent reception.

Symbol synchronisation: A digital communication system receiver must know the intervals (of duration T_s) of the incoming pulses in order to correctly set the start and end times of integration in the correlation detector or the sampling (i.e. decision) instants in the matched filter or noncoherent detector. To this end a clock signal is locally generated having period T_s and transitions matching the boundaries of the incoming symbols, and is employed to trigger the integration process or decision instants. The clock signal is usually conveyed from transmitter to receiver in the level transitions within the transmitted pulse stream and is recovered in a clock extraction circuit at the receiver. The synchronisation between transmitter and receiver thus achieved is known as *symbol* or *bit synchronisation*.

Frame synchronisation: Carrier and symbol synchronisations allow the recovery of a bit stream at the receiver. But these bits are usually grouped in some way at the transmitter. For example, a TDM multiplexer may have interleaved bits from several input tributaries into *frames*; a channel encoder may have transformed every k bits into one code word of n data bits; the bits may have been segmented into headed blocks (called frames) for synchronous transmission. The receiver must have precise knowledge of the demarcations of these bit groupings or frames in order to correctly process the bit stream in the channel and source decoders. This introduces a further synchronisation requirement, known as *frame alignment* or *frame synchronisation*, which allows the receiver to identify the start of each frame or group of bits. In this way, the channel decoder can pick out the code words or frames from the recovered bit stream for error control checks, or the demultiplexer can correctly distribute the bits in the TDM signal to their respective ports without the need for additional address information, etc. To aid frame synchronisation a special pattern of bits known as a *frame alignment word* (FAW) is usually inserted at the transmitter into the bit stream at regular intervals. This serves as a marker onto which the decoder is synchronised at the receiver as follows: The time offset between the FAW and the start of frame (SOF) is defined for the system. The decoder starts by searching the bit stream to locate the FAW and hence the SOF. Frame intervals in the bit stream are then maintained and regularly adjusted with the aid of the receiver clock signal and subsequent FAWs.

The synchronisations discussed above are all that is needed in point-to-point and point-to-multipoint communication systems. However, in multipoint-to-multipoint or *networked* systems a further synchronisation, known as *network synchronisation*, is required to co-ordinate events in various parts of the network. For example, transmit earth stations in a TDMA (time division multiple access) satellite network must be synchronised so that each sends burst transmissions that reach the satellite within time slots centrally allocated for that station.

1.3.8 Filtering operations

A number of filtering operations have been devised to mitigate the degrading effects of practical channels on information transmission. We introduce below three of the techniques — namely *equalisation, raised cosine* and *matched filtering* — that have become commonplace in digital communication systems and identify the target channel impairment of each process.

Equalisation: A practical channel is not distortionless, but will attenuate some frequency components more than others — giving rise to *attenuation distortion*, and will impose different values of propagation delay on different frequency components — causing *phase distortion*. These distortions may be sufficient to lead to occasional transmission errors where a particular symbol is so distorted by the channel that it is mistaken for another at the receiver. To combat this, the transfer function $H_c(f)$ of the channel is

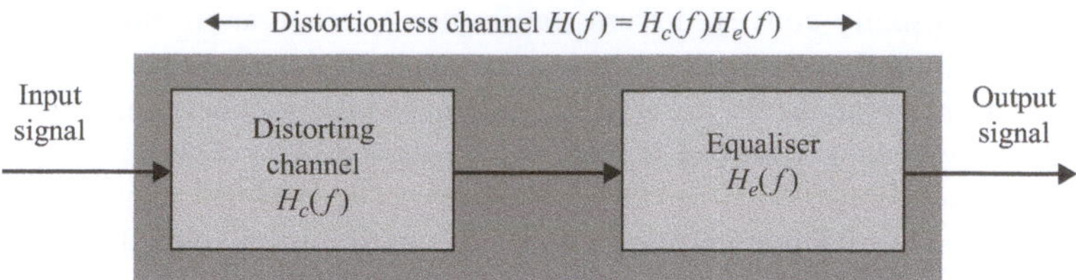

Fig. 1.22 Channel equalisation to minimise channel distortion

determined, which encapsulates the attenuation and delay distortions imposed by the channel on a signal transmitted through it. A filter, known as an *equaliser*, of transfer function $H_e(f)$ can then be designed for use at the receiver having at each frequency component a normalised gain equal to the loss of the channel, and advancing each frequency component by the relative delay of the channel. The former characteristic of the equaliser achieves *gain equalisation* whereas the latter yields *phase equalisation*. If a signal passes through *both* the channel and the equaliser — in whatever order, but in practice channel first (Fig. 1.22) — then every frequency component will be equally attenuated and equally delayed. Thus a cascade connection of the channel and equaliser equates to a distortionless channel. In practice equalisation is imperfect and is limited to a small frequency range of interest, but the use of an equaliser yields significant transmission improvement.

Pulse shaping: The transmission bandwidth of a practical channel is always limited. An equaliser will at best deliver an overall transmission channel having a *constant gain response* (i.e. flat pass band) and a *linear phase response* (i.e. constant delay) within a limited pass band. Such a channel will provide distortionless transmission only for *strictly band limited* signals of bandwidth less than the channel bandwidth. However, rectangular pulses and (in the limit) impulses — because of their sharp transitions — contain high-frequency components, which are blocked by the channel. Since by virtue of the inverse relationship between time and frequency a signal cannot be both band limited and duration limited, this action of the band limited channel will cause a rectangular pulse or impulse to spread out beyond its symbol interval so that it contributes significant energy into one or more adjacent symbol intervals as illustrated in Fig. 1.19. If left unchecked this *intersymbol interference* (ISI) will be a source of bit errors, as the receiver is unable to discriminate between energy in the current symbol and residual energy from previous symbols. We cannot prevent pulse spreading outright as this would require transmitting either infinite-duration pulses through *band limited* channels or finite-duration pulses through *band unlimited* channels. However, we can eliminate ISI by arranging for the path from the pulse generation point at the transmitter to the pulse detection point at the receiver to have an overall transfer function that *shapes* each pulse in such a way that, although it spreads out significantly into subsequent symbol intervals, it always has zero value at the crucial sampling or decision instants of those symbol intervals. This strategy is illustrated in Fig. 1.23 and requires the use of an additional filter known as a *raised cosine filter*.

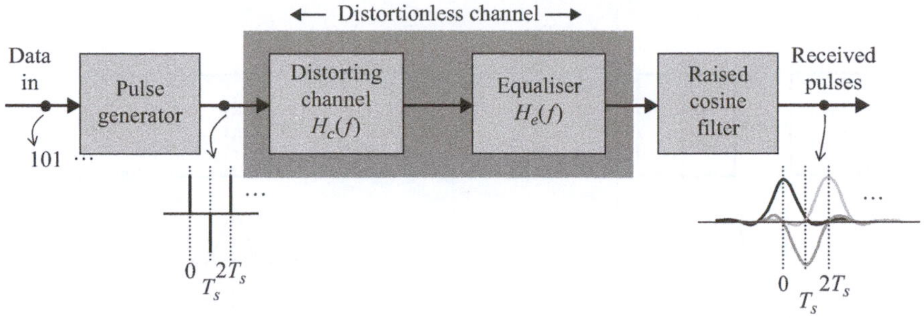

Fig. 1.23 Channel equalisation and raised cosine filtering to minimise channel distortion and ISI

Matched filter: Well-designed equalisers and raised cosine filters will combine effectively to allow error-free digital transmission through band limited distorting channels in the absence of noise. However, noise will be present in a practical system and is added to the transmitted signal along the path from the point of pulse generation at the transmitter to the point of pulse detection at the receiver. We will refer to this noise as *channel noise*, although it should be noted that the path encompasses more than the channel and extends from the *back end* of the transmitter to the *front end* of the receiver. The simplest type of channel noise to deal with is *additive white Gaussian noise* (AWGN). The noise is described as *additive* because it adds to the signal and is present in the same amount regardless of signal value. It is *white* because it contains all frequency components at the same amplitude (i.e. it has a uniform amplitude spectrum, or equivalently, a uniform power spectral density), just as *white light* contains equal amounts of all colours. Finally, it is described as *Gaussian* because it has a normal (i.e. Gaussian) probability density function. Our aim is to achieve optimum detection of each pulse (at the detection point) in the presence of channel noise. Recall that we are not interested in reconstructing the pulse shape. So at the detection point we may pass the noise-corrupted pulse signal through a filter whose characteristic is tailored specifically to the pulse shape we wish to detect in such a way that the filter output maximises contributions from the pulse while minimising contributions from noise. In a given symbol interval, the presence or absence of the pulse is indicated by whether the filter output taken at the interval's decision instant is above or below a specified threshold. Such a filter is called a *matched filter* since its characteristic is matched to the pulse shape to be detected. In fact we will show in Chapter 8 that the impulse response of the matched filter is a time-reversed and delayed version of the pulse. The use of matched filters is a highly effective technique that allows reliable detection of the presence of pulses even when their waveforms are visually unrecognisable due to degradation by noise. The matched filter effectively 'pulls' the pulse out of noise.

In summary, apart from the standard filtering operations used to pass or block a specified band of frequencies, namely *low pass, band pass, high pass* and *band stop filtering*, we need more sophisticated filtering operations for the optimum design of a digital communication system. These include (i) *equalisation* to mitigate channel

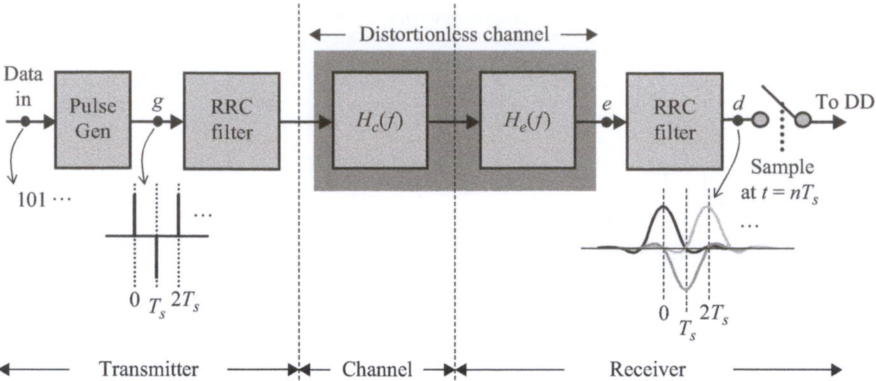

Fig. 1.24 Channel equalisation, raised cosine and matched filtering to minimise channel distortion, ISI and noise effects respectively (DD: decision device; RRC: root raised cosine)

distortions, (ii) *raised cosine filtering* to minimise ISI caused by the band limited nature of the channel, and (iii) *matched filtering* to cope with channel noise.

Fig. 1.24 shows an ingenious arrangement of filtering operations usually employed to mitigate all three degradations. The pulses (assumed for simplicity to be impulses) are filtered at the transmitter using a root-raised cosine (RRC) filter — a filter whose gain response is the square root of the gain response of a raised cosine filter. This yields RRC-shaped pulses that pass undistorted through the equalised channel and reach the point *e* in the receiver. For optimum detection in the presence of AWGN we must pass these pulses through a matched filter. From the symmetry of an RRC-shaped pulse this filter turns out to be another RRC filter. The output of the filter is sampled at time instants $t = nT_s$ and the samples are fed into a decision device that decides in favour of binary 1 if the sample is positive and in favour of binary 0 otherwise. Clearly this arrangement provides both channel equalisation and matched filtering. But note that the arrangement features two RRC filters in cascade, which yields a raised cosine filter. Thus the entire path (from pulse generator output point *g* at the transmitter to decision point *d* at the receiver) has a raised cosine filter response as required for minimising ISI. The arrangement of Fig. 1.24 therefore mitigates the three channel degradations discussed above.

1.4 Why digital communication?

1.4.1 The digital revolution

By the early 1970s advances in integrated circuit technology made digital transmission a cost-effective way of providing telephony, which was until then an analogue communication service. Pulse code modulation (PCM), a signal processing technique devised by British engineer Alec Reeves (1902–1971) back in 1937, was employed to digitise the analogue speech signal by converting it into a sequence of numbers. Digital exchanges and time division multiplexing (TDM) began effectively to replace analogue technology — i.e. analogue exchanges and frequency division multiplexing (FDM) — in the PSTN during the 1970s, although the first TDM telephony system had been installed as far back as 1962 by Bell

Laboratories in the United States. The momentum of digitalisation grew in the 1990s into what was described as a *digital revolution*: the telecommunication networks of some countries are now practically 100% digital — the only remaining analogue segment being in the *local loop* that connects a subscriber's telephone handset to the local exchange.

Furthermore the cellular mobile telephone concept, which was first demonstrated by Motorola in 1972 and implemented via analogue transmission in the first generation (1G) advanced mobile phone system (AMPS) of the 1980s, was quickly replaced in the 1990s by 2G systems (Digital AMPS and GSM) based on digital processing and transmission. And the small step into computer networking in 1971 through the Advanced Research Project Agency Network (ARPANET) evolved in the 1990s into an all-digital worldwide network called the *Internet*, which by 2010 facilitated ubiquitous multimedia communication, including audio and video streaming in addition to text, graphics and all forms of data exchange on a global scale. For example, every one of the numerous events of London Olympics 2012 could be watched live on the Internet — a first in Olympics history.

Digital audio broadcasting began in September 1995 with some field trials by the British Broadcasting Corporation (BBC). Although analogue AM and FM radio broadcast will continue into the foreseeable future, digital radio now provides the highest broadcast sound quality especially in the presence of multipath distortion. Furthermore, terrestrial digital television broadcast was launched in the United Kingdom in 1998 and provides a more efficient spectrum utilisation and wider service choice than its analogue counterpart. The adoption of digital TV gained momentum around the world, and in the UK analogue television broadcasting, which incidentally was started by the BBC in 1936, was eventually discontinued in 2012.

1.4.2 Advantages of digital over analogue communication

The above digital revolution was driven by the *overwhelming* advantages offered by digital communications. Consider the following benefits:

Noise immunity: The number of bit errors in the received data may be very small even in the presence of a significant amount of noise in the received signal. Digital communication transmits information by using a finite set of pulses or symbols to represent groups of bits. These symbols are then transmitted from source to destination. Although each symbol will be somewhat distorted by additive noise during transit, the change will very rarely be large enough to cause a misidentification of the symbol and hence of the bits represented.

Regeneration: Digital communication systems allow the possibility of regenerating (at sufficiently closely spaced regenerative repeaters) clean new symbols or pulses, which are free from all impairment effects and, barring misidentification of any of the symbols, are an exact replica of the original transmission. Thus, unlike analogue systems, noise does not accumulate from repeater to repeater and no further signal distortion occurs beyond that which was introduced at the ADC stage. Digital signals may also be stored in various storage media (e.g. optical or magnetic discs) and processed or re-transmitted later without loss of fidelity.

Error correction: It is possible to detect and even to correct errors in the received data by the use of various coding techniques, which generally insert some redundancy in the transmitted data. It really is a significant feat to be able to correct errors in a message whose content you do not know for sure — if you did know the content you wouldn't need any transmission. But we can do this in digital communication because the transmitted signal corresponds to a countable set of integers, some of which can be permanently excluded as forbidden code words. Error is declared (i.e. *detected*) if a forbidden code word is received. And it may be possible to *correct* the error if it is clear which legitimate code word is most likely to have been sent. Analogue transmission on the other hand corresponds to a continuum of real numbers. There are no forbidden values available and it is difficult to distinguish changes in the transmitted signal made by channel impairments from the legitimate variations imprinted at the transmitter by the (unknown) message signal.

Privacy and security: Increased reliance on telecommunication systems for private, business and military communications and the sale of entertainment and information services calls for *secrecy, authenticity* and *integrity*. The first requirement ensures that the information is received only by an authorised user, whereas the last two requirements assure the receiver that there has not been any impersonation of the sender and that the information has not been deliberately or accidentally altered in transit. Digital communication permits data encryption to be easily implemented on the information bit stream in order to satisfy these requirements.

Flexibility: The signal processing tasks of a digital communication system may be readily reconfigured simply by changing the software program, without any need to change the hardware. Modification of system functions can therefore be implemented more cheaply and speedily.

Integrated services: Voice, video and data can all be represented in a common bit stream format and transmitted simultaneously in a common communication system. Multimedia communication, the Internet and a host of other modern communication services are only feasible through digital technology.

Dynamic range: The dynamic range of a communication system refers to the amplitude ratio between the strongest and weakest signals that the system can process without significant distortion. Signals of a wider range of values (from very small to very large) than is possible with analogue systems can be accurately represented and transmitted with negligible distortion in digital communication systems. The dynamic range may be increased as much as desired by increasing the number of bits used to represent each sample of the analogue signal during the ADC process. There is of course a penalty, namely increased transmission bandwidth requirements.

Low cost: Inexpensive digital circuits may be used to implement the signal processing tasks required in digital communication. With continuing advances in semiconductor technology the cost of *very large scale integrated* (VLSI) circuits will drop even further, making digital communication systems cheaper than their analogue counterparts in spite of the simplicity of the latter. Cost is an important factor that determines whether a new technology is successfully assimilated into society. The digitalisation of telecommunications has been enabled to a large extent by the falling costs of digital circuits of increasing computational power.

1.4.3 Disadvantages of digital communication

Digital communication also has a number of seemingly disadvantageous features. However, in most cases, these disadvantages are under the system designer's control and their effects may be reduced as much as desired by making a suitable trade-off. It should therefore be noted that the following disadvantages are far outweighed by the advantages discussed above. Thus, the trend in telecommunication has been towards an all-digital network.

Large bandwidth: Digital communication systems may require more bandwidth than analogue systems. For example, a 4-kHz bandwidth is adequate for analogue speech transmission, whereas digital speech transmission using standard (64 kb/s) PCM requires a minimum bandwidth of 32 kHz — if binary transmission is employed. The bandwidth of all transmission media, especially radio, is very limited. Transmission techniques that minimise required bandwidth are therefore preferred in order to increase the number of users and services that can be accommodated. However, various low bit rate speech coding and data compression techniques have been devised to reduce the bandwidth requirements of digital audio and video transmission and storage. The end result is that digital television is in fact more bandwidth efficient than its analogue counterpart.

Complexity: Digital communication systems generally perform more complex processing operations on the input signal and require more sophisticated circuitry. Synchronisation usually has to be maintained between receiver and transmitter. However, advances in semiconductor technology make circuit complexity a less significant disadvantage — except that mastering the complex processing concepts involved in digital communications still poses an intellectual challenge. Most of the signal processing tasks may be performed in a single highly reliable and affordable VLSI unit, which can be easily replaced in the unlikely event of a malfunction. Furthermore, digital transmission techniques such as the asynchronous transfer mode (ATM) have been devised, which somewhat soften the criticality of synchronisation.

Quantisation distortion: Analogue signals such as speech must be converted to digital format prior to transmission or processing in a digital communication system. This conversion introduces an irreversible quantisation distortion — see Fig. 1.4d, where the quantised value at each sampling instant clearly differs from the value of the original signal. However, this distortion may be made as small as the system designer wishes by increasing the number of quantisation levels. The penalty for this improvement is increased bandwidth requirements, so a compromise has to be reached according to the practical constraints and priorities of the system design.

1.5 Summary

Digital communications is basically about the use of electrical means for the representation of information as integer numbers, the transmission of these numbers from one point to another, and the conversion of the numbers back to information at destination. The information signal originates from an *information source* and may be voice, music, sound, text, graphics, image or video. At the *transmitter* a *source coder* formats the information as integer numbers (usually in the form of a binary bit stream) aiming for an efficient representation by removing redundancy. The numbers may be scrambled in an *encryption* process to disguise the information. A *channel coder* inserts redundancy in a controlled manner to allow channel errors to be detected and/or corrected at destination. Thus the channel coder increases the number of integers transmitted and hence the system resources utilised. This overhead serves as an insurance against channel impairments. A *line coder* (in digital base band systems) or *modulator* (in digital modulated systems) then generates pulses or symbols selected from a finite set to represent the integers. These pulses are then shaped in a suitable filter to guard against *ISI*. The shaped pulses constitute the transmitted signal that is placed into the *channel* either directly or — in the case of *radio* and *optical fibre* media — indirectly through a transducer such as an *antenna* or a *laser diode*, which converts the signal into an electromagnetic waveform.

The channel provides a connection for the transmission. It may be an entire communication link (e.g. a wire pair linking a local subscriber device to a local exchange) or only a portion of a shared link resource (e.g. where *multiplexing* is employed). There is a host of ways to classify and analyse communication channels: *direct* or *switched*; *stable* or *time-varying*; *distorting* or *distortionless*; *linear* or *non-linear*; *simplex* or *duplex*; etc. The communication system itself may be *point-to-point, point-to-multipoint*, or *multipoint-to-multipoint*. A range of filtering operations including *equalisation* and *matched filtering* are employed at the receiver to counter the effects of channel distortions and noise. The channel delivers a weak and impaired received signal to the receiver that processes the signal in reverse order to proceedings at the transmitter, and delivers an estimate of the information to an information sink for presentation.

The main system resources consumed in digital transmission are *bandwidth, signal power* and *circuit complexity*. And the main aim of a digital communication system design is usually to provide *timely* data transmission at a high enough *bit rate* and an acceptably low *bit error rate*. In achieving this aim the system designer must make judicious use of all three resources, taking into account the constraints of the particular application. This is an exciting task and the main focus of this book will be to guide the reader towards its accomplishment.

Our overview of digital communications is now complete. In the next chapter we will build on this foundation with a discussion of the characterisation of channels and provision of tools for the analysis of linear transmission systems.

Chapter 2

Linear channels and systems

In Chapter Two

- ✓ Analysis of linear systems in the time and frequency domains. You will learn the concepts of impulse response, transfer function, gain response and phase response of linear time-invariant systems and how they are employed to model system effects and determine system output in response to an arbitrary input.
- ✓ Convolution operation. Our graphical introduction to the convolution operation is particularly illuminating for this process that many students struggle to grasp.
- ✓ Signal distortion. What constitutes signal distortion in a system and how do we ensure distortionless transmission.
- ✓ Multipath propagation. You will learn how to analyse multipath propagation and to calculate path loss and signal strength in free space and plane Earth radio channels.
- ✓ Mobile radio channel. Fast fading and median path loss in a terrestrial mobile radio environment.
- ✓ Worked Examples. Pay particular attention to the worked examples that should help you hone your skills in problem solving and extend your knowledge of the concepts discussed and their application in real world engineering.

2.1 Introduction

Communication systems can be characterised in both time and frequency domains. In the simplest cases, a system (which could be a simple circuit, a complex device or a communication channel) may be characterised in the time domain by its impulse response that gives the output of the system when excited by an impulse input. Such a system may equivalently be characterised in the frequency domain by its transfer function that stipulates how the system modifies a sinusoidal signal input. Both descriptions can be employed to determine system output, which consists exclusively of input frequencies, and hence the gain and distortions imposed by the system on a signal passed through it.

In the event of non-linearity, the action of the system on one input — impulse or sinusoid — is affected both by the magnitude of that input and by the presence of other inputs. It is then not possible to determine the response of such a system to an arbitrary input, which in the time domain comprises a train of weighted impulses and in the frequency domain is a sum of sinusoids, based on its response when excited by one impulse or by one sinusoid. The impulse response and transfer function are therefore not useful as analysis tools for non-linear systems and we must resort to the *transfer characteristic* of the system in determining system output, which inevitably includes new frequencies not present in the input. A discussion of non-linear systems behaviour and their analysis will be deferred to the next chapter.

2.2 Linearity and system classifications

Let us consider a transmission channel (henceforth referred to simply as a *system*) with an input signal $x(t)$ and a corresponding output signal $y(t)$ — also called the *response* of the system to the input $x(t)$, and let us denote this relationship as follows:

$$x(t) \xrightarrow{R} y(t)$$

which is read '$x(t)$ yields response $y(t)$'.

The system is a *linear time-invariant* system if it obeys the following rules:

Given that

$$x_1(t) \xrightarrow{R} y_1(t)$$

and

$$x_2(t) \xrightarrow{R} y_2(t) \tag{2.1}$$

then

$$a_1x_1(t) + a_2x_2(t) \xrightarrow{R} a_1y_1(t) + a_2y_2(t) \qquad \text{(i)}$$

$$x_1(t-\tau) \xrightarrow{R} y_1(t-\tau) \qquad \text{(ii)}$$

where a_1 and a_2 are arbitrary constants. Rule (i) is known as the *principle of superposition*, and states that the system produces an output by always doing the same thing to every input and then adding the results together. Rule (ii) expresses *time-invariance*: what the system does is not dependent on the time the input is applied. That is, the only change to the system output in response to a delayed input is a delay of exactly the same amount. In practice the system will not be *indefinitely time-invariant*, and it will be sufficient to treat the system as time-invariant if it is *short-term time-invariant* for the duration of the call or connection. In this case there will be a new challenge to obtain the channel characterisation (a task known as *channel estimation*) at the start of each call since the channel behaviour will vary from call to call. We will also assume that the system is *causal*, which means that there is no system response before an input is applied. That is, the present output of the system depends only on current and/or past inputs. In a *non-causal* system the output at any given instant also depends on or is influenced by future input values. In addition to these three, there are other basic system properties such as *memory* and *stability*. A system is classed as *memoryless* if its output at any instant t depends at most on the input values at the same instant. A system whose output at time t depends exclusively on the input at a past instant $t-\Delta t$ is also regarded as memoryless. To distinguish between these two, the first is described as *memoryless instantaneous-response* whereas the latter is referred to as *memoryless delayed-response*. If the output however depends on the current input as well as some of the past input values then the system is said to have *memory*. Employing relations from basic circuit theory, we illustrate in Fig. 2.1 that by these definitions a resistor is a memoryless system, whereas the capacitor and inductor both have memory. Finally, a system is *stable* if for every absolutely bounded input $|x(t)| < K_i$, the output is also absolutely bounded, $|y(t)| < K_o$.

When all three rules of linearity, time-invariance and causality are obeyed then we show below that such a system can be fully characterised in the time domain by its

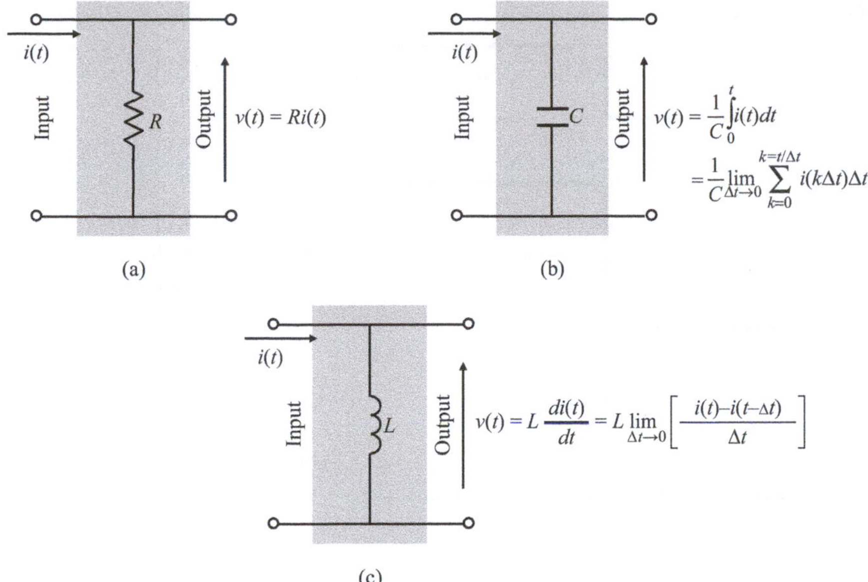

Fig. 2.1 Examples of simple systems with and without memory: (a) memoryless; (b) memory based on accumulation of all past input; (c) memory based on difference between current and immediate past input

impulse response $h(t)$ and in the frequency domain by its *transfer function* or *frequency response* $H(f)$. Before this, it would be helpful to familiarise yourself with Fig. 2.2 where we introduce various standard pulse signals that will feature at various points in this book. One of these is the *unit impulse* (also called *Dirac delta*) function $\delta(t)$ that is zero at all times except at $t = 0$ where its value is infinitely large such that the total area under the curve of the signal is unity. This area is often referred to as the *weight* of the function. The unit impulse is fully characterised by the two equations:

$$\begin{aligned} \delta(t) &= 0 \quad \text{for } t \neq 0 \\ \int_{-\infty}^{\infty} \delta(t)dt &= 1 \end{aligned} \tag{2.2}$$

The waveform of $\delta(t)$ is usually represented by an upward pointing vertical arrow or spike of unit height at $t = 0$ as shown in Fig. 2.2f. The unit impulse may be realised as a limiting case of a number of standard functions, one of which is the rectangular pulse defined in Fig. 2.2c, so that

$$\delta(t) = \lim_{\tau \to 0} \frac{1}{\tau} \text{rect}(t/\tau) \tag{2.3}$$

The unit impulse is an even function since, by the above definition

$$\delta(t) = \delta(-t) \tag{2.4}$$

and it obeys the *sifting property* (sometimes referred to as the *sampling property*) given by

$$\int_{-\infty}^{\infty} g(t)\delta(t - T_s)dt = g(T_s) \tag{2.5}$$

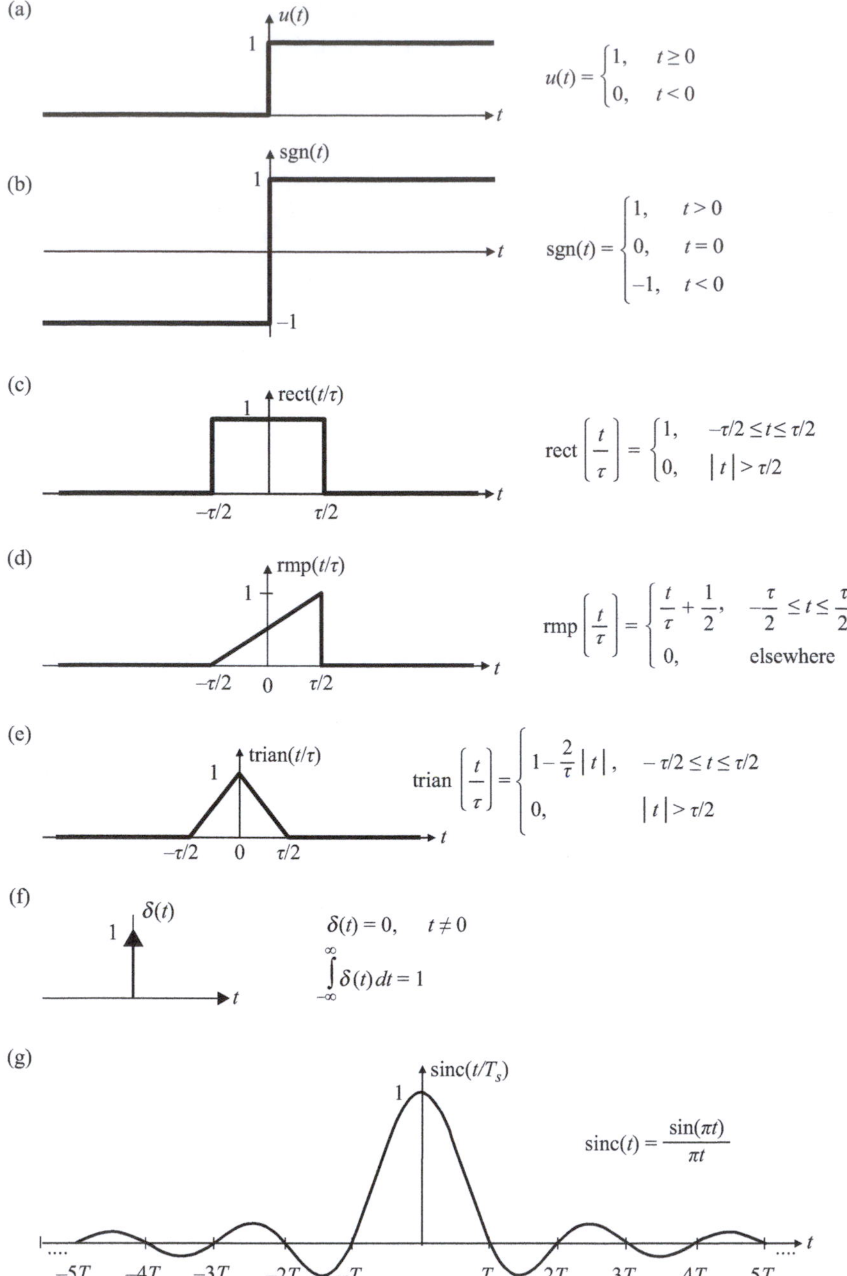

Fig. 2.2 Standard signals: (a) unit step function; (b) signum function; (c) rectangular pulse; (d) ramp pulse; (e) triangular pulse; (f) impulse function; (g) sinc function

The above property can be seen to hold by noting that $\delta(t - T_s)$ is zero everywhere except at $t = T_s$, which means that the only value of $g(t)$ that counts in the integral is $g(T_s)$ that does not depend on t and so can be taken out of the integral sign. By virtue of Eq. (2.2) the remaining integral then evaluates to unity, and hence the result in Eq. (2.5).

2.3 Time-domain characterisation of LTI systems

The impulse response $h(t)$ is defined as the response of an LTI system with zero-initial conditions (e.g. with all capacitors fully discharged) to a unit impulse function $\delta(t)$ applied at its input:

$$\delta(t) \xrightarrow{R} h(t) \tag{2.6}$$

We can obtain a general expression for the response $y(t)$ of this system to an arbitrary input $x(t)$ by taking the following steps, illustrated in Fig. 2.3:

(a) In view of rule (ii) of Eq. (2.1), a delayed impulse input $\delta(t-\tau)$ yields a delayed response $h(t-\tau)$.

(b) A sample (of thickness $\Delta\tau$) of an arbitrary input signal $x(t)$ taken at time $t=\tau$ has area $x(\tau)\Delta\tau$ and is equivalent to an impulse of weight $x(\tau)\Delta\tau$ located at $t=\tau$. In view of rules (i) and (ii) of Eq. (2.1) we obtain the following response to this scaled and delayed impulse input:

$$[x(\tau)\Delta\tau]\delta(t-\tau) \xrightarrow{R} [x(\tau)\Delta\tau]h(t-\tau)$$

(c) The entire signal $x(t)$ can be approximated as a train of samples, each of thickness $\Delta\tau$, taken at $t=k\Delta\tau$ where k takes on integer values from $-\infty$ to ∞, as necessary in order to span the arbitrarily long duration of $x(t)$. As in (b) the kth sample is a scaled and delayed impulse $[x(k\Delta\tau)\Delta\tau]\delta(t-k\Delta\tau)$. Because of rule (i), the response of the system to this sum of impulses is a sum of impulse responses:

$$\sum_{k=-\infty}^{\infty} [x(k\Delta\tau)\Delta\tau]\delta(t-k\Delta\tau) \xrightarrow{R} \sum_{k=-\infty}^{\infty} [x(k\Delta\tau)\Delta\tau]h(t-k\Delta\tau)$$

(d) In the limit $\Delta\tau \to 0$, the approximation of $x(t)$ becomes exact, the discrete $k\Delta\tau$ becomes a continuous variable τ, and the summation on the

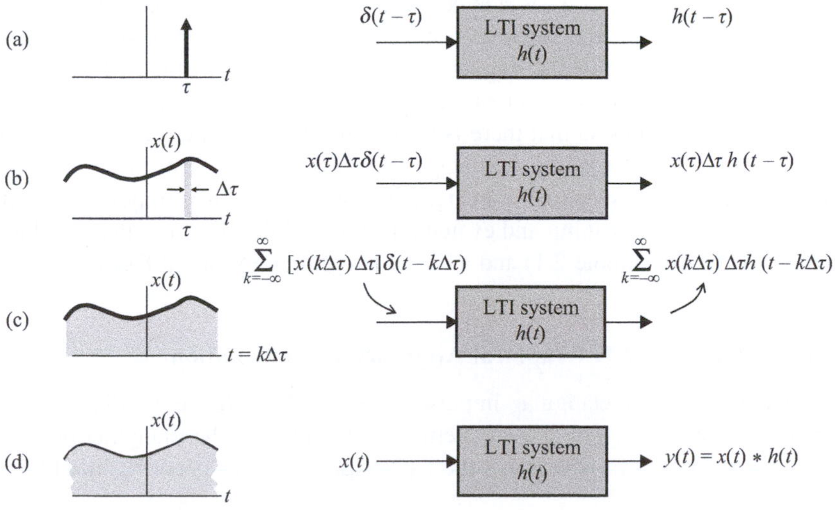

Fig. 2.3 Time-domain analysis of an LTI system

RHS of the above equation (and Fig. 2.3c) becomes an integration, so that we have

$$x(t) \xrightarrow{R} y(t) = \int_{-\infty}^{\infty} x(\tau)h(t-\tau)d\tau \equiv x(t) * h(t) \tag{2.7}$$

The above integral is called the *convolution integral*. Eq. (2.7) is a very important result. It tells us that the output $y(t)$ of a system in response to an arbitrary input $x(t)$ is obtained by convolving $x(t)$ with the impulse response $h(t)$ of the system. The convolution operation is usually denoted as above by an asterisk. By substituting $t-\tau$ for τ in the convolution integral, bearing in mind that t is a constant as far as the integration is concerned, we can show that the convolution operation is *commutative*. That is

$$y(t) = x(t) * h(t) = h(t) * x(t) \equiv \int_{-\infty}^{\infty} h(\tau)x(t-\tau)d\tau \tag{2.8}$$

Furthermore, since $h(t)$ is the response of the system to $\delta(t)$ — applied at time $t=0$ — it follows from the causality rule that $h(t)=0$ for $t<0$, and the integral in Eq. (2.7) can be confined to the interval $(0, \infty)$:

$$y(t) = \int_{0}^{\infty} h(\tau)x(t-\tau)d\tau = \int_{0}^{\infty} x(\tau)h(t-\tau)d\tau \tag{2.9}$$

In addition to being commutative, the convolution operation is *associative* as well as *distributive*. To summarise

$$\begin{aligned} &x(t) * h(t) = h(t) * x(t) && \text{Commutative law} \\ &x(t) * [h_1(t) * h_2(t)] = [x(t) * h_1(t)] * h_2(t) && \text{Associative law} \\ &x(t) * [h_1(t) + h_2(t)] = x(t) * h_1(t) + x(t) * h_2(t) && \text{Distributive law} \end{aligned} \tag{2.10}$$

These properties have important practical implications. For example, it implies that the order of arrangement of a cascade connection of LTI stages does not affect the overall output, and that the impulse response of a system comprising two LTI stages is the convolution of the impulse responses of the two stages. Note that a cascade connection means a series connection in which there is impedance matching between stages so that there is no signal reflection and the output of one stage is entirely delivered to the next stage and equals the input of that stage.

Convolution is a key operation in signal transmission so it is important to gain a good insight into its meaning and evaluation. We will do this through a graphical approach (Worked Example 2.1) and a direct integration (Worked Example 2.2).

Worked Example 2.1: Graphical Approach to Convolution

A system has the rectangular impulse response $h(t)$ shown in Fig. 2.4a. Determine the output $y(t)$ of this system when the input is the triangular pulse $x(t)$ shown. Discuss in what ways the input signal $x(t)$ has been modified by this system.

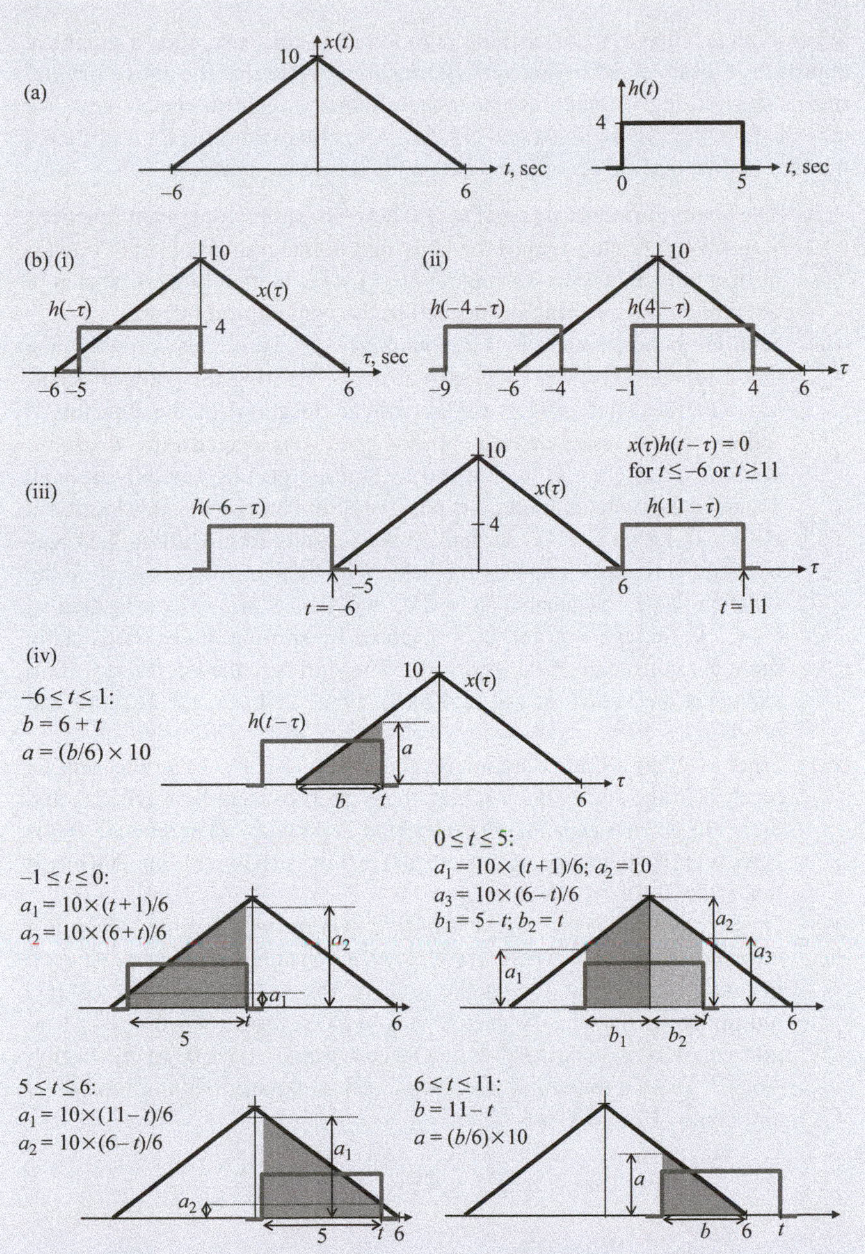

Fig. 2.4 Worked Example 2.1: (a) input signal x(t) and impulse response h(t); (b) solution steps (i)–(iv)

We wish to evaluate the convolution integral

$$y(t) = \int_{-\infty}^{\infty} x(\tau)h(t-\tau)d\tau$$

When both $x(t)$ and $h(t)$ are of finite duration as in this case, then a graphical approach is feasible and in fact very illuminating. Note that the above integral means that $y(t)$ is the total area under the product waveform $x(\tau)h(t-\tau)$. We take the following steps, illustrated in Fig. 2.4, to obtain this area for a sufficient number of different values of t, and hence the desired output $y(t)$:

(i) The operation needs $x(\tau)$ and $h(\tau)$. These are simply the given functions $x(t)$ and $h(t)$ plotted against the convolution integral variable τ. Flip $h(\tau)$ horizontally about $\tau=0$ to obtain $h(-\tau)$. This is merely the first step to obtaining $h(t-\tau)$, which is needed in the convolution integral.

(ii) Shift $h(-\tau)$ horizontally by t to obtain $h(t-\tau)$. To do this correctly note that a function $g(x)$ has value $g(0)$ at $x=0$. Treating the point of occurrence of the value $g(0)$ as the horizontal 'location' of the function, it follows that $g(x)$ is located at $x=0$, and $g(x-k)$ is located at $x-k=0$, i.e. at $x=k$. Thus $g(x-k)$ is obtained by shifting $g(x)$ horizontally through k units to the right (assuming k is positive). Similarly, $g(x+k)$ is located at $x+k=0$, i.e. at $x=-k$, so that $g(x+k)$ results from shifting $g(x)$ horizontally through k units to the left. With this in mind, we view the function $h(-\tau)$ as located at $\tau=0$, and hence $h(t-\tau)$ as located at $t-\tau=0$, i.e. at $\tau=t$, and thus obtained by shifting $h(-\tau)$ horizontally through t units to the right (for $t>0$). The shift is to the left if $t<0$. Two examples are shown in Fig. 2.4(b)(ii) for $t=\pm 4$, i.e. for $h(4-\tau)$ and $h(-4-\tau)$.

(iii) Since $x(\tau)$ has a finite duration (in this case $-6\leq\tau\leq 6$) a point will be reached in the shift value t where there is no overlap between $x(\tau)$ and $h(t-\tau)$, which means that the integrand is $x(\tau)h(t-\tau)$ and hence $y(t)$ is zero. Identify the range of t where $y(t)=0$. It is shown in Fig. 2.4(b)(iii) that $y(t)=0$ for $t\leq -6$ and $t\geq +11$.

(iv) Observe that $h(t-\tau)$ acts like a *sliding window* with gain G, which is a constant in this case where $h(t-\tau)$ has a constant height of 4. At each instant t, the output $y(t)$ is the area of the exposed portion of $x(\tau)$ multiplied by the *window gain* $G=4$. In the regions $t\leq -6$ or $t\geq 11$ no portion of $x(\tau)$ is exposed as noted above and $y(t)=0$. In the region $-6\leq t\leq -1$ a triangular portion of $x(\tau)$ is exposed having base $6+t$ and height $10\times(6+t)/6$. Thus

$$\begin{aligned} y(t) &= \frac{1}{2}\text{base}\times\text{height}\times 4 = \frac{1}{2}(6+t)\times\frac{10(6+t)}{6}\times 4 \\ &= \frac{10(6+t)^2}{3}, \qquad -6\leq t\leq -1 \end{aligned}$$

Similarly, in the region $-1\leq t\leq 0$ the exposed area is a trapezium of parallel sides $10(t+1)/6$ and $10(6+t)/6$, and base 5. In the region $0\leq t\leq 5$ the exposed area is two trapeziums of sides a_1, a_2 and a_3 and bases b_1 and b_2 as shown in Fig. 2.4(b)(iv) with their corresponding expressions. In the region $5\leq t\leq 6$ the exposed area is a trapezium with the indicated sides, and in the region $6\leq t\leq 11$ the exposed area is a triangle. It is a straightforward matter to evaluate each of these areas (not forgetting the window gain factor) to obtain the following result:

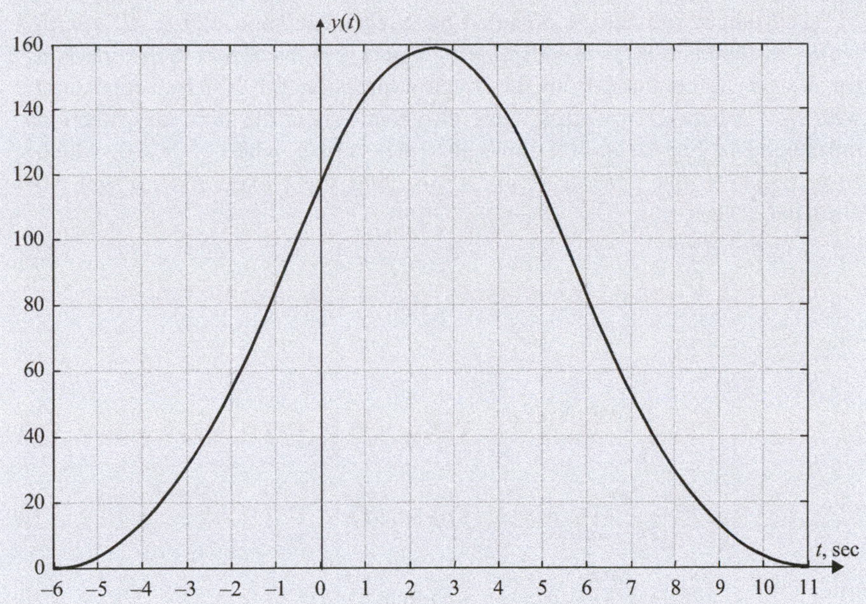

Fig. 2.5 Output of the system in Worked Example 2.1

$$y(t) = \begin{cases} 0, & t \leq -6 \\ \frac{10}{3}(6+t)^2, & -6 \leq t \leq -1 \\ \frac{50}{3}(2t+7), & -1 \leq t \leq 0 \\ \frac{10}{3}[35 - 2t(t-5)], & 0 \leq t \leq 5 \\ \frac{50}{3}(17-2t), & 5 \leq t \leq 6 \\ \frac{10}{3}(11-t)^2, & 6 \leq t \leq 11 \\ 0, & t \geq 11 \end{cases}$$

This output is plotted in Fig. 2.5. We see that the system has changed the input signal in several ways: (i) *Scaling*: the amplitude of $y(t)$ is about 16 times that of $x(t)$; (ii) *Delay*: the peak of $y(t)$ occurs 2.5 s after the peak of $x(t)$; (iii) *Smoothing*: a triangular-shaped $x(t)$ is smoothed into a bell-shaped $y(t)$; (iv) *Spreading*: the duration of $y(t)$ is 5 s longer than that of $x(t)$. Note that only the last two changes constitute a *distortion*.

Worked Example 2.2: Direct Evaluation of Convolution

A system has the impulse response $h(t) = \exp(-at)u(t)$, $a > 0$. Determine the output $y(t)$ of this system when the input is the sinusoidal signal $x(t) = \cos(\omega t)$. Discuss the modifications imposed on $x(t)$ and identify the type of system.

The output $y(t)$ can be obtained by direct evaluation of Eq. (2.7) with $x(\tau)=\cos(\omega\tau)$ and $h(t-\tau)=\exp[-a(t-\tau)]u(t-\tau)$. Note that the expression for $h(t-\tau)$ is obtained from the given expression for $h(t)$ by replacing t with $t-\tau$ wherever it occurs. Note also that $u(t)$ is the unit step function introduced in Fig. 2.2a. It follows that $u(t-\tau)=1$ when $t-\tau\geq 0$, which corresponds to the interval $-\infty<\tau\leq t$ along the τ-axis, and $u(t-\tau)=0$ elsewhere. Thus

$$\begin{aligned}
y(t) &= \int_{-\infty}^{\infty} \cos(\omega\tau)e^{-a(t-\tau)}u(t-\tau)d\tau = \int_{-\infty}^{t} \cos(\omega\tau)e^{-a(t-\tau)}d\tau \\
&= e^{-at}\int_{-\infty}^{t} \cos(\omega\tau)e^{a\tau}d\tau \\
&= e^{-at}\left[\frac{e^{a\tau}}{a^2+\omega^2}(a\cos\omega\tau+\omega\sin\omega\tau)\Big|_{\tau=-\infty}^{\tau=t}\right] \\
&= e^{-at}\left[\frac{e^{at}}{a^2+\omega^2}(a\cos\omega t+\omega\sin\omega t)-0\right] \\
&= \frac{a\cos\omega t+\omega\sin\omega t}{a^2+\omega^2} = \frac{1}{\sqrt{a^2+\omega^2}}\cos(\omega t+\phi); \quad \phi=-\tan^{-1}\left(\frac{\omega}{a}\right)
\end{aligned}$$

In the above, we carried out integration by parts in the second line and phasor combination of the sine and cosine terms in the last line. We see that the response of this system to a sinusoidal input is a sinusoidal signal of the same frequency but of different amplitude and phase. This describes the behaviour of LTI systems in general as further elaborated in the next section. In this case, the phase shift ϕ is given above, which corresponds to a time delay of $|\phi|/\omega$ through the system; and the attenuation A_{dB} experienced by the signal is

$$\begin{aligned}
A_{dB} &= 20\log\left(\frac{\text{Input amplitude}}{\text{Output amplitude}}\right) = 10\log(a^2+\omega^2) \\
&= 20\log a + 10\log\left[1+\left(\frac{\omega}{a}\right)^2\right]
\end{aligned}$$

Notice how the attenuation of the system increases with (angular) frequency ω. So this system is a *low pass filter* and its 3-dB bandwidth B_{3dB} or cut-off frequency f_2 (obtained by finding the value of ω at which A_{dB} is 3 dB more than its minimum value of $20\log a$ at $\omega=0$) occurs at a frequency where $\omega/a=1$. Thus

$$B_{3dB} \equiv f_2 = \frac{a}{2\pi} \quad \text{(Hz)}$$

2.4 Frequency-domain characterisation of LTI systems

The *frequency response* or *transfer function* $H(f)$ of an LTI system is defined as the ratio of system output to input when the input is a sinusoidal signal of frequency f. It indicates how the system alters the amplitude and phase of each sinusoid passed through it and is in general therefore a complex quantity that we may express in the form:

$$H(f) = |H(f)|\exp[j\phi_H(f)] \tag{2.11}$$

where $|H(f)|$ and $\phi_H(f)$ are respectively the magnitude and phase of $H(f)$. When a sinusoid of frequency f_k is passed through a system with transfer function $H(f)$, the sinusoid will have its amplitude multiplied by a factor $|H(f_k)|$ and its phase increased by $\phi_H(f_k)$ radians, and we may write

$$\cos(2\pi f_k t + \phi_k) \xrightarrow{R} |H(f_k)| \cos[2\pi f_k t + \phi_k + \phi_H(f_k)] \tag{2.12}$$

$|H(f)|$ therefore specifies gain as a function of frequency and is called the *amplitude response* or *gain response* of the system.

The phase of $H(f)$ is given by

$$\phi_H(f) = \tan^{-1}\left[\frac{\text{Im}\{H(f)\}}{\text{Re}\{H(f)\}}\right] \tag{2.13}$$

where $\text{Im}\{x\}$ and $\text{Re}\{x\}$ give the imaginary and real parts of x respectively. $\phi_H(f)$ specifies phase shift as a function of frequency and is called the *phase response* of the system.

We have already learnt how to determine the output of an LTI system using a time-domain characterisation of the system in terms of its impulse response $h(t)$. Given the transfer function $H(f)$ of an LTI system, which is a frequency-domain characterisation of the system, we can also obtain a general expression for the response of the system to an arbitrary input signal by taking the following steps, illustrated in Fig. 2.6:

(a) A single sinusoidal input of frequency f_k yields a sinusoidal response at the same frequency with amplitude gain $|H(f_k)|$ and phase shift $\phi_H(f_k)$, whatever the value of f_k.

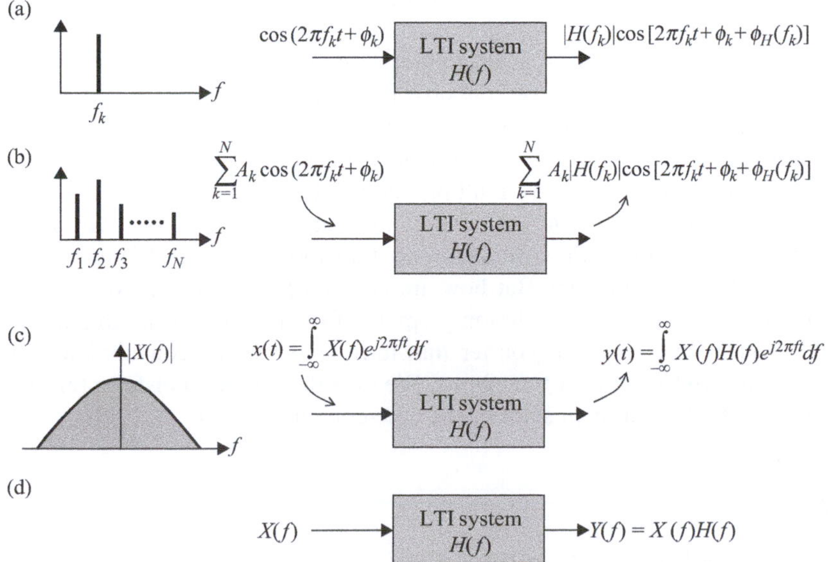

Fig. 2.6 Frequency-domain analysis of an LTI system

(b) In view of rule (i) — Eq. (2.1) — when the input is a group of sinusoids of frequencies f_k, $k = 1, 2, 3, \ldots, N$, the output will be the result of the system changing each sinusoidal signal in amplitude by a factor $|H(f_k)|$ and in phase by a shift $\phi_H(f_k)$, and then adding these modified sinusoids together. Note in particular that rule (i) — the principle of superposition — states that what the system does to any one of the sinusoidal signals is not affected in any way whatsoever by the presence of other signals.

(c) But an arbitrary signal $x(t)$ is in fact a sum of sinusoids expressed as

$$x(t) = \int_{-\infty}^{\infty} X(f)e^{j2\pi ft}df \tag{2.14}$$

where $X(f)$ is the Fourier transform of $x(t)$. Note that the integrand $[X(f)df]$ $\exp(j2\pi f)$ represents a sinusoid of frequency f and amplitude $|X(f)|df$. The RHS and hence $x(t)$ is therefore a *continuous sum* of sinusoids of all frequencies in the range $-\infty < f < +\infty$. So when we present $x(t)$ to our system the output $y(t)$ will, in view of rule (i), be the result of modifying each input sinusoid of frequency f by the complex factor $H(f)$, which includes both a gain factor and a phase shift, and then adding them all together in the same continuous (i.e. integral) fashion. That is

$$y(t) = \int_{-\infty}^{\infty} X(f)H(f)e^{j2\pi ft}df \tag{2.15}$$

(d) We see that Eqs. (2.14) and (2.15) have the same format. So since $X(f)$ in Eq. (2.14) is the Fourier transform of $x(t)$, it follows that $X(f)H(f)$ in Eq. (2.15) must be the Fourier transform of $y(t)$, denoted by $Y(f)$. Thus

$$Y(f) = H(f)X(f) \tag{2.16}$$

This is a very important result: The output spectrum $Y(f)$ for a transmission through an LTI system is the product of the input spectrum and the system's frequency response.

So we now have two approaches to the analysis of an LTI system. In the time-domain approach of Fig. 2.3 we convolve the input signal $x(t)$ with the system's impulse response $h(t)$ to give the output signal $y(t)$. We can then obtain the output spectrum $Y(f)$ if required by taking the Fourier transform of $y(t)$. In the frequency-domain approach of Fig. 2.6, we multiply the input signal's spectrum $X(f)$ by the system's transfer function $H(f)$ to obtain the output signal's spectrum $Y(f)$. If desired we may then take the inverse Fourier transform (IFT) of $Y(f)$ to obtain the output signal's waveform $y(t)$. But how are $h(t)$ and $H(f)$ related? To answer this question let us invoke the convolution property of the Fourier transform (Appendix C.1.11) which states that the Fourier transform (IFT) of the convolution of two signals is the product of their transforms. We take the Fourier transform (an operation denoted by $F[\ ]$) of both sides of the time-domain relation

$$y(t) = x(t) * h(t)$$

to obtain

$$\begin{aligned} F[y(t)] &= F[x(t) * h(t)] = F[x(t)]F[h(t)] \\ Y(f) &= X(f)F[h(t)] \end{aligned}$$

and equate the RHS of this equation and Eq. (2.16) to obtain

$$F[h(t)] = H(f)$$

The impulse response $h(t)$ and frequency response $H(f)$ therefore form a Fourier transform pair

$$h(t) \rightleftharpoons H(f) \qquad (2.17)$$

Eq. (2.17) completes the picture of interrelationship between the two analysis approaches. $H(f)$ can be determined theoretically by analysing the system's circuit diagram or the channel's signal propagation mechanisms. It may also be obtained experimentally by measuring the system's gain and phase shift for a sinusoidal input of frequency stepped through the range of interest. The impulse response $h(t)$ of the system is then obtained by taking the inverse Fourier transform of $H(f)$. The next worked example presents the circuit analysis method.

In deriving Eq. (2.16) for the output spectrum of an LTI system we followed an approach that emphasised the underlying principles of operation of the system. A more mathematical derivation can however be obtained starting with the convolution integral of Eq. (2.8):

$$y(t) = \int_{-\infty}^{\infty} h(\tau)x(t-\tau)d\tau$$

Expressing $x(t)$ in terms of $X(f)$ using the IFT formula

$$x(t) = \int_{-\infty}^{\infty} X(f)e^{j2\pi ft}df \qquad (2.18)$$

yields the following expression for $x(t-\tau)$:

$$x(t-\tau) = \int_{-\infty}^{\infty} X(f)e^{j2\pi f(t-\tau)}df = \int_{-\infty}^{\infty} e^{-j2\pi f\tau}X(f)e^{j2\pi ft}df$$

which we substitute for $x(t-\tau)$ in the above convolution integral for $y(t)$ to obtain

$$y(t) = \int_{-\infty}^{\infty} h(\tau)\left[\int_{-\infty}^{\infty} e^{-j2\pi f\tau}X(f)e^{j2\pi ft}df\right]d\tau$$

Integration is a linear operation so we may interchange the order of integration with respect to f and t. This gives

$$y(t) = \int_{-\infty}^{\infty}\left[X(f)e^{j2\pi ft}\left\{\int_{-\infty}^{\infty} h(\tau)e^{-j2\pi f\tau}d\tau\right\}\right]df = \int_{-\infty}^{\infty} \underline{X(f)H(f)}e^{j2\pi ft}df$$

$$\equiv \int_{-\infty}^{\infty} Y(f)e^{j2\pi ft}df$$

where we recognised that the term in curly brackets { } is by definition the FT of $h(t)$, denoted by $H(f)$. Notice that the second to last integral has the same form as the IFT integral of Eq. (2.18) so the underlined term must be the FT of $y(t)$, denoted by $Y(f)$. Thus, as earlier obtained

$$Y(f) = X(f)H(f)$$

Worked Example 2.3: LPF System Analysis

Determine the transfer function $H(f)$ and impulse response $h(t)$ of the simple RC low pass filter shown in Fig. 2.7a.

For $R = 1$ kΩ and $C = 79.58$ nF, determine

(a) The output spectrum when the input signal is an impulse function $\delta(t)$, and
(b) The output voltage $v_2(t)$ for an input voltage $v_1(t) = 10\cos(2000\pi t + 30°)$ V.

By definition $H(f)$ is the ratio of output to input when the input is a sinusoid of frequency f. The equivalent RC circuit for a sinusoidal input signal of frequency f is shown in Fig. 2.7b.

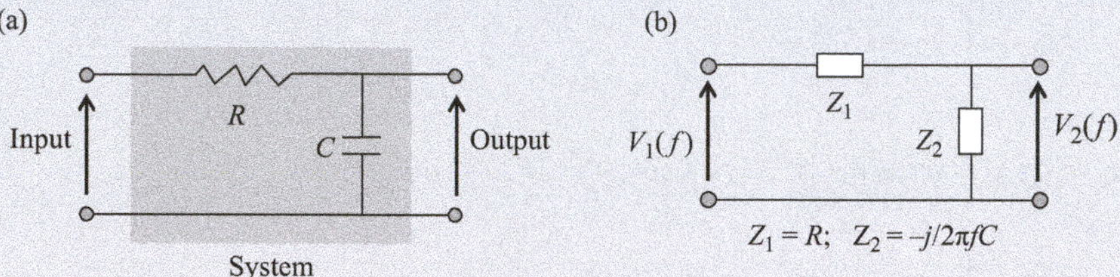

Fig. 2.7 RC low pass filter: (a) circuit diagram; (b) equivalent circuit for a sinusoidal input signal of amplitude V_1 and frequency f

A brief digression to briefly clarify the concept of impedance for those readers who have not done a course in circuit analysis is in order. The impedance Z of a circuit element (resistor, capacitor or inductor) is the ratio between the voltage V across the element and the current i flowing through the element. For a resistor of resistance R, Ohm's law states that $V = iR$, so that $V/i = R$, and therefore impedance $Z = R$.

For a capacitor of capacitance C, the voltage $V = Q/C$, where Q is the charge accumulated on the capacitor plates due to the flow of current. Assuming a sinusoidal current $i = I_p\cos(\omega t)$, we have

$$V = \frac{1}{C}Q = \frac{1}{C}\int_0^t I_p\cos(\omega t)dt = \frac{I_p\sin(\omega t)}{\omega C} = \frac{I_p\cos(\omega t - 90°)}{\omega C}$$
$$= \frac{I_p\cos(\omega t)}{\omega C}\cdot(-j) = \frac{i}{\omega C}\cdot(-j)$$

Thus, $Z = \frac{V}{i} = \frac{-j}{\omega C} = -j/2\pi fC$

Note that in the above, we replaced a phase shift of −90° by the factor $-j$. You can see why this is the case by noting that a phase shift of 180° corresponds to a factor −1, so a phase shift of 90° must correspond to a factor $\sqrt{-1}$ since we shift twice by 90° to obtain a 180° shift (or equivalently, we multiply twice by $\sqrt{-1}$ to get -1). The letter j denotes this imaginary number $\sqrt{-1}$. Note therefore that the presence of the factor $-j$ in the expression for the impedance of a capacitor is essential to capture the fact that the current i leads the voltage V by 90°.

Finally, for an inductor of inductance L

$$V = L\frac{di}{dt} = L\frac{d}{dt}\left[I_p\cos(\omega t)\right] = L\left[-I_p\omega\sin(\omega t)\right] = \omega LI_p\sin(\omega t + 180^\circ)$$
$$= \omega LI_p\cos(\omega t + 90^\circ) = j\omega LI_p\cos(\omega t) = j\omega Li$$

which yields $Z = \frac{V}{i} = j\omega L = j2\pi fL$.

So the impedance of an inductor L is ωL with the voltage leading the current by 90°.

Now back to the problem at hand: The output voltage is taken across the capacitance C and comes from a division of the input voltage $V_1(f)$ between the resistance R and the capacitance C according to the ratio of their respective impedances Z_1 and Z_2. Thus,

$$H(f) \equiv \frac{V_2(f)}{V_1(f)} = \frac{Z_2}{Z_1 + Z_2} = \frac{-j/2\pi fC}{R - j/2\pi fC}$$

$$= \frac{1}{1 + j2\pi fCR} \equiv |H(f)|\exp[j\phi_H(f)]$$

$$|H(f)| = \frac{1}{\sqrt{1 + 4\pi^2 f^2 C^2 R^2}} \equiv \text{Amplitude response}$$

$$\phi_H(f) = -\tan^{-1}(2\pi fCR) \equiv \text{Phase response}$$

We obtain the impulse response by taking the inverse Fourier transform of $H(f)$. Looking down the list of Fourier transform pairs in Appendix C.2, we find the pair

$$e^{-at}u(t) \rightleftharpoons \frac{1}{a + j2\pi f}$$

Casting $H(f)$ into the form of the RHS of this pair:

$$H(f) = \frac{1}{1 + j2\pi fCR} = \frac{1}{RC}\left[\frac{1}{1/RC + j2\pi f}\right]$$

it follows that the impulse response $h(t)$ is the LHS of the pair with $a = 1/RC$ and a scale factor $1/RC$. Thus

$$h(t) = \frac{\exp(-t/RC)}{RC}u(t)$$

The presence of the unit step function factor $u(t)$ indicates that $h(t) = 0$ for $t < 0$, as expected since the system is causal.

(a) Applying Eq. (2.16) and noting that the Fourier transform of a unit impulse is unity (Appendix C.2), we obtain the output spectrum

$$Y(f) = H(f)F[\delta(t)] = H(f)$$

This is a general result: Whenever the input is a unit impulse the spectrum of the output signal gives the transfer function of the system. Substituting the values of R and C in the amplitude and phase expressions for $H(f)$ yields the required amplitude and phase spectra:

$$|H(f)| = \frac{1}{\sqrt{1 + 4\pi^2 f^2 (79.58 \times 10^{-9})^2 (1000)^2}}$$
$$= \frac{1}{\sqrt{1 + (f/2000)^2}}$$
$$\phi_H(f) = -\tan^{-1}[2\pi fCR] = -\tan^{-1}(5 \times 10^{-4} f)$$

(b) $v_1(t)$ is a sinusoid of frequency $f = 1000$ Hz. The output is a sinusoid of the same frequency but with amplitude reduced by the factor $|H(f)|$ and phase increased by $\phi_H(f)$. Thus

$$v_2(t) = 10|H(f)|\cos[2000\pi t + 30^\circ + \phi_H(f)]\Big|_{f=1000}$$
$$= \frac{10}{\sqrt{1 + \left(\frac{1000}{2000}\right)^2}} \cos\left[2000\pi t + 30^\circ - \tan^{-1}(5 \times 10^{-4} \times 10^3)\right]$$
$$= 8.94 \cos[2000\pi t + 30^\circ - 26.6^\circ]$$
$$= 8.94 \cos(2000\pi t + 3.4^\circ) \text{ V}$$

The impulse response $h(t)$, gain response $|H(f)|$ and phase response $\phi_H(f)$ of this LPF are plotted in Fig. 2.8.

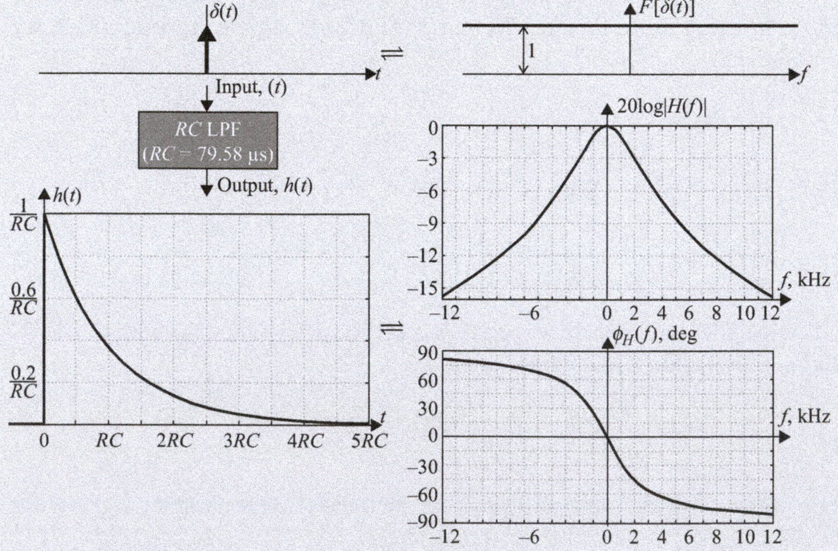

Fig. 2.8 Impulse response and transfer function of an RC low pass filter, shown for a time constant $RC = 79.58\ \mu s$

The practical significance of the Worked Example (2.2) should now be clearer: We evaluated the convolution integral with a unit-amplitude, zero-initial-phase sinusoid as the input signal. Under this condition, the amplitude of the output is the gain response $|H(f)|$ of the system and the phase of the output is the phase response $\phi_H(f)$. So the LTI system of Worked Example 2.2 has

$$|H(f)| = \frac{1}{\sqrt{a^2 + \omega^2}}; \qquad \phi_H(f) = -\tan^{-1}\left(\frac{\omega}{a}\right)$$

In fact, by setting $a = 1/RC$ in the above expressions, we see that the two systems are identical apart from a gain factor $1/RC$ in the system of Worked Example 2.3. So the two worked examples are simply two different approaches — time domain and frequency domain — of solving the same problem. Most engineers prefer the second approach because it avoids the evaluation of the convolution integral that often requires integration by parts.

2.5 Output spectral density of LTI systems

A physical interpretation of the Fourier transform $X(f)$ of a signal $x(t)$ — see Appendix C — is that $x(t)$ consists of sinusoidal signals, arranged in pairs having amplitude $|X(f)|df$ situated one in the interval $(f, f+df)$ and the other in $(-f, -(f+df))$, and spanning the entire frequency axis $-\infty < f < \infty$. If $x(t)$ is periodic of period T, then the frequency spacing $df \equiv f_o = 1/T$ is non-infinitesimal and we have a discrete spectrum; but if $x(t)$ is nonperiodic then $T \to \infty$ and we have an infinitesimal spacing of frequencies that results in a continuous spectrum. Whatever the type of $x(t)$ we can say that the signal power $P_{df}(f)$ in a frequency range df located at $\pm f$ comes from the two sinusoids of amplitude $|X(f)|df$ found in that range. Dividing $P_{df}(f)$ by df yields the power per unit frequency, called *power spectral density* (PSD) and denoted by $S_x(f)$. Thus

$$\begin{aligned} P_{df}(f) &= 2 \times \frac{|X(f)|^2 (df)^2}{2} = |X(f)|^2 (df)^2 \\ S_x(f) &= P_{df}(f)/df = |X(f)|^2 df \end{aligned} \tag{2.19}$$

For a periodic signal, substituting $df \equiv 1/T$ and $X(f) = C_n T$, where C_n is the amplitude of the nth harmonic in the exponential Fourier series defined in Appendix C, and replacing f by nf_o yields

$$\begin{aligned} &S_x(nf_o) = T|C_n|^2 \qquad \text{PSD of periodic signal} \\ &\text{where } C_n = \frac{1}{T}\int_{-T/2}^{T/2} x(t)e^{-j2\pi nf_o t}dt; \quad f_o = 1/T \end{aligned} \tag{2.20}$$

If the signal is nonperiodic there is no guarantee that $X(f)$ exists. However, we may determine $S_x(f)$ in terms of the Fourier transform $X_T(f)$ of a truncated and periodic

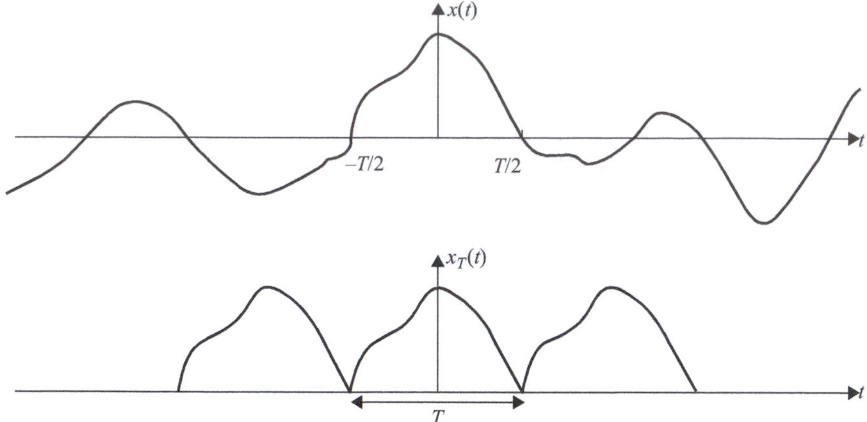

Fig. 2.9 A nonperiodic signal x(t) truncated in the interval (T/2, T/2) to form a truncated periodic version $x_T(t)$ of period T. In the limit T, x(t) and $x_T(t)$ are identical

version $x_T(t)$ of the signal $x(t)$ — as illustrated in Fig. 2.9, the interval of truncation T being gradually extended towards infinity in which limit $x_T(t) = x(t)$, so that Eq. (2.19) becomes (with $1/T$ replacing df)

$$S_x(f) = \lim_{T\to\infty} \frac{|X_T(f)|^2}{T}$$
$$\text{where } X_T(f) = \int_{-\infty}^{\infty} x_T(t)e^{-j2\pi ft}\,dt \qquad \text{PSD of nonperiodic signal} \qquad (2.21)$$

If the signal $x(t)$ is an energy signal (such as a realisable pulse of finite duration) then its Fourier transform $X(f)$ is guaranteed to exist, and since energy is the product of average power and the time duration T over which the power is expended, it follows that the energy per unit frequency of $x(t)$, called *energy spectral density* (ESD) and denoted by $\Psi_x(f)$, is given by

$$\Psi_x(f) = TS_x(f) = \lim_{T\to\infty}\left\{T\frac{|X_T(f)|^2}{T}\right\} = |X(f)|^2 \qquad (2.22)$$

which indicates that the ESD of a signal is simply the square of the signal's amplitude spectrum. The energy in an infinitesimally small frequency band $(f, f+df)$ is $\Psi_x(f)df$, and the total energy E of the signal is obtained by summing these contributions over the entire frequency axis $-\infty < f < \infty$ so that

$$E = \int_{-\infty}^{\infty} \Psi_x(f)df = \int_{-\infty}^{\infty} |X(f)|^2 df = 2\int_{0}^{\infty} |X(f)|^2 df \qquad (2.23)$$

We are able to take the last step above since the amplitude spectrum of a real signal has even symmetry so that the area under $|X(f)|^2$ in the negative frequency range $(-\infty, 0)$ equals the area in the positive range. Since the bandwidth of a signal is by definition the significant positive frequency range, we note therefore that energy per unit bandwidth is *double* the ESD. Similarly, the average power P of a signal is given in terms of the signal's PSD as

$$P = \int_{-\infty}^{\infty} S_x(f)df = 2\int_{0}^{\infty} S_x(f)df \tag{2.24}$$

and the power per unit bandwidth is $2S_x(f)$, i.e. double the PSD. Equating the frequency-domain expression for energy in Eq. (2.23) to the time-domain definition of the energy of $x(t)$ gives a special statement of *Parseval's theorem*:

$$\int_{-\infty}^{\infty} x^2(t)dt = \int_{-\infty}^{\infty} |X(f)|^2 df \tag{2.25}$$

It is useful to examine how the energy spectral density (ESD) of an energy signal and the power spectral density (PSD) of a power signal are affected by transmission through an LTI system. Squaring Eq. (2.16) yields

$$|Y(f)|^2 = |X(f)|^2|H(f)|^2$$

And employing Eq. (2.22) for ESD we obtain

$$\Psi_y(f) = \Psi_x(f)|H(f)|^2 \tag{2.26}$$

Since PSD is simply the average rate of ESD, a similar relationship will hold for power signals, namely

$$S_y(f) = S_x(f)|H(f)|^2 \tag{2.27}$$

So we see that the effect of the system is to modify — in other words *colour* — the signal's spectral density by the factor $|H(f)|^2$, which is the square of the system's gain response.

For an important example, let us consider white noise as the input signal. We know that this has a constant PSD $= N_o/2$. In view of Eq. (2.27) the noise at the system output will be *coloured noise* of PSD:

$$S_{cn}(f) = \frac{N_o}{2}|H(f)|^2 \tag{2.28}$$

Summing the above contribution over the entire frequency axis yields the total noise power at the output of the system:

$$P_n = \frac{N_o}{2}\int_{-\infty}^{\infty} |H(f)|^2 df = N_o\int_{0}^{\infty} |H(f)|^2 df \tag{2.29}$$

2.6 LTI system bandwidth

Every practical system is a band limited filter of some sort. It will pass only those frequencies that lie in a given range, called the *pass band* of the system. The amplitude of each passed frequency f is of course multiplied by the gain $|H(f)|$ of the system at that frequency. We discuss system bandwidth under two considerations. The first is to do with how the system affects a wanted signal whereas the second relates to the amount of noise admitted by the system.

2.6.1 3-*dB* and null-bandwidths

In respect of the passage of a wanted signal, the bandwidth of a system is the range of positive frequencies passed by the system without a significant relative reduction in amplitude. To express relative reduction, the gain $|H(f)|$ of the system is normalised so that $|H(f)|_{max} = 1$. The bandwidth of the system is then given by the frequency range over which $|H(f)|$ does not drop significantly below unity, the line (on what is significant) being usually drawn at $1/\sqrt{2}$. This defines the 3-*dB* or *half-power bandwidth* of the system. For example, the frequency response plotted in Fig. 2.8 shows that the RC filter of Worked Example 2.3 has a 3-dB bandwidth of 2 kHz.

The gain response of practical systems will usually not have a null, as in the amplitude spectrum of pulses. If however this is the case — see for example the popular *raised-cosine filter* introduced in Worked Example 2.5 — then the bandwidth of the system is expressed by a *null-bandwidth* definition rather than the 3-dB bandwidth. The null-bandwidth of such a low pass system is the frequency range from $f = 0$ to the first null in the gain response of the system. If the system is band pass, then the null-bandwidth is the width of the main lobe of its gain response.

Fig. 2.10 shows the normalised frequency response of ideal filters. The *low pass filter* (LPF) passes only frequencies below a cut-off frequency f_1, whereas a *high pass filter* (HPF) passes only frequencies above a cut-off frequency f_1. The *band pass filter* (BPF) passes only frequencies in the range from f_1 to f_2 whereas the *band stop filter* (BSF), also called *notch filter*, passes all frequencies except those in the range from f_1 to f_2. Notice how the gain of each filters drops suddenly from unity to zero at the cut-off frequency. There is a common boundary frequency between the *stop band*, where all frequencies are completely blocked and the pass band where they are passed unchanged. Such a *brick wall* filter response is unrealisable in real time.

A realisable Butterworth low pass filter response is shown in Fig. 2.11 for various values of filter order *n*. There are three regions in the response (marked in the diagram for the response curve of order $n = 3$), namely the pass band $0 \to f_1$ where

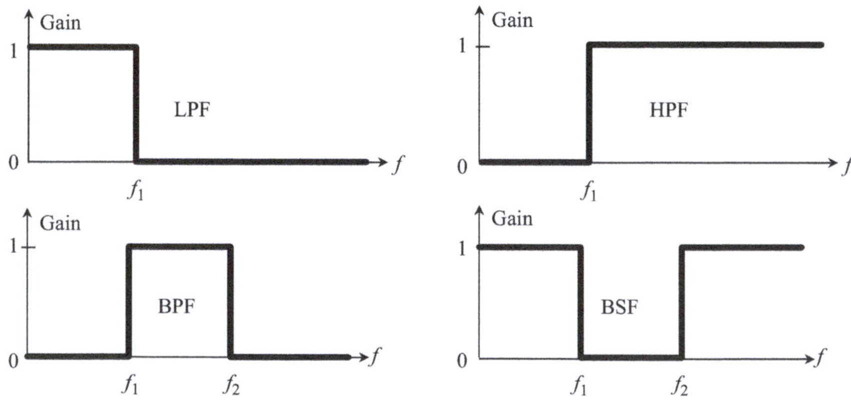

Fig. 2.10 Normalised frequency response of ideal filters

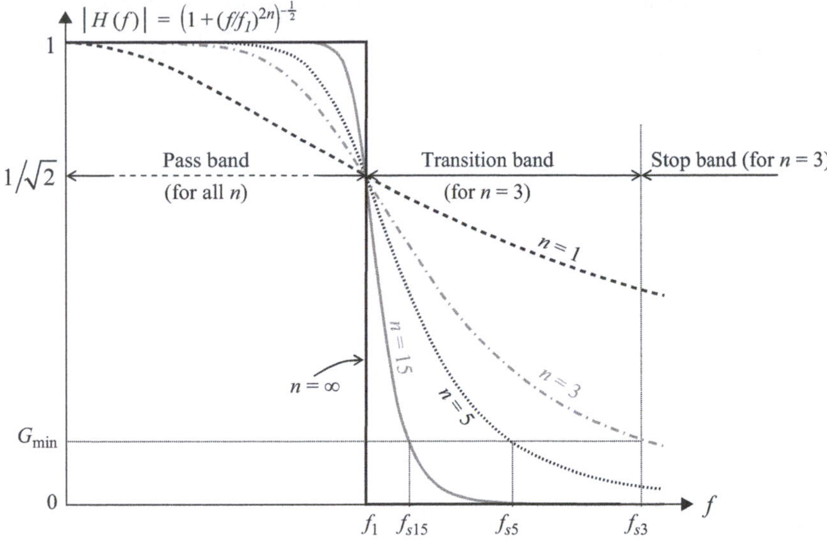

Fig. 2.11 Realisable Butterworth LPF

the minimum gain is $1/\sqrt{2}$, the *transition band* $f_1 \to f_s$ where the gain is dropping towards a specified threshold $G_{\min}$, and the *stop band* $f_s \to \infty$ where the gain is below $G_{\min}$. Observe that as filter order $n \to \infty$ the ideal response is approached where $f_s = f_1$ (i.e. the transition band disappears), and $G_{\min} = 0$, meaning that there is infinite attenuation in the stop band. Note from Fig. 2.11 that the pass band is the same ($0 \to f_1$) for all filter orders $n = 1, 2, 3, \ldots, \infty$, but the width of the transition band decreases as n increases. It reduces from a large value (beyond the range shown in Fig. 2.11) at $n = 1$, through $f_{s3} - f_1$ at $n = 3$, and all the way to zero at $n \to \infty$.

2.6.2 Noise-equivalent bandwidth

We are often interested in the noise power at the output of a system in order for example to determine signal-to-noise ratio (SNR), which gives a measure of transmission quality. Usually the system, through its frequency-dependent gain response $|H(f)|$, will pass more noise at some frequencies than at others. But as far as output noise power is concerned this system is equivalent to a system with a rectangular-shaped gain response that admits to an *equal* extent all the white noise falling within its finite pass band. As illustrated in Fig. 2.12 this *noise-equivalent* system has a bandwidth B and a constant gain response K (within its pass band) equal to the maximum value $|H(f)|_{\max}$ of the gain response of the original system.

We may formally define the *noise-equivalent bandwidth* of a system as *the bandwidth of a noiseless ideal brick wall filter, of constant gain response equal to the maximum gain response of the system, which passes the same amount of noise power through to its output as does a noiseless version of the system when both have white noise of equal PSD as input*. Thus following the method explained in Section 2.5, the noise power at the output of the equivalent system is as given in Fig. 2.12:

$$P_{ne} = K^2 N_o B \qquad \text{(Noise power at system output)} \tag{2.30}$$

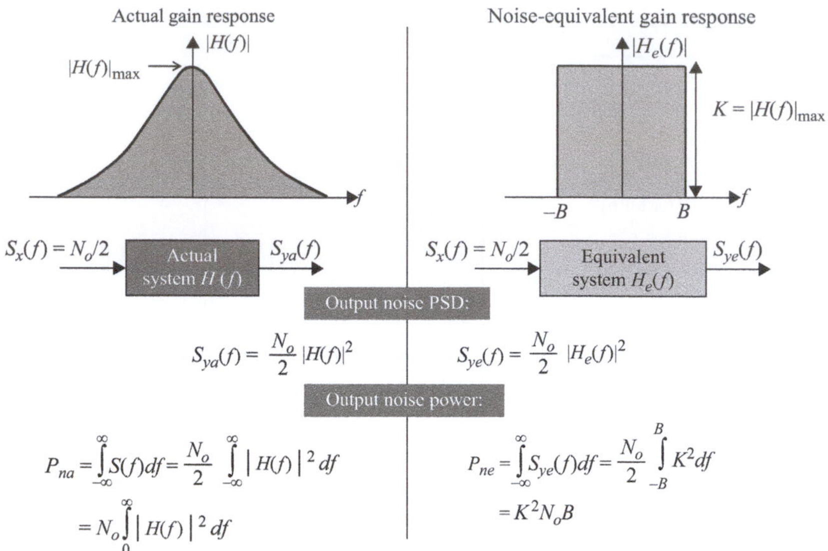

Fig. 2.12 Noise-equivalent bandwidth B

For equivalence this noise power must be equal to the noise power P_{na} at the output of the actual system — obtained earlier in Eq. (2.29) and repeated in Fig. 2.12. Equating the expressions for P_{ne} and P_{na} yields an expression for B:

$$B = \frac{1}{K^2}\int_0^\infty |H(f)|^2 df, \qquad \text{where } K = |H(f)|_{\max} \tag{2.31}$$

Eq. (2.31) defines the *noise-equivalent bandwidth* of the system. This is a very useful concept, which allows us to work with the more convenient ideal filter, as far as noise is concerned. In other words, we replace the realisable filter of frequency-dependent gain response $|H(f)|$ with an ideal brick wall filter of bandwidth B, given by Eq. (2.31). Once we know the noise-equivalent bandwidth of a system, then it is a matter of straightforward multiplication to obtain noise power using

$$P_n = N_o B \qquad \text{(Noise power referred to system input)} \tag{2.32}$$

for power referred to input, or Eq. (2.30) for output power. We will adopt this approach in all system noise calculations, and it will be taken for granted that B (except where otherwise indicated) refers to noise-equivalent bandwidth. If however this bandwidth is not known, then the 3-dB bandwidth of the system may be used in its place. This substitution underestimates noise power but the error will be small if the filter response has steep sides with a small transition width, as the following worked example illustrates.

Worked Example 2.4: Noise-Equivalent Bandwidth of Butterworth Filter

Determine the noise-equivalent bandwidth B of a Butterworth low pass filter of order n. How much error is incurred in noise power computation by using the 3-dB bandwidth in place of B?

The gain response of this filter is given in Fig. 2.11 as follows:

$$|H(f)| = 1/\sqrt{1+(f/f_1)^{2n}} \tag{2.33}$$

We note that $|H(f)|_{max} \equiv K = 1$ at $f = 0$, and apply Eq. (2.31) to obtain the noise-equivalent bandwidth

$$\begin{aligned} B &= \int_0^\infty |H(f)|^2 df = \int_0^\infty \frac{1}{1+(f/f_1)^{2n}} df = f_1 \int_0^\infty \frac{dx}{1+(x)^{2n}} \\ &= f_1 \frac{\pi}{2n} \text{cosec}\left(\frac{\pi}{2n}\right) \end{aligned}$$

It is clear from Eq. (2.33) that $|H(f)| = 1/\sqrt{2}$ at $f = f_1$ for all filter order n. So the 3-dB bandwidth is $B_{3dB} = f_1$, and it follows that B exceeds B_{3dB} by the factor

$$\frac{B}{B_{3dB}} = \frac{\pi}{2n} \text{cosec}\left(\frac{\pi}{2n}\right) \tag{2.34}$$

Using the 3-dB bandwidth in place of B therefore causes noise power to be underestimated by $10\log[\pi \text{cosec}(\pi/2n)/2n]$ dB. We see that the error incurred decreases rapidly as filter order n increases. It is 1.96 dB for $n = 1$ (which is the simple RC filter); 0.46 dB for $n = 2$; 0.20 dB for $n = 3$; 0.11 dB for $n = 4$; and 0.07 dB for $n = 5$.

Worked Example 2.5: Raised Cosine Filter

The raised cosine filter is universally employed in digital transmission systems to reduce inter-symbol interference. For transmission at symbol rate R_s symbols per second (a unit often called *baud*), the normalised gain response of the filter is

$$|H(f)| = \begin{cases} 1, & |f| \le f_1 \\ \frac{1}{2}\left[1 + \cos\left(\pi \frac{|f| - f_1}{f_2 - f_1}\right)\right], & f_1 \le |f| \le f_2 \\ 0, & |f| \ge f_2 \end{cases} \tag{2.35}$$
$$f_1 = (1-\alpha)R_s/2; \; f_2 = (1+\alpha)R_s/2; \; 0 \le \alpha \le 1$$

where α is known as the *roll-off factor* of the filter. Determine the null-bandwidth B_{null} and the noise-equivalent bandwidth B of the filter in terms of α and R_s.

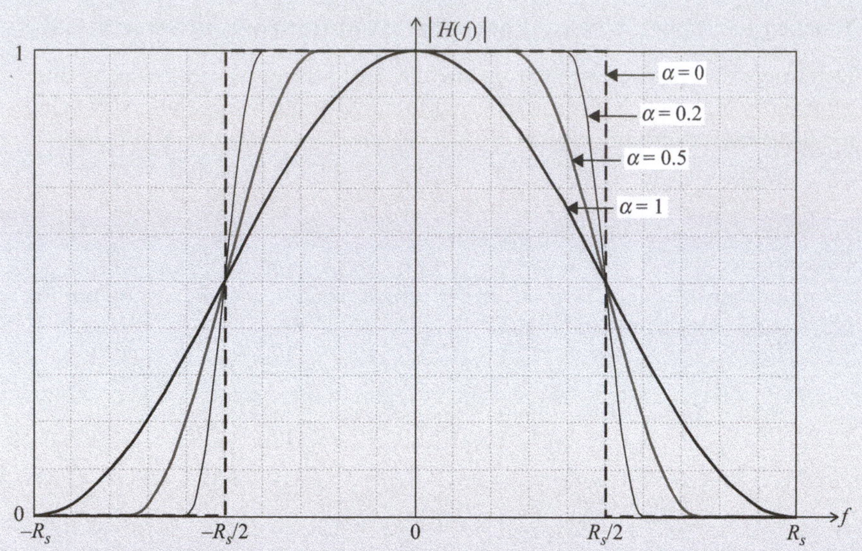

Fig. 2.13 Raised cosine filter of roll-off factor = 0, 0.2, 0.5, 1

From Eq. (2.35) and the gain response of the filter shown in Fig. 2.13 for various values of α, it is clear that the filter reaches its first and only null at $f = f_2 = (1+\alpha)R_s/2$. Thus, null-bandwidth

$$B_{\text{null}} = \begin{cases} (1+\alpha)\dfrac{R_s}{2}, & \text{Base band } (f \text{ centred at } 0) \\ (1+\alpha)R_s, & \text{Band pass } (f \text{ centred at } f_c \gg R_s) \end{cases} \tag{2.36}$$

with a minimum base band value $R_s/2$ when $\alpha = 0$ (which corresponds to an *ideal Nyquist channel* — unrealisable in real time), and a maximum value R_s when $\alpha = 1$ (which corresponds to a *full-cosine roll-off filter*). With a gain of zero at all frequencies above B_{null}, the raised cosine filter has the effect of limiting the spectrum of signals passed through it to an occupied bandwidth equal to B_{null}.

To determine the noise-equivalent bandwidth B of the raised cosine filter, we use Eq. (2.35) in (2.31) with $K = 1$, employing the substitution $\theta = \pi(f - f_1)/(f_2 - f_1)$ to evaluate the resulting integrals:

$$B = \int_0^\infty |H(f)|^2 df = \int_0^{f_1} 1^2 df + \frac{1}{4}\int_{f_1}^{f_2}\left[1 + \cos\left(\pi\frac{f-f_1}{f_2-f_1}\right)\right]^2 df + \int_{f_2}^{\infty} 0 \cdot df$$

$$= f_1 + \frac{1}{4}\int_{f_1}^{f_2}\left[\frac{3}{2} + 2\cos\left(\pi\frac{f-f_1}{f_2-f_1}\right) + \frac{1}{2}\cos\left(2\pi\frac{f-f_1}{f_2-f_1}\right)\right] df$$

$$= f_1 + \frac{3}{8}(f_2 - f_1) + \frac{1}{4}\frac{f_2-f_1}{\pi}\int_0^\pi 2\cos\theta d\theta + \frac{1}{4}\frac{f_2-f_1}{\pi}\int_0^\pi \frac{1}{2}\cos 2\theta d\theta$$

$$= f_1 + \frac{3}{8}(f_2 - f_1) + 0 + 0 = \frac{3}{8}f_2 + \frac{5}{8}f_1$$

$$= \frac{3}{8}(1+\alpha)\frac{R_s}{2} + \frac{5}{8}(1-\alpha)\frac{R_s}{2} = \frac{R_s}{2}(1 - \alpha/4)$$

Thus the noise-equivalent bandwidth of a raised cosine filter of roll-off factor α is

$$B = \begin{cases} (1-\alpha/4)\dfrac{R_s}{2}, & \text{Base band } (f \text{ centred at } 0) \\ (1-\alpha/4)R_s, & \text{Band pass } (f \text{ centred at } f_c \gg R_s) \end{cases} \tag{2.37}$$

It is common practice to assume ideal Nyquist filtering ($\alpha = 0$) when specifying B. This somewhat overestimates noise power but makes for a more reliable system design.

2.7 Distortionless transmission

We wish to specify the gain response and phase response that should be satisfied by a transmission system or communication channel in order to pass a signal without distortion so that its output signal is a scaled (boosted or attenuated) and delayed but otherwise exact copy of the input, as shown in Fig. 2.14. To determine the channel transfer function $H(f)$ that supports this *distortionless transmission*, we note that the channel output $y(t)$ is related to the input $x(t)$ by

$$y(t) = Kx(t - t_o) \tag{2.38}$$

We take the Fourier transform of this equation, applying the time shifting property (Appendix C), to obtain

$$Y(f) = KX(f)\exp(-j2\pi f t_o)$$

But, from Eq. (2.16), the RHS is the product of $X(f)$ and the channel transfer function $H(f)$, which must therefore be the expression

$$H(f) = K\exp(-j2\pi f t_o) \equiv |H(f)|\exp[j\phi_H(f)] \tag{2.39}$$

This is the required distortionless channel transfer function. We see that two conditions must be satisfied, as illustrated in Fig. 2.15:

(i) The amplitude response of the channel must be constant, i.e. $|H(f)| = K$.
(ii) The phase response of the channel must be linear, i.e. $\phi_H(f) = -2\pi t_o f$.

The transfer function of practical transmission channels will in general not satisfy the above conditions without additional filtering. Over the frequency band of interest, any departure of the amplitude response of the channel from a constant value K gives rise to *attenuation distortion*. Similarly, any departure of the phase response from a linear graph gives rise to *phase distortion*, also called *delay*

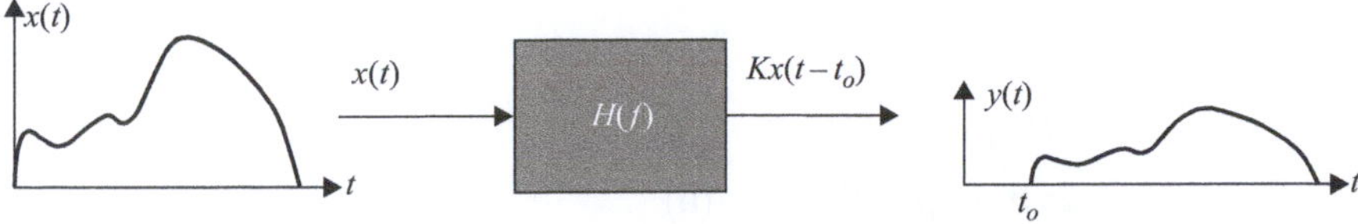

Fig. 2.14 Distortionless transmission system

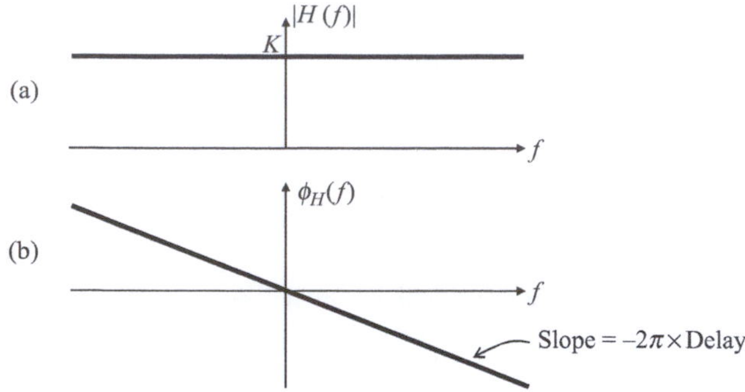

Fig. 2.15 (a) Amplitude and (b) phase response of a distortionless transmission system

distortion. A parameter known as the *group delay* τ_g of the channel is related to the slope of the channel's phase response as follows:

$$\tau_g(f) = -\frac{1}{2\pi}\frac{d\phi_H(f)}{df} \tag{2.40}$$

A channel without phase distortion has a phase response with a constant slope and hence its group delay $\tau_g(f) = t_o$, a constant. For all other channels, the phase response is non-linear, which means that group delay is a function of frequency, varying over the frequency band of interest. A frequency-domain measure of the phase distortion incurred in transmission through a distorting channel is usually given by *differential delay* τ_d, which is the difference between the maximum and minimum values of group delay within the frequency band of interest:

$$\tau_d = \tau_g(f)|_{\max} - \tau_g(f)|_{\min} \tag{2.41}$$

Distortionless transmission through a channel may be approximated over a desired frequency range by using a filter known as an *equaliser* at the output of the channel to compensate for the amplitude and phase distortions caused by the channel. The setup is as shown in Fig. 1.22. The overall system is distortionless, so we may write

$$\begin{aligned}
&H_c(f)H_e(f) = K\exp(-j2\pi ft_o)\\
&|H_c(f)|\exp[j\phi_{Hc}(f)]|H_e(f)|\exp[j\phi_{He}(f)] = K\exp(-j2\pi ft_o)\\
&|H_c(f)||H_e(f)|\exp\{j[\phi_{Hc}(f) + \phi_{He}(f)]\} = K\exp(-j2\pi ft_o)
\end{aligned}$$

Equating the amplitude and phase terms on both sides yields

$$\begin{aligned}
|H_c(f)||H_e(f)| &= K\\
\phi_{Hc}(f) + \phi_{He}(f) &= -2\pi ft_o
\end{aligned}$$

Thus

$$\begin{aligned}
&|H_e(f)| = K/|H_c(f)| && \text{(a)}\\
&\phi_{He}(f) = -[2\pi ft_o + \phi_{Hc}(f)] && \text{(b)}
\end{aligned} \tag{2.42}$$

Eq. (2.42) stipulates that

The amplitude response of the equaliser should be the inverse of the amplitude response of the distorting channel. This provides *gain* (or attenuation) *equalisation*.

The sum of the phase response of the equaliser and that of the distorting channel should be a linear function of frequency. This gives *phase* (or delay) *equalisation*.

Attenuation equalisation alone may be sufficient in some applications such as speech transmission. For example, the attenuation (in dB) of audio telephone lines increases as the square root of frequency. The amplitude response of the transmission medium in this case is

$$|H_c(f)| = K_1 \exp(-a\sqrt{f})$$

An equaliser with amplitude response

$$|H_e(f)| = K_2 \exp(+a\sqrt{f})$$

will adequately compensate for the attenuation distortion of the medium. The overall amplitude response of the combination of transmission medium and equaliser in tandem is flat and independent of frequency:

$$\begin{aligned} |H_c(f)||H_e(f)| &= K_1 \exp(-a\sqrt{f}) K_2 \exp(+a\sqrt{f}) \\ &= K_1 K_2 = K \text{ (a constant)} \end{aligned}$$

Before leaving this topic, it is worth pointing out that we could have taken a slightly more mathematical approach by first specifying the impulse response $h(t)$ of a distortionless transmission system and then taking its Fourier transform (FT) to obtain the transfer function $H(f)$. To be distortionless, the response of a system to a unit impulse $\delta(t)$ must be $K\delta(t-t_o)$, where K is a scale factor and t_o is a constant delay. Thus the system's impulse response is

$$h(t) = K\delta(t - t_o) \tag{2.43}$$

and hence the response $y(t)$ of the system to an arbitrary input signal $x(t)$ is

$$\begin{aligned} y(t) &= x(t) * h(t) = x(t) * K\delta(t - t_o) \\ &= K \int_{-\infty}^{\infty} x(\tau)\delta(t - t_o - \tau)d\tau \\ &= K \int_{-\infty}^{\infty} x(\tau)\delta(\tau - (t - t_o))d\tau \\ &= Kx(t - t_o) \end{aligned}$$

which is Eq. (2.38). Note that in the above steps, we made use of the even property of $\delta(t)$ — Eq. (2.4) — to obtain line 3, and the sifting property — Eq. (2.5) — to obtain the last line. An important observation to make on the above outcome is that the effect of convolving a signal $g(t)$ with a time delayed unit impulse $\delta(t-T_s)$ is to delay the signal by the same amount. That is

$$g(t) * \delta(t - T_s) = g(t - T_s) \tag{2.44}$$

Taking the FT of both sides of Eq. (2.43), using the FT definition in Eq. (C.6) (Appendix C) yields

$$H(f) = \int_{-\infty}^{\infty} K\delta(t - t_o)e^{-j2\pi ft}dt = Ke^{-j2\pi ft_o}\int_{-\infty}^{\infty} \delta(t - t_o)dt = Ke^{-j2\pi ft_o}$$

which is Eq. (2.39). Again, the sifting property was applied.

2.8 Multipath distortion

An important cause of attenuation and delay distortions is *multipath propagation*. This may occur in metallic cables as a result of multiple reflections from impedance irregularities, and improper terminations and imbalances at 2/4 wire interfaces that give rise to echoes. Multipath propagation is also encountered in multimode optical fibre mainly because the path length of the light within the fibre depends on incidence angle. In radio, HF transmission may reach a receiver after reflections at several ionospheric layers; and VHF and UHF reception in mobile communication may be through both direct transmission and reflections from objects. The received signal is the resultant of two or more (sinusoidal) signals of the same frequency but different amplitudes and phases combined at the receiver. It is therefore useful to briefly review the following systematic method of sinusoidal signal addition, which will help us later when we attempt to quantify path loss in a mobile radio channel.

Consider n multipath signals arriving at a receiver. They will in general be sinusoidal signals of the same frequency f but different phases ϕ_i and amplitudes A_i, $i = 1, 2, 3, \ldots, n$. The resultant signal will also be a sinusoid of the same frequency f and of amplitude A and phase ϕ obtained as follows in terms of ϕ_i and A_i:

$$\begin{aligned}
&\sum_{i=1}^{n}[A_i \cos(2\pi ft + \phi_i)] = A \cos(2\pi ft + \phi)\\
&\text{where} \qquad A = \sqrt{A_q^2 + A_I^2}\\
&\qquad\qquad \phi = \arctan(A_q/A_I)\\
&\qquad\qquad A_q = \sum_{i=1}^{n}(A_i \sin\phi_i)\\
&\qquad\qquad A_I = \sum_{i=1}^{n}(A_i \cos\phi_i)
\end{aligned} \tag{2.45}$$

The following steps are involved in applying Eq. (2.45):

(i) Make sure each sinusoid is in the standard cosine format so that you know its correct initial phase ϕ_i.

(ii) Express each sinusoid as a *phasor* — a line of length equal to the amplitude of the sinusoid and direction equal to the initial phase of the sinusoid. See Fig. 2.16a for an illustration.

(iii) Break each phasor into its *in-phase* component $A_{Ii} = A_i\cos(\phi_i)$ and *quadrature* component $A_{qi} = A_i\sin(\phi_i)$, as shown in Fig. 2.16a.

(a)

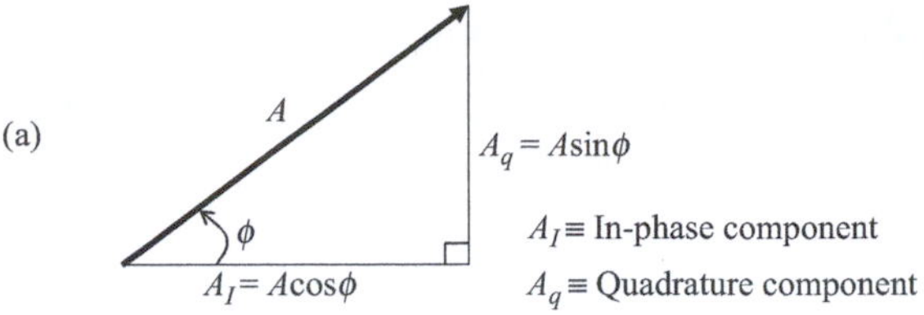

(b)

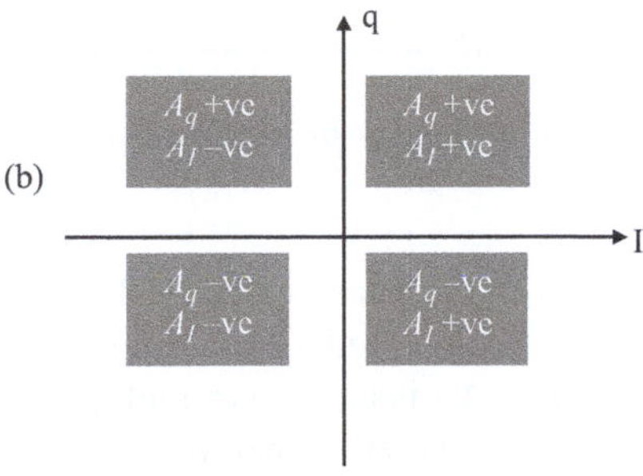

Fig. 2.16 (a) Phasor representation of Acos(2ft + ϕ); (b) quadrant of ϕ

(iv) Add all the in-phase components to obtain A_I — the in-phase component of the resultant sinusoid. Similarly, add all the quadrature components to obtain the resultant quadrature component A_q.

(v) Using Eq. (2.45), determine the amplitude A and phase ϕ of the resultant sinusoid from A_I and A_q. Care must be taken to assign the correct value to ϕ. A calculator will compute arctan(A_q/A_I) merely as a positive or negative angle of magnitude α between 0° and 90°. Fig. 2.16b shows the right quadrant for ϕ based on the signs of A_I and A_q. Thus the correct value of ϕ is obtained from A_I and A_q as follows:

$$\alpha = \tan^{-1}\left(\left|\frac{A_q}{A_I}\right|\right); \quad \phi = \begin{cases} 0^\circ, & A_q = 0,\ A_I = 0 \\ 90^\circ, & A_q > 0,\ A_I = 0 \\ -90^\circ, & A_q < 0,\ A_I = 0 \\ \alpha, & A_q \geq 0,\ A_I > 0 \\ 180 - \alpha, & A_q \geq 0,\ A_I < 0 \\ \alpha - 180, & A_q < 0,\ A_I < 0 \\ -\alpha, & A_q < 0,\ A_I > 0 \end{cases} \tag{2.46}$$

Note that *subtraction* of sinusoids may be accomplished by first incrementing by 180° the phase of each sinusoid that is to be subtracted and then proceeding with Eq. (2.45) exactly as in the case of addition.

Fig. 2.17 shows a simple plane Earth model of multipath propagation with a receiver of height h_r separated by distance d along a flat Earth from a transmitter of height h_t. A radio signal is received via two paths, one the direct line-of-sight of path length d_1, and the other through a secondary path of length d_2 involving a single reflection from a flat ground surface. If the radio signal is of frequency f and the two paths differ in propagation time by $\Delta\tau$, then there will be a phase difference of $\Delta\phi = \theta + 2\pi f\Delta\tau$ between the primary signal $x_1(t)$ and the secondary signal $x_2(t)$

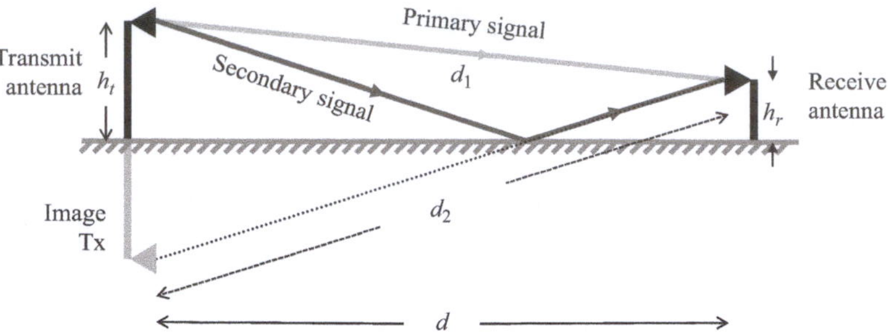

Fig. 2.17 Plane Earth multipath propagation model

arriving over the two paths, where θ is the phase of the reflection coefficient ρ, which means that upon reflection the wave is reduced in amplitude by a factor $|r|$ and shifted in phase through angle θ. If $\Delta\phi$ is an integer multiple of 2π then $x_1(t)$ and $x_2(t)$ add constructively to give a received signal $x(t)$ that has an enhanced amplitude. However, if the phase difference is an odd integer multiple of π, then the two components add destructively and the received signal is severely attenuated. If $x_1(t)$ and $x_2(t)$ have equal amplitude, then $x(t)$ is zero under this situation. In practice, the amplitude of the direct signal is larger than that of the reflected signal. The received signal amplitude therefore varies from a non-zero minimum under destructive interference to a maximum value under constructive interference.

For values of phase difference $\Delta\phi$ other than integer multiples of π, the amplitude and phase of the received signal $x(t)$ are determined according to the method of sinusoidal signal addition given in Eq. (2.45). From this and the geometry of the multipath scenario in Fig. 2.17, we can obtain the amplitude A and phase ϕ of the resultant in terms of the amplitude A_1 of the primary signal and amplitude $A_2 = |\rho|A_1$ of the secondary signal, bearing in mind that the phase of the primary signal is $\phi_1 = 0$ (since this is the reference path), and the phase of the secondary signal is $\phi_2 = \Delta\phi = \theta + 2\pi f\Delta\tau$:

$$\begin{aligned}
A_q &= A_1 \sin\phi_1 + A_2 \sin\phi_2 = A_2 \sin(\Delta\phi) \\
A_I &= A_1 \cos\phi_1 + A_2 \cos\phi_2 = A_1 + A_2 \cos(\Delta\phi) \\
A &= \sqrt{A_q^2 + A_I^2} = \sqrt{A_1^2 + A_2^2 + 2A_1A_2 \cos(\Delta\phi)} \\
&= A_1\sqrt{1 + |\rho|^2 + 2|\rho| \cos(\Delta\phi)} \\
\phi &= \tan^{-1}\left[\frac{A_q}{A_I}\right] = \tan^{-1}\left[\frac{\sin(\Delta\phi)}{A_1/A_2 + \cos(\Delta\phi)}\right] = \tan^{-1}\left[\frac{\sin(\Delta\phi)}{1/|\rho| + \cos(\Delta\phi)}\right] \\
\Delta\phi &\equiv \phi_2 = \theta + 2\pi f\Delta\tau
\end{aligned} \tag{2.47}$$

The difference Δt in propagation time between the two paths is given by the path difference $d_2 - d_1$ of the two paths divided by the speed of light $c = 3 \times 10^8$ m/s, which from Fig. 2.17 yields

$$\Delta\tau = \frac{1}{c}\left(\sqrt{(h_t + h_r)^2 + d^2} - \sqrt{(h_t - h_r)^2 + d^2}\right) \tag{2.48}$$

At large separation distances between transmitter and receiver such that $d \gg h_t, h_r$ we can employ the approximation

$$(1 \pm a)^m \approx 1 \pm ma \qquad \text{for } a \ll 1$$

to reduce Eq. (2.48) as follows:

$$\begin{aligned}
\Delta\tau &= \frac{1}{c}\left\{ d\left(1 + \left[\frac{h_t + h_r}{d}\right]^2\right)^{1/2} - d\left(1 + \left[\frac{h_t - h_r}{d}\right]^2\right)^{1/2} \right\} \\
&\approx \frac{1}{c}\left\{ \left(d + \frac{d}{2}\left[\frac{h_t + h_r}{d}\right]^2\right) - \left(d + \frac{d}{2}\left[\frac{h_t - h_r}{d}\right]^2\right) \right\} \\
&\approx \frac{1}{c}\cdot\frac{d}{2}\left\{ \frac{h_t^2 + h_r^2 + 2h_t h_r - (h_t^2 + h_r^2 - 2h_t h_r)}{d^2} \right\} \approx \frac{1}{c}\cdot\frac{d}{2}\left\{\frac{4h_t h_r}{d^2}\right\} \\
&\approx \frac{1}{c}\cdot\frac{2h_t h_r}{d}
\end{aligned} \tag{2.49}$$

We explore below several important consequences, namely (i) attenuation and phase distortion, (ii) fast fading and inverse 4th-power law dependence of signal strength on distance, and (iii) dispersion.

(i) The exact values of A and ϕ given by Eq. (2.47) depend on frequency f. Some frequencies may be severely attenuated whereas some are enhanced. This causes attenuation and phase distortion. An example of multipath-induced attenuation distortion based on the plane Earth propagation model of Fig. 2.17 is shown in Fig. 2.18. Notice how, for example, a signal frequency

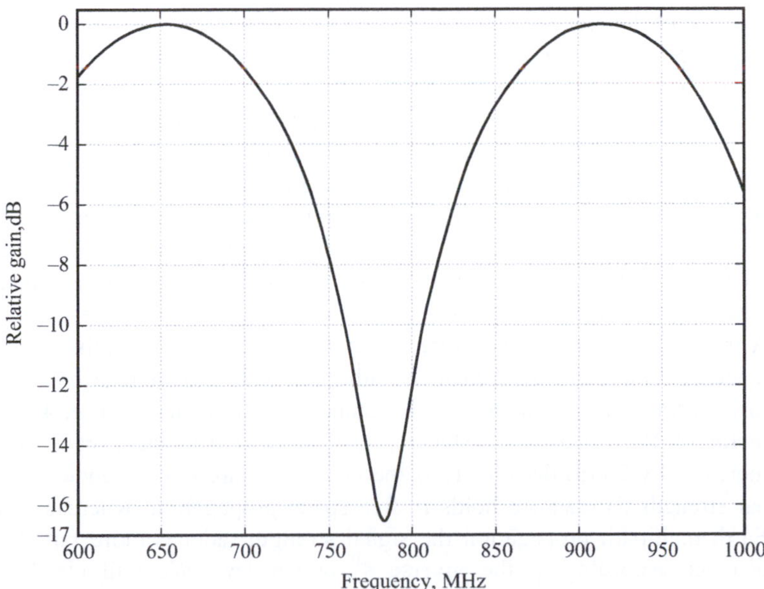

Fig. 2.18 Multipath-induced attenuation distortion due to a single reflection from a plane Earth. This is based on the geometry of Fig. 2.15 with $h_t = 30$ m, $h_r = 2$ m, $d = 100$ m, and reflection coefficient −0.75

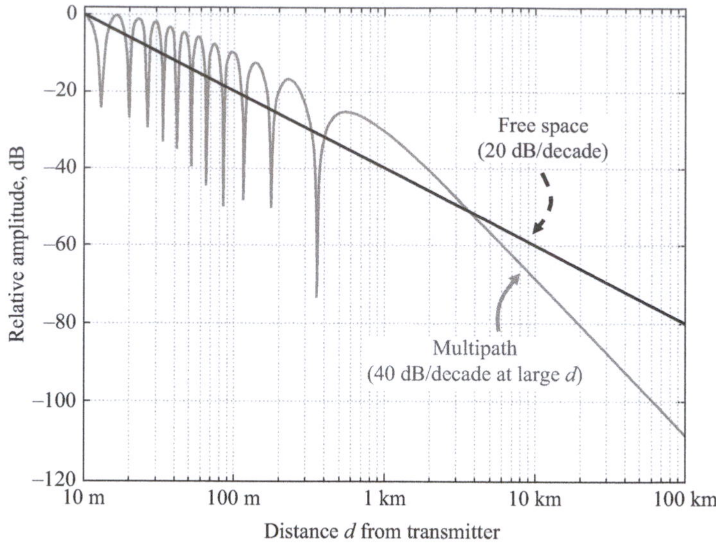

Fig. 2.19 Plane Earth path loss: $h_t = 30$ m, $h_r = 2$ m, $f = 900$ MHz, and reflection coefficient 1

of 783 MHz is attenuated by 4.5 dB more than the attenuation experienced at 800 MHz, and by 16.5 dB more than the attenuation at 650 MHz. A reflection coefficient of −0.75 was assumed, which means $A_2 = -0.75A_1$ in Eq. (2.47). That is, reflection reduces the signal's amplitude by a factor of 0.75 and changes its phase by 180°.

(ii) The amplitude and phase of the received signal depends on the propagation delay difference $\Delta\tau$ between the two paths, which in turn depends on the location of the receiver. If the receiver is not stationary then at distances not too far removed from the sources of multipath this dependence gives rise to rapid changes in the resultant amplitude, a phenomenon called *fast fading*. This is evident in Fig. 2.19, which shows plane Earth path loss as a function of distance and we can see that, close to the transmitter, there is rapid fluctuation of signal strength with distance: A small change in distance can result in a very significant change in received signal strength by several tens of dB. For example, for the system configuration used in Fig. 2.19, a receiver located at 30 m from the transmitter will experience 25 dB more attenuation when moved 3.5 m (equal to 10.5 wavelengths) closer to the transmitter. A mobile receiver will therefore experience time-selective fading as its distance from the transmitter changes with time. At large distances however this loss increases monotonically with distance by 40 dB/decade (i.e. 40 dB per factor of 10 increase in distance), compared to free space path loss that increases by 20 dB/decade. Thus the inverse-square law dependence of signal strength on distance holds in free space propagation, whereas in plane Earth multipath propagation the signal strength reduces more rapidly with distance according to the inverse 4th-power law. We will clarify these assertions in full in the next worked example.

(iii) In digital communications, multipath propagation over two or more differently delayed paths gives rise to pulse broadening or *dispersion*. One transmitted narrow pulse becomes a sequence of two or more narrow pulses at the receiver and this is received as one broadened pulse. Since pulse duration is the reciprocal of symbol rate, channel dispersion places a limit on the symbol rate that can be used without the overlap of adjacent pulses.

Worked Example 2.6: Free Space and Plane Earth Propagation Equations

We wish to derive expressions for

(a) The *free space propagation* equation (also called *Friis* equation) that gives the path loss (i.e. ratio between transmitted power P_t and received power P_r) of a radio wave propagating in free space, and

(b) The *plane Earth propagation* equation that gives the path loss of a radio wave in the propagation scenario depicted in Fig. 2.17 in which the transmitted signal reaches the receiver via a direct path and a secondary path created by reflection from a flat earth.

We will assume that the distance d between transmitter and receiver is much larger than the dimensions of the transmitter and receiver.

The expression we seek is a ratio of powers so we will work with normalised powers (where medium impedance Z is set to unity) since, provided we are consistent at both ends, the impedances cancel out and their particular value is immaterial.

(a) *Friis Equation*

Consider an isotropic antenna that radiates a (sinusoidal) radio wave of amplitude A_o equally in all directions. The (normalised) power of the transmitted wave is $P_t = A_o^2/2$ and at a distance d from the transmitter this power will be uniformly distributed over the surface of an imaginary sphere of area $4\pi d^2$ centred on the transmitter so that a receiver at this distance d will see a power flux density $P_{fd} = P_t/4\pi d^2$ and therefore will collect power $P_r = P_{fd}A$, where A is the effective aperture or collection area of the receive antenna, which for an isotropic antenna is given by [1], $A = \lambda^2/4\pi$. The received power is therefore

$$P_r = P_{fd}A = \frac{P_t}{4\pi d^2}\cdot\frac{\lambda^2}{4\pi} = \frac{P_t}{(4\pi d/\lambda)^2} \tag{2.50}$$

We see that the transmitted power is reduced by a factor $(4\pi d/\lambda)^2$ by the time it reaches the receiver. This factor is the *free space path loss* L_s, which, by substituting $\lambda = c/f$, may be more conveniently expressed in dB in terms of frequency f in GHz and distance d in km as

$$\begin{aligned} L_s &= \frac{P_t}{P_r} = (4\pi d/\lambda)^2 = 20\log\left(\frac{4\pi d}{\lambda}\right) \quad \text{dB} \\ &= 92.44 + 20\log f_{\text{GHz}} + 20\log d_{\text{km}} \quad \text{dB} \end{aligned} \tag{2.51}$$

Notice that the received power P_r is proportional to $1/d^2$ (i.e. it follows an inverse-square law dependence on distance). Notice also that the distance-dependent term in the logarithmic expression for L_s contributes dB values of 0, 20, 40,... at $d_{\text{km}} = 1, 10, 100, \ldots,$ km respectively, which indicates that L_s increases by 20 dB for every factor of 10 increase in distance. This is usually expressed as 20 dB/decade. Alternatively, the contribution is 0, 6, 12, 18,..., dB at 1, 2, 4, 8,..., km respectively, which is a loss of 6 dB per factor of 2 increase in distance or 6 dB/octave.

Expressing Eq. (2.50) in terms of transmitted and received amplitudes A_o and A_1 gives

$$\frac{P_r}{P_t} = \frac{A_1^2/2}{A_o^2/2} = \frac{1}{(4\pi d/\lambda)^2}$$

or

$$A_1 = A_o \cdot \frac{\lambda}{4\pi d} \tag{2.52}$$

which shows that the signal amplitude is inversely proportional to distance as expected if power follows an inverse-square law dependence on distance.

Finally, in practice the transmit and receive antennas have respective gains G_t and G_r in the direction along the link. The received power P_r is thereby increased by the factor G_tG_r so that Eq. (2.50) yields

$$\frac{P_r}{P_t} = G_tG_r\left(\frac{\lambda}{4\pi d}\right)^2 \tag{2.53}$$

which is the well-known *Friis equation* relating received power and transmitted power in free space propagation. In logarithmic units this reads

$$P_r = P_t + G_t + G_r - L_s = \text{EIRP} + G_r - L_s \quad \text{(in dBW)} \tag{2.54}$$

where G_t and G_r are transmit and receive antenna gains in dB, L_s is free space path loss in dB given by Eq. (2.51), P_t is the transmitted power in dBW, and EIRP $= P_t + G_t$ is the *effective isotropically radiated power* in dBW, which is the amount of input power that an isotropic antenna would require in order to emit the same power as radiated in its boresight direction by our directional antenna of gain G_t supplied with input power P_t. Note that dBm unit is sometimes used in place of dBW, and the two units are related by adding 30 to the dBW value to convert it into a dBm value. For example, 20 dBW = 50 dBm.

(b) *Plane Earth propagation equation*

The resultant amplitude A in this scenario is given by Eq. (2.47) where A_1 is the signal amplitude reaching the receiver at a distance $d_1 \approx d$ through the primary path, and A_1 is given in terms of the transmitted amplitude A_o by Eq. (2.52). Thus

$$\begin{aligned} A &= A_1\sqrt{1 + |\rho|^2 + 2|\rho|\cos(\theta + 2\pi f\Delta\tau)} \\ &= A_o \cdot \frac{\lambda}{4\pi d} \cdot \sqrt{1 + |\rho|^2 + 2|\rho|\cos(\theta + 2\pi f\Delta\tau)} \end{aligned}$$

At large d, Δt is given approximately by Eq. (2.49) and reflection occurs at near-grazing incidence so that the reflection coefficient $\rho \approx -1$, which means that $|\rho| \approx 1$ and $\theta \approx 180°$ in the above equation. Thus

$$\begin{aligned}\frac{A}{A_o} &= \frac{\lambda}{4\pi d} \cdot \sqrt{2 - 2\cos(2\pi f \Delta\tau)} = \frac{\lambda}{4\pi d} \cdot \sqrt{4\sin^2(\pi f \Delta\tau)} = \frac{\lambda \sin(\pi f \Delta\tau)}{2\pi d} \\ &\approx \frac{\lambda}{2\pi d} \sin\left(\pi f \frac{2h_t h_r}{cd}\right) \approx \frac{\lambda}{2\pi d} \sin\left(\frac{2\pi h_t h_r}{\lambda d}\right) \approx \frac{\lambda}{2\pi d} \cdot \frac{2\pi h_t h_r}{\lambda d} \\ &\approx \frac{h_t h_r}{d^2}\end{aligned}$$

where we used the trigonometric identities of Eq. (B.10) and (B.14) in Appendix B to simplify the first line, and used the approximation $\sin(\theta) \approx \theta$ for $\theta \to 0$ to eliminate the sine function in the second line. Noting (as in part (a) above) that the ratio of powers equals the square of the ratio of amplitudes, and including the increase in received power due to transmit and receive antenna gains G_t and G_r, we obtain

$$\frac{P_r}{P_t} = G_t G_r \cdot \frac{h_t^2 h_r^2}{d^4} \tag{2.55}$$

which in logarithmic units reads

$$\begin{aligned}P_r &= \text{EIRP} + G_r + 20\log(h_t) + 20\log(h_r) - 40\log(d) \\ &\equiv \text{EIRP} + G_r - L_p\end{aligned} \tag{2.56}$$

where

$$L_p\ (\text{dB}) = 10n\log(d) - 20\log(h_t) - 20\log(h_r), \quad n = 4 \tag{2.57}$$

is referred to as the plane Earth path loss, in analogy with L_s in Eq. (2.51), which represents free space path loss.

Notice that in this multipath scenario comprising a single reflection from a flat earth and reception at large distances from the transmitter, the received power decreases very rapidly with distance as $1/d^4$. In other words, the *path loss exponent* is $n = 4$, which represents a more rapid decline of signal strength with distance than in free space propagation where $n = 2$. As illustrated in Fig. 2.19, this means that path loss increases with distance by 40 dB/decade or 12 dB/octave in this multipath propagation scenario, compared to 20 dB/decade or 6 dB/octave in free space.

Worked Example 2.7: Multipath Propagation

A 1 GHz radio signal is received via a primary path and one secondary path. The primary signal power is −100 dBm; the secondary signal is attenuated by 3 dB and shifted in phase by −10° upon reflection, and the propagation time difference between the primary and secondary paths is 100.25 ns. Determine the received signal power.

Primary signal power $P_1 = -100\text{ dBm} = 10^{-(100/10)}\text{ mW} = 10^{-13}\text{ W}$

Assuming normalised power we obtain the amplitude A_1 of the primary signal as

$$A_1 = \sqrt{2P_1} = \sqrt{2 \times 10^{-13}} = 4.472 \times 10^{-7}\text{ V} = 0.4472\ \mu\text{V}$$

The secondary signal amplitude A_2 is 3 dB below A_1. Thus

$$A_2 = (10^{-3/20})A_1 = 3.166 \times 10^{-7}\text{ V} = 0.3166\ \mu\text{V}$$

Using the primary path as reference, the phase of the primary signal $\phi_1 = 0$; and the phase ϕ_2 of the secondary signal and amplitude A of the resultant follow from Eq. (2.47) with $\theta = -10°$, $\Delta\tau = 100.25$ ns, $f = 1$ GHz and A_1 and A_2 as above. Thus

$$\Delta\phi \equiv \phi_2 = -10 \times \frac{\pi}{180} + 2\pi \times 10^9 \times 100.25 \times 10^{-9} \equiv \frac{\pi}{2} - \frac{\pi}{18} = \frac{4}{9}\pi$$
$$A = \sqrt{A_1^2 + A_2^2 + 2A_1A_2\cos(\Delta\phi)} = 0.5911\ \mu\text{V}$$

And received power

$$P = \frac{A^2}{2} = 1.747 \times 10^{-13}\text{ W} = 10\log\left(\frac{1.747 \times 10^{-13}}{1 \times 10^{-3}}\right)\text{ dBm}$$
$$= -97.58\text{ dBm}$$

2.9 Terrestrial mobile radio channel

It is important to note that the plane Earth model of multipath propagation derived in the previous section will not yield reliable results in realistic mobile radio environments where, in addition to a ground surface that follows the curvature of the Earth (and hence is *non-plane*), there is often a multiplicity of other objects such as hills, trees, buildings, vehicles and other manmade structures that interact with the transmitted radio signal through various mechanisms including reflection, scattering, absorption and diffraction. Furthermore, a line-of-sight or direct path signal might not exist. The received signal at a given point is thus the resultant of a large number of waves each having random and statistically independent amplitude, phase and angle of arrival. The resultant signal strength will therefore vary significantly from point to point within a small area of a few wavelengths in diameter, and this translates into rapid temporal variation or *fast fading* in a moving receiver. In this practical scenario, a useful path loss model that predicts the *median* signal strength within each small area may be obtained by introducing into the plane Earth path loss formula of Eq. (2.57) an empirically determined factor (called *clutter factor*) β that depends on frequency and the type of environment, but not on distance from the transmitter. An example of such formulation is the *semi-empirical* path loss model by Ibrahim and Parsons cited in [2], derived from extensive measurements in London at frequencies 168, 445 and 896 MHz, which gives

$$L_p(dB) = 40\,\log d - 20\,\log h_t - 20\,\log h_r + \beta$$
$$\beta = 20 + \frac{f_{\text{MHz}}}{40} + 0.18L - 0.34H + K \qquad (2.58)$$
$$K = \begin{cases} 0.094U - 5.9, & \text{Highly urbanised areas} \\ 0, & \text{Otherwise} \end{cases}$$

where $d \leq 10{,}000$ m is the distance between base station antenna of height h_t and mobile antenna of height $h_r \leq 10$ m; L is the land usage factor that specifies the percentage of the 500 m $\times$ 500 m test square that is covered by buildings of whatever height; U is the degree of urbanisation factor that specifies the percentage of building site area in the test square occupied by buildings having four or more floors; H is the difference between the average ground heights of the test squares containing the transmitter and receiver; and f_{MHz} is the signal frequency in MHz.

2.10 Summary

Our study of the characterisation of linear signals and systems in *time* and *frequency* is now complete. By virtue of the *principle of superposition* being upheld, a *linear* system is fully characterised in the time domain by its *impulse response* $h(t)$ and in the frequency domain by its *transfer function* $H(f)$, both functions constituting a *Fourier transform pair*. The output of such systems may be determined in the time domain by *convolving* the input signal with $h(t)$ or in the frequency domain by *multiplying* the input spectrum by $H(f)$. The shape, or more specifically non-zero duration of $h(t)$ is indicative of *pulse spreading* in the system, whereas a non-flat shape of the magnitude of $H(f)$ and any non-linear change of its phase with frequency are indicative of *attenuation* and *phase distortions* respectively. These distortions can be minimised by connecting a suitably designed *equalisation filter* in cascade with the system in order to flatten the overall gain response and *linearise* the overall phase response within the frequency band of interest to the transmission.

We also discussed the various effects of multipath radio propagation in general and analysed in detail the special case of a radio channel in which multipath propagation arises due to the existence at the receiver of a primary signal received directly from the transmitter as well as a secondary signal arriving via reflection from a plane Earth surface. The phenomena of fast fading and inverse 4th power decrease in median signal strength with distance were delineated and a semi-empirical formulation presented for application in real environments containing multiple reflecting objects.

In the next chapter we turn our attention to the analysis of *non-linear* systems where the principle of superposition does not hold and as a result we are unable to determine system response to an arbitrary input based on its response to an impulse or a sinusoid.

2.11 References

[1] Kraus, J. D. (1988). *Antennas 2E*, New York, NY: McGraw-Hill. ISBN 0-07-100482-3

[2] Parsons, J. D. (2000). *The mobile radio propagation channel 2E*. Chichester: Wiley. ISBN 0-471-98857-X, pp. 88–91

[3] Cann, A. J. (1980). 'Nonlinearity model with variable knee sharpness', *IEEE Transactions on Aerospace and Electronic Systems*, 16(6), pp. 874–877

Chapter 3
Non-linear systems

In Chapter Three

- ✓ Analysis of non-linear systems. You will learn how to characterise non-linear systems and determine their output where, because the principle of superposition does not hold, we cannot employ the concepts of system impulse response or transfer function.
- ✓ Simulation of non-linear systems. Step-by-step computer simulation approach to obtain the output of a non-linear system in response to a narrow band input. The concept of quadrature and in-phase components of a band pass signal is introduced.
- ✓ Intermodulation distortion. You will also learn how interference is generated in these systems when multiple carriers are present and how to calculate the interference power and hence carrier-to-interference ratio.

3.1 Introduction

The linear systems that we examined in the previous chapter simply modify the amplitudes and phases of existing frequency components of an input signal — giving rise to attenuation and phase distortions (although distortionless transmission is possible as also discussed). In particular, linear systems do not create any new frequency components. Thus the *absolute bandwidth* of the output signal — defined as a frequency range encompassing all the signal's positive frequency components — of a linear system cannot exceed its input value. In fact output bandwidth is often smaller than input bandwidth due to components at one or both ends of the input frequency spectrum being blocked altogether by the typically band limited channel. In view of the inverse relationship between time and frequency, this bandwidth-narrowing action of the linear system manifests itself as pulse spreading in the time domain.

In the more general cases involving non-linearity, the action of the system on one input — impulse or sinusoid — is affected both by the magnitude of that input and by the presence of other inputs. It is then not possible to determine the response of such a system to an arbitrary input, which in the time domain comprises a train of weighted impulses and in the frequency domain is a sum of sinusoids, based on its response when excited by one impulse or sinusoid. The impulse response and transfer function are therefore not useful as analysis tools for non-linear systems and we must resort to the *transfer characteristic* of the system in determining system output, which inevitably includes new frequencies not present in the input.

Non-linear systems or channels cause non-linear distortions, which are characterised by the following:

(i) The principle of superposition, stated in Eq. (2.1), does not hold. Rather, what the system does to one signal is affected by the presence of other signals. The analysis steps in Figs. 2.3 and 2.6 that led to Eqs. (2.7) and (2.16) are therefore not applicable. That is, the concepts of *impulse response* and *transfer function* are no longer applicable as a systems analysis tool since what the system does to an arbitrary input signal is no longer simply equal to the sum of what it does to the signal's constituent impulses or sinusoids.

(ii) The output signal is no longer directly proportional to the input signal at all levels. Thus the gain of the non-linear system is not a constant value G but rather a function $G(A)$ of input signal amplitude A.

(iii) The output signal contains frequency components not present in the input signal. This *bandwidth expansion* by the non-linear system imposes a requirement for the output signal, if sampled without first being filtered to exclude the new frequencies, to be processed using a faster sampling rate than necessary for the input signal in order to avoid alias distortion. Second, the new frequency components may fall in the band occupied by other user signals in a multicarrier system and thereby cause interference.

In what follows we study the characterisation and analysis of two classes of non-linear systems, first memoryless systems and then those with memory. We then apply this insight to develop a simple procedure for computer simulation of non-linear systems and to compute the intermodulation distortions produced in multicarrier operation of non-linear systems.

3.2 Memoryless non-linear systems

The output $y(t)$ of a memoryless non-linear system at any time instant t depends exclusively on the input signal $x(t)$ applied at that instant. The functional relationship between $y(t)$ and $x(t)$ is usually referred to as the *transfer characteristic* of the system. This means that the output signal is always in step with the input signal and an excitation at the input is felt instantaneously (i.e. without delay) at the system output. Thus, a memoryless system does not cause any *phase shift* or *phase distortion*. The only effect of a memoryless non-linear system on a signal transmitted through it is *amplitude distortion*, which is manifested in the time domain by a distortion in the waveform of the output signal $y(t)$ when compared to the waveform of the input $x(t)$, and in the frequency domain by the appearance of new frequency components in the output spectrum $Y(f)$.

The transfer characteristic of simple non-linear systems or devices may be given in analytic form obtained through a theoretical derivation; an example of this being the diode equation that gives an exponential relationship between (output) current and applied (input) forward-bias voltage in a *p-n* junction diode (system), or the *full-wave rectifier* that converts a bipolar input signal into a unipolar output signal $y(t) = |x(t)|$. Another example is the *μ-law compressor*, illustrated in Fig. 3.1, which is usually implemented digitally in non-linear pulse code modulation to

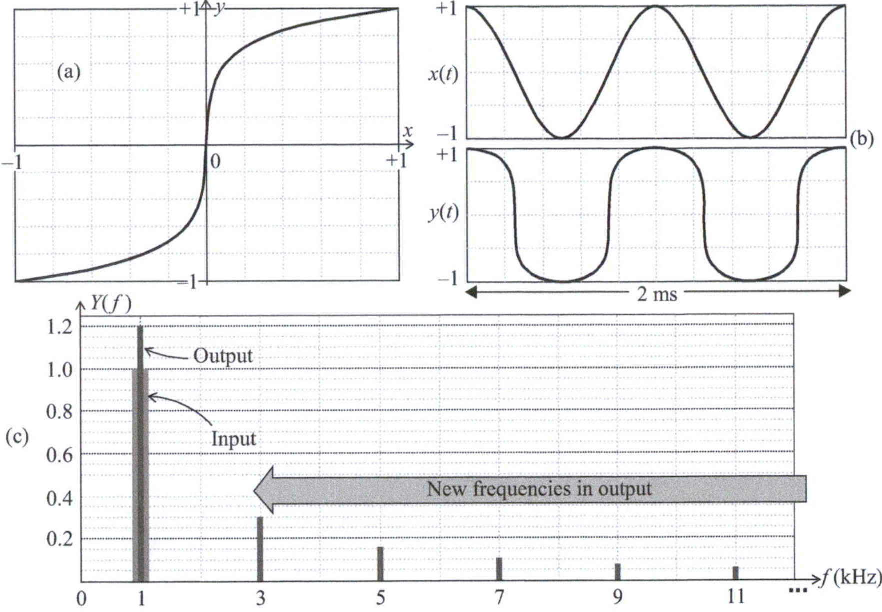

Fig. 3.1 *μ-Law compressor: (a) transfer characteristic; (b) response y(t) to a sinusoidal input x(t); (c) output spectrum Y(f)*

predistort the signal prior to sampling and quantisation. Using the signum function sgn(x) — defined in Fig. 2.2b — the compressor's transfer characteristic is given in terms of a normalised input by

$$y(t) = \frac{\ln[1 + \mu|x(t)|]}{\ln(1 + \mu)}\,\text{sgn}[x(t)]; \quad -1 \le x(t) \le 1, \quad \mu = 255 \tag{3.1}$$

Figs. 3.1a, 3.1b and 3.1c respectively show the transfer characteristic of the compressor, the output $y(t)$ when the input $x(t)$ is a unit-amplitude sinusoidal signal of frequency 1 kHz, and the single-sided output amplitude spectrum $Y(f)$. Notice that the output contains new frequencies at harmonics of the input frequency. The spectrum shown extends only up to harmonics with amplitude not less than 4% of the output amplitude at input frequency.

The transfer characteristic of practical non-linear systems such as travelling wave tube amplifiers (TWTA) used in satellite repeaters are usually obtained through measurements, with the readings laid out in tabular form featuring a column of input power levels versus a column of corresponding output power. In this case the response y to any input x that lies between two measurement points x_1 and x_2 with respective measured outputs y_1 and y_2 is obtained through *linear interpolation* as

$$y = y_1 + \left[\frac{y_2 - y_1}{x_2 - x_1}\right](x - x_1) \tag{3.2}$$

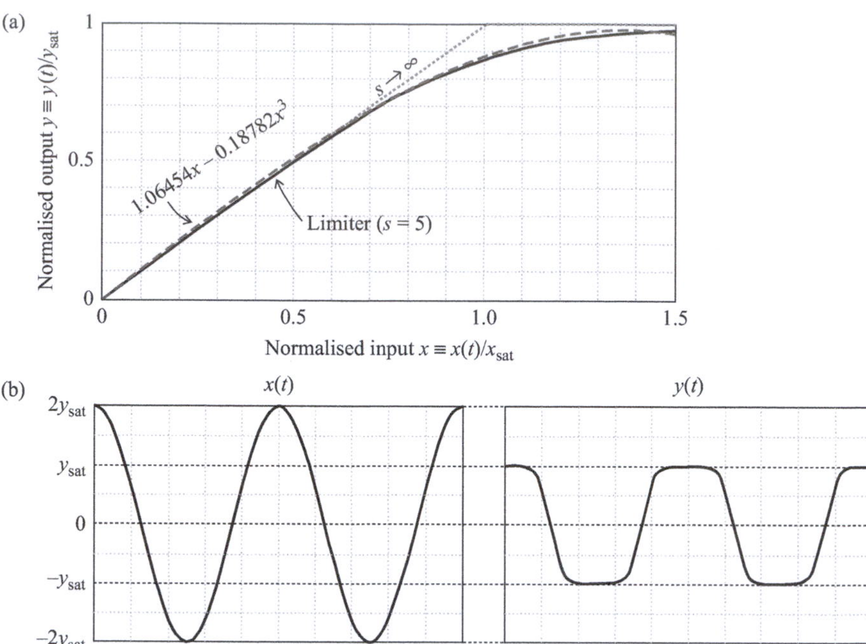

Fig. 3.2 (a) 3rd-Order polynomial fit to transfer characteristic of a limiter (knee sharpness parameter $s = 5$); (b) response y(t) to a sinusoidal input x(t)

Over the input range, it is often possible to approximate the transfer characteristic by a polynomial function of order N in x so that we write

$$y(t) \approx a_1 x(t) + a_2 x^2(t) + \cdots + a_N x^N(t) = \sum_{n=1}^{N} a_n x^n(t) \tag{3.3}$$

Fig. 3.2 shows an example of a 3rd-order polynomial fit to the transfer characteristic of a *limiter*, a parametric form of which is given by [1]

$$y(t) = \frac{y_{sat}}{\left[1 + (x_{sat}/|x(t)|)^s\right]^{1/s}} \operatorname{sgn}[x(t)] \tag{3.4}$$

where y_{sat} is the saturated output level, x_{sat} is the input saturation level and s is a knee sharpness parameter that governs how the output approaches saturation as input is increased. Note that $y(t)$ attains to y_{sat} at $x(t) = x_{sat}$ only for $s \to \infty$; whereas for finite values of s, y_{sat} is a limiting output level that is approached as input increases beyond x_{sat}.

To see how a non-linear system creates new frequency components, consider a simple input signal $x(t)$ consisting of only two sinusoids of zero initial phase at frequencies f_1 and f_2:

$$x(t) = A_1 \cos(2\pi f_1 t) + A_2 \cos(2\pi f_2 t) \tag{3.5}$$

Let this signal be transmitted through a non-linear system whose output $y(t)$ is a non-linear function of the input, represented by a 3rd-order polynomial:

$$y(t) = a_1 x(t) + a_2 x^2(t) + a_3 x^3(t) \tag{3.6}$$

Substituting Eq. (3.5) into Eq. (3.6) gives

$$
\begin{aligned}
y(t) = {} & \frac{1}{2}a_2\left(A_1^2 + A_2^2\right) + \left[a_1 + \frac{3}{4}a_3\left(A_1^2 + 2A_2^2\right)\right]A_1 \cos 2\pi f_1 t \\
& + \left[a_1 + \frac{3}{4}a_3\left(A_2^2 + 2A_1^2\right)\right]A_2 \cos 2\pi f_2 t \\
& + \frac{1}{2}a_2A_1^2 \cos[2\pi(2f_1)t] + \frac{1}{2}a_2A_2^2 \cos[2\pi(2f_2)t] \\
& + \frac{1}{4}a_3A_1^3 \cos[2\pi(3f_1)t] + \frac{1}{4}a_3A_2^3 \cos[2\pi(3f_2)t] \\
& + a_2A_1A_2 \cos[2\pi(f_1 + f_2)t] + a_2A_1A_2 \cos[2\pi(f_1 - f_2)t] \\
& + \frac{3}{4}a_3A_1^2A_2 \cos[2\pi(2f_1 + f_2)t] + \frac{3}{4}a_3A_2^2A_1 \cos[2\pi(f_1 + 2f_2)t] \\
& + \frac{3}{4}a_3A_1^2A_2 \cos[2\pi(2f_1 - f_2)t] + \frac{3}{4}a_3A_2^2A_1 \cos[2\pi(2f_2 - f_1)t]
\end{aligned} \tag{3.7}
$$

We note first of all that if in Eq. (3.3), $a_k = 0$, for all $k > 1$, then the system is linear and $y(t)$ is proportional to $x(t)$ through a constant gain factor $G = a_1$. If however these coefficients are non-zero then the system is non-linear and we observe in Eq. (3.7) the following distortions:

There is a DC component (i.e. $f = 0$), which was not present in the input. This happens whenever any of the even coefficients $a_2, a_4, \ldots$ in Eq. (3.3) is non-zero. However, $a_2 \ll a_3$ in non-linear systems such as amplifiers that are suitably modelled by the limiter family of Eq. (3.4), yielding a negligible DC component in such cases.

The system gain is no longer constant but varies with input amplitude. At frequency f_1, for example, an input amplitude A_1 yields output amplitude $\left[a_1 + \frac{3}{4}a_3\left(A_1^2 + 2A_2^2\right)\right]A_1$. The system gain is thus $G(A) = a_1 + \frac{3}{4}a_3\left(A_1^2 + 2A_2^2\right)$, and clearly now depends on input amplitude. We see that $G(A) \approx a_1$, a constant at low input levels that correspond to the *linear region* of operation of the non-linear device. In high power amplifiers (HPAs) typically $a_3 < 0$ and $a_2 \sim 0$, so that $G(A)$ has maximum value a_1 and decreases as input level increases until the output peaks at a maximum value referred to as the *saturated output* level y_{sat}. Beyond this saturation point any further increase in input level results in a reduction in output level. Notice also that the system gain is different when f_1 is present alone ($A_2 = 0$) than when f_1 is accompanied by another frequency component f_2 ($A_2 \neq 0$). Thus, what the system does to one signal is affected by the presence of other signals, a sign of non-linearity as earlier noted.

For each input frequency component f_k, there appear new frequency components in the output signal at mf_k, $m = 2, 3, \ldots, N$, where N is the order of the non-linearity ($= 3$ in this case). Since these new frequencies are harmonics of the input frequency, this type of distortion is termed *harmonic distortion*.

For any two input frequencies f_1 and f_2, there appear new components at $mf_1 \pm nf_2$, $|m| + |n| = 2, 3, \ldots, N$. These are the sum and difference of the harmonic frequencies. This type of distortion is termed *intermodulation distortion*. The frequency component at $mf_1 \pm nf_2$ is called an *intermodulation product* (IMP) of order $|m| + |n|$. The power in an IMP decreases with its order.

The phase of the output signal (at frequencies f_1 and f_2) is the same as the input phase at each frequency, since the coefficients a_1, a_2 and a_3, and the amplitudes A_1 and A_2 are all real numbers. Thus, as expected of memoryless systems, there is no phase shift. The introduction of a phase shift between input and output requires systems with memory that contain energy storage components such as capacitors and inductors. In reality, memoryless systems are an idealisation and all practical transmission systems have non-zero memory since there will always be some amount of residual capacitive and inductive effects in any conductors or semiconductors within the system.

Our discussion above has focused on what is in fact the memoryless instantaneous-response class of systems. There is in addition the memoryless delayed-response type of system in which the output $y(t)$ depends exclusively on a past instantaneous input $x(t-\Delta t)$. Here there is a fixed delay Δt between system input and its corresponding response, and the output is determined only by the input applied Δt seconds earlier. In the context of a transmission channel we may treat Δt as a fixed *propagation delay* and say that it takes an input signal this constant time duration to pass through the channel so that an input at time t is received (at the output) at time $t+\Delta t$. Noting that the angle of a sinusoid of frequency f increases by $2\pi f$ radians per second, we see that for a sinusoidal input the angle of the output at time t will be $2\pi f\Delta t$ radians larger than that of the input at the time of application, and it is tempting to associate a phase shift $\Delta\varphi = 2\pi f\Delta t$ with this system. That would however be misleading since the phase shift $\Delta\varphi$ is merely a consequence of the expiration of time Δt, rather than any particular action of the system. That is, the signal would have undergone the same phase shift simply by staying at the input for the same length of time Δt without being transmitted through the system. Thus discounting this extra phase advance incurred over the fixed delay interval Δt by all sinusoids of the same frequency (whether or not transmitted through the system), we see that there is in fact no phase shift due to system processing of the input signal. The discussion presented above for memoryless instantaneous-response systems is therefore equally applicable to memoryless delayed-response systems, provided we note that $x(t)$ actually predicts $y(t+\Delta t)$ so that the LHS of Eqs. (3.3), (3.6) and (3.7) is replaced by $y(t+\Delta t)$.

3.3 Non-linear systems with memory

When the output of the system depends on past as well as current input, we can describe the system action as combining multiple samples of the input signal in some non-linear fashion (dictated by the transfer characteristic of the system) that incorporates moving weighted averaging (hence low pass filtering) or differencing (hence high pass filtering) of the collated samples. The implication is that filtering is inherent in systems with memory, and therefore a non-linear system with memory will have a frequency-dependent transfer characteristic. We can identify the following transformational effects of a non-linear system with memory on a signal transmitted through it.

Harmonic and intermodulation distortion due to the creation of new frequency components as earlier discussed. This is an exclusive feature of all non-linear systems, whether memoryless or with memory.

Distortion of the amplitude spectrum within the frequency band common to both input and output as a result of individual input frequencies being differently scaled in amplitude by the system. This is an exclusive feature of all systems with memory, whether linear or non-linear.

Alteration of the phase spectrum due to a shift of the phase of each input frequency component by an amount that depends on both input amplitude and frequency. Phase alteration is an exclusive feature of systems with memory, but if the system is linear then the alteration is independent of amplitude. And if the alteration is not a linear function of frequency then we have what is termed *phase distortion*, also called *delay distortion*, as earlier elaborated in Section 2.7.

Scaling of each sinusoidal input of frequency f and amplitude A by a gain factor $G(f, A)$ to produce an output of amplitude $[G(f, A)]A$ at frequency f. This is a feature of all systems, both linear and non-linear, with and without memory. The distinctions lie in the functional dependence of G. If the gain G does not depend on amplitude A then the system is linear. If G is constant (i.e. independent of both f and A) then the system is linear and memoryless (and hence free from attenuation distortion — see Section 2.7). If G depends only on f, then the system is linear and with memory; but if G depends on both f and A then the system is non-linear and with memory.

For a narrow band input signal whose bandwidth is small compared to the non-linear system's bandwidth it is sufficient to characterise the system by a single set of measurements taken at the centre frequency f_c of the input signal. More specifically, we apply a sinusoidal input signal $x(t)$ to the system and measure the output $y(t)$, where

$$\begin{aligned} x(t) &= A\cos(2\pi f_c t) \\ y(t) &= A_o\cos[2\pi f_c t + \Phi(A)] \equiv [G(A)]A\cos[2\pi f_c t + \Phi(A)] \end{aligned} \tag{3.8}$$

and we repeat these measurements over the entire allowable range of input amplitudes for the system, leading to a table of values of system gain $G(A)$ and system phase shift $\Phi(A)$ versus A. Just as the plot of $y(t)$ versus $x(t)$, see Fig. 3.2, is the transfer characteristic of the system, so $G(A)$ as a function of A is termed *AM/AMcharacteristic*, and $\Phi(A)$ versus A the *AM/PM characteristic* of the non-linear system. There are two issues worthy of further elaboration.

First, $G(A)$ and $\Phi(A)$ provide exact characterisations of the system's performance only at the frequency f_c at which measurements were taken. However, on the reasonable assumption that these characteristics change slowly with frequency, the error can be considered negligible when $G(A)$ and $\Phi(A)$ are applied at all frequencies of a narrow band input signal with bandwidth (centred around f_c) much smaller than system bandwidth. If however the input is a wide band signal, or its bandwidth is comparable to that of the non-linear system, then the change in $G(A)$ and $\Phi(A)$ over the frequency range of the input signal will be significant. In such a case, multiple sets of measurements over a range of frequencies covering the input bandwidth will be necessary in order to more accurately characterise the system. Linear interpolation may then be employed to estimate system characteristics at frequency or amplitude values that fall between measurement points.

Second, we note that the AM/AM and AM/PM characteristics obtained through the measurements indicated by Eq. (3.8) model all of the effects outlined above except the harmonic and intermodulation distortions of the non-linear system. This shortcoming however does not in any way diminish the usefulness of these characteristics for determining the output of a non-linear transmission system as seen by a receiver tuned to the narrow frequency band or channel around f_c (called the *principal zone*) since the unrepresented distortion components will fall outside this zone. Knowledge of the distortion components is important however when we need to assess the level of interference in adjacent channels of the system, and this is the subject of Section 3.5.

3.4 Simulation of non-linear systems

Armed with a lookup table of measurements of $G(A)$ and $\Phi(A)$ at frequency f_c, extendable as necessary through linear interpolation, we may determine the response in the principal zone of a non-linear system to an arbitrary modulated narrow band signal of carrier frequency f_c by taking the steps illustrated in Fig. 3.3. The input signal $x(t)$ is given by

$$x(t) = a_x(t)\cos[2\pi f_c t + \varphi_x(t)] \tag{3.9}$$

where $a_x(t)$ is called the *natural envelope* of $x(t)$ and is a constant if $x(t)$ is not *amplitude modulated* (AM); $\varphi_x(t)$ is called the *phase* of $x(t)$ and is a constant if $x(t)$ is neither *frequency modulated* (FM) nor *phase modulated* (PM); and the bandwidth B of $x(t)$ satisfies the condition $B \ll f_c$. Note therefore that Eq. (3.9) encompasses all types of band pass signals, including AM, FM, PM or hybrid combinations of these as special cases. Given this input, the output $y(t)$ is obtained as follows.

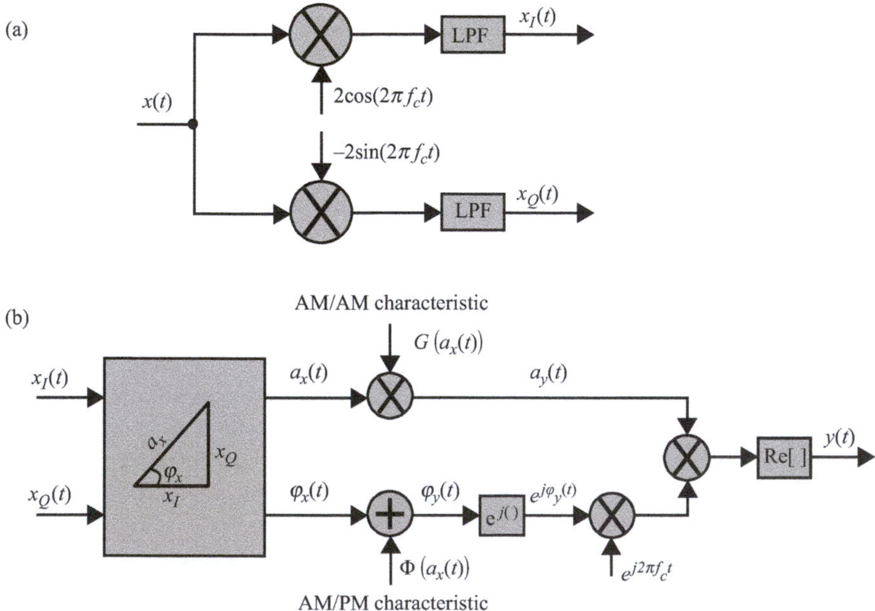

Fig. 3.3 Non-linear system simulation: (a) extracting in-phase and quadrature components of band pass input signal x(t); (b) computing output signal y(t)

First, compute the *in-phase* and *quadrature* components $x_I(t)$ and $x_Q(t)$ of the input signal:

$$\begin{aligned} x_I(t) &= a_x(t)\cos[\varphi_x(t)] \\ &= x(t)\cos(2\pi f_c t) + x\left(t+\frac{1}{4f_c}\right)\cos\left[2\pi f_c\left(t+\frac{1}{4f_c}\right)\right] \end{aligned} \tag{3.10}$$

The first line of Eq. (3.10) defines $x_I(t)$, whereas the second line gives an expression for calculating its value at any time instant t. This expression elegantly fulfils the carrier multiplication and low pass filtering indicated in Fig. 3.3a. You may verify that this is the case by inserting the expression for $x(t)$ from Eq. (3.9) into the second line of Eq. (3.10), expanding the result using trigonometric identities, and noting that

$$\cos\left[4\pi f_c\left(t+\frac{1}{4f_c}\right)\right] = \cos(4\pi f_c t + \pi) = -\cos(4\pi f_c t)$$

and that

$$a_x(t) \approx a_x\left(t+\frac{1}{4f_c}\right); \quad \varphi_x(t) \approx \varphi_x\left(t+\frac{1}{4f_c}\right)$$

since, if A_m denotes the peak value of the (base band) envelope $a_x(t)$, then in the small time interval $1/(4f_c)$, the change in $a_x(t)$ and $\varphi_x(t)$ is bounded by

$$\Delta a_x \leq \frac{1}{2}\pi A_m \frac{B}{f_c}; \quad \Delta\varphi_x \leq \frac{\pi^2}{2}\frac{B}{f_c}$$

both of which are negligibly small when $B \ll f_c$. This expression is indeed the average of $x_I(t)$ in the interval $(t, t+1/4f_c)$; but because f_c is much larger than the fluctuation rate in $x_I(t)$, the smoothing (and hence loss of detail) incurred by taking this average is negligible. Furthermore, the time instants t and $t+1/4f_c$ used for averaging are chosen to ensure that contributions from the high frequency component $(2f_c)$ exactly cancel out, hence achieving the desired low pass filtering. Similarly,

$$\begin{aligned} x_Q(t) &= a_x(t)\sin[\varphi_x(t)] \\ &= -x(t)\sin(2\pi f_c t) - x\left(t+\frac{1}{4f_c}\right)\sin\left[2\pi f_c\left(t+\frac{1}{4f_c}\right)\right] \end{aligned} \tag{3.11}$$

Second, using the in-phase and quadrature components determined above, compute the envelope $a_x(t)$ and phase $\varphi_x(t)$ of the band pass signal as

$$\begin{aligned} a_x(t) &= \sqrt{x_I^2(t) + x_Q^2(t)} \\ \varphi_x(t) &= \tan^{-1}\left[x_Q(t)/x_I(t)\right] \end{aligned} \tag{3.12}$$

Finally, from the lookup table of AM/AM and AM/PM characteristics of the system or their polynomial or other functional approximation, read or calculate the gain $G(a_x(t))$ and phase shift $\Phi(a_x(t))$ of the system at input value $a_x(t)$. The envelope and phase of the output signal are therefore given by

$$\begin{aligned} a_y(t) &= G(a_x(t))a_x(t) \\ \varphi_y(t) &= \varphi_x(t) + \Phi(a_x(t)) \end{aligned} \tag{3.13}$$

and hence the output signal $y(t)$ of the non-linear system in the principal zone is

$$y(t) = a_y(t)\cos\left[2\pi f_c t + \varphi_y(t)\right] \tag{3.14}$$

Note that for computer simulation purposes, the time instants t should be selected to be a discrete sequence $t = 0$, T_s, $2T_s$, $3T_s, \ldots$, where the *sampling interval* T_s is constrained only by the bandwidth B of the base band signal whose variation is encapsulated in the envelope $a_x(t)$ or phase $\varphi_x(t)$ of $x(t)$, and not at all by the carrier frequency f_c, according to the sampling theorem (discussed further in Chapter 4):

$$T_s \leq \frac{1}{2B} \tag{3.15}$$

3.5 Intermodulation distortion

Knowledge of the levels of harmonic and intermodulation components created by a non-linear system is important in order to assess the quality of service in adjacent channels, an assessment that requires estimation of the ratio between wanted signal power and the sum of all noise and interference within the channel. Non-linearity in transmission channels arises predominantly due to high power amplification using active devices. A prime example is the transparent satellite repeater (called transponder), a block diagram of which is shown in Fig. 3.4a, where operations of RF signal reception, frequency downconversion and high power amplification are performed. We will restrict our consideration to this satellite communication

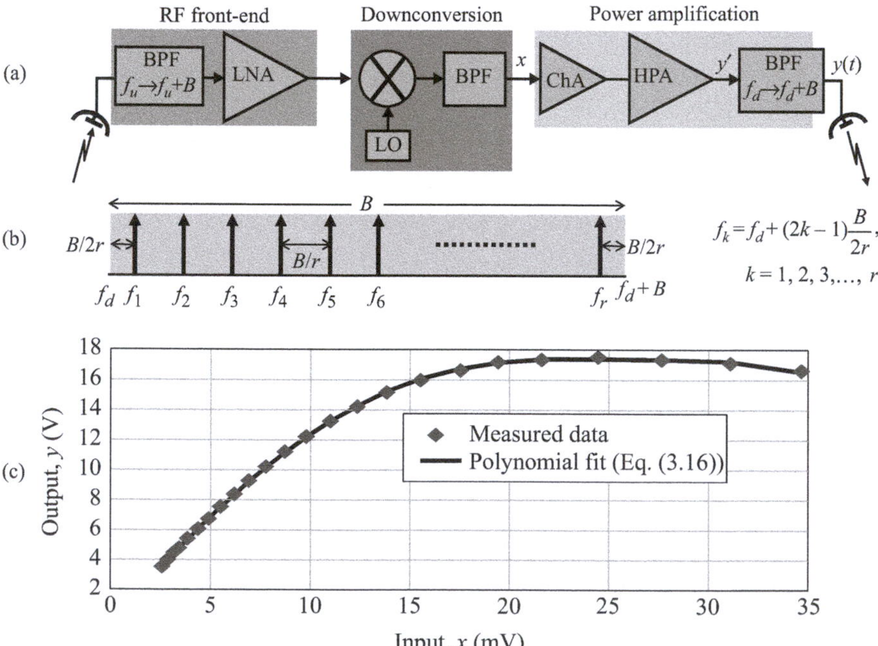

Fig. 3.4 (a) Satellite transponder of bandwidth B; (b) HPA input comprising r equally spaced equal-amplitude carriers; (c) measured TWTA transfer characteristic at 12.2 GHz

channel scenario and seek to determine which harmonic and intermodulation products will constitute interference during multicarrier usage of the transponder, quantify this interference, and hence obtain an expression for the carrier-to-intermodulation ratio in each channel (i.e. sub-band) of the transponder.

Radio transmission systems usually incorporate a band pass filter (BPF), as in Fig. 3.4a, to block out-of-band frequency components and ensure that operation is confined within the allocated band as required by regulatory authorities. Therefore if the transponder is of bandwidth B and operates in the band $f_d \to f_d + B$, then only modulation and intermodulation products in this band will 'escape' and contribute to interference in a receiver tuned to one of the channels of the transponder. Typically, $B \ll f_d$; for example, $B \sim 36$ MHz and $f_d >\sim 4000$ MHz. The non-linear channel behaviour resides in the HPA whose characteristic may be accurately modelled by a 7th-order polynomial consisting exclusively of odd powers (to retain the odd function feature of the transfer characteristic) of the form

$$y(t) = Gx(t) + bx^3(t) + cx^5(t) + dx^7(t) \tag{3.16}$$

where G is the *small-signal* (or *linear-region*) gain of the amplifier, and the coefficients alternate in sign, with G and c positive and b and d negative. Fig. 3.4c shows the excellent fit achieved by Eq. (3.16) on measured data of a Ku-band transponder recorded at 12.2 GHz. In this example, $G = 1414$; $b = -1.928 \times 10^6$; $c = 1.558 \times 10^9$; $d = -5.01 \times 10^{11}$ and the output reaches a maximum $y_{sat} = 17.4$ V, called *output saturated level*, at a *saturated input* level $x_{sat} = 23.8$ mV, which indicates that system voltage gain starts at a maximum of 1414 or 63 dB at $\sim$0 V, dropping by 5.7 dB to 731 or 57.3 dB at saturation.

Before proceeding any further, an important caveat is in order: Eq. (3.16) is a good empirical fit that applies only within a limited input range up to $|x| \approx 1.4x_{sat}$. Results will be unreliable and misleading if the model is employed outside this range of validity. For example, the above model yields accurate results up to input $x \approx 1.4x_{sat}$. Beyond this point the model's output falls rapidly, reaching zero at $1.74873x_{sat}$ and going monotonically increasingly negative for larger input (e.g. $y = -62$ V at $x = 2x_{sat}$ and $y = -3270$ V at $x = 3x_{sat}$) — results that bear no relation whatsoever to actual measurement. The analysis that follows is therefore valid on the condition (usually satisfied in practice) that the non-linear amplifier *is never driven significantly beyond its saturation point.*

There is no intermodulation distortion when the input is a single carrier. So the response $y_1(t)$ of the system to a single sinusoidal input $A\cos(2\pi f_c t)$ will be the wanted signal obtained by substituting $x_1(t) = A\cos(2\pi f_c t)$ into Eq. (3.16). Thus,

$$\begin{aligned}
y_1'(t) &= GA\cos(2\pi f_c t) + bA^3\cos^3(2\pi f_c t) + cA^5\cos^5(2\pi f_c t) + dA^7\cos^7(2\pi f_c t) \\
&= GA\cos(2\pi f_c t) \\
&\quad + \frac{bA^3}{4}[\cos(2\pi \cdot 3f_c \cdot t) + 3\cos(2\pi f_c t)] \\
&\quad + \frac{cA^5}{16}[\cos(2\pi \cdot 5f_c \cdot t) + 5\cos(2\pi \cdot 3f_c \cdot t) + 10\cos(2\pi f_c t)] \\
&\quad + \frac{dA^7}{64}[\cos(2\pi \cdot 7f_c \cdot t) + 7\cos(2\pi \cdot 5f_c \cdot t) + 21\cos(2\pi \cdot 3f_c \cdot t) \\
&\quad + 35\cos(2\pi f_c t)]
\end{aligned}$$

where we have used standard trigonometric identities to expand the powers of $\cos(2\pi f_c t)$. Eliminating the components at frequencies $3f_c$, $5f_c$ and $7f_c$, which are blocked because they are outside the system pass band (see Fig. 3.4a), we obtain the system output for a single carrier as

$$y_{s,1}(t) = \left[GA + \frac{3}{4}bA^3 + \frac{5}{8}cA^5 + \frac{35}{64}dA^7\right]\cos(2\pi f_c t) \tag{3.17}$$

Let us now consider an input $x_r(t)$ comprising r unmodulated carriers of equal amplitude A and frequencies $f_1, f_2, \ldots, f_r$, equally spaced within the frequency band $f_d \rightarrow f_d + B$ as shown in Fig. 3.4b. That is

$$x_r(t) = A\sum_{k=1}^{r}\cos(2\pi f_k t) \tag{3.18}$$

The response $y_r(t)$ to this multi-carrier input will consist of the wanted signal output $y_{s,r}(t)$ of the form of Eq. (3.17) at each input frequency as well as intermodulation products (IMPs) $y_{I,r}(t)$ arising from a mixing of the input frequencies caused by non-linearity. That is

$$y_r(t) = y_{s,r}(t) + y_{I,r}(t)$$

where

$$y_{s,r}(t) = \left[GA + \frac{3}{4}bA^3 + \frac{5}{8}cA^5 + \frac{35}{64}dA^7\right]\sum_{k=1}^{r}\cos(2\pi f_k t) \tag{3.19}$$

To determine $y_{I,r}(t)$ we substitute Eq. (3.18) into Eq. (3.16) and expand the powers of the series as follows:

$$\begin{aligned} bA^3\left(\sum_{k=1}^{r}\cos(2\pi f_k t)\right)^3 &= bA^3\Bigg[\sum_{k=1}^{r}\cos^3(2\pi f_k t) + \sum_{k=1}^{r}\cos^2(2\pi f_k t)\sum_{j\neq k}^{r}\cos(2\pi f_j t) \\ &\quad + \sum_{k=1}^{r}\cos(2\pi f_k t)\sum_{j\neq k}^{r}\cos(2\pi f_j t)\sum_{i=1}^{r}\cos(2\pi f_i t)\Bigg] \end{aligned} \tag{3.20}$$

$$\begin{aligned} cA^5\left(\sum_{k=1}^{r}\cos(2\pi f_k t)\right)^5 &= cA^5\Bigg[\sum_{k=1}^{r}\cos^5(2\pi f_k t) + \sum_{k=1}^{r}\cos^4(2\pi f_k t)\sum_{j\neq k}^{r}\cos(2\pi f_j t) \\ &\quad + \sum_{k=1}^{r}\cos^3(2\pi f_k t)\sum_{j\neq k}^{r}\cos(2\pi f_j t)\sum_{i=1}^{r}\cos(2\pi f_i t) + \sum_{k=1}^{r}\cos^2(2\pi f_k t) \\ &\quad \times\sum_{j\neq k}^{r}\cos(2\pi f_j t)\sum_{i=1}^{r}\cos(2\pi f_i t)\sum_{m=1}^{r}\cos(2\pi f_m t) + \sum_{k=1}^{r}\cos(2\pi f_k t) \\ &\quad \times\sum_{j\neq k}^{r}\cos(2\pi f_j t)\sum_{i=1}^{r}\cos(2\pi f_i t)\sum_{m=1}^{r}\cos(2\pi f_m t)\sum_{n=1}^{r}\cos(2\pi f_n t)\Bigg] \end{aligned} \tag{3.21}$$

$$
\begin{aligned}
dA^7\left(\sum_{k=1}^{r}\cos(2\pi f_k t)\right)^7
&= dA^7\left[\sum_{k=1}^{r}\cos^7(2\pi f_k t) + \sum_{k=1}^{r}\cos^6(2\pi f_k t)\sum_{j\neq k}^{r}\cos(2\pi f_j t) + \sum_{k=1}^{r}\cos^5(2\pi f_k t)\right.\\
&\times \sum_{j\neq k}^{r}\cos(2\pi f_j t)\sum_{i=1}^{r}\cos(2\pi f_i t) + \sum_{k=1}^{r}\cos^4(2\pi f_k t)\sum_{j\neq k}^{r}\cos(2\pi f_j t)\\
&\times \sum_{i=1}^{r}\cos(2\pi f_i t)\sum_{m=1}^{r}\cos(2\pi f_m t) + \sum_{k=1}^{r}\cos^3(2\pi f_k t)\sum_{j\neq k}^{r}\cos(2\pi f_j t)\\
&\times \sum_{i=1}^{r}\cos(2\pi f_i t)\sum_{m=1}^{r}\cos(2\pi f_m t)\sum_{n=1}^{r}\cos(2\pi f_n t) + \sum_{k=1}^{r}\cos^2(2\pi f_k t)\\
&\times \sum_{j\neq k}^{r}\cos(2\pi f_j t)\sum_{i=1}^{r}\cos(2\pi f_i t)\sum_{m=1}^{r}\cos(2\pi f_m t)\sum_{n=1}^{r}\cos(2\pi f_n t)\sum_{p=1}^{r}\cos(2\pi f_p t)\\
&+ \sum_{k=1}^{r}\cos(2\pi f_k t)\sum_{j\neq k}^{r}\cos(2\pi f_j t)\sum_{i=1}^{r}\cos(2\pi f_i t)\sum_{m=1}^{r}\cos(2\pi f_m t)\\
&\left.\times \sum_{n=1}^{r}\cos(2\pi f_n t)\sum_{p=1}^{r}\cos(2\pi f_p t)\sum_{q=1}^{r}\cos(2\pi f_q t)\right]
\end{aligned}
\tag{3.22}
$$

As previously shown in the derivation of Eq. (3.17), each first term in square brackets on the RHS of the above expansions reduces to a component of $y_{s,r}(t)$ that we have already included in Eq. (3.19). The rest of the terms reduce to IMPs due to non-linearity-induced mixing of different frequencies, but only those whose frequency falls within the filter's pass band (illustrated in Fig. 3.4b) given by

$$
f_d \to f_d + B \equiv f_1 - \frac{1}{2r}B \to f_1 + \left(1 - \frac{1}{2r}\right)B \tag{3.23}
$$

will 'survive' to contribute to $y_{I,r}(t)$. A necessary albeit insufficient condition for an IMP to 'survive' is that it must have $\sum m_i = 1$, which is the absolute value of the sum of the positive and negative integer factors in the linear combination that yields the IMP frequency. For example, the IMP $2f_1 - f_2$ has $\sum m_i = |2-1| = 1$, and the IMP $f_1 - f_4 - f_2$ has $\sum m_i = |1-1-1| = 1$, whereas the 3rd-order IMP $f_1 + f_2 + f_3$ has $\sum m_i = |1+1+1| = 3$ and therefore does not survive. You may wish to verify that, in view of the narrow band condition $f_d \gg B$, every IMP with $\sum m_i > 1$ has absolute frequency that exceeds the upper limit in Eq. (3.23), whereas every IMP with $\sum m_i = 0$ falls below the lower limit, so that both are blocked by the filter.

However, not all IMPs with $\sum m_i = 1$ will fall in the system pass band and contribute to interference in adjacent channels. For example, considering only 3rd-order IMPs for now, a multi-carrier system with $r = 2$ uniformly spaced carriers has $f_1 = f_d + B/4$, $f_2 = f_d + 3B/4$, and the two possible 3rd-order IMPs both fall outside the pass band since

$$
\begin{aligned}
2f_1 - f_2 &= 2(f_d + B/4) - (f_d + 3B/4) = f_d - B/4 \qquad &&(\text{less than } f_d)\\
2f_2 - f_1 &= 2(f_d + 3B/4) - (f_d + B/4) = f_d + 5B/4 \qquad &&(\text{greater than } f_d + B)
\end{aligned}
$$

Similarly, if there are 3 uniformly spaced carriers $f_1 = f_d + B/6$, $f_2 = f_d + B/2$, $f_3 = f_d + 5B/6$, then only $2f_2 - f_1$, $2f_2 - f_3$, and $f_1 + f_3 - f_2$ are in-band IMPs, the rest being out-of-band. The number of in-band IMPs grows rapidly with the number of carriers. The condition that a 3rd-order IMP frequency must satisfy in order to be in-band is quite straightforward, namely

$$\begin{array}{lll} f_d \le 2f_j - f_k \le f_d + B & j,\ k = 1,\ 2, \ldots,\ r; & j \ne k \\ f_d \le f_i + f_j - f_k \le f_d + B & i,\ j,\ k = 1,\ 2, \ldots,\ r; & i \ne j \ne k \end{array}$$

which upon using the expression, given in Fig. 3.4b, for carrier frequency in terms of f_d and B yields

$$\begin{array}{ll} \frac{1}{2}k + \frac{1}{4} \le j \le \frac{1}{2}(k + r) + \frac{1}{4}, & j \ne k \\ k + \frac{1}{2} \le i + j \le k + r + \frac{1}{2}, & i \ne j \ne k \end{array} \tag{3.24}$$

To understand the above conditions, recall that there are r carriers numbered $1, 2, \ldots, r$, having uniformly spaced frequencies $f_1, f_2, \ldots, f_r$ within a bandwidth B. The first line of Eq. (3.24) stipulates that an IMP formed by doubling the jth carrier frequency before subtracting another (the kth frequency) will be in-band only if the frequency that is doubled is neither so large that subtraction cannot bring the result back down into the band, nor so small that subtraction takes the result below the band. Note that if there are only two carriers (i.e. $r = 2$), then only the first line applies (since the second line requires 3 distinct carriers), and (you may wish to verify for yourself that) the condition of this first line cannot be satisfied when $r = 2$; so there are no in-band 3rd-order IMPs. Similarly, the second line of Eq. (3.24) stipulates the condition for combining three distinct carriers to form an in-band 3rd-order IMP: The two frequencies summed must be neither so large that when a 3rd frequency is subtracted it is insufficient to bring the result back down into the band, nor so small that such subtraction takes the result below the band.

Following the considerations outlined above, we apply standard trigonometric identities to carefully evaluate all but the first terms on the RHS of Eqs. (3.20), (3.21) and (3.22), retain only those IMPs that not only have $\sum m_i = 1$ but are also in-band, and combine the amplitudes of coincident (i.e. coherent) products. For a dual-carrier system $r = 2$, with two input frequencies f_1 and f_2 uniformly spaced within the transponder bandwidth, the IMP output is simple enough to be written in equation form and is given by

$$y_{I,2}(t) = \left(\frac{3}{2}bA^3 + \frac{45}{8}cA^5 + \frac{595}{32}dA^7\right)(\cos 2\pi f_1 t + \cos 2\pi f_2 t) \tag{3.25}$$

where b, c and d are the polynomial coefficients of the transfer characteristic of the transponder (Eq. (3.16)), and A is the input amplitude of each carrier. However, for $r \ge 3$, the number of IMPs is too large to be conveniently expressed in equation form. The result for $r = 4$ is summarised in Table 3.1, which gives the amplitude and the number of in-band products of each class of IMP. Note that the single-frequency IMP denoted f_k in Table 3.1 results from the mixing of 3 or more carriers. For example, the 3rd-order IMP $f_k - f_j + f_i$, the 5th-order IMP $f_k - 2f_j + 2f_i$ and the 7th-order IMP $f_k - 3f_j + 3f_i$ each reduces to IMP f_k when $j = i$ in the respective Eqs. (3.20) to (3.22). In a 4-carrier system, there are four possible values of f_k, $k = 1, 2, 3, 4$; hence the number of in-band products of this IMP is $N_{w=1} = 4$ (see Table 3.1). The amplitude A_{Iw} of each product given in Table 3.1 is obtained by collating all contributions from the 3rd-order non-linearity of Eq. (3.20), the 5th-order non-linearity of Eq. (3.21) and the 7th-order non-linearity of Eq. (3.22), which have respective factors bA^3, cA^5 and dA^7.

Table 3.1 Amplitude and number of in-band IMPs in a narrow band multicarrier nonlinear system with 4 carriers of equal amplitude A

Row, *w*	**IMP**	**Number of products, N_w**	**Amplitude, A_{Iw}**
1	f_k	4	$\frac{9}{2}bA^3 + \frac{315}{8}cA^5 + \frac{11865}{32}dA^7$
2	$2f_k - f_j$	4	$\frac{3}{4}bA^3 + \frac{85}{8}cA^5 + \frac{8715}{64}dA^7$
3	$f_k - f_j + f_i$	6	$\frac{3}{2}bA^3 + \frac{75}{4}cA^5 + \frac{3465}{16}dA^7$
4	$3f_k - 2f_j$	2	$\frac{5}{8}cA^5 + \frac{1085}{64}dA^7$
5	$3f_k - f_j - f_i$	4	$\frac{5}{4}cA^5 + \frac{1365}{32}dA^7$
6	$f_k - 2f_j + 2f_i$	6	$\frac{15}{8}cA^5 + \frac{1365}{32}dA^7$
7	$f_k - f_j - f_i + 2f_m$	6	$\frac{15}{4}cA^5 + \frac{1155}{16}dA^7$
8	$f_k + f_j - 2f_i + f_m$	2	$\frac{15}{4}cA^5 + \frac{1155}{16}dA^7$
9	$4f_k - f_i - 2f_i$	6	$\frac{105}{64}dA^7$
10	$f_k - 3f_j + 3f_i$	4	$\frac{35}{16}dA^7$
11	$f_k - f_i + 3f_i - 2f_m$	4	$\frac{105}{16}dA^7$
12	$2f_k - 3f_j + 2f_i$	4	$\frac{105}{32}dA^7$
13	$f_k + f_j - 3f_i 2f_m$	2	$\frac{105}{16}dA^7$
14	$2f_k - f_j + 2f_i - 2f_m$	4	$\frac{315}{32}dA^7$

So what have we achieved thus far? We now have the wanted signal output $y_{s,r}(t)$ given by Eq. (3.19) and the IMPs output $y_{I,r}(t)$ given by Eq. (3.25) for $r=2$ and Table 3.1 for $r=4$ and similar tabular compilations for other values of r. So we may now proceed to determine the average ratio between signal power and intermodulation power, called carrier-to-intermodulation ratio (C/I), which assumes that IMPs are uniformly distributed across the bandwidth B. We will then discuss the adjustment necessary to obtain the actual C/I in individual channels of the transponder, which takes into account the non-uniform distribution of IMP locations.

Working with normalised powers (related to amplitude by $A^2/2$), the total carrier power P_c in the bandwidth B is obtained by summing the power of each of the r carriers in Eq. (3.19):

$$P_c = \frac{r}{2}\left(GA + \frac{3}{4}bA^3 + \frac{5}{8}cA^5 + \frac{35}{64}dA^7\right)^2 \tag{3.26}$$

And total intermodulation power P_I in the same bandwidth is obtained by summing the powers of the IMPs as follows:

$$P_I = \begin{cases} \left(\frac{3}{2}bA^3 + \frac{45}{8}cA^5 + \frac{595}{32}dA^7\right)^2, & r = 2 \\ \sum_{w=1}^{\text{All IMP types}} \left(\frac{1}{2}N_w A_{Iw}^2\right), & r \geq 3 \text{ (Using e.g. Table 3.1)} \end{cases} \tag{3.27}$$

where the summation covers all rows of Table 3.1, and A_{Iw} and N_w are respectively from the last and penultimate columns of the table. Average C/I is thus

$$(C/I)_{\text{avg}} = \frac{P_c}{P_I} = \frac{r\left(GA + \frac{3}{4}bA^3 + \frac{5}{8}cA^5 + \frac{35}{64}dA^7\right)^2}{\sum_{w=1}^{\text{All IMP types}} N_w A_{Iw}^2} \tag{3.28}$$

Given that x_{sat} is the saturated input voltage amplitude of the transponder, the saturated input power is $x_{\text{sat}}^2/2$ for the entire transponder and $x_{\text{sat}}^2/2r$ per carrier, so that the saturated amplitude of each carrier is

$$A_{\max} = x_{\text{sat}}/\sqrt{r}$$

This is the amplitude of each carrier that will drive the transponder into saturation. The ratio $A_{\max}/A$ expresses the factor by which each input carrier amplitude A has been backed off from its saturated level, a value of unity (or 0 dB) indicating that the transponder is operated at saturation. In terms of this input back-off factor, defined in dB as

$$\text{IBO} = 20 \log_{10}(A_{\max}/A) \text{ dB}$$

the input amplitude A of each carrier in Eq. (3.28)

$$A = \frac{x_{\text{sat}}}{\sqrt{r}} 10^{-\text{IBO}/20} \tag{3.29}$$

Substituting this into Eq. (3.28) and evaluating at various values of IBO yields graphs of $(C/I)_{\text{avg}}$ versus back-off IBO in dB shown in Fig. 3.5. The number of

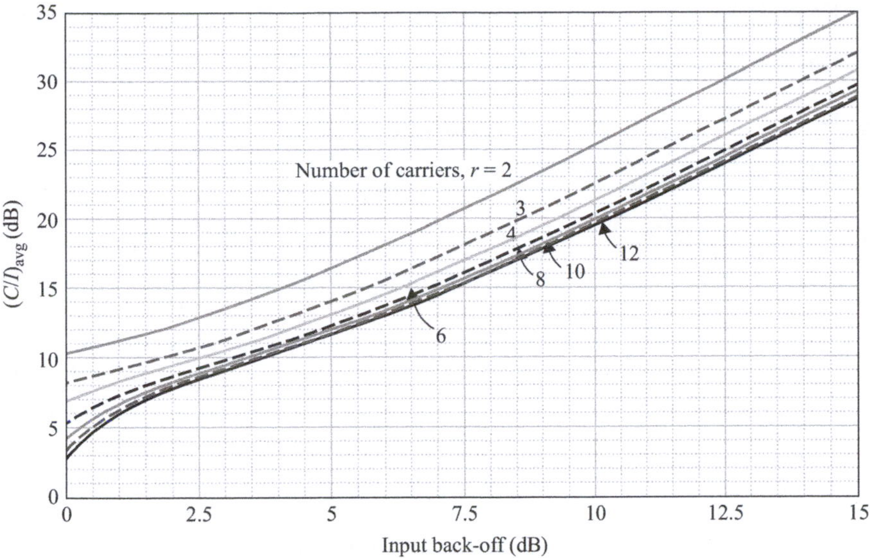

Fig. 3.5 Average carrier-to-intermodulation ratio versus input power back-off for various numbers of carriers, r = 2 to 12

IMPs increases rapidly with the number of carriers. However, in view of Eq. (3.29), for any given back-off the power of an IMP arising from 3rd-order non-linearity (NL) decreases with r as $r^{3/2}$ whereas the 5th- and 7th-order IMPs decrease in power as $r^{5/2}$ and $r^{7/2}$ respectively so that for $r >\approx 10$, the average C/I of this particular transponder (having a 7th-order polynomial transfer characteristic given by Eq. (3.16) with coefficients $G = 1414$; $b = -1.928 \times 10^6$; $c = 1.558 \times 10^9$; $d = -5.01 \times 10^{11}$) tends to the approximate relation

$$(C/I)_{avg}\ (\text{dB}) = 3.9 + 1.598 \times \text{IBO}, \quad r \gg 1 \tag{3.30}$$

We see that if operated at saturation with a large number of carriers, the output signal power of the non-linear transponder will be only about 3.9 dB higher than that of IMPs, giving rise to significant distortions. Fig. 3.5 shows that a dual-carrier transponder operated at saturation fares better, having average C/I of ~10.5 dB. Note that, unlike noise effects, this distortion cannot be mitigated by increasing signal power. This is because the intermodulation power at the output increases more rapidly than output signal power as input signal power is increased, leading to a rapid decline in C/I. To minimise distortions due to intermodulation during multi-carrier operation, the amplifier must be used in its more linear region, and this requires a significant back-off from saturation. For example, Fig. 3.5 shows that an input back-off of 10 dB delivers around 15 dB improvement in C/I.

The above evaluation of C/I assumed a uniform distribution of IMP products over the bandwidth B of the non-linear system. However, more IMPs will fall in the central sub-bands than in the outer ones. Fig. 3.6 shows the distribution of 3rd-order IMPs in a transponder divided into 20 equal sub-bands. Specifically, the number of 3rd-order IMPs in the kth sub-band is

$$N_{3,k} = \lfloor 3r(r-2)/8 - [(k-1-r/2)(k-r/2)-1]/2 \rfloor, \quad k = 1, 2, 3, \ldots, r \tag{3.31}$$

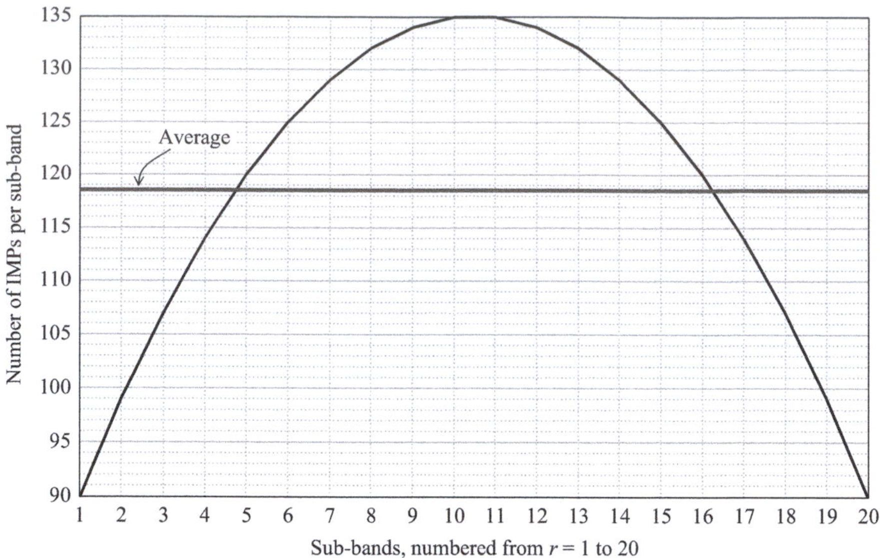

Fig. 3.6 Number of 3rd-order IMPs in each of 20 equally spaced sub-bands of a transponder. The average count per sub-band is a also shown

where the brackets $\lfloor z \rfloor$ denote the integer part of z. For r even, Eq. (3.31) simplifies to

$$N_{3,k} = 3r(r-2)/8 - (k-1-r/2)(k-r/2)/2, \quad r \text{ even} \tag{3.32}$$

which is maximum in the two central sub-bands $k=r/2$ and $k=1+r/2$, and is minimum in the two outermost sub-bands $k=1$ and $k=r$, with respective values

$$N_{3,k\,\max} = \frac{3}{8}r(r-2); \quad N_{3,k\,\min} = \frac{1}{4}r(r-2), \quad r \text{ even} \tag{3.33}$$

For r odd, the central sub-band is $k=(r+1)/2$, and the corresponding expressions are

$$N_{3,k\,\max} = \left\lfloor \frac{3}{8}[r(r-2)+1] \right\rfloor; \quad N_{3,k\,\min} = \left\lfloor \frac{1}{4}r(r-2) + \frac{1}{2} \right\rfloor, \quad r \text{ odd} \tag{3.34}$$

C/I is lowest in the sub-band or channel at the centre of the transponder pass band, and increases outwards to a maximum in the two channels at either end of the pass band. However, when contributions from all IMPs (3rd, 5th and 7th orders) are included, the variation in C/I across the channels is not as large as indicated merely by a count of 3rd-order IMPs in the channels. It is found that maximum channel C/I is within ~0.25 dB of the average C/I that we determined earlier, and the lowest C/I is within ~0.45 dB of $(C/I)_{\text{avg}}$, depending on the number of carriers.

Worked Example 3.1: Memoryless Non-linear Device Operation

MEASUREMENT of a TWT amplifier at 4 GHz yields a transfer characteristic represented by a polynomial $y = Gx + bx^3 + cx^5 + dx^7$ in the interval $-1.3x_{sat} \leq x \leq 1.3x_{sat}$, where $G = 893.5$, $b = -1.1568 \times 10^6$, $c = 9.0961 \times 10^8$, $d = -2.84223 \times 10^{11}$ and saturated input voltage $x_{sat} = 30$ mV. Assuming normalised power (i.e. power is measured across a 1 Ω impedance), and that the characteristic of the amplifier is frequency independent over a narrow bandwidth around 4 GHz, determine

---

(a) The saturation output power P_{o1sat} of the amplifier in single carrier operation
(b) The saturation power gain G_{sat} of the device in dB
(c) The small-signal or linear region power gain G_{ss} of the amplifier in dB
(d) The AM/AM conversion coefficient ρ_{am} (in dB per dB) at input back-off (IBO) 20 dB, and at IBO = 1 dB
(e) The point of compression to 1 dB
(f) The AM/PM conversion factor ρ_{pm} (in degrees per dB) at IBO = 20 dB and IBO = 1 dB, given that AM/PM characteristic is represented (in degrees) by the function

$$\Phi(\text{IBO}) = \frac{-59.32 \times \text{IBO}^2 + 2734 \times \text{IBO} - 30970}{\text{IBO}^2 + 1.353 \times \text{IBO} + 633}$$

in the interval $-1.5 \leq \text{IBO} \leq 20$ dB
(g) The carrier-to-intermodulation power ratio (C/I) when the amplifier is used in a 2-carrier mode of operation with IBO = 8 dB

(a) Saturation output power P_{o1sat} is the power at the output of the amplifier when a saturated input power P_{i1sat} or saturated input voltage $x_{sat} = 30$ mV (given) is applied. We use the given transfer characteristic to obtain the saturated output voltage y_{sat}, and hence P_{o1sat} as follows:

$$\begin{aligned} y_{sat} &= Gx_{sat} + bx_{sat}^3 + cx_{sat}^5 + dx_{sat}^7 \\ &= 893.5 \times 30 \times 10^{-3} - 1.1568 \times 10^6 \times (30 \times 10^{-3})^3 \\ &\quad + 9.0961 \times 10^8 \times (30 \times 10^{-3})^5 \\ &\quad - 2.84223 \times 10^{11} \times (30 \times 10^{-3})^7 \\ &= 26.805 - 31.2336 + 22.103523 - 6.215957 = 11.459 \text{ V} \end{aligned}$$

$$P_{o1sat} = y_{sat}^2/2 = 65.654 \text{ W} = 18.17 \text{ dBW}$$

(b) Saturation power gain

$$G_{sat} = 20 \log_{10}\left(\frac{y_{sat}}{x_{sat}}\right) = 20 \log_{10}\left(\frac{11.459}{30 \times 10^{-3}}\right) = 51.64 \text{ dB}$$

(c) Small signal gain

$$G_{ss} = 20 \log_{10}(G) = 20 \log_{10}(893.5) = 59.02 \text{ dB}$$

(d) The AM/AM conversion coefficient ρ_{am} is the slope of the curve of output power P_{o1} in dB relative to P_{o1sat} versus input power P_{i1} in dB relative to P_{i1sat}. It gives the dB increase in output power per dB increase in input power. In general ρ_{am} is unity at very large input back-off (small signal region), decreasing to zero at saturation. We may determine ρ_{am} at a specified IBO as the difference × 10 between P_{o1} at IBO − 0.05 dB and P_{o1} at IBO + 0.05 dB. This is a 0.1 dB input power interval, which is a compromise between the need to avoid averaging the slope of the curve at IBO over a large interval and the need to minimise rounding errors in the calculation when a very small interval is used. From Eq. (3.29), under single carrier operation ($r = 1$) the input voltage x_{IBO} at input back-off IBO (dB) is $x_{IBO} = 10^{-IBO/20}x_{sat}$. Thus, for IBO = 20 dB

$$x_{IBO\pm0.05} = 10^{-(20\pm0.05)/20} \times 30\ \text{mV} = 2.98278,\ 3.01732\ \text{mV}$$

Substituting these into the transfer characteristic as in (a) yields the corresponding output voltages $y_{IBO\pm0.05}$ and hence output powers $P_{o1,IBO\pm0.05}$:

$$\begin{aligned} y_{IBO\pm0.05} &= 2.63463,\ 2.66442\ \text{V} \\ P_{o1,IBO\pm0.05} &= (y_{IBO\pm0.05})^2/2 = 5.4041,\ 5.5018\ \text{dBW} \end{aligned}$$

$P_{o1,IBO-0.05} - P_{o1,IBO-0.05}$ is the dB change in output power when the input power changes by 0.1 dB around IBO, so we multiply this by a factor of 10 to obtain the change in output corresponding to a 1 dB change in input, which is the required AM/AM conversion coefficient:

$$\begin{aligned} \rho_{am} &= 10\left(P_{o1,IBO-0.05} - P_{o1,IBO+0.05}\right) = 10(5.5018 - 5.4041) \\ &= 0.977\ \text{dB/dB} \end{aligned}$$

Similarly, for IBO = 1 dB

$$\begin{aligned} x_{IBO\pm0.05} &= 10^{-(1\pm0.05)/20} \times 30\ \text{mV} = 26.58406,\ 26.89189\ \text{mV} \\ y_{IBO\pm0.05} &= 11.429896,\ 11.432978\ \text{V} \\ P_{ol,IBO\pm0.05} &= 18.150546,\ 18.152888\ \text{dBW} \\ \rho_{am} &= 10\left(P_{ol,IBO-0.05} - P_{ol,IBO+0.05}\right) = 0.023\ \text{dB/dB} \end{aligned}$$

(e) Notice from solutions (b) and (c) that the system gain reduces from $G_{ss} = 59.02$ dB in the small signal region to $G_{sat} = 51.64$ dB at saturation. The point (P_{i1}, P_{o1}) corresponding to a 1 dB reduction in gain from its maximum value G_{ss} is known as the 1 dB compression point. Plotting $\Delta G = 20\log_{10}(y_{IBO}/x_{IBO}) - G_{ss}$ versus IBO as shown in Fig. 3.7, we read the value of IBO at which $\Delta G = -1$ dB and hence calculate the corresponding input and output power pair, which constitutes the 1 dB compression point. We see that $\Delta G = -1$ dB at IBO = 10 dB, so that

$$x_{IBO=10} = 10^{-10/20}x_{sat} = 9.4868\ \text{mV}$$

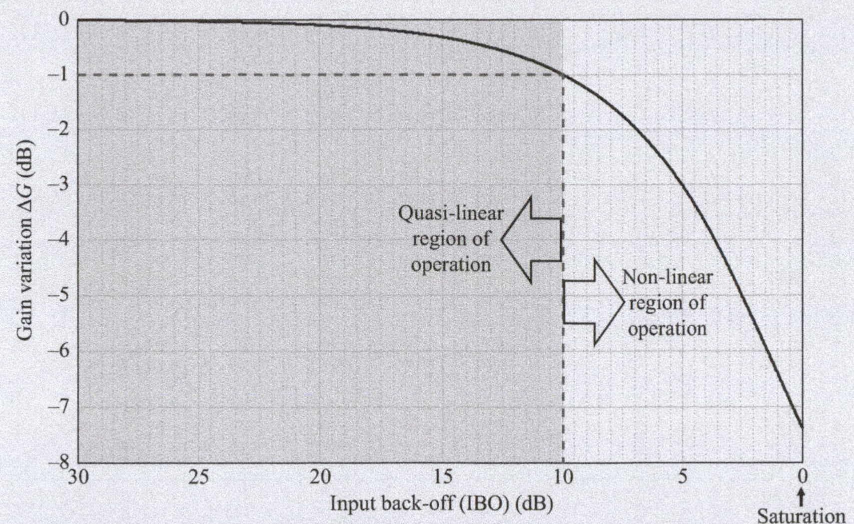

Fig. 3.7 Worked Example 3.1: *Gain variation versus input back-off*

which upon substituting into the transfer characteristic yields corresponding output voltage $y_{\mathrm{IBO}=10} = 7.5567$ V. The 1 dB compression point is thus

$$(P_{i1}, P_{o1}) = \left(x^2_{\mathrm{IBO}=10}/2, y^2_{\mathrm{IBO}=10}/2\right) = (-43.47, 14.56)\ \mathrm{dBW}$$

Alternatively, we may express the 1 dB compression point as a pair of output back-off (OBO) and input back-off (IBO) pair. The above output power ($P_{o1} = 14.56$ dBW) is below the saturation output power ($P_{o1\mathrm{sat}} = 18.17$ dBW) by OBO = 3.61 dB. Thus the 1 dB compression point is also (IBO, OBO) = (10, 3.61) dB. This point is often used to define the quasi-linear region of operation of the amplifier as illustrated in Fig. 3.7. In this example, a roughly linear operation is obtained in the shaded region of the figure with maximum total input power = −43.47 dBW or minimum IBO of 10 dB, which gives maximum output power = 14.55 dBW. Beyond this point (in the unshaded region of the graph) the device exhibits non-linear behaviour.

(f) The AM/PM conversion factor ρ_{pm} is the slope of the Φ(IBO) versus IBO curve. Fig. 3.8 shows this curve for the worked example at hand. Taking the derivative of the given function of Φ(IBO) with respect to IBO:

$$\Phi(\mathrm{IBO}) = \frac{-59.32 \times \mathrm{IBO}^2 + 2734 \times \mathrm{IBO} - 30970}{\mathrm{IBO}^2 + 1.353 \times \mathrm{IBO} + 633}$$

$$\rho_{\mathrm{pm}} = \frac{d\Phi(\mathrm{IBO})}{d(\mathrm{IBO})} = \frac{1772524 - 2814.26 \times \mathrm{IBO}^2 - 13159.12 \times \mathrm{IBO}}{\left(\mathrm{IBO}^2 + 1.353 \times \mathrm{IBO} + 633\right)^2}$$

At IBO = 20 dB and 1 dB, we obtain $\rho_{\mathrm{pm}} = 0.34°/\mathrm{dB}$ and $4.35°/\mathrm{dB}$ respectively. In general ρ_{pm} increases as we move towards saturation (i.e. as IBO decreases).

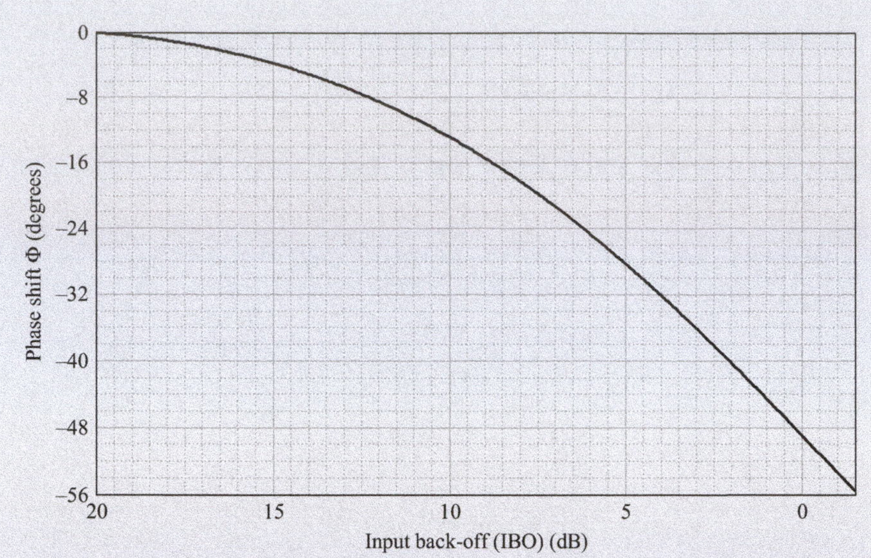

Fig. 3.8 Worked Example 3.1: Phase shift versus input back-off

(g) With $r = 2$, IBO = 8 dB and $x_{\text{sat}} = 30$ mV, Eq. (3.29) gives the amplitude of each of the two carriers as

$$A = \frac{30}{\sqrt{2}} 10^{-8/20} = 8.45 \text{ mV}$$

Eqs. (3.19) and (3.25) give the output signal amplitude A_s and output IMP amplitude A_I in each of the two channels as

$$A_s = GA + \frac{3}{4}bA^3 + \frac{5}{8}cA^5 + \frac{35}{64}dA^7$$

$$A_I = \frac{3}{2}bA^3 + \frac{45}{8}cA^5 + \frac{595}{32}dA^7$$

and hence C/I as

$$C/I = 10\log_{10}\left(\frac{A_s^2/2}{A_I^2/2}\right) = 18.5 \text{ dB}$$

In concluding this section, we should point out that non-linearity is not always a nuisance in communication systems, rather it does find extensive application in signal processing tasks such as modulation, demodulation, frequency down- and up-conversions, frequency multiplication, and so on. For example, if we add a carrier signal of frequency say f_2 to a message signal of frequency say f_1 and then pass the sum signal through a non-linear device having transfer characteristic $y = x + 0.6x^2 + 0.12x^3$ we obtain the results shown in Fig. 3.9. We see in this figure that the output signal contains the carrier f_2 and side frequencies $f_2 - f_1$ and $f_2 + f_1$, which constitute an amplitude modulated carrier and can be readily separated from the other 'unwanted' frequency components using a BPF. Non-linear devices used in this way include diodes and transistors.

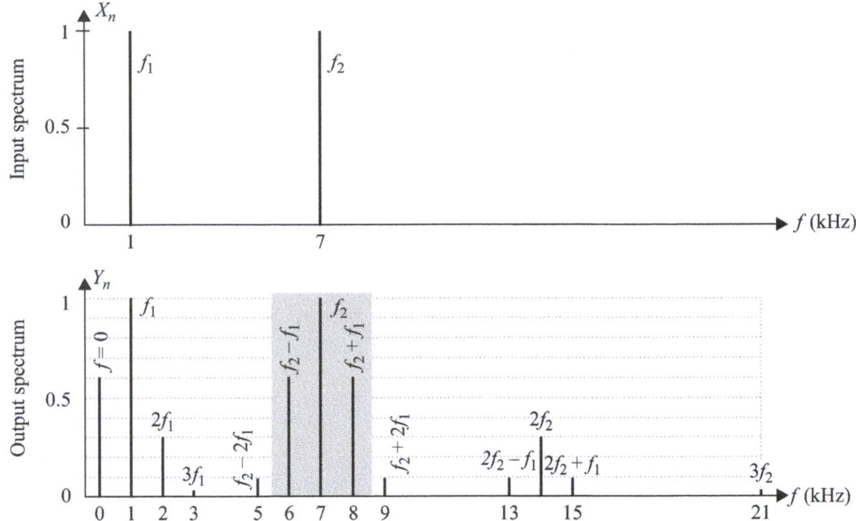

Fig. 3.9 Harmonic and intermodulation products in a non-linear system of transfer characteristic $y = x + 0.6x^2 + 0.12x^3$. Shaded portion of output spectrum represents amplitude modulation of f_2 (carrier) by f_1 (message)

3.6 Summary

The principle of superposition does not hold in *non-linear* systems where response to one input is modified in the presence of other inputs. So we are unable to determine the response of this system to an arbitrary input signal based on its response to an impulse or a sinusoid. We may however employ the system's *transfer characteristic*, which gives its output as a polynomial function of input over a specified range. In so doing we find that non-linearity not only distorts the wanted signal, but it also introduces *harmonic* and *intermodulation* products into the output, and these may interfere with other signals in adjacent channels or frequency bands. This is undesirable in transmission channels and must be minimised by ensuring that HPA operations are linear, for example by using a larger (and heavier) HPA than necessary and employing sufficient input power back-off to set its operating point within the linear region.

In the next chapter we begin an in-depth study of the efficient representation of digital signals for transmission, and explore how the conflicting goals of minimising transmission bandwidth requirement and maintaining high fidelity with the source signal are achieved in practice.

3.7 Reference

[1] Cann, A. J. (1980). 'Nonlinearity model with variable knee sharpness', *IEEE Transactions on Aerospace and Electronic Systems*, 16(6), pp. 874–877

Chapter 4
Sampling of base band and band pass signals

In Chapter Four

- ✓ An insightful and graphical introduction to sampling using both time-domain and frequency-domain perspectives.
- ✓ Base band signal sampling and the quantification of alias distortion.
- ✓ A lucid discussion of two methods of band pass signal sampling, namely uniform sampling and quadrature sampling. You will be able to avoid the common pitfalls and confusion associated with this important topic.
- ✓ An easy-to-follow presentation of the mathematical foundation of sampling that validates conclusions reached through an intuitive graphical approach and gives you a more complete understanding of the subject of sampling.
- ✓ Two methods of non-instantaneous sampling, namely natural sampling and flat-top sampling, their shortcomings such as aperture effect and practical mitigation measures.
- ✓ Anti-alias filter specification to reduce alias distortion to tolerable levels.
- ✓ Worked Examples.

4.1 Introduction

To exploit the benefits and advantages of digital communications enumerated in Section 1.4.2, a message signal such as voice or video that is originally in analogue form must first be converted into a digital signal in the form of a regular sequence of numbers. It is this sequence of numbers that constitutes the input bit stream of a digital communication system that processes it in a variety of ways to gain various advantages, such as bandwidth reduction, privacy, power saving, protection against channel-induced errors and distortion, before conveying it to the intended destination where, after further processing, it is ultimately presented in its original analogue format to a consumer.

This chapter deals in complete detail with the first two steps illustrated in Fig. 1.6 involved in converting an analogue signal in to a sequence of binary numbers (called *bit stream*), specifically the processes of *anti-alias filtering* and *sampling*. We take an insightful and easy-to-grasp look at sampling from both time-domain and frequency-domain perspectives, and learn how the *sampling rate* and the parameters of the *anti-alias filter* are chosen to set any desired maximum level for *alias distortion*. Of particular note is the treatment of the often confused or misleadingly presented subject of band pass sampling.

In what follows we first introduce the sampling concept and then examine the sampling of sinusoidal signals. Through a graphical approach we verify the sampling theorem and discover the penalty incurred for flouting it. We next apply these findings to a detailed study of the sampling of low pass and band pass signals and provide a quantification of alias distortion through a parameter known as the *signal-to-alias distortion ratio* (SADR). The mathematical basis of sampling is then presented and some of the practical issues involved in sampling are discussed, including anti-alias filter design and aperture effect due to non-instantaneous sampling. The chapter includes carefully selected worked examples to help deepen your understanding and develop your problem-solving competence in the subject.

4.2 The sampling concept and theorem

An analogue signal $g(t)$ is continuous in time and value, and (except for random signals with an impulsive autocorrelation function) has inherent redundancy since values of the signal taken at close time instants are related in some way. It is of interest to consider whether $g(t)$ can be faithfully reconstructed from a sequence of its values taken at regular intervals $T_s > 0$, and to establish what constraints are set on the size of T_s by the characteristics of $g(t)$. We know for example that if $g(t)$ is constant then one sample is sufficient and $T_s = \infty$; and if $g(t)$ is linear then only two samples are needed for reconstruction and T_s can be any non-zero value. But the case of practical interest is when $g(t)$ is an arbitrary information-bearing signal. The process of taking values of $g(t)$ at a discrete set of regular time instants is referred to as *sampling*. It converts the continuous-value continuous-time signal $g(t)$ in to a continuous-value discrete-time signal (or sequence) $g(nT_s)$, where $n = 0, 1, 2, 3, \ldots$; T_s is the *sampling interval*, and its reciprocal $f_s = 1/T_s$ is known as the *sampling frequency* or *sampling rate* in hertz (Hz).

Sampling is hugely important, and in fact indispensable in digital communication of voice, audio, video and sensor signals. First of all, sampling is an essential step in the digitisation process as illustrated in Fig. 1.6. Second, the gaps between samples of one signal can be exploited to simultaneously accommodate without any overlap multiple signals inside one composite signal — a technique known as *time division multiplexing* (TDM) that facilitates the efficient utilisation of communication system resources such as the bandwidth capacity of the transmission link or medium. TDM is achieved through interleaving, as illustrated in Fig. 4.1 for a 3-channel system ($N = 3$), whereby the start of the regular sampling instants of N independent signals, numbered 0 to $N - 1$, is staggered so that the nth sample of the kth signal $g_k(t)$ is taken at time instant t given below and all samples are connected to a single outlet and conveyed as one composite (TDM) signal:

$$t = \left(n + \frac{k}{N}\right)T_s; \quad k = 0, 1, 2, \ldots, N - 1; \quad n = 0, 1, 2, 3, \ldots \tag{4.1}$$

The rule governing the selection of an appropriate sampling interval T_s or sampling rate $f_s = 1/T_s$ is given by the *sampling theorem*, which may be stated as follows:

> *A finite-energy band limited analogue signal that has no frequency components outside a bandwidth B (Hz) may be perfectly reconstructed from its samples taken at regular intervals at the rate $f_s \geq 2B$ (samples per second, or Hz).*

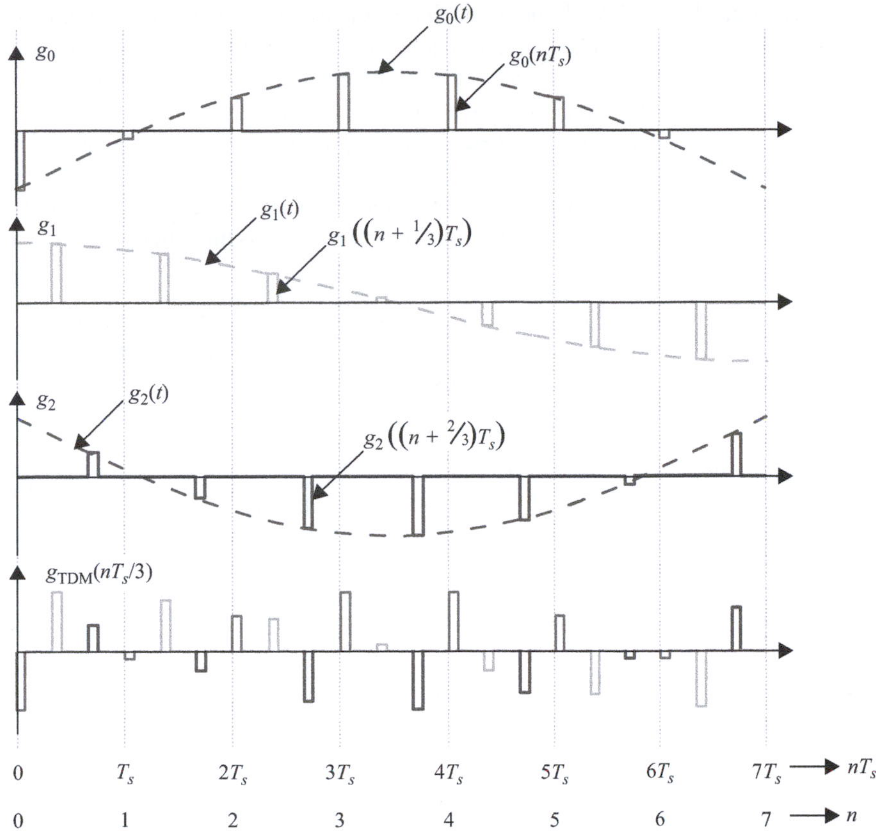

Fig. 4.1 Interleaving of the samples of three signals $g_0(t)$, $g_1(t)$ and $g_2(t)$ to form one multiplexed signal $g_{TDM}(t)$

There are a number of important points to note about the sampling theorem as stated above:

The analogue signal must have a finite bandwidth B. Signals such as white noise with an infinite bandwidth cannot be reconstructed without distortion from samples taken at a regular interval $T_s > 0$. To comply with the sampling theorem, it is usually necessary in practice to artificially restrict the bandwidth of the signal to a value $\leq f_s/2$ prior to sampling at rate f_s.

For a low pass analogue signal the bandwidth B equals the maximum frequency component f_m of the signal. The theorem therefore stipulates that the sampling rate of a low pass signal must be at least twice the maximum frequency component of the signal. This requires that at least two samples are taken during each period of the highest frequency sinusoid in the signal.

For band pass signals where bandwidth is centred on a positive (carrier) frequency $f_c > B/2$, the maximum frequency component f_m of the signal exceeds bandwidth B, sometimes by many orders of magnitude. Perfect reconstruction is clearly possible if the signal is treated as if it were low pass and sampling is done at a rate that is at least $2f_m$. The theorem nevertheless indicates that sampling at the lower rate $f_s \geq 2B$ is also possible. Beware, not every rate above $2B$ will guarantee distortion-free reconstruction in *uniform sampling* of a band pass signal. There is however an alternative method known as *quadrature sampling* that ensures perfect reconstruction of a band pass signal from its samples taken at any rate $\geq 2B$. Section 4.5 gives a lucid treatment of this often confused but important subject of band pass sampling.

Under the above conditions, the original signal $g(t)$ can be recovered from its samples $g(nT_s)$, with absolutely no distortion, by passing them through a *reconstruction filter.*

The minimum sampling rate $f_{s\,\min}$ specified by the sampling theorem is referred to as the *Nyquist rate* or *Nyquist frequency*, with its reciprocal called the *Nyquist sampling interval*, which is the maximum sampling interval $T_{s\,\max}$ allowed by the theorem. Thus

$$\begin{aligned} &\text{Nyquist rate:} && f_{s\,\min} = 2B \\ &\text{Nyquist interval:} && T_{s\,\max} = \frac{1}{2B} \end{aligned} \tag{4.2}$$

4.3 Sampling a sinusoidal signal

By virtue of the Fourier theorem (Appendix C) every realisable information-bearing signal is a (discrete or continuous) sum of sinusoids, so it is only logical that we first consider the effect of sampling on a sinusoidal signal; for if a sinusoid can be reconstructed from its samples then, in view of the linearity of the sampling process, the same rule can be extended to any arbitrary signal. Consider then Fig. 4.2a, which shows a 0.5 ms segment of a sinusoidal signal $g(t)$ of frequency $f_m = 4$ kHz. This analogue signal unambiguously represents a unique frequency f_m

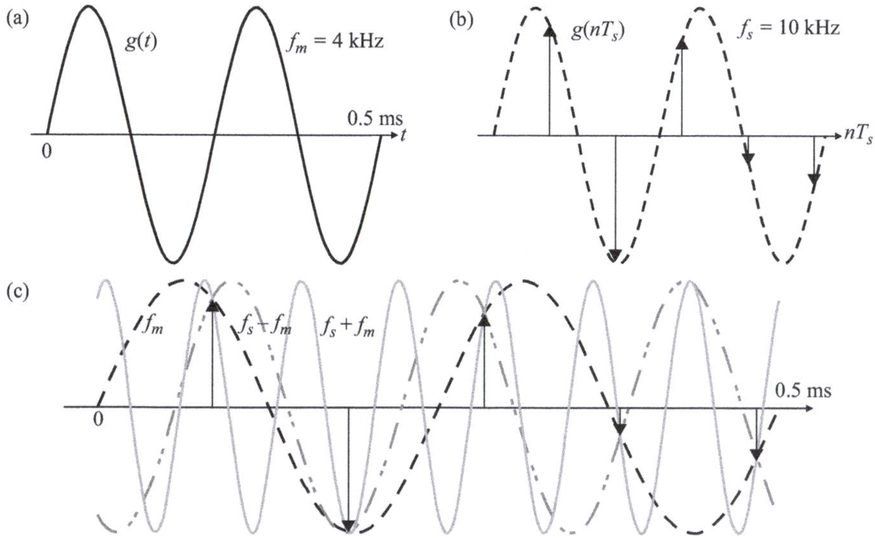

Fig. 4.2 (a) Sinusoidal signal g(t); (b) samples g(nT_s) taken at rate $f_s = 1/T_s$; (c) the samples g(nT_s) could have come from sinusoids of frequencies $nf_s \pm f_m$, $n = 1, 2, 3, \ldots$

in the sense that no other frequency will fit this waveform and there can therefore be no confusion at the receiver as to which frequency was sent. Consider next the sampled signal $g(nT_s)$ shown in Fig. 4.2b obtained by taking instantaneous samples of $g(t)$ at a rate f_s ($= 1/T_s$) of 10 kHz, which yields 5 samples in a 0.5 ms segment as shown. So are we able to recover $g(t)$ from $g(nT_s)$? In other words, would a receiver know that the sequence definitely came from frequency f_m and none other? Unsurprisingly we find that, unlike $g(t)$, the sequence no longer represents a unique frequency; rather it fits (i.e. contains) an infinite set of frequencies f_m, $f_s - f_m$, $f_s + f_m$, $2f_s - f_m$, $2f_s + f_m$, $3f_s - f_m$, $3f_s + f_m, \ldots$ That is, sampling any of these other sinusoids (of frequencies $nf_s \pm f_m$, $n = 1, 2, 3, \ldots$), described as *image frequencies*, would have yielded exactly the same sequence $\{g(nT_s)\}$ as obtained by sampling at rate f_s the sinusoid $g(t)$ of *base band frequency* f_m. Fig. 4.2c shows this fit for f_m, $f_s - f_m$ and $f_s + f_m$, whereas Fig. 4.3a shows how the sum of the base band and image frequencies up to $n = 20$ approaches $g(nT_s)$.

The single-sided amplitude spectrum of the sinusoid $g(t)$ of amplitude A contains a single spectral line of height A located at $f = f_m$, whereas the preferred double-sided amplitude spectrum, denoted by $|G(f)|$ and shown in Fig. 4.3b, contains two spectral lines of height $A/2$ located at $f = \pm f_m$ — obtained simply by exploiting the identity

$$A\cos(2\pi f_m t) \equiv \frac{A}{2}\cos(2\pi f_m t) + \frac{A}{2}\cos[2\pi(-f_m)t]$$

Similarly, since the sampled signal $g(nT_s)$ contains frequencies f_m, $f_s - f_m$, $f_s + f_m$, $2f_s - f_m$, $2f_s + f_m, \ldots$, its amplitude spectrum is as shown in the bottom half of Fig. 4.3b. We see therefore that sampling a sinusoidal signal of frequency f_m has the effect of replicating (without distortion) the base band frequency f_m at intervals of f_s. As illustrated in Fig. 4.3b, the frequency f_m and hence the original signal $g(t)$

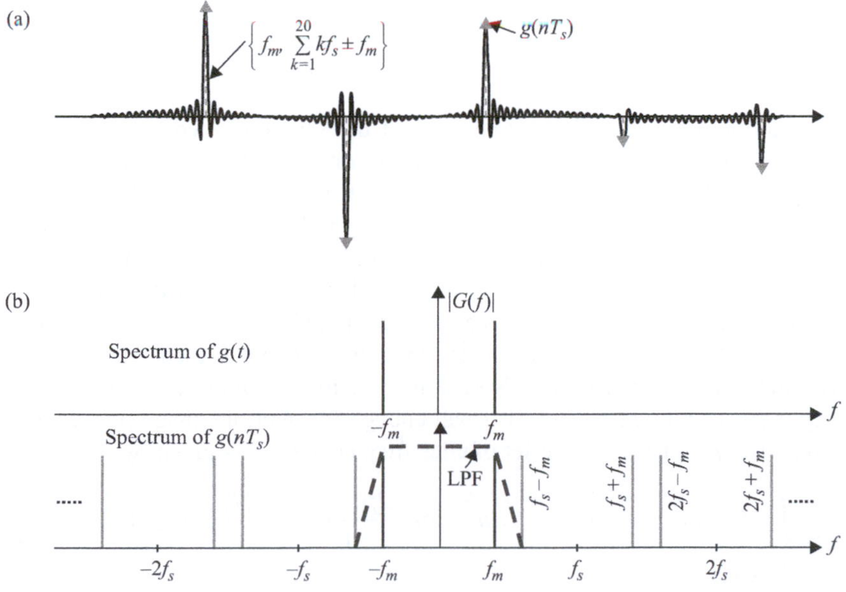

Fig. 4.3 (a) Sum of base band and image frequencies yields $g(nT_s)$; (b) spectra of $g(t)$ and $g(nT_s)$

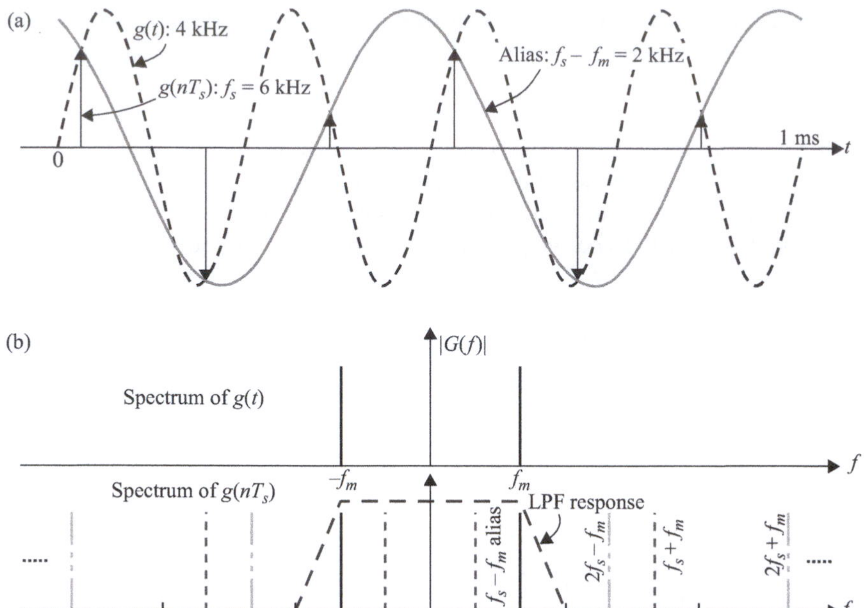

Fig. 4.4 Sampling a sinusoid g(t) of frequency $f_m = 4$ kHz at rate $f_s = 6$ kHz ($<2f_m$): (a) waveforms of g(t) (dashed) and alias (bold); (b) spectra of g(t) and g(nT_s)

can be recovered from $g(nT_s)$ by passing the samples through a low pass filter (LPF) — known as a *reconstruction filter* — which passes only f_m and blocks all the image frequencies. Since the lowest image frequency is $f_s - f_m$, this clearly requires that f_m should fall within the filter pass band whereas $f_s - f_m$ is above the pass band. That is, $f_s - f_m \geq f_m$, or

$$f_s \geq 2f_m \tag{4.3}$$

This condition happens to be a stipulation of the sampling theorem for low pass signals. So what happens if we use a sampling rate $f_s < 2f_m$ in direct violation of Eq. (4.3)? Fig. 4.4 illustrates the effect of this choice both in the time domain (a) and in the frequency domain (b). In this example, $f_m = 4$ kHz and $f_s = 6$ kHz. The lowest image frequency $|f_s - f_m|$ that also exactly fits the sequence is now smaller than f_m and is known as an *alias frequency* f_a since the LPF will recover this frequency in place of, or (depending on the LPF bandwidth) in addition to, f_m. Thus, the reconstructed signal will contain a frequency component f_a not present in the original signal, and is therefore distorted. Note that there may be more than one alias frequency if the sampling frequency is small enough for multiple image frequencies to be lower than f_m. The complete set of alias frequencies $f_{a,k}$ is given by

$$f_{a,k} = \begin{cases} |kf_s - f_m|, \quad k = 1, 2, \ldots, n; & \text{for } \dfrac{2f_m}{n+1} < f_s < \dfrac{2f_m}{n}, \quad n \geq 1 \\ \text{None} & \text{for } f_s \geq 2f_m \end{cases} \tag{4.4}$$

To clarify, if $f_s < 2f_m$ but $f_s > f_m$ (i.e. $n = 1$ in the above equation) as in Fig. 4.4a, then only one image frequency $|f_s - f_m|$ is lower than f_m and there is therefore only

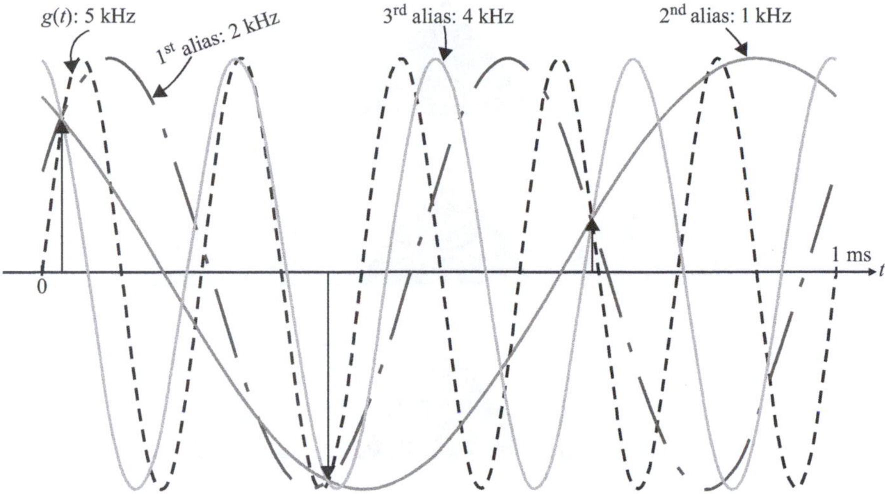

Fig. 4.5 Sampling a sinusoid g(t) of frequency $f_m = 5$ kHz at rate $f_s = 3$ kHz creates three alias frequencies at 1, 2 and 4 kHz

one alias; but if $\frac{2}{3}f_m < f_s < f_m$ (i.e. $n = 2$) then both $|f_s - f_m|$ and $|2f_s - f_m|$ are lower than f_m resulting in two alias frequencies. Fig. 4.5 shows the result of sampling a sinusoid of frequency $f_m = 5$ kHz at a sampling rate $f_s = 3$ kHz — which corresponds to $n = 3$ in Eq. (4.4). We see that 3 alias frequencies are created at 1, 2 and 4 kHz. That is, there are 3 sinusoids having frequencies less than f_m that exactly fit the samples. Note that this distortion, known as *alias distortion*, is the result of *under-sampling* and is not in any way a deficiency of sampling, which on this occasion does exactly what it always does, i.e. it replicates the spectrum of $g(t)$ at a regular frequency spacing f_s (as shown in Fig. 4.4b). It is just that if f_s is small enough to cause an overlap of one or more image (i.e. replicated) frequencies into base band then distortion will occur. Increasing f_s to a value that satisfies Eq. (4.3) eliminates this overlap and hence prevents alias distortion.

4.4 Base band signal sampling and alias distortion

Sampling is a linear process since the principle of superposition (Eq. (2.1)) holds — the sample of a sum of signals is the sum of the samples of the constituent signals taken individually. So when a base band signal $g(t)$ having spectrum $G(f)$ is sampled at rate f_s, then each frequency component of $G(f)$ is replicated as discussed above at intervals of f_s. The result is that the entire base band $G(f)$ is replicated at intervals of f_s and *image bands* are formed at nf_s, for $n = 1, 2, 3, \ldots$ If $g(t)$ is band limited with bandwidth f_m as shown in Fig. 4.6a, then the replicated bands do not overlap provided $f_s \geq 2f_m$ in line with Eq. (4.3). Under this condition, $G(f)$ and hence $g(t)$ can be recovered without any distortion from $g(nT_s)$ using an LPF as illustrated in Fig. 4.6b. The minimum sampling rate that satisfies Eq. (4.3) is $f_s = 2f_m$, called the *Nyquist rate*. Although it is theoretically possible to recover $g(t)$ without distortion from samples taken at Nyquist rate, an ideal brick wall reconstruction filter would be required as shown in Fig. 4.6c to keep out the lowest image band in its entirety.

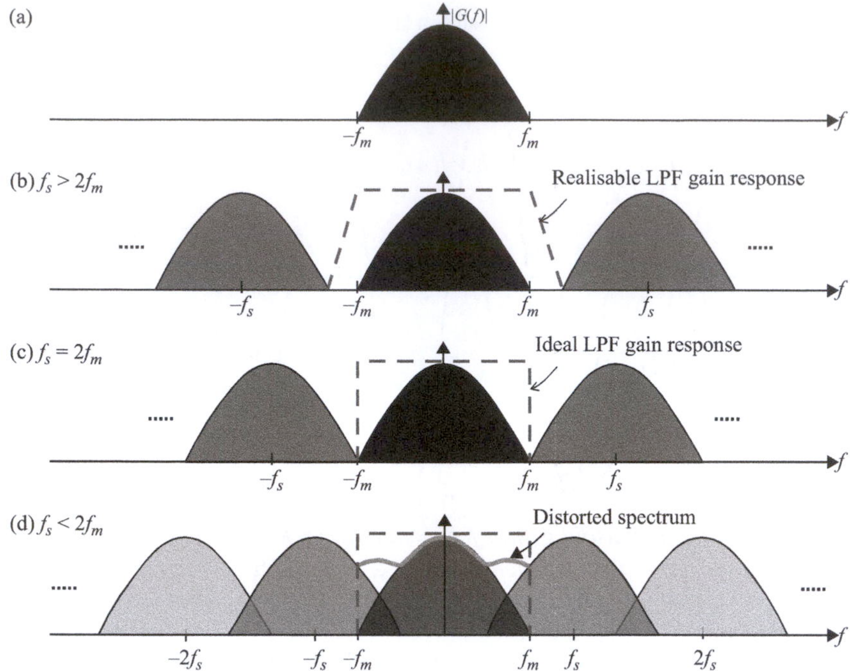

Fig. 4.6 (a) Spectrum of band limited base band signal g(t), and (b)–(d) spectra of its samples $g(nT_s)$ at various sampling rates f_s

A sampling rate $f_s < 2f_m$ (Fig. 4.6d) causes alias distortion due to an overlap of one or more image bands into base band. A useful measure of the amount of alias distortion is provided by the ratio between signal power and alias power at the output of the reconstruction filter, referred to as signal-to-alias distortion ratio (SADR). To obtain a general expression for SADR, consider employing a filter of gain response $|H(f)|$ to reconstruct an analogue signal of power spectral density $S_g(f)$ from its instantaneous samples. The signal power comes from the base band PSD $S_g(f)$ extracted by (i.e. passed through) $|H(f)|$, whereas the alias power is due to contributions from the infinite set of image PSDs $S_g(f \pm nf_s)$, $n = 1, 2, 3, \ldots$ transmitted through $|H(f)|$. Using the expression for output PSD given in Eq. (2.27), we take the ratio of the total of each of the two powers in the frequency range $-\infty < f < \infty$ and obtain SADR as

$$\text{SADR} = \frac{\int\limits_{-\infty}^{\infty} S_g(f)|H(f)|^2 df}{\sum\limits_{n=1}^{\infty} \int\limits_{-\infty}^{\infty} S_g(f \pm nf_s)|H(f)|^2 df} \tag{4.5}$$

For an ideal brickwall reconstruction filter of bandwidth $f_s/2$, the situation is as shown in Fig. 4.7. Note that a sinc-shaped PSD is used for illustration purposes

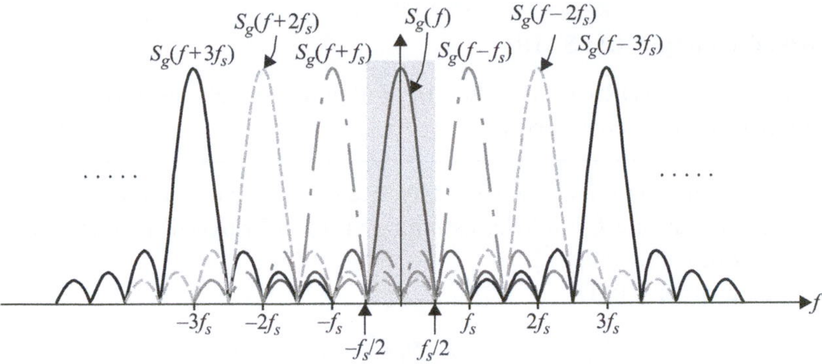

Fig. 4.7 Alias distortion in an ideal brickwall reconstruction filter (pass band shaded), illustrated for a signal g(t) with a sinc-shaped PSD

only and is not a requirement. Normalising $|H(f)|$ to unity in the filter pass band, the output signal power P_s is

$$P_s = \int_{-f_s/2}^{f_s/2} S_g(f)|H(f)|^2 df = 2\int_0^{f_s/2} S_g(f)df$$

To compute alias power P_a, we sum contributions from image bands along the negative side of the frequency axis and double this to include the identical contribution from the positive side. Thus

$$P_a = 2\left[\int_{-f_s/2}^{f_s/2} S_g(f+f_s)df + \int_{-f_s/2}^{f_s/2} S_g(f+2f_s)df + \int_{-f_s/2}^{f_s/2} S_g(f+3f_s)df + \cdots\right]$$

Substituting f for $f+nf_s$, $n=1, 2, 3, \ldots$ in to the above integrals, the respective limits of integration become $(2n-1)f_s/2$ to $(2n+1)f_s/2$ so that

$$P_a = 2\left[\int_{f_s/2}^{3f_s/2} S_g(f)df + \int_{3f_s/2}^{5f_s/2} S_g(f)df + \int_{5f_s/2}^{7f_s/2} S_g(f)df + \cdots\right] = 2\int_{f_s/2}^{\infty} S_g(f)df$$

Thus, SADR in this special case of an ideal brickwall reconstruction filter with bandwidth equal to half the sampling rate is given by

$$\begin{aligned}\text{SADR} &= \frac{P_s}{P_a} = \int_0^{f_s/2} S_g(f)df \bigg/ \int_{f_s/2}^{\infty} S_g(f)df \\ &= 10\log_{10}\left[\int_0^{f_s/2} S_g(f)df \bigg/ \int_{f_s/2}^{\infty} S_g(f)df\right] \quad \text{dB}\end{aligned} \tag{4.6}$$

Worked Example 4.1: SADR

Determine the SADR (in dB) when the following signals are reconstructed from their instantaneous samples.

(a) A rectangular pulse train of duty cycle $d = ¼$ and period $T = 1$ ms sampled at rate $f_s = 8$ kHz, assuming an ideal brickwall reconstruction filter of bandwidth $f_s/2$.

(b) A signal with PSD

$$S_g(f) = \begin{cases} [1 + \cos(\pi|f|/R_s)]^2, & 0 \le |f| \le R_s \\ 0, & |f| \ge R_s \end{cases}$$

sampled at rate $f_s = 1.6R_s$, assuming an ideal brickwall reconstruction filter of bandwidth $f_s/2$.

(c) The signal in (b) recovered using a 3rd-order Butterworth filter of 3-dB bandwidth $f_s/2$.

(a) A rectangular pulse train of amplitude A, duty cycle d and period $T \equiv 1/f_0$ has a sinc-shaped discrete amplitude spectrum given by the Fourier series

$$g(t) = Ad \sum_{n=-\infty}^{\infty} \text{sinc}(nd)\cos(2\pi nf_0 t) \equiv \sum_{n=-\infty}^{\infty} A_n \cos(2\pi nf_0 t) \quad (4.7)$$

and shown in Fig. 4.8a over the frequency range ±12 kHz for $d = ¼$, $T = 1$ ms, and $f_0 = 1/T = 1$ kHz. Sampling at rate $f_s = 8$ kHz will result in

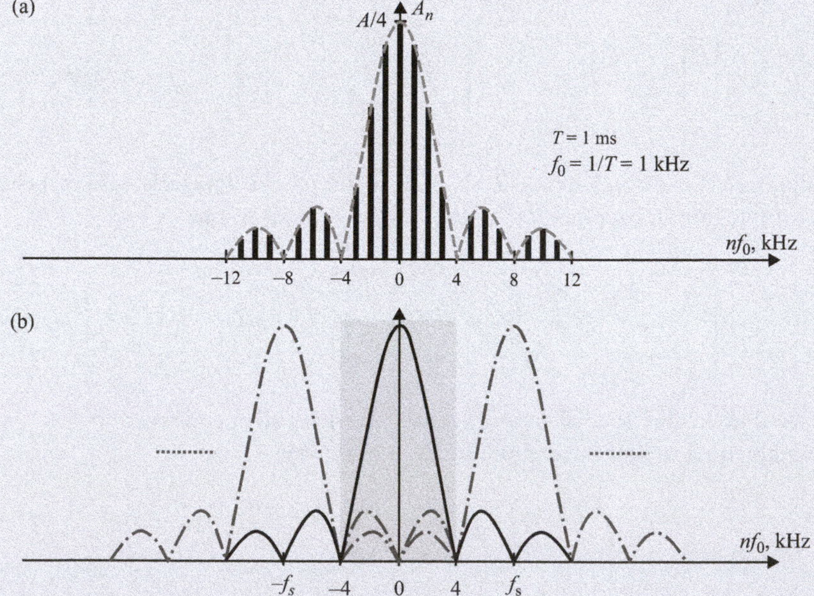

Fig. 4.8 (a) Amplitude spectrum of rectangular pulse train g(t) of period T; (b) amplitude spectrum of $g(nT_s)$, with sampling rate $f_s = 1/T_s =$ 8 kHz

an overlap of the tails of image bands into the pass band of the ideal reconstruction filter (as shown in Fig. 4.8b with the filter pass band shaded). This gives rise to alias distortion. A discrete version of Eq. (4.6) applicable to this problem may be written as

$$\mathrm{SADR} = 10\log_{10}\left[\frac{A_0^2 + \sum_{n=1}^{4} A_n^2/2}{\sum_{n=5}^{\infty} A_n^2/2}\right] \text{ dB}$$

We substitute $A_n = Ad\,\mathrm{sinc}(nd) = \dfrac{Ad\sin(n\pi d)}{n\pi d} = \dfrac{A}{n\pi}\sin\left(\dfrac{n\pi}{4}\right),\ n \geq 1,$ and $A_0 = A/4$, and manipulate to obtain

$$\mathrm{SADR} = \frac{\dfrac{A^2}{16} + \sum_{n=1}^{4} \dfrac{A^2}{n^2\pi^2}\sin^2(n\pi/4)}{\sum_{n=5}^{\infty} \dfrac{A^2}{n^2\pi^2}\sin^2(n\pi/4)}$$

$$= \frac{\dfrac{1}{16} + \dfrac{1}{\pi^2}\sum_{n=1}^{4}\dfrac{\sin^2(n\pi/4)}{n^2}}{\dfrac{1}{\pi^2}\sum_{n=1}^{\infty}\dfrac{\sin^2(n\pi/4)}{n^2} - \dfrac{1}{\pi^2}\sum_{n=1}^{4}\dfrac{\sin^2(n\pi/4)}{n^2}}$$

But

$$\frac{1}{\pi^2}\sum_{n=1}^{4}\frac{\sin^2(n\pi/4)}{n^2} = \frac{1}{\pi^2}\left(\frac{1}{2} + \frac{1}{4} + \frac{1}{18} + 0\right) = \frac{29}{36\pi^2}$$

and

$$\frac{1}{\pi^2}\sum_{n=1}^{\infty}\frac{\sin^2(n\pi/4)}{n^2} = \frac{1}{2\pi^2}\left[\sum_{n=1}^{\infty}\frac{1}{n^2} - \sum_{n=1}^{\infty}\frac{\cos n\pi/2}{n^2}\right]$$

which upon using the series relations

$$\sum_{n=1}^{\infty}\frac{1}{n^2} = \frac{\pi^2}{6}; \quad \sum_{n=1}^{\infty}\frac{\cos n\pi/2}{n^2} = -\frac{\pi^2}{48}$$

reduces to 3/32. Thus

$$\mathrm{SADR} = \frac{1/16 + 29/36\pi^2}{3/32 - 29/36\pi^2} = 11.8811 = 10.75 \text{ dB}$$

There is therefore significant alias distortion, with signal power only ~11 dB above alias power. This would be unacceptable in practice. A method for reducing alias distortion by filtering prior to sampling is discussed in Section 4.8. As an exercise, try and rework this problem with a higher sampling rate $f_s = 16$ kHz in order to calculate SADR improvement due to this doubling of sampling rate. What is the engineering trade-off involved (i.e. what price do you pay) in this improvement?

(b) SADR in this case is obtained through a straightforward application of Eq. (4.6), with $f_s = 1.6R_s$ (and hence $f_s/2 = 4R_s/5$), $|f|$ in the expression for $S_g(f)$ replaced by f since the integration is in the positive range of the frequency axis where f is always positive, and the upper limit of integration in the denominator reduced from ∞ to R_s, since $S_g(f)$ is zero beyond $f = R_s$. The situation at hand is depicted in Fig. 4.9a, which shows the image bands and their slight incursion into the pass band of the ideal LPF. Thus

$$\text{SADR} = \frac{\int_0^{4R_s/5} [1+\cos(\pi f/R_s)]^2 df}{\int_{4R_s/5}^{R_s} [1+\cos(\pi f/R_s)]^2 df}$$

$$= \frac{\int_0^{4R_s/5} \left[3/2 + \frac{1}{2}\cos(2\pi f/R_s) + 2\cos(\pi f/R_s)\right] df}{\int_{4R_s/5}^{R_s} \left[3/2 + \frac{1}{2}\cos(2\pi f/R_s) + 2\cos(\pi f/R_s)\right] df}$$

$$= \frac{\left[\frac{3}{2}f + \frac{R_s}{4\pi}\sin(2\pi f/R_s) + \frac{2R_s}{\pi}\sin(\pi f/R_s)\right]\Big|_0^{4R_s/5}}{\left[\frac{3}{2}f + \frac{R_s}{4\pi}\sin(2\pi f/R_s) + \frac{2R_s}{\pi}\sin(\pi f/R_s)\right]\Big|_{4R_s/5}^{R_s}}$$

$$= \frac{R_s[6/5 + \sin(8\pi/5)/4\pi + 2\sin(4\pi/5)/\pi] - 0}{3R_s/2 - R_s[6/5 + \sin(8\pi/5)/4\pi + 2\sin(4\pi/5)/\pi]}$$

$$= 1008 = 30 \text{ dB}$$

Note that the denominator in the penultimate line above involves the subtraction of two numbers that are close in values, and so care needs to be taken to avoid rounding errors. SADR is high, an indication that alias distortion is small. In fact in this type of idealised situation in which the signal is strictly band limited — note in the problem statement that PSD = 0 for $f \geq R_s$ — and an ideal brickwall reconstruction filter is used, it is possible to completely eliminate alias distortion (so that SADR = ∞) by sampling at Nyquist rate, which in this case is $f_s = 2R_s$, rather than at the rate of $1.6R_s$ specified in the problem. In (c) below, we reconstruct this same sampled signal using a realisable filter and observe what effect this has on SADR.

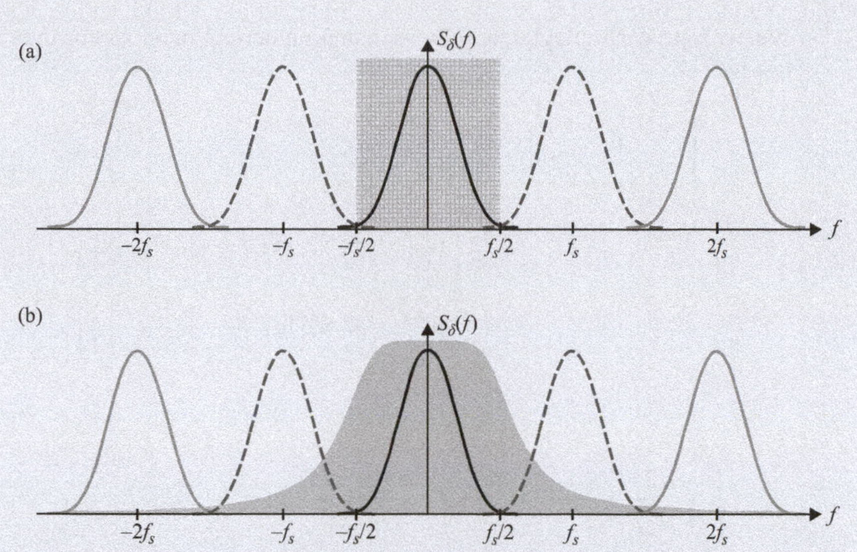

Fig. 4.9 Power spectral density of sampled signal (sampling rate $f_s = R_s$) in Worked Example 4.1b with reconstruction filter response shaded for (a) ideal brickwall filter, and (b) 3rd-order Butterworth filter

(c) The new scenario is as shown in Fig. 4.9b, with the gain response of the filter shaded and the more significant incursion of image bands through the filter clearly evident. First, let us determine the recovered signal power. This is the power in the base band spectrum that is admitted through the 3rd-order Butterworth filter of gain response $|H(f)| = 1/\sqrt{1+(f/f_1)^6}$ (see Fig. 2.11) and is given by the numerator of Eq. (4.5), with the limits of integration restricted to the frequency range $-R_s \le f \le R_s$ since $S_g(f) = 0$ outside this range, and the filter cut-off frequency $f_1 = f_s/2 = 4R_s/5$:

$$
\begin{aligned}
P_s &= \int_{-\infty}^{\infty} S_g(f)|H(f)|^2 df \\
&= 2\int_0^{R_s} \left[3/2 + \tfrac{1}{2}\cos(2\pi f/R_s) + 2\cos(\pi f/R_s)\right]|H(f)|^2 df \\
&= 3\int_0^{R_s} \frac{1}{1+(f/f_1)^6} df + \int_0^{R_s} \frac{\cos(2\pi f/R_s)}{1+(f/f_1)^6} df + 4\int_0^{R_s} \frac{\cos(\pi f/R_s)}{1+(f/f_1)^6} df
\end{aligned}
$$

We evaluate each integral as follows, using numerical methods for the last two:

$$3\int_0^{R_s}\frac{1}{1+(f/f_1)^6}df = \frac{12}{5}R_s\int_0^{5/4}\frac{dx}{1+(x)^6}$$

$$= \frac{R_s}{5}\left[2\tan^{-1}(2x-\sqrt{3}) + 2\tan^{-1}(2x+\sqrt{3}) + 4\tan^{-1}(x) + \sqrt{3}\ln\left(\frac{x^2+\sqrt{3}x+1}{x^2-\sqrt{3}x+1}\right)\right]\Bigg|_{x=0}^{x=5/4}$$

$$= 2.3721R_s$$

$$\int_0^{R_s}\frac{\cos(2\pi f/R_s)}{1+(f/f_1)^6}df = \frac{4R_s}{5}\int_0^{5/4}\frac{\cos(8\pi x/5)}{1+(x)^6}dx = -0.0797R_s$$

$$4\int_0^{R_s}\frac{\cos(\pi f/R_s)}{1+(f/f_1)^6}df = \frac{16R_s}{5}\int_0^{5/4}\frac{\cos(4\pi x/5)}{1+(x)^6}dx = 0.6606R_s$$

Summing the three results gives

$$P_s = 2.953R_s$$

Next, we turn our attention to the denominator of Eq. (4.5), the amount of image band power that is admitted by the filter. This constitutes alias power P_a. In view of the even symmetry of both $|H(f)|$ and $S_g(f)$, the contributions of image bands on either side of the frequency axis are identical, so we only need to add contributions from image bands on the RHS of the axis and then double this result. Thus, the nth image band on either side of the frequency axis contributes alias power given by

$$P_{a,n} = 2\int_{-\infty}^{\infty}S_g(f-nf_s)|H(f)|^2 df, \quad n = 1, 2, 3, \ldots$$

$$= 2\int_{-\infty}^{\infty}\left[1+\cos\left(\pi\frac{|f-nf_s|}{R_s}\right)\right]^2\frac{1}{1+(f/f_1)^6}df$$

$$= \frac{8R_s}{5}\int_{2n}^{2n+5/4}\left[1+\cos\left(4\pi\frac{x-2n}{5}\right)\right]^2\frac{1}{1+x^6}dx$$

The above integral was evaluated numerically, and results are tabulated in Table 4.1 for the first 10 image bands. It can be seen that only the first

Table 4.1 Worked Example 4.1c: Contributions to alias power by image bands in Fig. 4.9b

Image band (n)	**Alias power ($P_{a,n}$)**	**Relative level (dB)**	**Cumulative SADR (dB)**
1	$2.40522 \times 10^{-2} R_s$	0	19.94
2	$5.05049 \times 10^{-4} R_s$	−16.78	19.85
3	$4.9611 \times 10^{-5} R_s$	−26.86	19.84
4	$9.3765 \times 10^{-6} R_s$	−34.09	19.84
5	$2.5518 \times 10^{-6} R_s$	−39.74	19.84
6	$8.7675 \times 10^{-7} R_s$	−44.38	19.84
7	$3.5422 \times 10^{-7} R_s$	−48.32	19.84
8	$1.61235 \times 10^{-7} R_s$	−51.74	19.84
9	$8.042 \times 10^{-8} R_s$	−54.76	19.84
10	$4.312 \times 10^{-8} R_s$	−57.46	19.84
Total	$2.4620316 \times 10^{-2} R_s$	0.10	19.84

image band ($n = 1$) makes a significant contribution to alias power; the contributions of the 2nd, 3rd and 4th image bands being respectively 16.78, 26.86 and 34.09 dB below that of the first. Adding these contributions gives total alias power $P_a = 2.462 \times 10^{-2} R_s$, and hence SADR (using the signal power P_s obtained earlier)

$$\text{SADR} = \frac{P_s}{P_a} = \frac{2.953 R_s}{2.462 R_s} = 96.35 = 19.84 \text{ dB}$$

Comparing the SADR obtained here to that in (b) we see that there is a 10 dB reduction attributable entirely to the design of the reconstruction filter — more specifically to the steepness of the slope of the filter's gain response beyond its pass band. Furthermore, it is important to examine Table 4.1 and observe that calculating SADR based on alias power from only the first image band would give SADR = 19.94 dB, an error of only 0.1 dB, whereas adding alias powers from the 4th and higher image bands makes no difference to SADR expressed correct to two decimal places. The width of the transition band (see Fig. 2.11) of the reconstruction filter and the sampling frequency are important factors in determining the significance of contributions to aliasing from image bands other than the first. To investigate this, it is suggested that you undertake the exercise of reworking this same problem using a 1st-order Butterworth reconstruction filter. Is alias power from the second image band up to ~17 dB lower than that of the first band as was found here with a 3rd-order filter?

It is clear from Worked Example 4.1 above that the SADR at the output of a realisable reconstruction filter of gain response $|H(f)|$ that has a sharp roll-off beyond its pass band (or equivalently a narrow transition band) can be determined to a good level of accuracy by including alias contributions from only the first image band. Taking this observation into account simplifies Eq. (4.5) to

$$\text{SADR} = \frac{\int_{0}^{\infty} S_g(f)|H(f)|^2 df}{\int_{-\infty}^{\infty} S_g(f - f_s)|H(f)|^2 df} \tag{4.8}$$

4.5 Band pass signal sampling

It is sometimes necessary to sample a band pass signal $g_{bp}(t)$ of bandwidth B whose positive frequencies are centred on $f_c \gg 0$, as shown in Fig. 4.10a. An obvious way of doing this would be to treat $g_{bp}(t)$ as a (base band or low pass) signal that contains frequency components from zero up to the highest frequency f_m. We would then ensure the non-overlap of replicated bands in the sampled signal $g_{bp}(n/f_s)$, and hence the possibility of recovering without distortion the original signal $g_{bp}(t)$ from the sampled sequence $g_{bp}(n/f_s)$, by choosing a sampling rate $f_s \geq 2f_m$. This is illustrated in Fig. 4.9b for $f_s = 2f_m$. Note that the spectrum $G_{bp\delta}(f)$

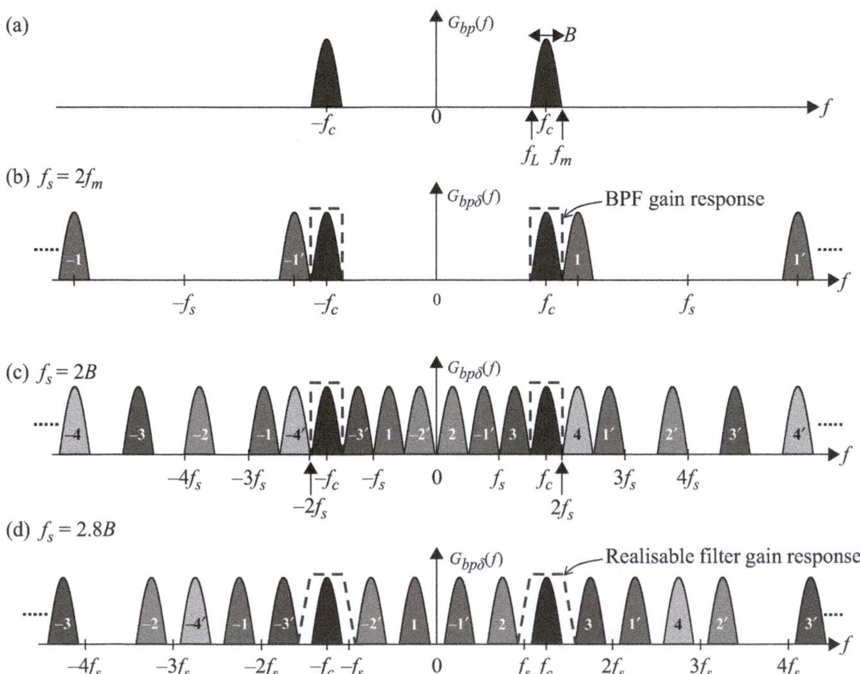

Fig. 4.10 (a) Spectrum of band pass signal $g_{bp}(t)$ of bandwidth B located at (f_L, f_m), where $f_L = 3B$, and (b) – (d) spectra of $g_{bp}(n/f_s)$ for various sampling rates f_s

of the sampled signal results from the replication of the band $f_L \rightarrow f_m$ of $G_{bp}(f)$, denoted by (f_L, f_m), at LHS and RHS locations

$$\begin{aligned}\text{Band}_n &= \pm\{nf_s \pm (f_L, f_m)\}, \quad n = 1, 2, 3, \ldots \\ &= \pm\{(nf_s - f_m, nf_s - f_L), \quad (nf_s + f_L, nf_s + f_m)\}\end{aligned} \tag{4.9}$$

Fig. 4.10b shows the above image bands for $n = 1$ (with the LHS band labelled 1 and the RHS band labelled 1′), which (for $f_s = 2f_m$) are at locations

$$\text{Band}_1 = \pm\{(f_m, f_m + B), \quad (2f_m + f_L, 2f_m + f_L + B)\}$$

We see that a band pass reconstruction filter of bandwidth B centred at f_c is required to recover the band pass signal $g_{bp}(t)$, whereas the filter employed to reconstruct base band signals is low pass as discussed in the previous section.

The biggest drawback of sampling a band pass signal at the rate $f_s = 2f_m$ suggested above is that this rate can be prohibitively high since f_c is often quite large, sometimes many orders of magnitude larger than B. It is however not necessary to sample at such a high rate if we observe that large swathes of $G_{bp}(f)$ are empty, with occupancy only within the two intervals $\pm(f_L, f_m)$, so distortion-free reconstruction ought to be possible at a much lower sampling rate if we can find one that simply ensures that none of the replicated bands of Eq. (4.9) overlaps into any of the two intervals $\pm(f_L, f_m)$. This is indeed the case for a sampling rate f_s in the range

$$\frac{2f_m}{p} \leq f_s \leq \frac{2f_L}{p-1}, \quad 1 \leq p \leq \left\lfloor \frac{f_m}{B} \right\rfloor \tag{4.10}$$

where p is an integer and $\lfloor x \rfloor$ denotes the integer part of a positive real number x. Let us elaborate on two special cases of Eq. (4.10) to shed more light on its interpretation and use:

If $f_m = B$, meaning that the signal is a base band signal, which by definition has a bandwidth B equal to its highest frequency component f_m, then $p = 1$ and the applicable sampling rate falls in the range

$$\frac{2f_m}{1} \leq f_s \leq \frac{2f_L}{1-1}; \quad \Rightarrow 2f_m \leq f_s \leq \infty; \quad \Rightarrow f_s \geq 2f_m$$

which is simply Eq. (4.3) for base band signal sampling. That is, alias-free reconstruction is achievable using a sampling rate f_s that is at least twice the bandwidth $B \equiv f_m$.

If $f_L = qB$, where q is an integer, it means that the band (f_L, f_m) of $g_{bp}(t)$ starts at a location that is an integer number of bandwidths. Note that $q = 0$ corresponds to a start at the origin, which is the base band case discussed above, so here we consider only the band pass cases where $q \geq 1$. Substituting $f_L = qB$ and $f_m = qB + B = (q+1)B$ into Eq. (4.10) yields the applicable range of sampling rates as

$$\frac{2(q+1)B}{p} \leq f_s \leq \frac{2qB}{p-1}, \quad 1 \leq p \leq q+1$$

Let us take a moment to examine the possibilities for f_s. First, when $q = 1$, then there are only two possible values for p, namely $p = 1$ and $p = 2$, which respectively yield the following ranges for f_s:

$$q = 1,\ p = 1: \quad \frac{4B}{1} \le f_s \le \frac{2B}{1-1}; \ \Rightarrow f_s \ge 4B$$

$$q = 1,\ p = 2: \quad \frac{4B}{2} \le f_s \le \frac{2B}{2-1}; \ \Rightarrow f_s = 2B$$

Thus, when $q = 1$ (which means that $f_L = B$ and $f_m = 2B$) then alias-free sampling is possible at any rate not lower than $4B$ or at the specific rate $f_s = 2B$. The first option ($f_s \ge 4B$) corresponds to sampling at a rate that is at least twice the maximum frequency component of the band pass signal. This has been earlier explored and is illustrated in Fig. 4.10b. So it is the second option ($f_s = 2B$) that delivers a saving in sampling rate by taking advantage of the fact that the region $f < f_L$ is unoccupied in a band pass signal to allow sampling to replicate the band (f_L, f_m) at a regular spacing f_s that might be much less than $2f_m$ without an overlap of any of the image bands into $\pm(f_L, f_m)$. Next, examining the situation for $q = 2$ (when p can take on values 1, 2 and 3), $q = 3$ (when p can be 1, 2, 3, 4) and so on, we find that all cases of integer band positioning (i.e. $f_L = qB$, $q = 0, 1, 2, 3, \ldots$) allow *alias-free* sampling (i.e. no overlap between image and original bands) at a minimum rate $f_s = 2\text{B}$. In addition there are windows of other non-minimum rates that support alias-free sampling. For example, for a band pass signal in which the lowest frequency component f_L is 3 times its bandwidth (i.e. $q = 3$ and $f_m = 4\text{B}$), there are four increasingly narrower windows from which f_s can be selected for alias-free sampling, namely

Window 1: $f_s \ge 8B$

Window 2: $4B \le f_s \le 6B$

Window 3: $\frac{8}{3}B \le f_s \le 3B$

Window 4: $f_s = 2B$

In general, for all values of integer band positions q, the smallest value of p (=1) gives the sampling rates window $f_s \ge 2f_m$, the largest value of p ($=q + 1$) gives the minimum sampling rate $f_s = 2B$, and the values of p in between gives a total of $q - 1$ windows of allowed sampling rates between $2B$ and $2f_m$. Fig. 4.10c and d illustrate alias-free sampling at rate $f_s = 2B$ and $f_s = 2.8B$ respectively for a band pass signal in which $f_L = 3B$. The pair of image bands replicated at nf_s is labelled n and n' in the figures. Notice in Fig. 4.10c how the first three image bands fall below f_L, whereas the 4th (and higher) image bands fall above f_m, so that none of them overlaps into the original band (f_L, f_m), which can therefore be extracted using a BPF to reconstruct the original signal $g_{bp}(t)$ without any distortion. A non-minimum rate (such as $f_s = 2.8B$ in this example) allows a realisable reconstruction filter to be used as illustrated in Fig. 4.10d.

It is important to recognise the fact that allowed sampling rates for a band pass signal fall in disjoint windows. This means that one cannot increase sampling rate willy-nilly above an allowed minimum value in an attempt for example to insert a guard band between original and replicated bands to permit the use of realisable reconstruction filters. The following general guidelines should be borne in mind when sampling a band pass signal of bandwidth B and frequency band (f_L, f_m).

You cannot sample at any rate less than $2B$, otherwise there will be alias distortion.

You can sample at any rate $\geq 2f_m$ without alias distortion.

You can use the theoretical minimum rate $f_s = 2B$ only if the pass band is located at an integer number of bandwidths from the origin, i.e. if $f_m = (q+1)B$, where q is a positive integer.

There are some rates between $2B$ and $2f_m$ that will allow alias-free sampling, but these rates must satisfy Eq. (4.10). Any rate not satisfying this equation will result in an overlap of a replicated band into the original band thereby causing alias distortion.

To provide a transition band Δf on either side of (f_L, f_m) for a realisable reconstruction filter, apply Eq. (4.10) to the augmented band $(f_L - \Delta f/2, f_m + \Delta f/2)$ and determine the minimum allowed sampling rate for this augmented band. When the true signal of band (f_L, f_m) is sampled at the minimum rate so determined, there will be a guaranteed guard band $\geq \Delta f$ on either side of (f_L, f_m) separating it from adjacent image bands. These guidelines are further clarified in the following worked example.

Worked Example 4.2: Band pass Signal Sampling

(a) Determine the sampling rate f_s needed to sample a band pass signal of bandwidth 20 kHz centred on carrier frequency 3012.5 kHz so that there is a transition band 25% of bandwidth on either side to allow alias-free reconstruction using a realisable band pass filter.

(b) Explain whether aliasing would be enhanced or reduced by using a sampling rate of 77 kHz for this band pass signal.

(c) What would be the minimum sampling rate for alias-free reconstruction of this signal if no provision is made for a transition band? Comment on the practical difficulties associated with the use of this minimum sampling rate.

(a) The positive frequency band of the signal is given by the interval

$$\begin{aligned}(f_L, f_m) &= (f_c - B/2, f_c + B/2) = (3012.5 - 20/2, \ 3012.5 + 20/2) \\ &= (3002.5, \ 3022.5) \text{ kHz}\end{aligned}$$

Required transition band Δf = 25% of B = 5 kHz, so the augmented band is

$$\begin{aligned}(f_L', f_m') &= (3002.5 - \Delta f/2, \ 3022.5 + \Delta f/2) \\ &= (3002.5 - 2.5, \ 3022.5 + 2.5) \\ &= (3000, \ 3025) \text{ kHz}\end{aligned}$$

which has bandwidth $B' = 25$ kHz that starts at 3000 kHz, a location that is an integer number ($q = 3000/25 = 120$) of bandwidths from the origin. This is therefore a case of integer band positioning so the minimum sampling rate applicable is

$$f_{s\,\text{min}} = 2B' = 50 \text{ kHz}$$

Applying this sampling rate to sample the band pass signal (3002.5, 3022.5) kHz will ensure that there is a gap of 5 kHz between it and adjacent image bands on either side. To verify that this is indeed the case, we focus on the positive frequency axis since, due to symmetry, the two halves of the axis have identical spectra. For each $n > 1$, the RHS image band (labelled n' in Fig. 4.10) lies above (f_L, f_m) so it is the LHS image bands given in Eq. (4.9) as $(nf_s - f_m,\ nf_s - f_L)$, $n = 1, 2, 3, \ldots$ that need checking for overlap with (f_L, f_m). To avoid overlap the lower limit $nf_s - f_m$ of the nth band must exceed f_m and the upper limit of the next lower image band $(n-1)f_s - f_L$ must be below f_L, where n is the number of the highest image bands closest to (f_L, f_m) and is given by

$$n = \left\lceil \frac{2f_m}{f_s} \right\rceil = \left\lceil \frac{2 \times 3022.5}{50} \right\rceil = \lceil 120.9 \rceil = 121$$

where $\lceil x \rceil$ denotes the smallest integer larger than or equal to the positive real number x. The lower limit of image band $nf_s - f_m$ for $n = 121$ is

$$121 \times 50 - 3022.5 = 3027.5 \text{ kHz}$$

which is 5 kHz above the band (3002.5, 3022.5) kHz of the band pass signal, and the upper limit of the next lower image band $(n-1)f_s - f_L$ is

$$(121 - 1) \times 50 - 3002.5 = 2997.5 \text{ kHz}$$

which is 5 kHz below the band (3002.5, 3022.5) kHz. Thus, we see that not only is there no overlap, but also the required 5 kHz guard band has been created around the band (3002.5, 3022.5) kHz to allow a realisable reconstruction filter to successfully block all image bands.

(b) So what would happen if we increased this sampling rate from 50 to 77 kHz? On the surface it would appear that this increases the gap between original and image bands and therefore reduces aliasing. However, that is the case only for low pass signals, and it is a common error to extend this thinking to the band pass signal scenario. Let us therefore take a closer look at the spectrum of the signal sampled at $f_s = 77$ kHz. The index n of the image band immediately above the band (3002.5, 3022.5) kHz is now

$$n = \left\lceil \frac{2f_m}{f_s} \right\rceil = \left\lceil \frac{2 \times 3022.5}{77} \right\rceil = \lceil 78.51 \rceil = 79$$

so that the lower limit of the higher adjacent image band and the upper limit of the lower adjacent image band are

$$f_{n,L} = 79 \times 77 - 3022.5 = 3060.5 \text{ kHz}$$
$$f_{n-1,m} = (79 - 1) \times 77 - 3002.5 = 3003.5 \text{ kHz}$$

We see that sampling at 77 kHz creates a situation where the 78th image band, which extends up to 3003.5 kHz, overlaps into the wanted band (3002.5, 3022.5) kHz thereby causing alias distortion. This serves to emphasise the point made earlier on the contrast between band pass and low pass signal sampling: Raising sampling rate in low pass sampling guarantees a reduction in alias distortion, whereas the same action (in this case increasing f_s from 50 to 77 kHz) could actually increase aliasing in band pass sampling.

(c) The minimum sampling rate applicable to this signal (having $B = 20$ kHz, $f_m = 3022.5$ kHz) lies in the lowest sampling rate window given by Eq. (4.10) when p takes on its maximum value:

$$p = \lfloor f_m/B \rfloor = \lfloor 3022.5/20 \rfloor = 151$$

Thus

$$\frac{2f_m}{151} \le f_s \le \frac{2f_L}{151 - 1}; \;\Rightarrow \frac{2 \times 3022.5}{151} \le f_s \le \frac{2 \times 3002.5}{150}$$
$$\Rightarrow 40.0331 \le f_s \le 40.0333$$

So assuming an ideal brickwall reconstruction band pass filter, the minimum sampling rate that ensures alias-free reconstruction is 40.0331 kHz, fractionally larger than $2B$. Again, to check that this minimum sampling rate $f_{s\,\min} = 40.0331$ kHz does allow alias-free operation, we examine the limits of adjacent image bands:

$$\begin{aligned} f_{n,L} &= \lceil 2f_m/f_{s\,\min} \rceil \times f_{s\,\min} - f_m \\ &= 151 \times 40.0331 - 3022.5 = 6045 - 3022.5 \\ &= 3022.5 \text{ kHz} \\ f_{n-1,m} &= (\lceil 2f_m/f_{s\,\min} \rceil - 1) \times f_{s\,\min} - f_L \\ &= (151 - 1) \times 40.0331 - 3002.5 = 6004.97 - 3002.5 \\ &= 3002.47 \approx 3002.5 \text{ kHz} \end{aligned}$$

It can be seen that the image bands just touch the pass band (3002.5, 3022.5) kHz without overlapping. However, operating at this theoretical minimum sampling rate is fraught with practical implementation challenges. First, the allowed window is extremely narrow (40.0331 to 40.0333 kHz in this case) and any slight drift in sampling rate will cause an excursion outside this window resulting in aliasing. Second, there is no guard band whatsoever, so only an unrealisable ideal filter would successfully keep out adjacent image bands in order to prevent alias distortion.

The method of band pass sampling discussed so far may be described as *uniform band pass sampling* since it produces a single sequence taken at a uniform interval T_s of size that lies within a set of allowed windows. There is a different approach to band pass signal sampling, called *quadrature sampling*, which accepts any rate $f_s \geq 2B$, irrespective of band location and without the need to use Eq. (4.10) to establish allowed sampling rate windows. If the band pass signal $g(t)$ is centred on a frequency f_c that is much larger than its bandwidth B, then we can employ Eqs. (3.10) and (3.11), adapted below for convenience, to obtain the in-phase and quadrature components of $g(t)$, denoted by $g_I(t)$ and $g_Q(t)$, respectively, where

$$\begin{aligned} g_I(t) &= g(t)\cos(2\pi f_c t) + g(t+\tau)\cos[2\pi f_c(t+\tau)] \\ g_Q(t) &= -g(t)\sin(2\pi f_c t) - g(t+\tau)\sin[2\pi f_c(t+\tau)] \end{aligned}; \quad \tau = \frac{1}{4f_c} \tag{4.11}$$

$g_I(t)$ and $g_Q(t)$ are base band signals that together constitute the so-called *complex envelope* of $g(t)$ in the form

$$\tilde{g}(t) = g_I(t) + jg_Q(t) \tag{4.12}$$

obtained by translating without distortion the LHS $G_L(f) \equiv -(f_c - B/2, f_c + B/2)$ and RHS $G_R(f) \equiv (f_c - B/2, f_c + B/2)$ bands of $g(t)$ from their locations at $\pm f_c$ to be co-located at the origin, as illustrated in Fig. 4.11a. This topic is treated in more detail in Section 9.4 under Complex Representation of Signals, but the basis of Eq. (4.11) was discussed in Section 3.4. The signals $g_I(t)$ and $g_Q(t)$ occupy the positive frequency band (0, $B/2$) and hence have bandwidth $B/2$. Together they contain all the information in the band pass signal $g(t)$. The processing steps illustrated in Fig. 4.11b yield the sequence $g(nT_s)$ by interleaving the in-phase and quadrature samples $g_I(2nT_s)$ and $g_Q(2nT_s)$ taken at a uniform interval $2T_s$

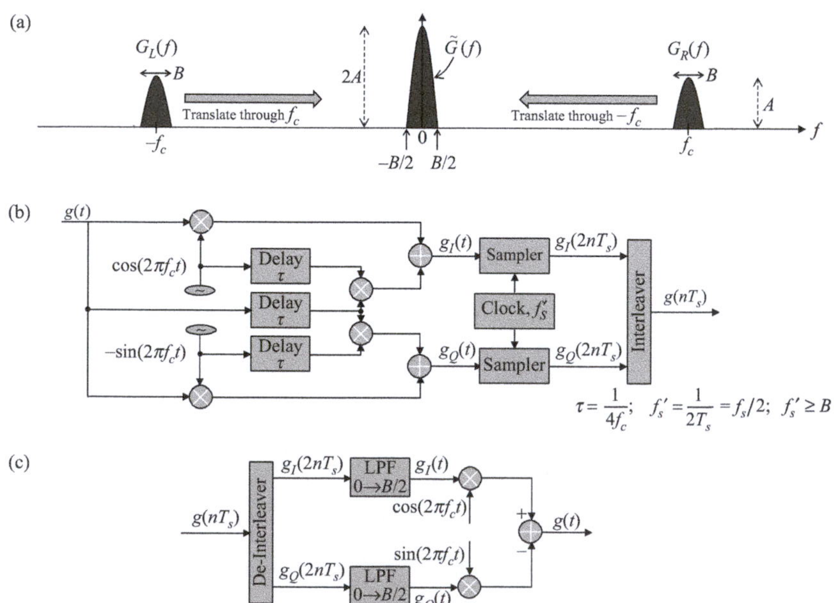

Fig. 4.11 *(a) Spectrum of complex envelope of g(t) obtained by translating LHS and RHS spectra of g(t) to the origin; (b) quadrature sampling of g(t); (c) reconstruction of g(t) from quadrature samples*

corresponding to a sampling rate $f_s/2 = 1/(2T_s)$, where $f_s \geq 2B$. This method of *quadrature sampling* is a special case of the more general *second-order sampling* approach discussed in [1]. Fig. 4.11c shows the steps involved in reconstructing the original band pass signal $g(t)$ from the sequence $g(nT_s)$. A de-interleaver is first employed to extract the sequences $g_I(2nT_s)$ and $g_Q(2nT_s)$, which are then passed through separate low pass reconstruction filters of bandwidth $B/2$ to recover $g_I(t)$ and $g_Q(t)$, and these are combined to give $g(t)$ as

$$g(t) = g_I(t)\cos(2\pi f_c t) - g_Q(t)\sin(2\pi f_c t) \tag{4.13}$$

This approach transforms a potentially problematic uniform band pass sampling into a more straightforward sampling of two half-bandwidth low pass signals, which lends itself to the simpler base band sampling procedure of Section 4.4 and is also better suited to system simulation as discussed in Section 3.4. However, it should be emphasised that Eq. (4.11) and hence the method used in Fig. 4.11b to obtain the in-phase and quadrature components of $g(t)$ are applicable only if $f_c \gg B$, a condition usually satisfied in practice.

Worked Example 4.3: Quadrature Sampling

We wish to demonstrate the applicability and accuracy of the quadrature sampling procedure presented in Fig. 4.11b and c in the following band pass signals:

(a) An amplitude-modulated signal $g_{am}(t) = [A_c + A_m\cos(2\pi f_m t)]\cos(2\pi f_c t)$, the result of amplitude modulating a carrier of frequency $f_c = 100$ kHz and amplitude $A_c = 20$ V by a sinusoidal message of frequency $f_m = 1$ kHz and amplitude $A_m = 10$ V.

(b) A frequency-modulated signal $g_{nbfm}(t) = A_c\cos[2\pi f_c t + \beta\sin(2\pi f_m t)]$, the result of frequency modulating a 100 kHz carrier frequency f_c of amplitude $A_c = 20$ V by a sinusoidal message signal of frequency $f_m = 1$ kHz at a modulation index $\beta = 0.1$.

By expanding the AM signal expression it may be readily seen that it contains 3 spectral lines at f_c and $f_c \pm f_m$ (= 100, 99 and 101 kHz) with respective amplitudes $A_c = 20$ V and $A_m/2 = 5$ V, and hence is of bandwidth $B = 2f_m =$ 2 kHz. The other signal is a narrow band FM, since $\beta \ll 1$, and therefore has bandwidth $B = 2f_m$ featuring three spectral lines f_c and $f_c \pm f_m$ with amplitudes A_c and $\beta A_c/2$. This also can be demonstrated through expansion of the expression for $g_{nbfm}(t)$ using trigonometric identities. Note that the two signals have different phase spectra and so their waveforms are quite different in spite of this similarity in amplitude spectra.

(a) To determine the in-phase and quadrature components of $g_{am}(t)$, we compare its given expression to the canonical form of a band pass signal given in Eq. (4.13):

$$\begin{aligned} g_{am}(t) &= [A_c + A_m\cos(2\pi f_m t)]\cos(2\pi f_c t), && \text{Given signal} \\ &\equiv g_I(t)\cos(2\pi f_c t) - g_Q(t)\sin(2\pi f_c t), && \text{Canonical form} \end{aligned}$$

It follows that

$$g_I(t) = A_c + A_m\cos(2\pi f_m t); \quad g_Q(t) = 0 \tag{4.14}$$

These are base band signals of bandwidth f_m $(=B/2)$, each of which can therefore be sampled at minimum rate $2f_m$ $(=B)$, or uniform interval $1/B$ $(\equiv 2T_s) = 0.5$ ms, to generate the sequences $g_I(2nT_s)$ and $g_Q(2nT_s)$, which are interleaved to give the sequence $g_{am}(nT_s)$ that fully represents the information in the AM signal $g_{am}(t)$. For example, starting at $t=0$ the samples would be taken at time instants (in ms) $2nT_s = 0.5n$, $n = 0, 1, 2, 3, \ldots$ to produce the following sequences

$2nT_s$ (ms):	0,	0.5,	1.0,	1.5,	2.0,	2.5,	3.0,	...
$g_I(2nT_s)$:	30,	10,	30,	10,	30,	10,	30,	...
$g_Q(2nT_s)$:	0,	0,	0,	0,	0,	0,	0,	...
$g_{am}(nT_s)$:	30, 0,	10, 0,	30, 0,	10, 0,	30, 0,	10, 0,	30, 0,	...

Note that $g_{am}(nT_s)$ contains 2 samples per 0.5 ms interval, which is a sampling rate $f_s = 2/(0.5\text{ ms}) = 4\text{ kHz} \equiv 2B$. The process for reconstructing $g_{am}(t)$ from $g_{am}(nT_s)$ is as shown in Fig. 4.11c. First $g_{am}(nT_s)$ is de-interleaved into the two sequences $g_I(2nT_s)$ and $g_Q(2nT_s)$. Passing each sequence through an LPF will successfully recover $g_I(t)$ and $g_Q(t)$ since the sampling interval of 0.5 ms ensures that two samples have been taken per period T_m of the maximum frequency component f_m ($= 1$ kHz, and hence $T_m = 1/f_m = 1$ ms) in the signal, in keeping with the sampling theorem. Once we have $g_I(t)$ and $g_Q(t)$, the AM signal $g_{am}(t)$ is obtained as stipulated in Eq. (4.13).

It is therefore clear that this quadrature sampling steps would lead to the successful recovery of $g_{am}(t)$. What remains is to verify that the steps of Fig. 4.11b, i.e. Eq. (4.11), would generate the same sequences as obtained above for $g_I(2nT_s)$ and $g_Q(2nT_s)$. Substituting $t = 0, 0.5, 1, 1.5, \ldots$ ms, and $\tau = 1/(4f_c) = 2.5$ µs in to Eq. (4.11) yields,

for $t = 0$ ms,

$$\begin{aligned}
g_I &= g_{am}(0)\cos(0) + g_{am}(2.5\times10^{-6})\cos\left[2\pi\times10^5\times2.5\times10^{-6}\right]\\
&= g_{am}(0) + g_{am}(2.5\times10^{-6})\cos(\pi/2) = g_{am}(0) = A_c + A_m = 30\\
g_Q &= -g_{am}(0)\sin(0) - g_{am}(2.5\times10^{-6})\sin\left[2\pi\times10^5\times2.5\times10^{-6}\right]\\
&= -g_{am}(2.5\times10^{-6})\sin(\pi/2) = -\left[A_c + A_m\cos\left(5\pi/10^3\right)\right]\cos(\pi/2) = 0
\end{aligned}$$

for $t = 0.5$ ms,

$$\begin{aligned}
g_I &= g_{am}(5\times10^{-4})\cos(100\pi) + g_{am}(502.5\times10^{-6})\cos(\pi/2)\\
&= [A_c + A_m\cos(\pi)]\cos(100\pi) = A_c - A_m = 10\\
g_Q &= -g_{am}(5\times10^{-4})\sin(100\pi) - g_{am}(502.5\times10^{-6})\sin(\pi/2)\\
&= -[A_c + A_m\cos(201\pi/200)]\cos(\pi/2) = 0
\end{aligned}$$

and so on, which is the same sequence as obtained earlier using Eq. (4.14). We therefore conclude that Fig. 4.11b does successfully generate the quadrature samples of the band pass signal $g_{am}(t)$.

(b) To obtain expressions for $g_I(t)$ and $g_Q(t)$ of the FM signal, we expand its waveform expression using the trigonometric identity $\cos(A+B) = \cos A\cos B - \sin A\sin B$ and then compare the resulting expression with the canonical form in Eq. (4.13):

$$\begin{aligned} g_{nbfm}(t) &= A_c\cos[2\pi f_c t + \beta\sin(2\pi f_m t)] \\ &= A_c\cos[\beta\sin(2\pi f_m t)]\cos(2\pi f_c t) - A_c\sin[\beta\sin(2\pi f_m t)]\sin(2\pi f_c t) \\ &\equiv g_I(t)\cos(2\pi f_c t) - g_Q(t)\sin(2\pi f_c t) \end{aligned}$$

It follows that

$$g_I(t) = A_c\cos[\beta\sin(2\pi f_m t)]; \quad g_Q(t) = A_c\sin[\beta\sin(2\pi f_m t)] \tag{4.15}$$

When $\beta \ll 1$ (called *narrow band FM*, as in this case), then

$$\cos[\beta\sin(2\pi f_m t)] \approx 1, \quad \sin[\beta\sin(2\pi f_m t)] \approx \beta\sin(2\pi f_m t)$$

so that the above expressions simplify to

$$\begin{aligned} g_I(t) &\approx A_c \\ g_Q(t) &\approx \beta A_c\sin(2\pi f_m t) \end{aligned} \tag{4.16}$$

These are base band signals of bandwidth f_m that can be sampled as in (a) at rate $2f_m$, corresponding to intervals $1/(2f_m) = 0.5$ ms to yield the sequences (assuming that sampling starts at $t = 0.25\text{ ms} \equiv T_s$)

$$\begin{aligned} g_I((2n+1)T_s) &= \{20,\ 20,\ 20,\ 20, \ldots\} \\ g_Q((2n+1)T_s) &= \{2,\ -2,\ 2,\ -2, \ldots\} \end{aligned}$$

which are interleaved to give

$$g_{nbfm}(nT_s) = \{20,\ 2,\ 20,\ -2,\ 20,\ 2,\ 20,\ -2, \ldots\} \tag{4.17}$$

a sequence that is processed as previously discussed in (a). Finally, we also check whether Fig. 4.11b generates the same values as above at corresponding time instants t (ms) = 0.25, 0.75, 1.25, ...

At $t = 0.25$ ms

$$\begin{aligned} g_I &= g_{nbfm}(2.5\times10^{-4})\cos(50\pi) + g_{nbfm}(2.525\times10^{-4})\cos(50\pi+\pi/2) \\ &= g_{nbfm}(2.5\times10^{-4}) = 19.9 \\ g_Q &= -g_{nbfm}(2.5\times10^{-4})\sin(50\pi) - g_{nbfm}(2.525\times10^{-4})\sin(50\pi+\pi/2) \\ &= -g_{nbfm}(2.525\times10^{-4}) = 1.9964 \end{aligned}$$

At $t = 0.75$ ms

$$g_I = g_{nbfm}(7.5\times10^{-4})\cos(150\pi) + g_{nbfm}(7.525\times10^{-4})\cos(150\pi+\pi/2)$$
$$= g_{nbfm}(7.5\times10^{-4}) = 19.9$$
$$g_Q = -g_{nbfm}(7.5\times10^{-4})\sin(150\pi) - g_{nbfm}(7.525\times10^{-4})\sin(150\pi+\pi/2)$$
$$= -g_{nbfm}(7.525\times10^{-4}) = -1.9964$$

and so on. Interleaving these values produces the sequence

$$g_{nbfm}(nT_s) = \{19.9,\ 1.9964,\ 19.9,\ -1.9964, \ldots\}$$

This is roughly the same sequence as in Eq. (4.17). The small difference is due to the approximation made in going from Eq. (4.15) to (4.16) on which the previous sequence was based. Sampling the exact $g_I(t)$ and $g_Q(t)$ expressions of Eq. (4.15) would have produced the sequence

$$g_{nbfm}(nT_s) = \{19.9,\ 1.9967,\ 19.9,\ -1.9967, \ldots\}$$

at $t = 0.25$ ms, 0.75 ms,..., which closely matches the sequence based on Fig. 4.11b. It is important to understand the significance of the approximation in Eq. (4.16). This eliminates (i.e. filters out) the higher harmonics (at $2f_m$, $3f_m$, ...) present in the exact expressions for $g_I(t)$ and $g_Q(t)$ given by Eq. (4.15), which would introduce side frequencies at $f_c \pm nf_m$, $n = 1, 2, 3, \ldots$ in the band pass (FM) signal. For $\beta \ll 1$, these side frequencies are insignificant beyond $n = 1$, allowing the FM signal to be constituted by a carrier and one pair of side frequencies covering a bandwidth of $2f_m$ without significant distortion. However, for $\beta > \sim 0.25$, the side frequencies become increasingly significant leading to several consequences. First, Eq. (4.16) is no longer valid and the in-phase and quadrature components are given only by Eq. (4.15). Second, the bandwidth of $g_I(t)$ and $g_Q(t)$ exceeds f_m by a factor that depends on β. More specifically, $g_I(t)$ consists of a DC component plus a number of even harmonics at $2f_m$, $4f_m$, ..., whereas $g_Q(t)$ consists exclusively of odd harmonics at f_m, $3f_m$, ... This point is clearly illustrated in Fig. 4.12 where it can be seen that at $\beta = 0.1$ Eq. (4.16) gives a good match to the exact base band waveforms of Eq. (4.15); but at $\beta = 5$ several harmonics (suppressed in Eq. (4.16)) must be included in order to faithfully represent $g_I(t)$ and $g_Q(t)$.

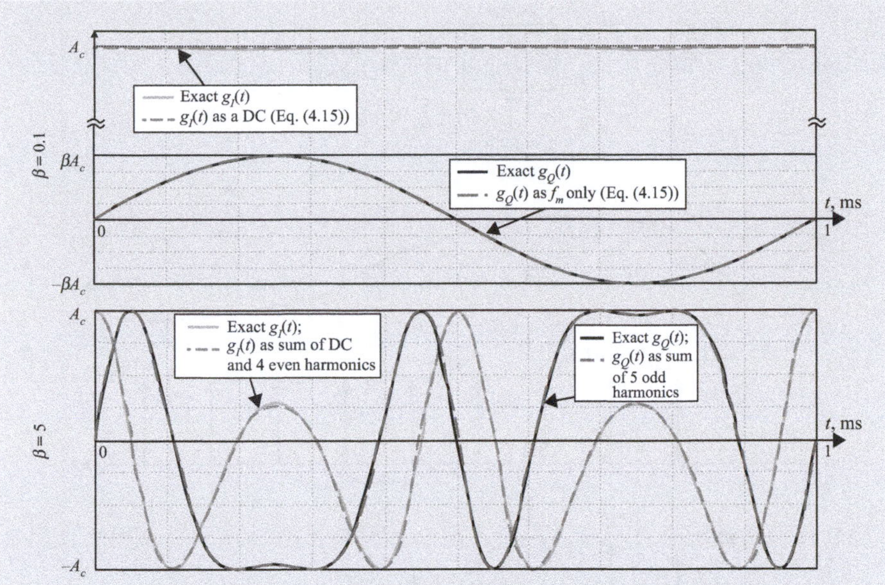

Fig. 4.12 In-phase and quadrature components of an FM signal carrying a sinusoidal message of frequency $f_m = 1$ kHz. Eq. (4.15) applies only at $\beta \ll 1$. At large β (e.g. $\beta = 5$), $g_I(t)$ and $g_Q(t)$ contain even and odd harmonics of f_m, respectively

4.6 Mathematical basis of sampling

An instantaneously sampled signal $g_\delta(t)$ is the product of an analogue signal $g(t)$ and an impulse train $\delta_{T_s}(t)$ of period T_s as illustrated in Fig. 4.13. The multiplication process is achieved by using the impulse train as a switching signal to switch the output terminal from ground (0 V) to the input terminal ($g(t)$) at the regular instants of an impulse. At these instants, the analogue input signal $g(t)$ is passed through to the output $g_\delta(t)$. In the absence of an impulse, there is no signal path from input to output and the output terminals remain grounded so that the output signal is zero. We may therefore write

$$\begin{aligned} g_\delta(t) &= g(t) \times \delta_{T_s}(t) \\ &= g(t)\sum_n \delta(t - nT_s) = \sum_{n=-\infty}^{\infty} g(nT_s)\delta(t - nT_s) \end{aligned} \tag{4.18}$$

We used the sifting property of the impulse function to reach the last expression which indicates that (as shown in Fig. 4.13d) the sampled signal is a sequence of impulses located at the sampling instants $t = nT_s$, each of weight equal to the value of $g(t)$ at that instant. The impulse train $\delta_{T_s}(t)$ is a limiting case of a rectangular pulse train having amplitude $1/\tau$, period T_s and duty cycle $d = \tau/T_s$ in the limit $\tau \to 0$ with switching period T_s kept fixed so that d also tends to zero. This rectangular pulse

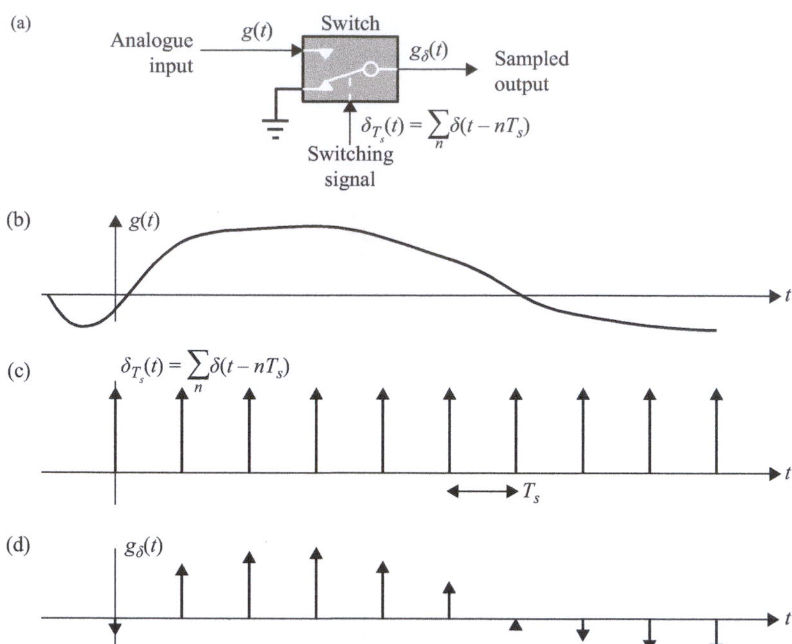

Fig. 4.13 Instantaneous sampling

train shown in Fig. 4.14a may be expressed in terms of the Fourier series in Eq. (4.7), with $f_0 \equiv f_s = 1/T_s$, having n^{th} harmonic amplitude given by

$$A_n = Ad\,\text{sinc}(nd)$$

as shown in Fig. 4.14b. This amplitude decreases with frequency according to a sinc envelope, with the first null at harmonic number $n = 1/d$ (corresponding to frequency $nf_s = f_s/d$). In the limit $d \to 0$ (when the rectangular pulse train becomes an impulse train), the first null is at $f \to \infty$, so the envelope is practically flat with an infinite number ($= 1/d$) of harmonics in the main and only lobe, and all harmonic components have equal amplitude

$$A_n = Ad = \frac{1}{\tau}\frac{\tau}{T_s} = \frac{1}{T_s} = f_s$$

The impulse train may therefore be written as a Fourier series

$$\sum_n \delta(t - nT_s) = \sum_{n=-\infty}^{\infty} A_n \cos(2\pi n f_s t) = f_s \sum_{n=-\infty}^{\infty} \cos(2\pi n f_s t)$$

with amplitude spectrum shown in Fig. 4.14c so that Eq. (4.18) gives the sampled signal as

$$\begin{aligned} g_\delta(t) &= g(t)\left[f_s \sum_{n=-\infty}^{\infty} \cos(2\pi n f_s t)\right] \\ &= f_s\left[g(t) + 2g(t)\sum_{n=1}^{\infty} \cos(2\pi n f_s t)\right] \end{aligned} \tag{4.19}$$

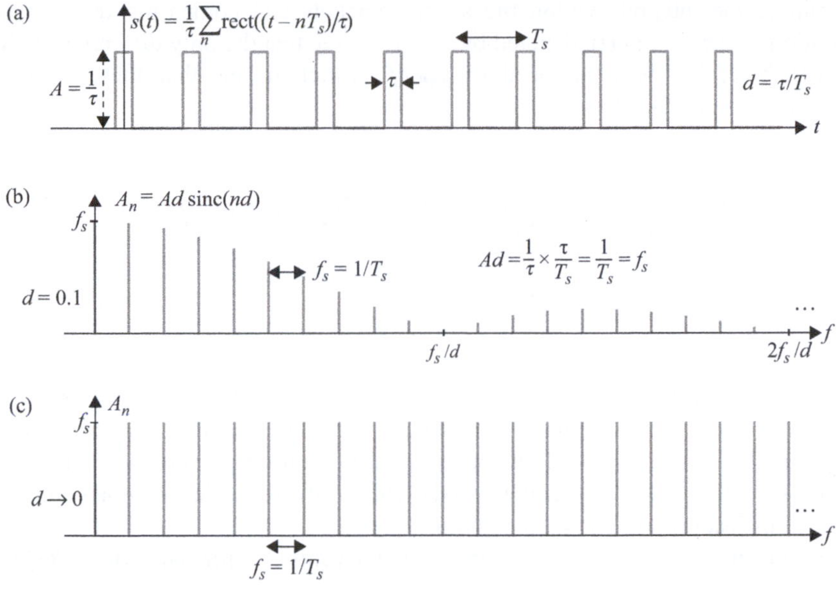

Fig. 4.14 Keeping the period T_s fixed, then in the limit $d\to 0$ the switching rectangular pulse train s(t) in (a) (with amplitude spectrum shown in (b) for $d=0.1$) becomes the unit impulse train of Fig. 4.13c with a flat amplitude spectrum as in (c)

where we have taken out $n=0$ from the general summation and combined $\cos(-n\theta)$ with $\cos(n\theta)$ to give $2\cos(n\theta)$. Recalling that the effect of multiplying $g(t)$ by $2\cos(2\pi n f_s t)$ is merely to shift its spectrum $G(f)$ to two frequency locations at $\pm n f_s$, we therefore see that (apart from a scale factor f_s that may be eliminated by using in Eq. (4.18) an impulse train of weight $1/f_s$) the sampled signal $g_\delta(t)$ consists of the original signal $g(t)$ plus the results of shifting without modification the spectrum $G(f)$ of $g(t)$ to frequency locations $f=\pm n f_s$, $n=1, 2, 3, \ldots$ To prove that this assertion is correct, we take the Fourier transform (FT) of Eq. (4.19) using the FT definition of Eq. (C.6) in Appendix C, denoting the FT of $g_\delta(t)$ on the LHS by $G_\delta(f)$ and using the Euler relation $\cos\theta = ½(e^{j\theta}+e^{-j\theta})$ to replace the cosine terms on the RHS with exponentials:

$$
\begin{aligned}
G_\delta(f) &= f_s \times FT\left[g(t) + 2g(t)\sum_{n=1}^{\infty}\cos(2\pi n f_s t)\right] \\
&= f_s \times \left\{G(f) + \sum_{n=1}^{\infty} FT\left[g(t)e^{j2\pi n f_s t} + g(t)e^{-j2\pi n f_s t}\right]\right\} \\
&\equiv f_s \times \left\{G(f) + \sum_{n=1}^{\infty}\int_{-\infty}^{\infty}\left[g(t)e^{j2\pi n f_s t} + g(t)e^{-j2\pi n f_s t}\right]e^{-j2\pi n f t}dt\right\} \\
&= f_s \times \left\{G(f) + \sum_{n=1}^{\infty}\left[\int_{-\infty}^{\infty} g(t)e^{-j2\pi(f-nf_s)t}dt + \int_{-\infty}^{\infty} g(t)e^{-j2\pi(f+nf_s)t}dt\right]\right\}
\end{aligned}
$$

Examining the integrals within the square brackets and comparing them to the definition of the FT of $g(t)$, denoted by $G(f)$, we see that the only difference is that the variable f has been replaced by $f \pm nf_s$, which means that these integrals represent $G(f \pm nf_s)$. Thus

$$\begin{aligned} G_\delta(f) &= f_s \times \left\{ G(f) + \sum_{n=1}^{\infty} G(f - nf_s) + \sum_{n=1}^{\infty} G(f + nf_s) \right\} \\ &\equiv f_s \times \left\{ G(f) + \sum_{n=1}^{\infty} G(f \pm nf_s) \right\} \end{aligned} \tag{4.20}$$

Eq. (4.20) states that (apart from a scale factor that may be dealt with as earlier explained) the spectrum of the instantaneously sampled signal $g_\delta(t)$ consists of the spectrum $G(f)$ of the original analogue signal $g(t)$ plus replications of $G(f)$ at a regular spacing f_s along the frequency axis. This confirms earlier non-mathematical discussions leading to Figs. 4.3b and 4.6.

If the original signal $g(t)$ is strictly band limited with bandwidth $B \le f_s/2$ so that

$$|G(f)| = 0, \quad \text{for } |f| \geq \frac{f_s}{2} \tag{4.21}$$

then $G(f \pm nf_s) = 0$ in the frequency interval $(-B, B)$ and Eq. (4.20) gives

$$G(f) = \frac{1}{f_s} G_\delta(f), \quad -B \le f \le B \tag{4.22}$$

From the time-shifting property of FT (Appendix C.1.2) and the Table of Fourier transforms (Appendix C.2), which states that $\text{FT}[\delta(t)] = 1$, it follows that

$$\text{FT}[\delta(t - nT_s)] = 1 \times e^{-j2\pi fnT_s} = e^{-j2\pi fnT_s}$$

so we may take the FT of Eq. (4.18) to obtain $G_\delta(f)$ as

$$G_\delta(f) = \sum_{n=-\infty}^{\infty} g(nT_s) e^{-j2\pi fnT_s} \tag{4.23}$$

Substituting this expression for $G_\delta(f)$ in to Eq. (4.22) yields

$$G(f) = \frac{1}{f_s} \sum_{n=-\infty}^{\infty} g(nT_s) e^{-j2\pi fnT_s}, \quad -B \le f \le B \tag{4.24}$$

Eq. (4.24) indicates that $G(f)$, and hence $g(t)$, is wholly determined in terms of the samples $g(nT_s)$. Note that the condition stated in Eq. (4.21) is essential for Eq. (4.24) to hold. To recover $g(t)$ from these samples, we take the inverse Fourier transform (IFT) of Eq. (4.24) and limit the integration to the frequency range ($-f_s/2$, $f_s/2$) since $G(f)$ is zero outside this range (see first line below), and interchange the order of integration and summation (second line) to obtain

$$\begin{aligned} g(t) &= \mathrm{IFT}[G(f)] \equiv \int_{-\infty}^{\infty} G(f)e^{j2\pi ft}df = \int_{-f_s/2}^{f_s/2} [G(f)]e^{j2\pi ft}df \\ &= \int_{-f_s/2}^{f_s/2} \left[\frac{1}{f_s}\sum_{n=-\infty}^{\infty} g(nT_s)e^{-j2\pi fnT_s}\right] e^{j2\pi ft}df \\ &= \frac{1}{f_s}\sum_{n=-\infty}^{\infty} g(nT_s)\int_{-f_s/2}^{f_s/2} e^{j2\pi f(t-nT_s)}df = \frac{1}{f_s}\sum_{n=-\infty}^{\infty} g(nT_s)\frac{e^{j2\pi f(t-nT_s)}}{j2\pi(t-nT_s)}\Bigg|_{-f_s/2}^{f_s/2} \\ &= \frac{1}{f_s}\sum_{n=-\infty}^{\infty} g(nT_s)\frac{e^{j\pi f_s(t-nT_s)} - e^{-j\pi f_s(t-nT_s)}}{j2\pi(t-nT_s)} \end{aligned}$$

Recognising that $\sin\theta = (e^{j\theta} - e^{-j\theta})/2j$, and introducing the unit factor f_s/f_s in order to complete the $\sin \pi x/\pi x \equiv \mathrm{sinc}(x)$ format yields

$$g(t) = \frac{1}{f_s}\sum_{n=-\infty}^{\infty} g(nT_s)\frac{\sin[\pi f_s(t-nT_s)]}{\pi f_s(t-nT_s)}f_s$$

from which it follows that

$$g(t) = \sum_{n=-\infty}^{\infty} g(nT_s)\,\mathrm{sinc}\left(\frac{t-nT_s}{T_s}\right) \tag{4.25}$$

This is an important result. It states that on condition, given by Eq. (4.21), that the bandwidth of the signal is less than or equal to half the *sampling frequency* f_s, we can reconstruct the original signal $g(t)$ from its samples $g(nT_s)$ taken at a regular *sampling interval* $T_s = 1/f_s$. You will recognise this as the sampling theorem that was more completely stated at the start of our discussion of sampling. The reconstruction is achieved as follows using an *interpolation function* $\mathrm{sinc}(t/T_s)$: The nth sample $g(nT_s)$ is multiplied by the interpolation function $\mathrm{sinc}(t/T_s)$ delayed by nT_s. All the waveforms obtained in this manner for $-\infty < n < \infty$ are summed to give the original signal $g(t)$ for $-\infty < t < +\infty$. To reconstruct $g(t)$ within a finite interval requires the inclusion of not only the interpolation functions with peaks falling within the interval, but also those functions on either side of the segment having significant tails within the segment.

But what is the significance of this so-called interpolation function? To answer this important question, observe that the input to the reconstruction process (i.e. device) is the sampled signal $g_\delta(t)$ in Eq. (4.18) and the output is the analogue signal $g(t)$ in Eq. (4.25) as illustrated in Fig. 4.15a, which means that to every input $g(nT_s)\delta(t-nT_s)$ there is a corresponding response

$$g(nT_s)\mathrm{sinc}\left(\frac{t-nT_s}{T_s}\right)$$

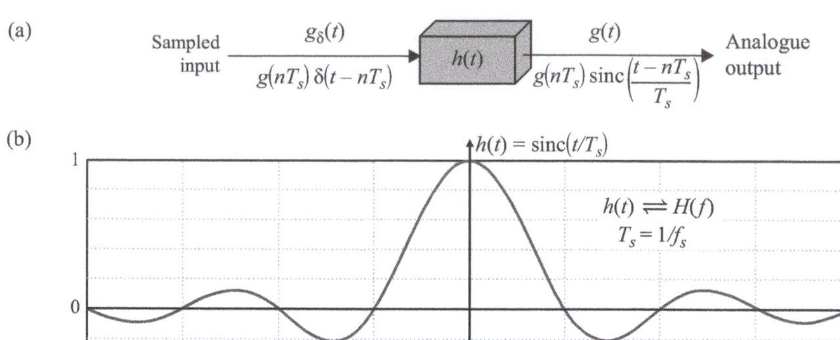

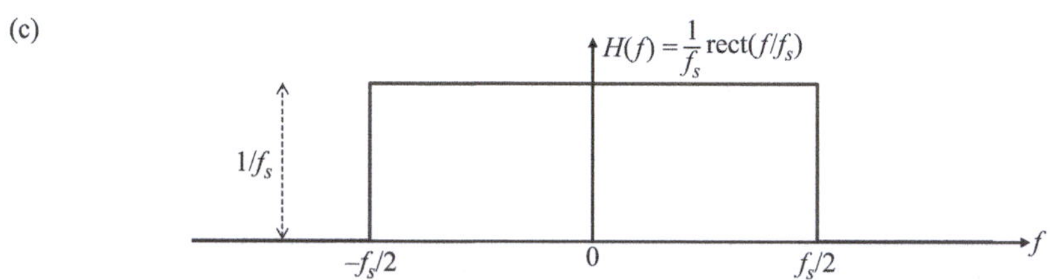

Fig. 4.15 Signal reconstruction from instantaneous samples: (a) block diagram; (b) impulse response of reconstruction filter; (c) transfer function of reconstruction filter

But the input is an impulse $\delta(t)$ delayed by nT_s and weighted by the factor $g(nT_s)$, so it follows from our discussion in Section 2.3 that the response to this input is the impulse response $h(t)$ of the device weighted by the same factor $g(nT_s)$ and delayed by the same amount nT_s. Therefore, the interpolation function in Eq. (4.25) is an indication that the reconstruction is implemented in a device with impulse response

$$h(t) = \mathrm{sinc}(t/T_s) \tag{4.26}$$

The FT of this impulse response gives the transfer function $H(f)$ of the device. From the Table of FT pairs in Appendix C.2, we obtain

$$H(f) = T_s\,\mathrm{rect}(fT_s) = \frac{1}{f_s}\,\mathrm{rect}\left(\frac{f}{f_s}\right) \tag{4.27}$$

Please refer to Fig. 2.2c for a definition of the rect function if in any doubt. The above impulse response $h(t)$ and transfer function $H(f)$ are plotted in Fig. 4.15b and c. We see that the reconstruction device is an ideal brickwall low pass filter of bandwidth $f_s/2$. This confirms the conclusions reached in the previous sections using an intuitive graphical approach.

Worked Example 4.4: Sinc Interpolation Function

We may now revisit the low pass sampling discussion presented earlier, employing the mathematically rigorous approach of this section. We will employ Eq. (4.25) to demonstrate the signal reconstruction process for 3 different sequences:

(a) Samples of a sinusoidal signal taken in keeping with the sampling theorem
(b) Samples of a sinusoidal signal taken at a rate below Nyquist sampling frequency
(c) Samples of a non-sinusoidal band limited low pass signal

(a) Given the samples

$$g(nT_s) = \ldots, 5, -9.1355, 9.7815, -6.6913, 1.0453, 5, -9.1355, 9.7815, -6.6913, 1.0453, 5, -9.1355, 9.7815, -6.6913, 1.0453, 5, \ldots$$

obtained at the corresponding sampling instants

$$nT_s\,(\text{ms}) = \ldots,\ -0.5,\ -0.4,\ -0.3,\ -0.2,\ -0.1,\ 0,\ 0.1,\ 0.2,\ 0.3,\ 0.4,\ 0.5,\ 0.6,\ 0.7,\ 0.8,\ 0.9,\ 1.0, \ldots$$

the sampling interval $T_s = 0.1$ ms $= 10^{-4}$ s, so the nth interpolation function is

$$w_n(t) = g(nT_s)\text{sinc}\left(\frac{t - nT_s}{T_s}\right) = g(nT_s)\text{sinc}\left(10^4(t - nT_s)\right)$$

For example, the first sample given above is at $n = -5$ with a value $g(nT_s) = 5$, so its interpolation function is $w_{-5}(t) = 5\,\text{sinc}(10^4 t + 5)$. The next interpolation function is $w_{-4}(t) = -9.1355\,\text{sinc}(10^4 t + 4)$, and so on. The value of the reconstructed signal $g(t)$ at any time instant t is given by the sum of the values of these interpolation functions at that instant. For example

$$\begin{aligned} g(t)|_{t=0} &= \sum_{n=-\infty}^{n=\infty} w_n(0) \approx \sum_{n=-5}^{n=10} w_n(0) \\ &= 5\,\text{sinc}(5) - 9.1355\,\text{sinc}(4) + \cdots + 5\,\text{sinc}(0) + \cdots + 5\,\text{sinc}(-10) \\ &= 5 \end{aligned}$$

$$\begin{aligned}g(t)|_{t=30\ \mu s} &= \sum_{n=-\infty}^{n=\infty} w_n(30\times 10^{-6}) \approx \sum_{n=-5}^{n=10} w_n(30\times 10^{-6})\\ &\approx 5\,\mathrm{sinc}(5.3) - 9.1355\,\mathrm{sinc}(4.3)\\ &\quad + \cdots + 5\,\mathrm{sinc}(0.3) + \cdots + 5\,\mathrm{sinc}(-9.7)\\ &\approx -0.2429 - 0.5471 + \cdots + 4.2920 + \cdots - 0.1327\\ &\approx -2.5865\end{aligned}$$

Note that the above value is approximate because an exact result requires use of all interpolation functions (and hence samples) preceding and following the time instant t of interest. Proceeding as above, we obtain $g(t)$ at sufficiently close values of time t to give a smooth plot of $g(t)$ versus t. Fig. 4.16 shows a plot of the reconstructed signal over the segment $0 \le t \le 0.5$ ms, based on samples from $n = -10$ to $n = 15$. In other words, the summation includes not only the six interpolation functions whose peaks fall in the segment (0, 0.5) ms, but also 10 preceding and 10 following functions whose tails are significant in the segment. These interpolation functions have been plotted in Fig. 4.16 with labels $w_0(t)$ to $w_5(t)$ placed on the ones whose peaks are within the segment.

We can now reveal that the samples given above came from the sinusoidal signal

$$g(t) = 10\cos(2\pi \times 4000t + \pi/3) \tag{4.28}$$

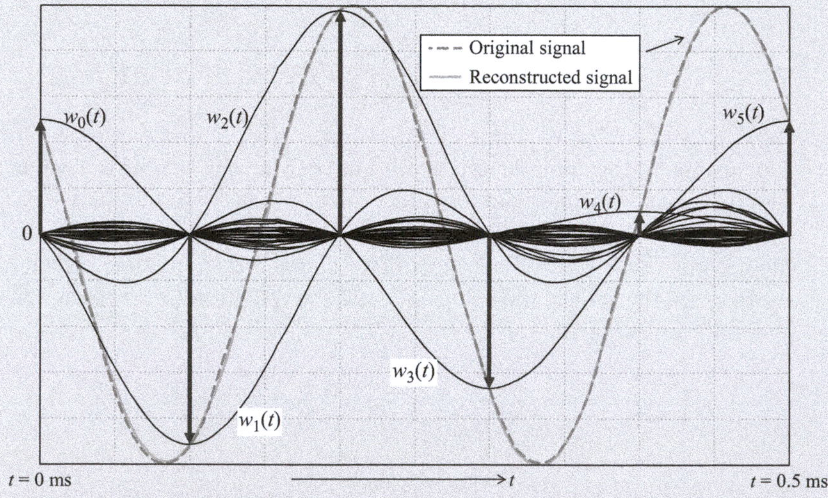

Fig. 4.16 Sinc interpolation to reconstruct a 4 kHz sinusoid from its samples taken at 10 kHz. The samples (thick arrow), original signal (thick dashed), reconstructed signal (thick solid) and interpolation waveforms (thin solid, some labelled $w_n(t)$, $n = 0, 1, 2, \ldots, 5$) are all shown over a 0.5 ms segment

The receiver of course did not have this information; however, the process of sinc interpolation followed here led to a successful reconstruction as can be seen in Fig. 4.16, where the above original signal is also plotted for comparison. The sampling interval $T_s = 0.1$ ms corresponds to sampling rate $f_s = 1/T_s = 10$ kHz that obeys the sampling theorem, hence the successful recovery of the original signal.

(b) Consider next if the above sinusoid — Eq. (4.28) — is sampled at 6 kHz (in clear violation of the sampling theorem) to produce the samples

$$g(nT_s) = \ldots, 5, -10, 5, 5, -10, 5, 5, -10, \ldots$$

at the corresponding sampling instants

$$nT_s\,(\text{ms}) = \ldots, -5/6, -4/6, -3/6, -2/6, -1/6, 0, 1/6, 2/6, \ldots$$

Following the procedure discussed in (a), we obtain the reconstructed signal using sinc interpolation. A 1 ms segment $(0 \leq t \leq 1)$ ms of the reconstructed signal is shown in Fig. 4.17 along with the original signal from which the samples were taken. Plots of the interpolation functions summed are also shown. We see that the reconstructed signal is a 2 kHz sinusoid, rather than the 4 kHz sinusoid that was sampled. This is the *aliasing error* due to under-sampling, which was discussed at length in the previous sections.

(c) Finally, Fig. 4.18 illustrates the application of the sinc interpolation procedure to reconstruct a non-sinusoidal band limited signal from its

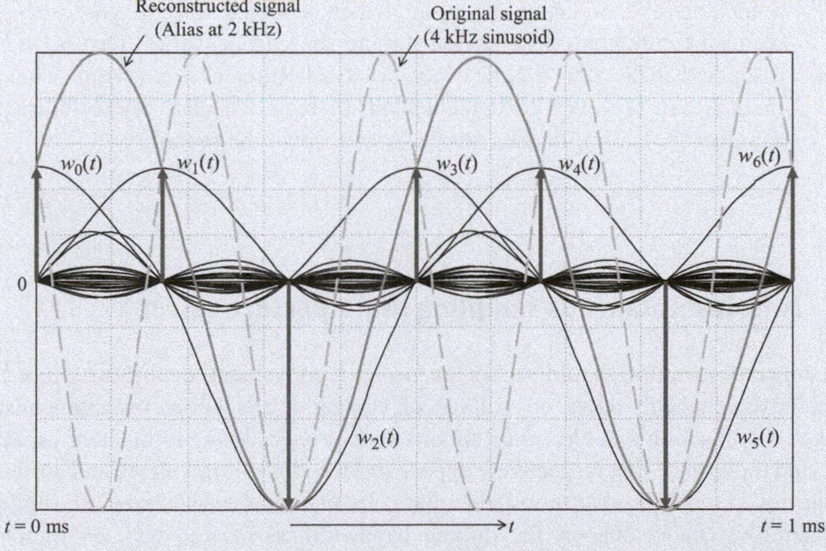

Fig. 4.17 Sinc interpolation to reconstruct a 4 kHz sinusoid from its samples taken at 6 kHz. The reconstructed signal is a 2 kHz alias. The samples (thick arrow) and interpolation waveforms (some labelled $w_n(t)$, $n = 0, 1, 2, \ldots, 6$) are also plotted over a 1 ms segment

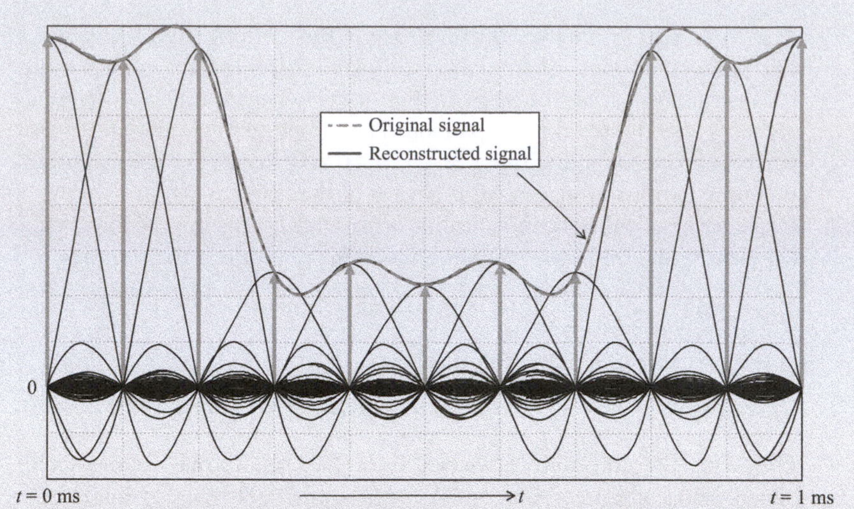

Fig. 4.18 Sinc interpolation to reconstruct a signal of bandwidth 5 kHz from its samples taken at 10 kHz. The samples (thick arrow) and interpolation waveforms (thin curves) are also plotted over a 1 ms segment

samples taken at Nyquist rate. Unsurprisingly, the reconstruction is perfect, being performed by an ideal brickwall low pass filter. From the above discussion it should now be clear why this type of filter is *unrealisable* (i.e. cannot be made to operate in *real-time*): It is based on a method of sinc interpolation in which samples from a yet-to-be-reached future contribute to determining the current value. That is, as discussed in Section 2.2, an ideal brickwall filter is *non-causal*. The method can however be implemented on recorded data where future samples are accessible data stored further down the record.

4.7 Non-instantaneous sampling and aperture effect

Instantaneous sampling should be seen as merely a convenient idealisation since its switching mechanism relies on voltage or current levels being instantaneously changed — an action that electronic circuits cannot execute due to inherent capacitance and inductance that respectively oppose sudden changes in voltage and current. To put this in more familiar terms, electronic circuits are always bandwidth-limited and therefore cannot support the infinite bandwidth associated with an impulse (such as shown in Fig. 4.14c). The inevitable curtailment of bandwidth by practical electronic circuits causes an expansion of pulse duration (by virtue of the inverse relationship between time and frequency), which converts an impulse in to a rectangular pulse at best.

Consider then the switch of Fig. 4.13a operating as previously described, but this time under the control of a rectangular pulse train (of amplitude A, duty cycle d

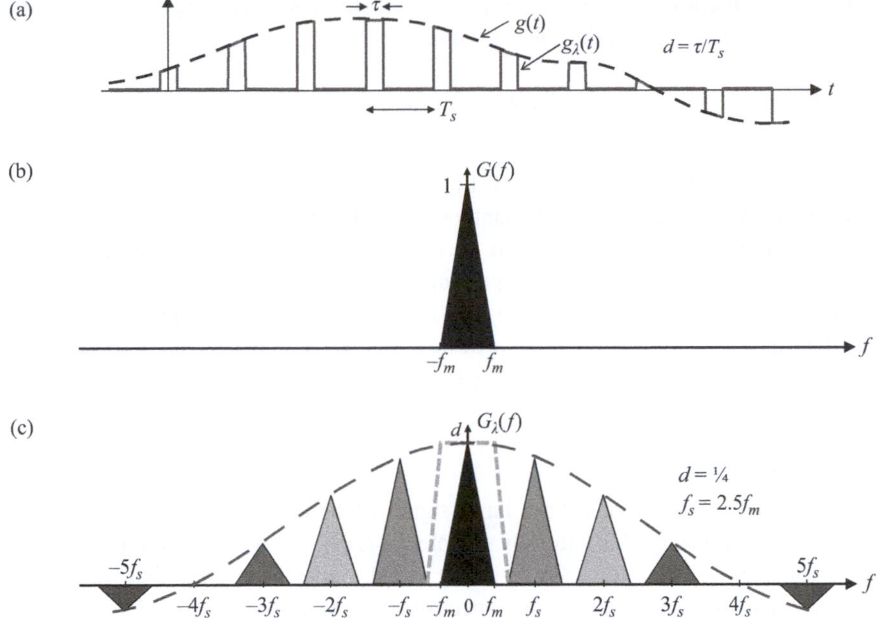

Fig. 4.19 Natural sampling: (a) original g(t) and sampled g(t) waveforms; (b) illustrative spectrum of g(t); (c) spectrum of g(t) for d = ¼, $f_s = 2.5f_m$

and period T_s) as the switching signal. The sampled signal $g_\lambda(t)$ is as shown in Fig. 4.19a and follows the analogue input $g(t)$ for a fraction d of the switching period (when the switching pulse is on) and is at 0 V for the remaining fraction $(1 - d)$ of the period when the pulse is off. This type of sampling is known as *natural sampling*. So how does the spectrum $G_\lambda(f)$ of a naturally sampled signal differ from the spectrum of an instantaneously sampled signal? To answer this question, let us write $g_\lambda(t)$ as the product of $g(t)$ and the rectangular pulse train $\text{rect}_{T_s}(t/\tau)$ expressed in terms of its Fourier series:

$$\begin{aligned} g_\lambda(t) &= g(t) \times \text{rect}_{T_s}(t/\tau) \\ &= g(t) \times Ad \sum_{n=-\infty}^{\infty} \text{sinc}(nd)\cos(2\pi n f_s t) \end{aligned}$$

Evaluating the $n = 0$ term using $\text{sinc}(nd) = 1$ and noting that the function under the summation sign is even, which means that the same result is obtained by doubling the sum taken over only positive values of n, we obtain

$$g_\lambda(t) = Ad\left[g(t) + 2g(t)\sum_{n=1}^{\infty} \text{sinc}(nd)\cos(2\pi n f_s t)\right] \tag{4.29}$$

Ignoring the scale factor (which can be normalised to unity), this expression for $g_\lambda(t)$ is similar to Eq. (4.19) except that in Eq. (4.29) the nth multiplying sinusoid is of amplitude sinc(nd). Spectral implications will therefore be similar to Eq. (4.20) and to the conclusions drawn in relation to Eq. (4.19), except for a factor sinc(nd) on the nth image band. That is, the sampled signal $g_\lambda(t)$ consists of the original signal $g(t)$ plus the results of shifting the spectrum $G(f)$ of $g(t)$ scaled by the factor sinc(nd) to frequency locations $f = \pm nf_s$, $n = 1, 2, 3, \ldots$ Normalising Ad to unity, the spectrum $G_\lambda(f)$ of $g_\lambda(t)$ is therefore given by

$$G_\lambda(f) = G(f) + \sum_{n=1}^{\infty} \mathrm{sinc}(nd) G(f \pm nf_s) \tag{4.30}$$

You may wish to prove this result by following the same steps that led to Eq. (4.20). It can be seen that natural and instantaneous sampling share a common feature, namely image bands in the sampled spectrum are replicated without distortion. However, in the naturally sampled spectrum the nth image band is reduced in size by the factor sinc(nd). Since sinc(nd) = 0 at non-zero integer values of nd, it follows that the following image bands will be scaled by a factor of zero and therefore will be missing from the spectrum of $g_\lambda(t)$:

$$n = \frac{1}{d}, \frac{2}{d}, \frac{3}{d}, \ldots$$

An illustrative spectrum of a base band signal of bandwidth f_m and the resulting spectrum when this signal is naturally sampled at $f_s = 2.5 f_m$ using a rectangular pulse train of duty cycle $d = ¼$ are shown in Fig. 4.19b and c. It is clear that, just as for instantaneous sampling, the original spectrum $G(f)$ and hence signal $g(t)$ can be recovered without distortion by passing $g_\lambda(t)$ through a low pass filter of bandwidth f_m. Furthermore, natural sampling will yield a higher SADR than is obtainable with instantaneous sampling at the same rate since the amplitude of alias components introduced by the nth image band is reduced by a factor of sinc(nd). For example, at $d = ½$, alias signal power from the first image band is reduced by ~4 dB, and alias from the second image band is completely eliminated. Taking this into account in Eq. (4.5) yields the following expression for SADR when an analogue signal of PSD $S_g(f)$ that was naturally sampled at rate f_s using a rectangular pulse train of duty cycle d is reconstructed using a low pass filter of gain response $|H(f)|$:

$$\mathrm{SADR}_{\text{natural sampling}} = \frac{\int_{-\infty}^{\infty} S_g(f)|H(f)|^2 df}{\sum_{n=1}^{\infty} \mathrm{sinc}^2(nd) \int_{-\infty}^{\infty} S_g(f \pm nf_s)|H(f)|^2 df} \tag{4.31}$$

However, in spite of these advantages natural sampling is rarely used in practice because it places a limitation on the maximum frequency that can be accurately digitised following sampling. To uncover the parameters involved in setting this limitation, let us consider a sinusoidal signal $A_m \sin(2\pi f_m t)$. Its maximum rate of change is

$$\frac{d}{dt}(A_m \sin 2\pi f_m t)\Big|_{\max} = 2\pi f_m A_m \cos 2\pi f_m t\Big|_{\max} = 2\pi f_m A_m$$

If the duration of the digitisation process is τ, then the maximum change in the signal during this interval will be $2\pi f_m A_m \tau$, and this must be no more than one step size in order to ensure that quantisation error does not exceed half a step size Δ. That is

$$2\pi f_m A_m \tau \le \Delta$$

If the quantiser has range $2A_m$ (i.e. it is fully loaded by the input signal) and employs k bits/sample, then there are 2^k quantisation intervals, each of size $\Delta = 2A_m/2^k$. Substituting this in the above condition yields

$$f_m \leq \frac{1}{2^k \pi \tau} \tag{4.32}$$

This equation may also be used to stipulate the ADC conversion time limit in order to digitise a given naturally sampled signal. For example, to digitise a naturally sampled 4 kHz voice signal in a 12-bit quantiser requires a conversion time less than 19.43 ns in order to achieve the required conversion accuracy of half a quantiser step size.

The method of non-instantaneous sampling usually implemented in practice by an integrated circuit (IC) is known as *flat-top sampling* in which the instantaneous value of the analogue signal at each sampling instant is held constant for the duration of the sample digitisation process. Fig. 4.20a shows a block diagram for implementing the *sample-and-hold* operation required in flat-top sampling. Switch S1 is normally open and S2 is normally closed. A rectangular pulse train of switching period T_s, pulse duration τ and duty cycle $d = \tau/T_s$ is used as the switching signal to actuate the two switches. S1 is turned on momentarily by the positive-going edge of the pulse while S2 is turned off at the same instant and held off for the entire duration τ of the pulse. The capacitor C charges rapidly (through the negligible resistance path of 'S1 on') to the value of $g(t)$ at the instant that S1 is turned on and holds that voltage for the entire pulse duration τ (during which S2 is off) since it has negligible discharge through the high resistance path of 'S2 off'. At the end of the pulse, S2 returns to its normally closed state and C discharges rapidly

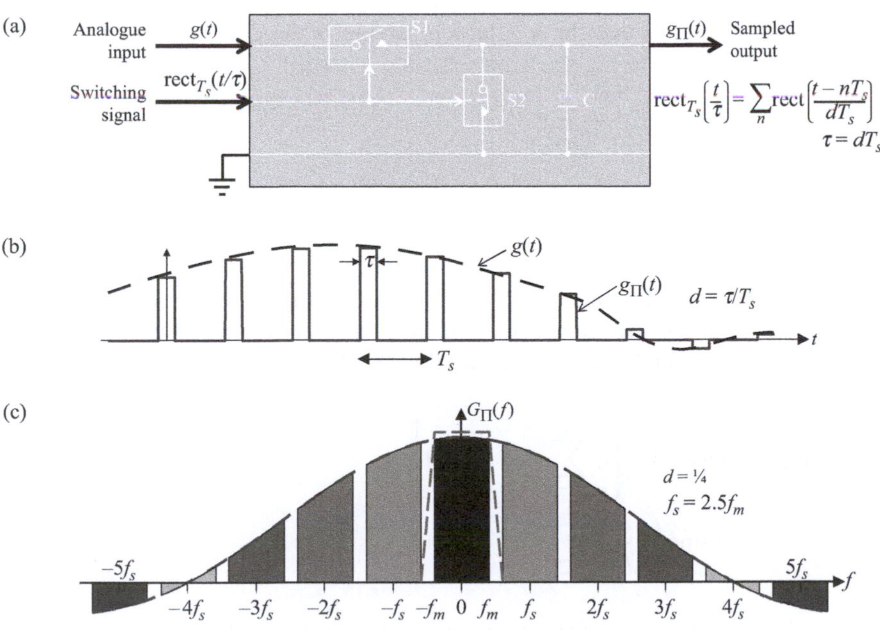

Fig. 4.20 Flat-top sampling: (a) circuit implementation; (b) original g(t) and sampled g(t) waveforms; (c) spectrum of g(t) based on rectangular-shaped spectrum of g(t)

through S2 to zero and remains there until the next pulse causes S1 to turn momentarily on thereby charging C to the voltage level of $g(t)$ at this next sampling instant. This carries on repeatedly so that the capacitor voltage gives a flat-top-sampled version of $g(t)$. The input signal $g(t)$ is usually connected to S1 through an operational amplifier (opamp, not shown) and the output $g_\Pi(t)$ is taken from across the holding capacitor C through another opamp (also not shown). Fig. 4.20b shows an illustrative analogue waveform and the resulting flat-top sampled output for $d = ¼$.

Two important clarifications are in order here. First, in some literature $g_\Pi(t)$ is called a PAM signal (for *pulse amplitude modulation*). However, that description is unnecessary and is avoided in this book in order to reserve the term 'modulation' for the type of signal processing in which a parameter of a higher frequency signal (called *carrier*) is varied in proportion to the instantaneous values of a lower frequency *message* signal resulting in a *modulated* signal that has no frequency components near zero. Second, the sampled waveform shown in Fig. 4.20b is very much a cleaned up version of what obtains in practice. In reality, $g_\Pi(t)$ will follow (i.e. track) $g(t)$ while S1 is (non-instantaneously) on. However, the crucial requirement that is fulfilled here, which sets this technique apart from natural sampling, is that $g_\Pi(t)$ stays at a constant level (while S2 is off) long enough for the sample to be digitised (typically through repeated comparison to fixed quantisation levels). Even then there will be a slight drop in $g_\Pi(t)$ during each 'constant level' phase due to a small discharge of the capacitor C through the large but non-infinite resistance of 'S2 off'.

To determine the spectral impact of flat-top sampling, we note that the nth flat-top sample in Fig. 4.20b may be obtained by delaying a unit rectangular pulse rect(t/τ) by nT_s and multiplying the delayed pulse by the factor $g(nT_s)$. Since convolving a function with $\delta(t - nT_s)$ simply delays the function by nT_s — see Eq. (2.44) — it follows that the nth flat-top sample is the convolution of rect(t/τ) with $g(nT_s)\delta(t - nT_s)$. Hence, the entire flat-top sampled signal $g_\Pi(t)$ is the convolution of rect(t/τ) with $\sum_{n=-\infty}^{\infty} g(nT_s)\delta(t - nT_s)$. But this sum signal is simply $g_\delta(t)$ as given in Eq. (4.18). Thus

$$g_\Pi(t) = \text{rect}\left(\frac{t}{\tau}\right) * g_\delta(t) \tag{4.33}$$

As convolution in the time domain corresponds to multiplication in the frequency domain, we obtain the spectrum $G_\Pi(f)$ of flat-top sampling as the product of the FT of rect(t/τ) listed in Appendix C.2 and the FT of $g_\delta(t)$ given by Eq. (4.20):

$$\begin{aligned} G_\Pi(f) &= \tau\,\text{sinc}(f\tau)\left[f_s \sum_{n=-\infty}^{\infty} G(f - nf_s)\right] \\ &= d\,\text{sinc}\left(f\frac{d}{f_s}\right) \sum_{n=-\infty}^{\infty} G(f - nf_s) \end{aligned} \tag{4.34}$$

where we have used $\tau = d/f_s$. This spectrum is shown in Fig. 4.20c (for $d = ¼$) based on a rectangular-shaped $G(f)$ to make spectral distortion more discernible. Notice that there is spectral band replication just as in instantaneous and natural sampling. However, unlike in those two there is now a distortion of the original and replicated

(a)

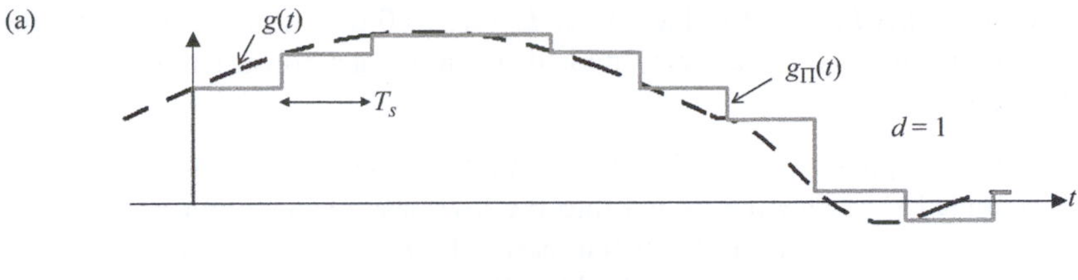

(b)

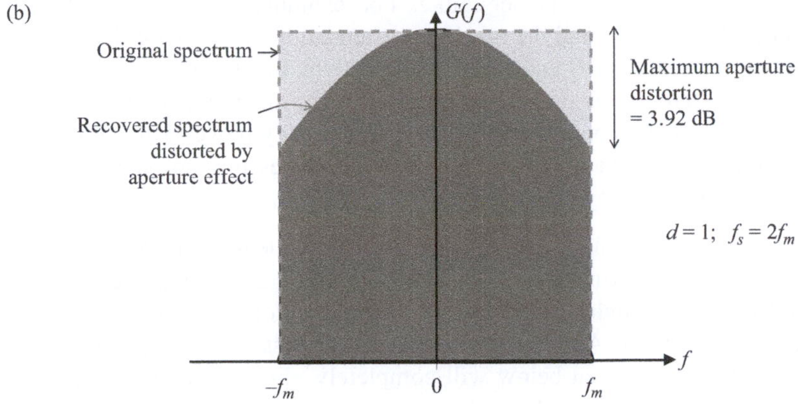

Fig. 4.21 Worst case scenario for aperture effect occurs when switching signal duty cycle d = 1 and sampling is at Nyquist rate $f_s = 2f_m$: (a) waveforms of original g(t) and sampled g(t) signals; (b) original spectrum (illustrative rectangular shape) of g(t), and recovered spectrum

bands by a sinc envelope factor. This amplitude distortion is known as *aperture effect* and is akin to the distortion observed in images obtained using a finite scanning spot where details in the image that are smaller than the scanning aperture are lost. It will occur in a reconstructed signal at a receiver whenever the input to the low pass reconstruction filter is $g_\Pi(t)$, a sequence of weighted rectangular pulses rather than $g_\delta(t)$, a sequence of weighted impulses. The worst case scenario is when the width τ of the sampling pulse equals the sampling interval T_s (i.e. $d = 1$) so that $g_\Pi(t)$ is a staircase waveform as shown in Fig. 4.21a.

The source of aperture distortion is the factor sinc(fd/f_s), which has a maximum value of 1 at $f = 0$ and gradually decreases to 0 at $f = f_s/d$, so a measure of aperture distortion L_{ad} at a frequency f within the recovered original band ($-f_m, f_m$) is given by this factor expressed in dB:

$$L_{ad} = -20\log_{10}\left[\text{sinc}\left(f\frac{d}{f_s}\right)\right], \quad 0 \le f \le f_s/2 \tag{4.35}$$

The maximum attenuation $L_{ad\,\max}$ occurs at the highest frequency component f_m in the analogue signal. Expressing sampling frequency f_s as a multiple M of f_m

$$f_s = Mf_m, \quad M \ge 2$$

it follows from Eq. (4.35) that the maximum attenuation due to aperture effect is

$$L_{ad\,\max} = -20\log_{10}[\text{sinc}(d/M)], \quad M \ge 2 \tag{4.36}$$

The fact that sinc(d/M) ≈ 1 (which gives $L_{ad\,\max} \approx 0$ dB) when $d/M \ll 1$ suggests two different ways to minimise amplitude distortion in a reconstructed signal due to aperture effect:

Use narrow pulses (with $d \ll 1$) to constitute the signal $g_\Pi(t)$ which carries the sampled sequence that is fed into the low pass reconstruction filter. When $d \ll 1$, then even in the worst case of Nyquist rate sampling ($M = 2$) amplitude distortion is negligible. For example, at $d = 0.1$ and $M = 2$, sinc(d/M) = 0.996 and the maximum attenuation is only 0.04 dB.

Use over-sampling ($f_s \gg f_m$) so that $M \gg 2$ in Eq. (4.36). For example, if $M = 10$ then even in the worst case of $d = 1$ the maximum attenuation due to aperture effect is only 0.14 dB. There is a penalty however for this solution as transmission bandwidth requirement increases proportionately with the sampling frequency employed.

When neither $d \ll 1$ nor $M \gg 2$, amplitude distortion due to aperture effect will be significant (up to a maximum of 3.92 dB at $d = 1$, $M = 2$) and may be reduced by using a filter that doubles up as both a reconstruction filter and an equaliser against aperture effect. A compensated ideal low pass filter having normalised gain response $|H(f)|$ given below will completely eliminate aperture effect and reconstruct an identical signal $g(t)$ from an input sequence of weighted rectangular pulses $\{g_\Pi(t)\}$ as would be reconstructed from an input sequence of weighted impulses $\{g_\delta(t)\}$:

$$|H(f)| = \begin{cases} \dfrac{1}{\text{sinc}(fd/f_s)}, & 0 \le f \le f_m \\ 0, & f > f_m \end{cases} \tag{4.37}$$

4.8 Anti-alias filter

In Section 1.2.3 and Fig. 1.6, the first step in the analogue-to-digital conversion (ADC) process was identified as low pass filtering to exclude subjectively insignificant high frequency components and limit the maximum frequency component in the analogue low pass signal to f_m prior to taking samples at a rate $f_s \ge 2f_m$. Alias distortion in a signal reconstructed from these samples is thereby prevented, and so such a filter is called an *anti-alias filter*. The value of f_m very much depends on the type of signal and communication service. In telephony for example, good subjective speech quality having the desired features of speaker recognition, intelligibility and naturalness but lacking in 'the presence' can be achieved with the maximum speech frequency limited to $f_m = 3.4$ kHz. In high fidelity audio that includes voice and instrumental sound, all audible frequency components up to $f_m = 20$ kHz must be faithfully represented. And in video, the maximum frequency component f_m ranges from 6.75 to 74.25 MHz depending on the vertical refresh rate and the total number of pixels per screen in the TV or PC graphics standard. Before bringing our discussion of sampling to a close, it is important to examine the specification of the anti-alias filter.

Ideally the anti-alias filter should have a normalised gain response that (expressed as a ratio) is flat and unity up to frequency f_m, and zero beyond f_m. However, for reasons stated in Worked Example 4.4c (Section 4.6), this ideal

Table 4.2 Anti-alias filter specification

Parameter	Ideal filter	Realistic filter
Pass band	$0 \rightarrow f_m$ (Width $= f_m$)	$0 \rightarrow f_m$ (Width $= f_m$)
Transition band	$f_m \rightarrow f_m$ (Width $= 0$)	$f_m \rightarrow f_s - f_m$ (Width $= f_s - 2f_m$)
Stop band	$f_m \rightarrow \infty$	$f_s - f_m \rightarrow \infty$
Stop band attenuation	Infinite	$\geq$ SQNR + 3 (dB) at $f = f_s - f_m$

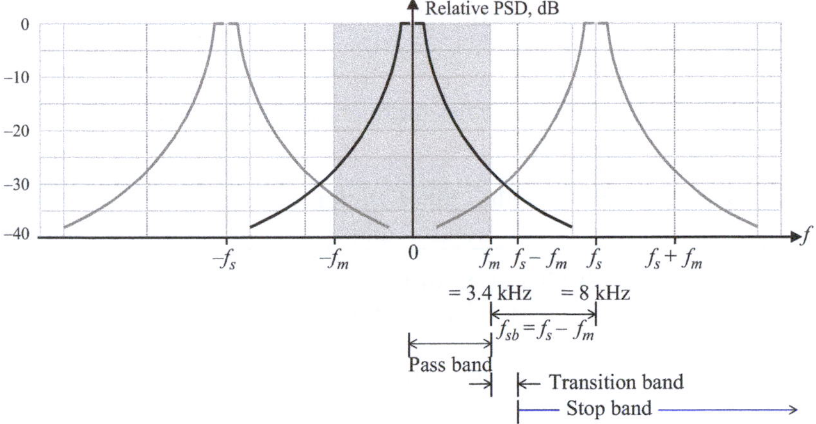

Fig. 4.22 Model of unfiltered speech spectrum sampled at $f_s = 8$ kHz showing two image bands either side of the original base band

performance cannot be implemented in real time. So we must be content to use a realistic filter having the specifications given in Table 4.2, with the ideal filter parameters listed alongside for comparison. Fig. 4.22 illustrates how the transition and stop bands are determined. The concept is applicable to any signal and sampling rate, but this figure uses the example of voice signals the PSD of which is typically roughly flat from 100 to 500 Hz followed by a roll-off of ~–10 dB/octave from 500 Hz to 7 kHz. It can be seen that if the original signal is sampled at rate f_s and reconstructed with a low pass filter of pass band $(0, f_m)$, then frequency components beyond $f_{sb} = f_s - f_m$ in the first image band will overlap into the wanted base band $(0, f_m)$ at the reconstruction filter output thereby causing significant alias distortion. Therefore, prior to sampling, we must filter the signal using a low pass filter of bandwidth f_m (which by definition means that the filter attenuation = 3 dB at $f = f_m$). Furthermore, the filter's attenuation must build up rapidly in the transition band (f_m, f_{sb}) to reach a minimum attenuation A_{min} at $f = f_{sb}$, which ensures that alias amplitude is negligible when compared to the wanted signal amplitude at $f = f_m$.

So how large should the minimum stop band attenuation A_{min} be? The complexity and cost of a filter increase with the tightness of its transition band and magnitude of its stop band attenuation, so we must allow a reasonable transition width $f_s - 2f_m$ and set only the minimum satisfactory A_{min} in order to avoid over-specifying the anti-alias filter. First of all, this means that sampling must be at above Nyquist rate $2f_m$. For example, in voice telephony $f_m = 3.4$ kHz and $f_s =$ 8 kHz, which provides a transition bandwidth of 1.2 kHz; and in high fidelity audio,

$f_m = 20$ kHz and $f_s = 44.1$ kHz, which gives a transition bandwidth of 4.1 kHz. Second, since sampling is usually accompanied by digitisation that unavoidably introduces quantisation noise, it is sufficient to ensure that $A_{\min}$ is set at the level required to make the alias amplitude of frequency f_{sb} to be as much below the amplitude of wanted frequency f_m as the latter will be above quantisation noise after digitisation. Given that f_m is attenuated by 3 dB by the filter, it follows that the filter should attenuate f_{sb} by an additional amount equal to SQNR dB, where SQNR is the *signal-to-quantisation noise ratio* discussed in the next chapter. Thus, the stop band attenuation of the anti-alias filter is set at

$$\begin{aligned} A_{\min} &= A|_{f=f_{sb}} = \text{SQNR} + 3 \text{ dB} \\ f_{sb} &= f_s - f_m \end{aligned} \tag{4.38}$$

The following worked example clarifies the application of these concepts.

Worked Example 4.5: Anti-alias Filter Design

A low pass Butterworth filter of bandwidth $f_m = 3.4$ kHz is to be used as an anti-alias filter for voice signal prior to sampling at $f_s = 8$ kHz. The sampled signal will be digitised in a PCM codec that has a signal-to-quantisation noise ratio (SQNR) of 38.6 dB. Determine the order n of the filter necessary to sufficiently minimise alias distortion.

The gain response of this filter is given in Eq. (2.33) from which it follows that the attenuation A of the filter as a function of frequency f is

$$A = 10 \log_{10}\left[1 + (f/f_m)^{2n}\right] \text{ dB} \tag{4.39}$$

where $f_m = 3.4$ kHz is the cut-off frequency and 3-dB bandwidth of the filter and n is the filter order. Manipulating to make n the subject of the above equation, and substituting $A = A_{\min}$ at $f = f_{sb}$ leads to

$$n = \left\lceil \frac{\log\left[10^{A_{\min}/10} - 1\right]}{2 \log(f_{sb}/f_m)} \right\rceil \tag{4.40}$$

where the brackets $\lceil\ \rceil$ indicate that n must be an integer larger than or equal to the enclosed value. From Eq. (4.38), the filter stop band starts at frequency

$$f_{sb} = f_s - f_m = 8 - 3.4 = 4.6 \text{ kHz}$$

at which the attenuation must be at least

$$A_{\min} = \text{SQNR} + 3 = 38.6 + 3 = 41.6 \text{ dB}$$

Substituting these values in to Eq. (4.40) yields

$$\begin{aligned} n &= \left\lceil \frac{\log\left[10^{41.6/10} - 1\right]}{2 \log(4.6/3.4)} \right\rceil = \left\lceil \frac{4.15997}{0.26256} \right\rceil = \lceil 15.844 \rceil \\ &= 16 \end{aligned}$$

To check, we use $n = 16$ in Eq. (4.39) to obtain the attenuation at $f = 4.6$ kHz as

$$A = 10\log_{10}\left[1 + (4.6/3.4)^{32}\right] = 42 \text{ dB}$$

which is at least 41.6 dB as required, thus ensuring sufficient attenuation of the voice components at frequencies 4.6 kHz and above. At a sampling rate of 8 kHz, these are the only frequencies that will overlap into the pass band of a low pass reconstruction filter set at (0, 3.4) kHz.

4.9 Summary

This concludes our detailed treatment of sampling, a process that converts a continuous-value continuous-time (analogue) signal into a continuous-value discrete-time (analogue) signal. The sampling theorem that governs distortion-free sampling can only be obeyed in strictly band limited signals that have no frequency components outside a bandwidth B. However, most naturally occurring analogue signals are not strictly band limited in the sense that they do contain stray frequency components outside the bandwidth B. In that case alias distortion is unavoidable and the design goal is to reduce this distortion to a negligible level in order to allow a signal of acceptable quality to be reconstructed from samples even when the original signal is not strictly band limited. This and other issues to do with the sampling of base band and band pass signals as well as measures to mitigate practical constraints and distortions have been discussed in full. The next stage in the digitisation process is known as *quantisation*, which goes an irreversible step further and converts the sampled signal into a discrete-value discrete-time (*digital*) signal. This along with PCM is the subject of the next chapter.

4.10 Reference

[1] Vaughan, R. G., Scott, N. L., and Rod White, D. (1991). 'The theory of band pass sampling', *IEEE Transactions on Signal Processing*, 39(9), pp. 1973–1984

Chapter 5

Quantisation and PCM

In Chapter Five

- ✓ Quantisation: An in-depth study of scalar quantisation involving an introduction to the different types of quantisation, detailed discussion of uniform quantisation, its performance measure, design parameters, trade-offs and drawbacks, and a comprehensive specification of logarithmic non-uniform quantisation that delivers a constant signal-to-quantisation noise ratio (SQNR) across all input signal levels. The benefit and price of non-uniform quantisation are also quantified in terms of companding gain and companding penalty.
- ✓ PCM: A comprehensive treatment of A-law and μ-law PCM source coding standards, including derivation of the standards from the ideal logarithmic compression curve through piece-wise linear approximations, detailed SQNR analysis, and performance comparisons with linear ADC.
- ✓ Lossy data compression: A brief overview highlighting various categories of low bit rate speech coding, namely waveform coders, vocoders and hybrid coders. The discussion particularly focuses on speech quality measures and trade-offs in low bit rate speech coding, and on differential quantisation, linear prediction and various special cases of bit rate reduction, including ADPCM, ADM and LPC-10.

5.1 Introduction

This chapter completes the treatment of the digitisation or, more appropriately, source coding of analogue signals started in the previous chapter. It deals in detail with the processes involved in converting a sampled analogue signal into a sequence of binary numbers (called *bit stream*) as illustrated in Fig. 1.6. We discuss the quantisation process with an emphasis on the performance metric of signal-to-quantisation noise ratio (SQNR), and the measures and trade-offs involved in maximising SQNR as well as ensuring its consistency across the full range of input signal amplitudes. The two standardised methods for binary representation of non-uniformly quantised signals, namely A-law and μ-law pulse code modulation (PCM), are then presented and their bandwidth savings and SQNR improvement for weak signals (called *companding gain*) when compared to linear ADC are evaluated. As always there is a price to pay because this improvement afforded to weak signals is achieved at the expense of a coarser quantisation of stronger signals and hence a reduction in their SQNR, which we quantify as *companding penalty*.

First we discuss the process of uniform quantisation, explain the subtle differences between the mid-rise and mid-tread types, calculate quantisation noise and explore various measures for reducing this noise power and the costs involved.

It emerges from our evaluation that uniform quantisation is unsuitable especially for voice communication due to its excessive bandwidth requirement and the inconsistency of signal quality when input signal amplitudes are spread over a wide range as is typically the case for speech signals. We then set about developing a method of non-uniform quantisation that fully remedies these deficiencies and find that this can be achieved through a coarser quantisation of large input signals and finer quantisation of small inputs using signal compression prior to uniform quantisation. The transfer characteristic of the compressor device is derived and is shown to obey a logarithmic function. A piecewise linear approximation of this theoretical curve to overcome practical implementation difficulties is discussed and expressions are derived to assess the impact of the approximation on SQNR which is no longer strictly constant but varies by about 2.5 dB over a 36 dB input amplitude range.

A 64 kb/s PCM is the toll quality benchmark for telephone speech communication, but this bit rate is excessive in most storage and wireless communication applications. We therefore include a brief discussion of bit rate reduction measures for telephone speech coding, known generally as *lossy data compression* or more specifically as *low bit rate speech coding*. There are a number of important worked examples that serve to further clarify and extend the concepts discussed.

5.2 Concept and classes of quantisation

The sampled signal $g(nT_s)$, obtained by sampling an analogue signal $g(t)$ at regular intervals T_s as discussed in the previous chapter, is still an analogue signal since it can take on any value in a continuum from the smallest to the largest value of $g(t)$. To transform $g(nT_s)$ into a digital signal, which by definition has discrete values taken from a finite set of numbers, we must replace (i.e. approximate) each value of $g(nT_s)$ by its nearest neighbour in the finite set of allowed values or levels. This is the process of quantisation. Unlike sampling, quantisation is an irreversible process that introduces irrecoverable errors (called *quantisation noise*) because once a sample value has been approximated by its nearest allowed level we lose the detail and information that would allow us to recover (i.e. return to) the original value. However, human audio-visual perception is limited in its sensitivity, so the quantised signal will be subjectively identical to the original signal if the approximations (or loss of detail or added quantisation noise) are kept small. Although we can make this noise as small as we wish by setting a large number of closely spaced allowed levels, the design goal is always to achieve a *subjectively acceptable quality* through a measure known as the *signal-to-quantisation noise ratio* (SQNR), rather than to aim for an *objectively perfect quality* at prohibitive costs.

Quantisation can be classified as *uniform* or *nonuniform*, and *mid-tread* or *mid-rise*. In uniform quantisation the quantisation levels to which sample values are approximated are equally spaced, whereas in non-uniform quantisation the spacing is unequal, being smaller near zero and progressively larger away from the origin. Mid-tread (also known as mid-step) quantisation has the value zero as a quantisation level, whereas in mid-rise quantisation the value zero is at the boundary between two quantisation intervals. The description 'mid' is a reference to the fact that zero is the middle value of the quantiser input range from say $-C$ to $+C$, and it is one of the levels or steps or treads in mid-tread, whereas in mid-rise quantisers zero is a point of transition where you 'rise' from one quantisation interval to another.

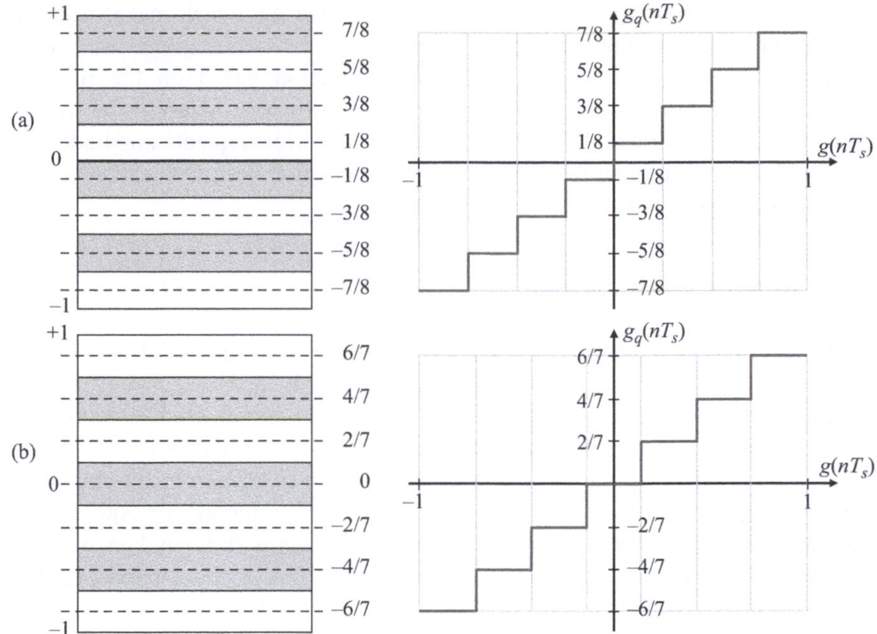

Fig. 5.1 Quantisation types: (a) uniform mid-rise; (b) uniform mid-tread

Fig. 5.1 illustrates uniform quantisation, based on $N = 8$ quantisation intervals and a normalised input range $(-1, +1)$, of the mid-rise type in (a) and the mid-tread type in (b). Two formats have been adopted for this illustration. The RHS shows the *transfer characteristic* of each quantiser, i.e. graph of quantised output $g_q(nT_s)$ versus sampled input $g(nT_s)$, whereas the LHS features alternately shaded quantisation intervals covering the (normalised) quantiser range -1 to $+1$. The dashed line bisecting each interval is the quantiser output for all inputs in that interval, i.e. it is the quantisation level to which all sample values falling in the interval are approximated. Both types of uniform quantisers have *odd symmetry*, but there are significant differences.

First of all, on an idle line with $g(nT_s) = 0$ V, thermal noise would cause the output of a mid-rise quantiser to jump randomly between two steps (in this case between $-1/8$ V and $+1/8$ V) resulting in *noise chatter* on idle lines. On the other hand, in a mid-tread quantiser this idle condition produces a steady 0 V quantised output thus eliminating noise chatter. Second, mid-rise quantisers have $N/2$ quantisation intervals above zero and $N/2$ intervals below, making a total of N intervals; whereas mid-tread quantisers have $N/2 - 1$ intervals above and below the one interval containing zero, making a total of $N - 1$ intervals. This means that to cover the same quantiser range, say $(-C, +C)$, a mid-tread quantiser has to use a larger step size Δ than a mid-rise quantiser, and hence incurs larger approximation errors or quantisation noise. The difference in step size is $2C/(N-1) - 2C/N \approx 2C/N^2$, which is quite small for large N. Mid-tread quantisation is used in the μ-law PCM standard, whereas A-law PCM employs mid-rise quantisation. In both standards the impact of the drawbacks highlighted here is minimised in that $N = 256$, which is large enough to make the difference in step size negligible, and non-uniform quantisation is used with a very small starting step size Δ_0, which makes idle line noise chatter $\pm\Delta_0/2$ insignificant.

In what follows, we discuss uniform quantisation based on the mid-rise structure and derive the design parameters and performance measures of quantisation in general. We then outline the drawbacks of uniform quantisation and find that a satisfactory solution lies in non-uniform quantisation designed with step sizes that increase exponentially away from the origin.

5.3 Uniform quantisation

Consider a quantiser of range $(-C, +C)$ divided into N equal quantisation intervals as shown in Fig. 5.2. The quantised value Q_j, $j = 0, 1, 2, 3, \ldots, N-1$, of each interval is the *mean* of all input samples that fall in that interval. On the assumption that the samples in each interval are *equally likely to be located anywhere within the interval*, in other words that they have a *uniform distribution*, the quantised value Q_j is the midpoint of the jth interval. Since these intervals, or equivalently quantised values Q_j, will be numbered using k binary digits (bits) and each output identified by this number, the number of intervals N has to be a power of 2 to ensure that all 2^k combinations of the k bits are utilised. That is

$$N = 2^k, \quad \Rightarrow \quad k = \log_2 N \tag{5.1}$$

The size of each interval, called *quantiser step size*, is

$$\Delta = \frac{2C}{N} = \frac{C}{2^{k-1}} \tag{5.2}$$

A quantiser is characterised by the number N of its quantisation levels or the number k of bits per sample, which is the number of bits needed to uniquely identify each of the N levels. We may also characterise a quantiser by specifying its

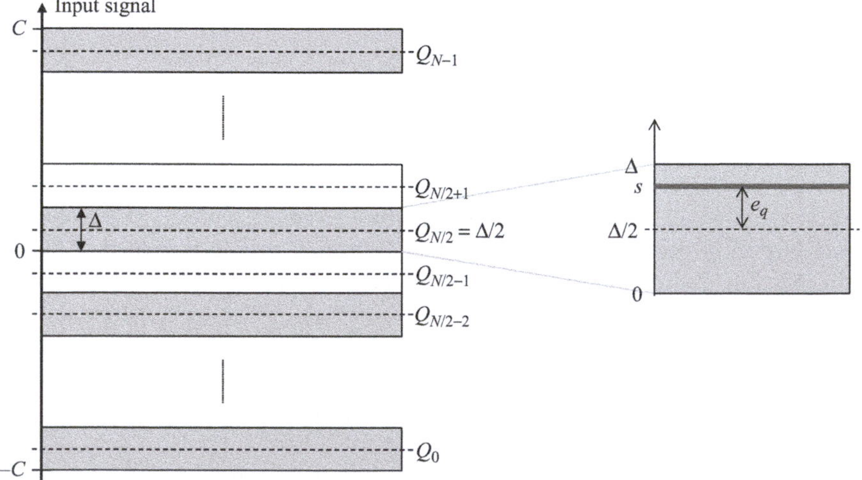

Fig. 5.2 Analysis of uniform quantisation. The interval just above zero is blown up on the RHS and shows the quantisation error e_q incurred when input sample s in this interval is quantised

dynamic range, which is the ratio between the largest amplitude A_2 of a sinusoidal signal that avoids clipping by the quantiser and the largest amplitude A_1 of a sinusoid whose variations are confined entirely to one interval of the quantiser and hence go undetected. Noting that $A_2 = C$ and $A_1 = \Delta/2$, we use Eq. (5.2) to obtain

$$\begin{aligned}\text{Dynamic range} &= \frac{C}{\Delta/2} = 2^k \\ &= 6k \text{ (dB)}\end{aligned} \tag{5.3}$$

Thus the dynamic range of a quantiser depends only on the number of bits per sample and increases by 6 dB per extra bit used to represent each sample.

The error incurred when an input sample s that falls in the jth interval is represented by the quantised value Q_j of that interval is

$$e_q = s - Q_j \tag{5.4}$$

Ignoring the sign, e_q reaches a maximum $e_{q\,\max}$ when s is just below the upper limit of the interval (i.e. $s = Q_j + \Delta/2$) or s is just above the lower limit of the interval (i.e. $s = Q_j - \Delta/2$). Thus maximum quantisation error is half the quantiser step size:

$$e_{q\,\max} = \frac{\Delta}{2} = \frac{C}{2^k} \tag{5.5}$$

With the sample values in each interval equally likely to be above or below the interval's quantised level, e_q has a mean of zero and therefore its variance and mean square value (which defines normalised power) are identical. If the input signal lies entirely within the quantiser range, i.e. it does not overload the quantiser, the error statistics are the same across all the uniform quantiser intervals spanned by the input. Therefore to determine *mean square quantisation error* (MSQE), which gives quantisation noise power, we focus on the interval just above zero (shown blown out on the RHS of Fig. 5.2) and sum over the entire interval ($0 \rightarrow \Delta$) the product of $e_q^2 = (s - \Delta/2)^2$ and the probability ds/Δ that the sample s lies in an infinitesimal interval ds around s. This gives

$$\begin{aligned}\text{MSQE} &= \int_0^\Delta e_q^2 ds/\Delta = \frac{1}{\Delta}\int_0^\Delta (s - \Delta/2)^2 ds \\ &= \frac{1}{\Delta}\left[\left(\frac{s^3}{3} - \frac{s^2\Delta}{2} + \frac{s\Delta^2}{4}\right)\Bigg|_0^\Delta\right] \\ &= \Delta^2/12\end{aligned} \tag{5.6}$$

It can be seen that quantisation error depends only on step size. Assuming small Δ this error will have the same effect as thermal noise on the quantised signal and is therefore referred to as *quantisation noise* with rms value:

$$e_{\text{qrms}} = \sqrt{\text{MSQE}} = \frac{\Delta}{2\sqrt{3}} \tag{5.7}$$

We can make this noise due to quantisation as small as desired by sufficiently reducing Δ, which, from Eq. (5.2), is achieved by increasing k, the number of bits per sample. Clearly, this increases the bit rate $R_b = kf_s$ (where f_s is sampling rate) of the resulting digital signal and hence the transmission bandwidth required.

The quality of the quantised signal is measured by the *signal-to-quantisation noise ratio* (SQNR), which is the ratio between signal power and quantisation noise power:

$$\text{SQNR} = \frac{\text{Signal power}}{\text{Quantisation noise power}} = \frac{P_s}{\text{MSQE}} \tag{5.8}$$

Let the input signal has peak value A_p, rms value A_{rms}, and peak-to-rms ratio $R = A_p/A_{rms}$. It follows that $P_s = A_{rms}^2 = (A_p/R)^2$, and hence, using Eqs. (5.6) and (5.2), that

$$\begin{aligned} \text{SQNR} &= \frac{A_p{}^2/R^2}{\Delta^2/12} = \frac{12A_p{}^2 \times (2^{k-1})^2}{R^2C^2} \\ &= \frac{3A_p{}^2}{R^2C^2}2^{2k} \end{aligned} \tag{5.9}$$

Expressing this in dB yields

$$\text{SQNR} = 4.77 + 6.02k - 20\log(R) + 20\log(A_p/C)\ \text{dB} \tag{5.10}$$

Recall that it is assumed and in fact required that the quantiser is never overloaded so that the peak value of the input signal $A_p \leq C$, which means that SQNR is reduced by the last term in the above equation. SQNR therefore reduces at the rate of 20 dB per decade reduction in input signal amplitude. The highest SQNR is experienced by a signal that fully loads the quantiser so that $A_p = C$ and $20\log(A_p/C) = 0$, which yields

$$\text{SQNR}_{\text{max}} = 4.77 + 6.02k - 20\log(R)\ \text{dB} \tag{5.11}$$

The highest SQNR achievable for different signals through uniform quantisation can be determined if we insert the value of R for that signal into Eq. (5.11). For example, sinusoidal signals have $R = \sqrt{2}$; speech signals have $R \approx 9$ dB; and a signal with uniform probability density function (pdf), i.e. the signal is equally likely to have any value in the range $(-C, C)$, has $R = \sqrt{3}$; so Eq. (5.11) gives

$$\text{SQNR}_{\text{max}} = \begin{cases} 6.02k + 1.76\ \text{dB}, & \text{Sinusoidal signal} \\ 6.02k - 4.23\ \text{dB}, & \text{Speech signal} \\ 6.02k\ \text{dB}, & \text{Signal with uniform pdf} \end{cases} \tag{5.12}$$

5.3.1 Quantisation design parameters and trade-offs

It is worth taking a moment to examine the parameters and trade-offs involved in achieving the usual design goal of maximising SQNR. These are encapsulated by Eqs. (5.9) and (5.10).

- SQNR increases exponentially with k, which, other factors being equal, is directly proportional to the transmission bandwidth required by the resulting digital signal. More specifically SQNR increases by 6 dB per extra bit used to represent each quantised sample. If information about the quantised levels is subsequently conveyed to the receiver without error then no further degradation will be introduced into the signal and this SQNR will be the signal to noise ratio (SNR) of the analogue signal recovered at the receiver, assuming that aperture distortion due to flat-top pulses at the reconstruction filter is avoided as discussed in the previous chapter. Therefore quantisation (and ultimately PCM) provides an effective mechanism for trading bandwidth for improvement in SNR. If the signal-to-noise ratio is SNR_1 at bandwidth B_1 and we increase the number of bits per sample by a factor $n > 1$, which means bandwidth also increases by the same factor to nB_1 (other factors being equal), then SNR increases exponentially to $(\text{SNR}_1)^n$, assuming the factor $3A_p^2/R^2C^2$ in Eq. (5.9) equals unity, which is the case if the signal fully loads the quantiser and $R = \sqrt{3}$.

- SQNR is inversely proportional to the square of the quantiser range $2C$ needed to fully accommodate the input signal without clipping. Thus reducing C by a factor $n > 1$ increases SQNR by a factor n^2 for the same number of bits per sample. Alternatively, reducing C by a factor $n > 1$ allows us to achieve the same SQNR using $\log_2 n$ fewer bits per sample. For example, reducing C by a factor of 16 allows the same SQNR to be achieved using $k = 4$ bits per sample as is realised with $k = 8$. You are right to wonder how, given an input signal with amplitude $A_p = C$, we may reduce C without causing severe clipping distortion. The trick is to recognise that it is possible to avoid direct quantisation of the input signal samples by exploiting the correlation that exists between neighbouring samples of information-bearing signals. Basically, we use the past p samples ($p \geq 1$) to obtain a prediction $\hat{g}(nT_s)$ of the current sample $g(nT_s)$ and obtain the prediction error $\varepsilon(nT_s) = g(nT_s) - \hat{g}(nT_s)$. We then quantise and transmit $\varepsilon(nT_s)$ rather than $g(nT_s)$. The receiver makes its own local prediction (based on the same algorithm used by the transmitter) and then simply adds the received error to obtain the original sample value. If the sampling rate is sufficiently high and the prediction algorithm is well designed, the errors are very small so that the range of $\varepsilon(nT_s)$ is much smaller than the range of $g(nT_s)$, and this allows us to reduce quantiser range accordingly and hence to realise the benefits discussed above. This is the principle behind various types of *differential quantisers* (e.g. differential pulse code modulation, DPCM).
- SQNR increases by 20 dB per decade increase in the ratio r ($= A_p/C$) between the peak-to-peak amplitude of the input signal and the quantiser input range, up to a maximum at $r = 1$. This maximum SQNR is given by Eq. (5.11) for general signals and by Eq. (5.12) for three special cases. So SQNR is maximised by ensuring that the input signal fully loads the quantiser. This is okay for signals such as a sinusoid that have a low peak-to-rms ratio R. However, for signals such as speech that have higher R in addition to dynamic variations featuring a mixture of high-amplitude and low-amplitude sub-intervals, SQNR during weak passages will be significantly lower than the maximum SQNR attained in strong passages. Note that this problem cannot be tackled by scaling the weak signals to fully load the quantiser since this would cause clipping distortion during the strong passages that intermittently follow.

5.3.2 Shortcomings of uniform quantisation

Uniform quantisation has two significant drawbacks that make it unsuitable for digital transmission systems although it is widely used for linear analogue to digital conversion (ADC) in digital signal processing (DSP) systems.

- First of all, in order to faithfully represent crucial but small variations in the sampled signal, such as consonants in speech which carry intelligibility but are usually of weak amplitudes, the quantiser step size Δ must be sufficiently small, and this requires a large k (bits/sample) and hence a large transmission bandwidth and storage capacity for the resulting digital signal. Take the example of speech with a dynamic range of ~60 dB, which gives the ratio between highest and lowest sample magnitude as $10^{60/20} = 1000$. What this means is that if the strongest sub-intervals within the speech signal have peak values as high as 1 V then the weakest significant sub-intervals (not silence periods) may have peak values as low as 1 mV. If we are to faithfully quantise these small samples then the step size must be no more than 1 mV, which leads to 1000 quantisation intervals in the positive range and a further 1000 in the negative range, making a total of 2048 intervals (to the next power of 2). This requires $k = \log_2(2048) = 11$ bits/sample, which combined with a sampling rate of 8000 samples per second (i.e. 8 kHz) produces a bit rate of 88 kb/s and imposes an unacceptably high transmission bandwidth requirement.
- Second, uniform quantisation produces an SQNR that varies across the input signal range. Near the origin the maximum quantisation error Δ/2 is comparable in magnitude to the small sample values A_L in this region and therefore the distortion is perceptible, whereas near the top end of the quantiser range this error is negligible compared to the large sample values A_H. So $A_L^2/\Delta^2 \ll A_H^2/\Delta^2$ and SQNR at the bottom end is much lower than SQNR at the top end. Another way of looking at this problem is that the small samples A_L underload the quantiser when compared to the large samples A_H and therefore their SQNR is $20 \log_{10}(A_H/A_L)$ dB lower.

It is desirable, especially in telephony, to maintain a high and consistent SQNR and service quality across all regimes of significant signal amplitudes while making judicious use of bandwidth. An excellent solution is provided through non-uniform quantisation designed with step sizes that increase exponentially away from the origin so that large input samples are coarsely quantised using larger step sizes whereas smaller input samples are more finely quantised using smaller step sizes. This will be dealt with in the next section after the following worked examples.

Worked Example 5.1: Quantiser Parameters

Given a 4-bit uniform quantiser (also called linear ADC) that covers the input range −16 V to 16 V, we wish to determine the following: (a) quantiser step size Δ; (b) quantisation noise power; (c) rms quantisation noise voltage; (d) signal-to-quantisation noise ratio (SQNR) when the quantiser is used to digitise a sinusoidal signal of amplitude $A_m = 16$ V; (e) SQNR when the input is a sinusoid of amplitude 1 V.

(a) Quantiser peak $C = 16$ V and range $= 2C = 32$ V. The quantiser uses $k = 4$ bits per sample. Eq. (5.2) yields step size

$$\Delta = \frac{C}{2^{k-1}} = \frac{16}{2^3} = 2\ \text{V}$$

(b) Quantisation noise power is given by MSQE in Eq. (5.6) as

$$\text{MSQE} = \frac{\Delta^2}{12} = \frac{2^2}{12} = \frac{1}{3}\ \text{W}$$

(c) rms quantisation noise voltage is of course the square root of noise power:

$$e_{\text{qrms}} = \sqrt{\text{MSQE}} = \sqrt{1/3} = 0.5774\ \text{V}$$

(d) Signal power $P_s = A_m^2/2 = 16^2/2 = 128$ W. Thus

$$\text{SQNR} = \frac{P_s}{\text{MSQE}} = \frac{128}{1/3} = 384 = 25.84\ \text{dB}$$

Note that the above result can be obtained using Eq. (5.12) for a sinusoid that fully loads a quantiser (as is the case here): SQNR $= 6.02 \times 4 + 1.76 = 25.84$ dB.

(e) Signal power $P_s = A_m^2/2 = 1^2/2 = 1/2$ W. Thus

$$\text{SQNR} = \frac{P_s}{\text{MSQE}} = \frac{1/2}{1/3} = 1.5 = 1.76\ \text{dB}$$

Again note that this could have been solved using Eq. (5.10), with $k = 4$, $A_p = 1$ V, $C = 16$ V, $R = \sqrt{2}$ (for a sinusoid) to obtain

$$\begin{aligned}\text{SQNR} &= 4.77 + 6.02k - 20\log(R) + 20\log(A_p/C)\ \text{dB}\\ &= 4.77 + 6.02 \times 4 - 20\log(\sqrt{2}) + 20\log(1/16)\ \text{dB}\\ &= 25.84 - 24.08\ \text{dB}\\ &= 1.76\ \text{dB}\end{aligned}$$

The difference in SQNR between (d) and (e) serves to highlight the problem with uniform quantisation. A strong input signal in (d) enjoys a reasonable SQNR, whereas a weak input signal in (e) with amplitude 24 dB below that of signal (d) experiences a poor SQNR that is lower than the SQNR in (d) by the same dB.

Worked Example 5.2: Dynamic Range

An analogue signal of range −8 V to +8 V and rms value 6 V fully loads the quantiser of a linear ADC that has a dynamic range of 42 dB. We wish to determine the signal-to-quantisation noise ratio (SQNR) in dB of the ADC output.

From Eq. (5.3) the number of bits (per sample) of the ADC is

$$k = \frac{\text{Dynamic range}}{6} = \frac{42}{6} = 7$$

The peak-to-rms ratio of the input signal is $R = 8/6 = 4/3$.

Substituting these values into Eq. (5.11), since the quantiser is fully loaded, yields

$$\begin{aligned} \text{SQNR} &= 4.77 + 6.02k - 20\log(R) \text{ dB} \\ &= 4.77 + 6.02 \times 7 - 20\log(4/3) \text{ dB} \\ &= 44.4 \text{ dB} \end{aligned}$$

5.4 Non-uniform quantisation

The main goal of non-uniform quantisation is to achieve a constant SQNR across all levels of input signal amplitude. This requires coarse quantisation of large input samples using large step sizes and finer quantisation of small input samples using smaller step sizes. In other words, a wider range of large input samples than that of small input samples is mapped to one quantised output level, meaning that the larger samples are effectively compressed. And since these output levels are then sequentially (i.e. uniformly) numbered irrespective of the size of the gap between adjacent levels, it follows that non-uniform quantisation is equivalent to a two-stage process of *compression* followed by uniform quantisation (or linear ADC) as illustrated in Fig. 5.3 for $N = 8$ quantisation intervals. This equivalent view involves the following steps at the transmitter and receiver.

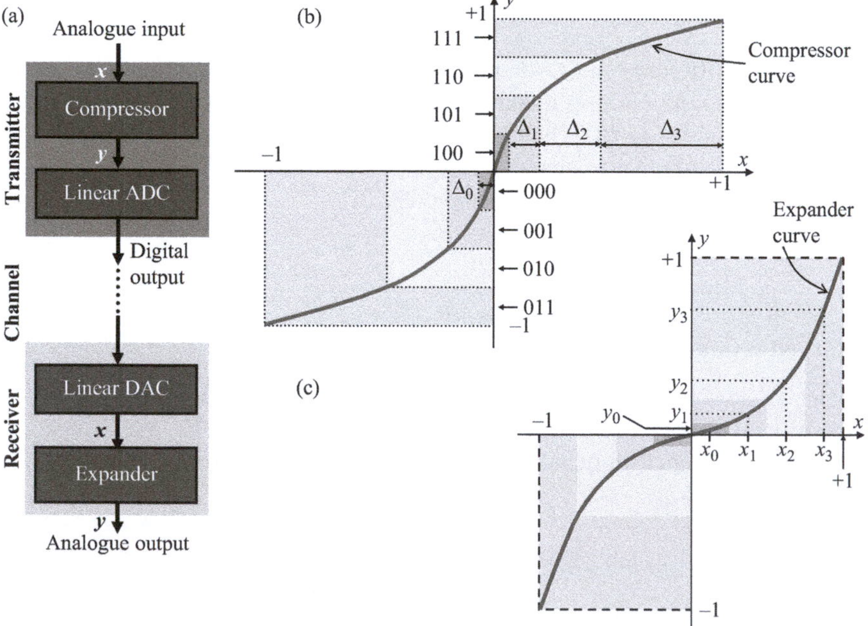

Fig. 5.3 Non-uniform quantisation: (a) equivalent steps; (b) compressor; (c) expander

- At the transmitter, first use a compressor to compress the analogue input samples in order to reduce high sample values relative to low sample values. Next, quantise the compressed signal using a uniform quantiser (i.e. linear ADC). Notice that Fig. 5.3b embodies these two steps. The input x is first converted into output y through the compressor curve and then the output y is divided into numbered uniform intervals. The end result is that we have achieved non-uniform quantisation of the input x by using progressively larger step sizes $\Delta_0, \Delta_1, \ldots$, as shown along the x-axis of the graph.
- At the receiver, recover the signal using a linear DAC (digital to analogue converter). Then expand the recovered signal using an expander having a transfer characteristic that is the exact inverse of the compressor. This perfectly removes the distortion due to compression at the transmitter. Notice in Fig. 5.3c how the expander curve maps equal intervals along the x-axis (expander input) to unequal intervals along the y-axis (expander output) that are progressively larger the further you are from the origin, just like the intervals along the x-axis of Fig. 5.3b.

This combined process of signal compression at the transmitter and expansion at the receiver is known as *companding* and does deliver tangible benefits the most important of which is the so-called *companding gain* defined in Section 5.4.5.

5.4.1 Ideal log-companding

To determine the function $y = f(x)$ that gives the required compressor characteristic, consider Fig. 5.4 which shows a compressor curve that maps an input interval of size Δ_x centred at x along the horizontal axis into its corresponding output interval along the vertical axis. The input and output are both normalised to the range

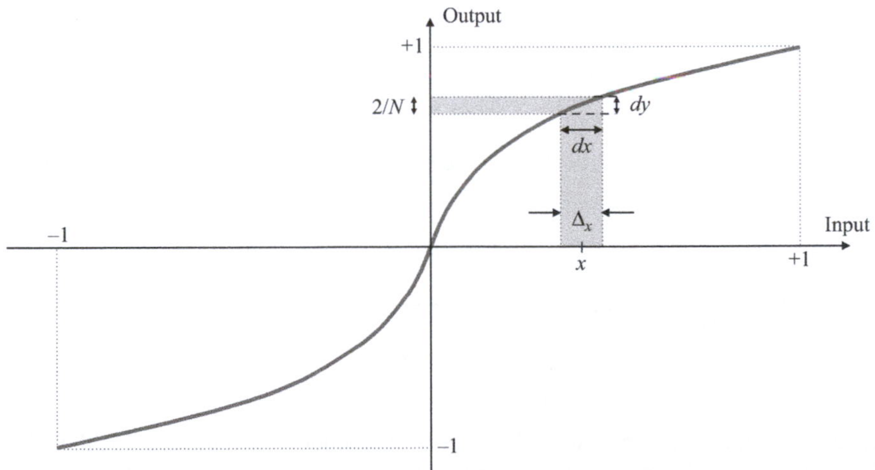

Fig. 5.4 Compressor characteristic, $y = f(x)$. The range $-1 \rightarrow +1$ along the y-axis is divided into N equal quantisation intervals, so each interval is of size $2/N$ as shown. The range $-1 \rightarrow +1$ along the x-axis is divided into N unequal quantisation intervals. One interval of size Δ_x centred at x is shown and is mapped by the compressor curve into an interval of size $2/N$ along the y-axis

$(-1, +1)$, and the quantiser has N intervals, which means that the step size along the vertical axis is $2/N$. For large N, the intervals $2/N$ and Δ_x represent small changes denoted by dy and dx, respectively and their ratio gives the slope of the compressor curve at x. That is

$$\frac{dy}{dx} = \frac{2/N}{\Delta_x} \tag{5.13}$$

Or

$$\Delta_x = \frac{2/N}{\text{Slope of compressor curve}} = \frac{2}{N}\frac{dx}{dy} \tag{5.14}$$

If the input signal has probability density function $p(x)$, then the probability that its sample falls in an interval of size dx centred at x is $p(x)dx$, and its mean square value or signal power is obtained by summing the product of this probability and x^2 over the entire input range $(-1, 1)$:

$$P_s = \int_{-1}^{1} x^2 p(x)dx \tag{5.15}$$

Eq. (5.6) gives the quantisation noise power in a uniform quantiser where step sizes are equal. To apply this to a non-uniform quantiser we multiply $\Delta_x^2/12$ by the probability $p(x)dx$ that the input sample falls in the interval x of step size Δ_x and sum this product over the entire input range to obtain

$$\begin{aligned}\text{MSQE} &= \int_{-1}^{1} \frac{\Delta_x^2}{12} p(x)dx \\ &= \frac{1}{3N^2} \int_{-1}^{1} \left(\frac{dx}{dy}\right)^2 p(x)dx\end{aligned} \tag{5.16}$$

where we made use of Eq. (5.14) for Δ_x. The ratio between Eqs. (5.15) and (5.16) yields

$$\text{SQNR} = \frac{P_s}{\text{MSQE}} = 3N^2 \frac{\int_{-1}^{1} x^2 p(x)dx}{\int_{-1}^{1} \left(\frac{dx}{dy}\right)^2 p(x)dx} \tag{5.17}$$

Since our goal is to achieve an SQNR that is independent of input level x, the RHS of the above equation must be independent of x, and this can only be the case if in the denominator

$$\left(\frac{dx}{dy}\right)^2 = K^2 x^2 \tag{5.18}$$

where K is some constant, so that the integrals cancel out leaving

$$\text{SQNR} = \frac{3N^2}{K^2} \tag{5.19}$$

which is independent of input signal level x as desired. Thus the compressor characteristic must be a function $y = f(x)$ that satisfies Eq. (5.18). Taking the square root of both sides of this equation, and making dy the subject of the equation and integrating yields

$$y = \frac{1}{K}\ln(x) + Y_1$$

where Y_1 is a constant that we choose in order to make $(x, y) = (1, 1)$ a point on the curve, as is obvious in Fig. 5.4, since the normalised maximum input is compressed to the normalised maximum output. Thus $Y_1 = 1$, and the desired compressor characteristic is

$$y = \frac{1}{K}\ln(x) + 1 \tag{5.20}$$

Taking the derivative with respect to x gives the slope of the compressor curve as

$$\frac{dy}{dx} = \frac{1}{Kx} \equiv \frac{\text{Constant interval (along } y\text{-axis)}}{\text{Quantiser step size (along } x\text{-axis)}} \tag{5.21}$$

Eq. (5.20) gives a complete specification of a compressor curve that can be used to compress the input signal x to give an output y, which when uniformly quantised achieves the desired fine quantisation of small input values and coarse quantisation of larger input values, and delivers an SQNR that is exactly the same across all input sample amplitudes, no matter how small. If this sounds too good to be true, it really is, because there is a practical problem with implementing this compression function. Its slope is infinite at $x = 0$, so the required quantiser step sizes become vanishingly small near the origin (as $x \to 0$), which means that you need an infinite number of intervals to be able to implement the curve down to $x = 0$. To circumvent this problem, we are in practice forced to abandon the logarithmic function in the region $x \to 0$ and to adopt a linear function in this region, say $|x| \leq 1/A$, where A is some constant ≥ 1. The effect of this pragmatic solution is that SQNR is constant in the region $1/A \leq |x| \leq 1$ where the logarithmic compression curve is followed, whereas in the region $-1/A \leq x \leq 1/A$ where uniform quantisation is followed, SQNR decreases with input amplitude as discussed in Section 5.3. Notice therefore that the constant A sets the boundaries of the logarithmic and linear compression regions. For example, if $A = 1$ the entire curve is linear and there is no compression (which corresponds to uniform quantisation) and if $A = \infty$ the compression curve is entirely logarithmic down to $x = 0$ (which is not feasible). Because the compression curve is a logarithmic function, digitisation based on non-uniform quantisation is sometimes referred to as *log-PCM*.

Worked Example 5.3: Logarithmic Versus Linear Compression

Given a quantiser, of normalised range $(-1, +1)$, that has $N = 256$ quantisation intervals, we wish to determine SQNR in the following cases.

(a) The intervals are uniform (which corresponds to a linear compressor) and the input is a sinusoid of amplitude 1 V.

(b) The quantiser incorporates a logarithmic compressor having $K = 1 + \ln(A)$, where $A = 87.6$, so that the intervals are non-uniform, and the input is a sinusoid of amplitude 1 V.

(c) The quantiser is linear as in (a) and the input is a sinusoid of amplitude 10 mV.

(d) The quantiser is logarithmic as in (b) and the input is a sinusoid of amplitude 10 mV.

(a) The sinusoid fully loads this linear quantiser that has $N=256$ intervals which corresponds to $k=\log_2 N=8$ bits/sample, so SQNR is as given by Eq. (5.12):

$$\text{SQNR} = 6.02k + 1.76 = 6.02 \times 8 + 1.76 = 49.92 \text{ dB}$$

(b) The SQNR of this logarithmic quantiser is given by Eq. (5.19) irrespective of input signal amplitude. Thus

$$\text{SQNR} = \frac{3N^2}{K^2} = \frac{3 \times 256^2}{(1 + \ln 87.6)^2} = 6564.2 = 38.2 \text{ dB}$$

(c) We apply Eq. (5.10) with $A_p = 10$ mV, $C = 1$ V and $R = \sqrt{2}$ (for a sinusoid) to obtain

$$\begin{aligned}\text{SQNR} &= 4.77 + 6.02k - 20\log(R) + 20\log(A_p/C) \text{ dB}\\ &= 4.77 + 6.02 \times 8 - 10\log(2) + 20\log(10/1000) \text{ dB}\\ &= 49.92 - 40 = 9.92 \text{ dB}\end{aligned}$$

Notice that this 10 mV sinusoid underloads the 1 V quantiser by 40 dB. Hence its SQNR is down by 40 dB from the value in (a) for an input that fully loads the quantiser.

(d) As noted in (b) the SQNR of the logarithmic quantiser is independent of input signal level, so SQNR for this 10 mV sinusoidal input is 38.2 dB, exactly the same as for the 1 V sinusoid in (b).

It can be seen that SQNR in (a) is higher than in (b). In general, for input signals at the top end, i.e. for inputs that approximately fully load the quantiser, a linear quantiser will always deliver a higher SQNR than a log-quantiser of the same number of bits per sample. However, the SQNR of log-quantisers might be modest but they are consistent for all input amplitudes, as can be seen by comparing the results in (c) and (d) where the linear quantiser has a very poor SQNR of 9.92 dB for the 10 mV sinusoid whereas the SQNR of the log-quantiser is still 38.2 dB.

The ITU-T has standardised two practicable logarithmic compressor characteristics, the A-law in Europe, and the μ-law in North America and Japan.

5.4.2 A-law companding

The A-law compressor characteristic is based on Eq. (5.20) with the following settings and modifications:

- The constant $K = 1 + \ln(A)$, with $A = 87.6$, so that the logarithmic compressor function takes the form

$$y_{\log} = \frac{1 + \ln(Ax)}{1 + \ln(A)} \tag{5.22}$$

- The logarithmic function $y_{\log}$ is followed down to $|x| = 1/A$ at which point it is abandoned and a linear characteristic $y_{\text{lin}} = mx + c$ is followed down to $x = 0$. The linear characteristic y_{lin} must satisfy two conditions. First, it must pass through the origin in order that the compressor characteristic retains its odd symmetry. This means that $c = 0$. Second, y_{lin} must be equal to $y_{\log}$ at $x = 1/A$ in order to maintain continuity, and this means that

$$m = \frac{A}{1 + \ln(A)}$$

which gives

$$y_{\text{lin}} = \frac{Ax}{1 + \ln(A)} \tag{5.23}$$

The A-law compressor characteristic may therefore be expressed as follows:

$$y = \begin{cases} \dfrac{1 + \ln(A|x|)}{1 + \ln(A)} \operatorname{sgn}(x), & \dfrac{1}{A} \le |x| \le 1 \\[2ex] \dfrac{Ax}{1 + \ln(A)}, & -\dfrac{1}{A} \le x \le \dfrac{1}{A} \end{cases} \tag{5.24}$$

The signum function expresses the odd symmetry feature of the compressor. To determine compressor output, apply the above function to the magnitude (i.e. absolute value) of the input, and then assign to the output the sign of the input. Subsequently we will restrict our analysis to the positive input range of a logarithmic compressor knowing that a negative input is processed exactly as if it were positive to obtain an output that is then negated.

5.4.3 μ-Law companding

The μ-law compressor is also based on Eq. (5.20) with the following settings and modifications:

- The constant K is set to $\ln(1 + \mu)$, where μ is a positive constant usually 255. The logarithmic compressor function therefore takes the form

$$y = \frac{\ln(x)}{\ln(1 + \mu)} + 1 = \frac{\ln(x) + \ln(1 + \mu)}{\ln(1 + \mu)} = \frac{\ln(x + \mu x)}{\ln(1 + \mu)}$$

- Next, the function is modified by replacing $\ln(x + \mu x)$ in the numerator by $\ln(1 + \mu x)$. This gives the μ-law compressor, which we express with the aid of the signum function (see the comments following Eq. (5.24)) as

$$y = \frac{\ln(1 + \mu|x|)}{\ln(1 + \mu)} \operatorname{sgn}(x), \quad -1 \le x \le 1 \tag{5.25}$$

The modification carried out in the second step above is crucial in producing a practicable logarithmic compressor. To see this, note that in the region $|x| \to 1$, the term $1 + \mu|x| \approx \mu|x|$, since $\mu \gg 1$ so that the compressor function is logarithmic

$$y_{\log} = \frac{\ln(\mu|x|)}{\ln(1 + \mu)} \operatorname{sgn}(x), \quad |x| \to 1 \tag{5.26}$$

And in the region $|x| \to 0$, the term $\ln(1 + \mu|x|) \approx \mu|x|$ so that the compressor function is linear

$$y_{\text{lin}} = \frac{\mu x}{\ln(1+\mu)}, \qquad |x| \to 0 \tag{5.27}$$

Therefore the μ-law compressor of Eq. (5.25) exhibits a gradual transition from a logarithmic function $y_{\log}$ in Eq. (5.26) in the region $|x| \to 1$ to a linear function y_{lin} in Eq. (5.27) in the region $|x| \to 0$. The A-law compressor on the other hand has a sudden transition from logarithmic to linear at the point $|x| = 1/A$. Fig. 5.5 shows the A-law and μ-law compressor characteristics for various values of A and μ. At $A = 1$ and $\mu \to 0$, both curves are linear and pass through the origin (0, 0) as well as the saturation points $(-1, -1)$ and (1, 1), so that input x equals output y at all points and there is no compression, which means that quantisation is uniform or linear. As the positive constants A and μ increase, compression increases until there is 'infinite compression' at A, $\mu = \infty$, whereby all non-zero input values are compressed into one output interval. Compression that is suitable for speech digitisation lies between these two extremes of zero compression at one end and infinite compression at the other.

5.4.4 *Specification of companding*

It can be seen from Fig. 5.5 that the A-law and μ-law compressor curves are almost identical at their standardised values of $A = 87.6$ and $\mu = 255$. With the aid of the horizontal and vertical graph lines, you should be able to read these two standard curves in Fig. 5.5 to observe the following approximate mappings: The top half $(0.5 \to 1)$ of input is compressed into the top $1/8^{\text{th}}$ of output $(0.875 \to 1)$; the next ¼ of input $(0.25 \to 0.5)$ is compressed into the next $1/8^{\text{th}}$ of output $(0.75 \to 0.875)$; the next $1/8^{\text{th}}$ of input $(0.125 \to 0.25)$ is mapped into the next $1/8^{\text{th}}$ of output $(0.625 \to 0.75)$; the next $1/16^{\text{th}}$ of input $(0.0625 \to 0.125)$ is expanded into the next $1/8^{\text{th}}$ of

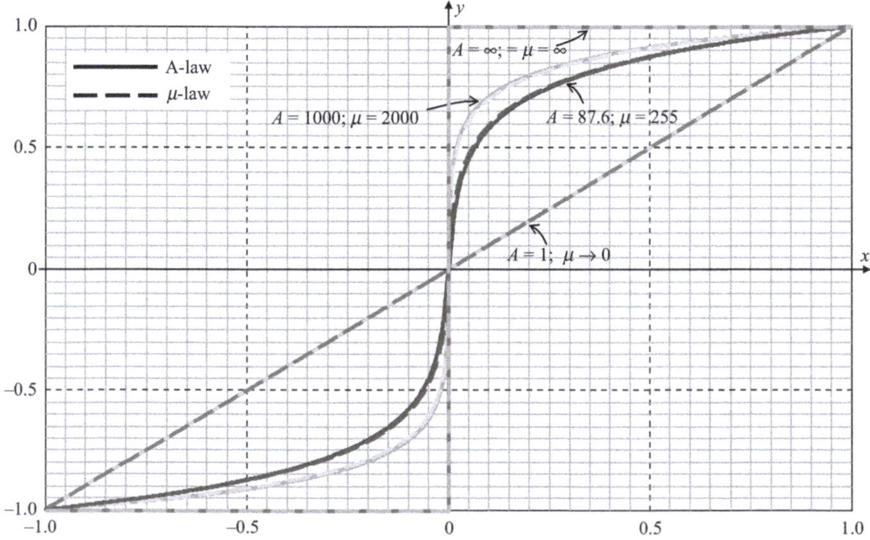

Fig. 5.5 A-law and μ-law compression curves at various values of A and μ

output (0.5 → 0.625); and the bottom 1/16th of input (0 → 0.0625) is expanded into the bottom half of output (0 → 0.5). These curves therefore map input intervals whose sizes reduce progressively by a factor of 2 (starting at the top end of input) into equal intervals at the output, and this achieves the desired compression of large input values and expansion of the smaller ones.

The amount of expansion of small input values and compression of large input values in each law is controlled by the constants A and μ in the respective laws. These constants can be set to achieve any specified compression as follows. If you want to

> *compress the top half of input* $0.5 \leq x \leq 1$ *into the top* $(1/2m)^{\text{th}}$ *of output* $(2m-1)/2m \leq y \leq 1$ *while expanding the bottom* $(1/2^m)^{\text{th}}$ *of input* $0 \leq x \leq 1/2^m$ *into the bottom half of output* $0 \leq y \leq 0.5$

then the value of A to use in the A-law expression of Eq. (5.24) is

$$A = 2^{2m}/e; \quad m \geq \log_2 e, \quad e = 2.718281828459\cdots \tag{5.28}$$

and the value of μ to use in Eq. (5.25) is

$$\mu = 2^{2m} - 2^{m+1}; \qquad m > 1 \tag{5.29}$$

Eq. (5.28) is derived in the next worked example and gives an exact match of the compression and expansion stated above, whereas Eq. (5.29) only gives an approximate match that improves with m, the maximum error being less than 0.25% for $m \geq 4$ and no more than 3.75% at any $m > 1$.

An indication of the extent of compression is given by the ratio between the maximum step size Δ_{max}, which occurs at $x = 1$, and the minimum step size Δ_{min} that occurs at $x = 0$. Using Eq. (5.14) we obtain

$$\frac{\Delta_{max}}{\Delta_{min}} = \left(\left.\frac{dy}{dx}\right|_{x=0}\right) \Big/ \left(\left.\frac{dy}{dx}\right|_{x=1}\right) \tag{5.30}$$

For the A-law, the applicable curves are Eq. (5.23) at $x = 0$ and Eq. (5.22) at $x = 1$. Taking their derivatives we obtain

$$\left.\frac{dy}{dx}\right|_{x=0} = \frac{A}{1+\ln A}; \quad \left.\frac{dy}{dx}\right|_{x=1} = \left.\frac{1/x}{1+\ln A}\right|_{x=1} = \frac{1}{1+\ln A} \tag{5.31}$$

Similarly, taking the derivative of the μ-law function in Eq. (5.25) written simply as

$$y = \frac{\ln(1+\mu x)}{\ln(1+\mu)}$$

we obtain

$$\frac{dy}{dx} = \frac{\mu}{1+\mu x} \cdot \frac{1}{\ln(1+\mu)}$$

and hence

$$\left.\frac{dy}{dx}\right|_{x=0} = \frac{\mu}{\ln(1+\mu)}; \quad \left.\frac{dy}{dx}\right|_{x=1} = \frac{\mu}{(1+\mu)\ln(1+\mu)} \tag{5.32}$$

Therefore the ratio between maximum and minimum step sizes is

$$\frac{\Delta_{max}}{\Delta_{min}} = \begin{cases} A, & \text{A-law} \\ 1+\mu, & \mu\text{-law} \end{cases} \tag{5.33}$$

If $\Delta_{max} = \Delta_{min}$, it means that step sizes are equal across the entire input range, which is the special case of uniform quantisation. We see from Eq. (5.33) that this happens in A-law when $A = 1$, and in μ-law when $\mu = 0$. Clearly, the constant A cannot be less than +1 and μ cannot be less than 0, otherwise Δ_{max} would be less than Δ_{min}, which is contradictory.

Worked Example 5.4: A-Law Specification

We wish to derive Eq. (5.28), giving an expression for the constant A in terms of m so that the A-law compressor will compress the top half of input into the top $(1/2m)^{\text{th}}$ of output while expanding the bottom $(1/2^m)^{\text{th}}$ of input to fill the bottom half of output.

The top $(1/2m)^{\text{th}}$ of output starts at $y = 1 - 1/2m = (2m - 1)/2m$, so the compressor specification given above corresponds to $y = (2m - 1)/2m$ at $x = 1/2$, and $y = 1/2$ at $x = 1/2^m$. Let us set the A-law function (logarithmic portion) to the second condition and solve for A:

$$\frac{1}{2} = \frac{1+\ln(A/2^m)}{1+\ln(A)}$$

$$1+\ln(A) = 2 + 2\ln(A) - \ln(2^{2m})$$

$$\ln A = \ln(2^{2m}) - 1 = \ln(2^{2m}) - \ln e = \ln(2^{2m}/e)$$

$$A = 2^{2m}/e$$

Next, we verify that this value for A satisfies the first condition. That is, we verify that if A is set to $2^{2m}/e$ in the A-law function (logarithmic portion) then y will be $(2m - 1)/2m$ at $x = 1/2$. Substituting $A = 2^{2m}/e$ and $x = 1/2$ into the A-law yields

$$\begin{aligned} y &= \frac{1+\ln(0.5\times 2^{2m}/e)}{1+\ln(2^{2m}/e)} = \frac{1+\ln(2^{2m-1}/e)}{1+\ln(2^{2m}/e)} \\ &= \frac{1+(2m-1)\ln 2 - \ln e}{1+2m\ln 2 - \ln e} = \frac{1+(2m-1)\ln 2 - 1}{1+2m\ln 2 - 1} \\ &= \frac{(2m-1)\ln 2}{2m\ln 2} = \frac{2m-1}{2m} \end{aligned}$$

Therefore the value $A = 2^{2m}/e$ ensures that the A-law function compresses the (normalised) input range $1/2 \le x \le 1$ into the output range $(2m-1)/2m \le y \le 1$ while expanding the bottom end input range $0 \le x \le 1/2^m$ into the output range $0 \le y \le 1/2$. Since the above derivation assumes that x lies in the logarithmic portion of the A-law function, it is essential that the top of this bottom input range, i.e. $x = 1/2^m$, lies within this logarithmic portion that only extends down to $x = 1/A$. This means that we must have

$$\frac{1}{2^m} \ge \frac{1}{A} = \frac{e}{2^{2m}}$$

which upon manipulation as follows yields the restriction on m stated in Eq. (5.28):

$$\begin{aligned}
&2^m \leq 2^{2m}/e \\
&\log_2(2^m) \leq \log_2(2^{2m}/e) = \log_2(2^{2m}) - \log_2 e \\
&m\log_2 2 \leq 2m\log_2 2 - \log_2 e \\
&m \leq 2m - \log_2 e \qquad\qquad (\text{since } \log_2 2 = 1) \\
&m \geq \log_2 e
\end{aligned}$$

5.4.5 Companding gain and penalty

The benefit of non-uniform quantisation is measured in terms of the improvement that it delivers to the SQNR of small samples. *Companding gain* G_c is one such measure defined as the ratio between the SQNR of small input values in a non-uniform quantiser and the SQNR of the same signal type when using a uniform quantiser of the same number of bits/sample. A non-uniform quantiser employs the smallest step size $\Delta_{\min}$ in the region around $x = 0$. The SQNR of the non-uniform quantiser in this region is therefore

$$\frac{P_s}{\text{MSQE}} = \frac{P_s}{\Delta_{\min}^2/12}$$

A uniform quantiser would employ a constant step size of $2/N$ (see Fig. 5.4), where N is the number of uniform intervals, so that the SQNR of the uniform quantiser in this small input region is

$$\frac{P_s}{(2/N)^2/12}$$

Dividing the former SQNR by the latter yields companding gain as

$$G_c = \left(\frac{2/N}{\Delta_{\min}}\right)^2 = \left(\left.\frac{dy}{dx}\right|_{x=0}\right)^2$$

where we have used Eq. (5.13) to express the gain as the square of the derivative of the compressor curve evaluated at the origin. These derivatives were computed in Eqs. (5.31) and (5.32) for the A-law and μ-law curves, so we may write

$$G_c = \begin{cases} \left(\dfrac{A}{1+\ln A}\right)^2, & \text{A-law} \\ \left(\dfrac{\mu}{\ln(1+\mu)}\right)^2, & \mu\text{-law} \end{cases}$$

In dB these expressions become

$$G_c = \begin{cases} 20\log_{10}\left(\dfrac{A}{1+\ln A}\right)\text{dB}, & \text{A-law} \\ 20\log_{10}\left(\dfrac{\mu}{\ln(1+\mu)}\right)\text{dB}, & \mu\text{-law} \end{cases} \tag{5.34}$$

Putting the standard values $A = 87.6$ and $\mu = 255$ into the above equations yields $G_c = 24.1$ dB for A-law and 33.3 dB for μ-law. Since one extra bit per sample yields a 6 dB improvement in SQNR, it follows that a 24 dB improvement due to companding is equivalent to an extra four coding bits. What this means is that an 8-bit A-law quantiser achieves the same SQNR for small inputs as a 12-bit uniform quantiser. Thus by using A-law quantisation the required number of bits per sample is reduced from 12 to 8, which represents a 33% reduction in bit rate and the same percentage reduction in bandwidth. In practice a piecewise linear approximation is adopted in the implementation of the A-law and μ-law standards, as discussed in Section 5.5.1. This leaves the A-law curve unchanged in the region around $x=0$ where the law is already governed by a linear curve, so that practical A-law companding gain remains at 24 dB as computed above. However, the piecewise linear approximation alters the μ-law curve in the region around $x=0$ so that practical μ-law companding gain is 30 dB, rather than the 33.3 dB obtained above, which corresponds to an extra five coding bits. That is, an 8-bit μ-law quantiser achieves the same SQNR as a 13-bit uniform quantiser in the small signal region, which represents a saving of 38.5% in bit rate and transmission bandwidth for the same quality of small input signals.

It is important to recognise that the improvement in SQNR enjoyed by the small input samples is at the expense of the SQNR of large input samples. There really is no free lunch! Let us therefore define *companding penalty* L_c as the ratio between the SQNR of large input values in a uniform quantiser and the SQNR of the same signal type when using a non-uniform quantiser of the same number of bits per sample. A non-uniform quantiser uses the largest step size $\Delta_{\max}$ at the top end of the input range, whereas a uniform quantiser uses the step size $2/N$ in this region. By a similar argument as for G_c, we obtain

$$
\begin{aligned}
L_c &= \left(\frac{\Delta_{\max}}{2/N}\right)^2 = \left(\frac{1}{dy/dx}\bigg|_{x=1}\right)^2 \\
&= \begin{cases} (1+\ln A)^2, & \text{A-law} \\ \left[\dfrac{(1+\mu)\ln(1+\mu)}{\mu}\right]^2, & \mu\text{-law} \end{cases}
\end{aligned}
$$

where we have made use of Eqs. (5.31) and (5.32) for dy/dx evaluated at $x=1$. Expressed in dB, the companding penalty is

$$
L_c\,(\text{dB}) = \begin{cases} 20\log_{10}(1+\ln A), & \text{A-law} \\ 20\log_{10}\left[\dfrac{(1+\mu)\ln(1+\mu)}{\mu}\right], & \mu\text{-law} \end{cases} \tag{5.35}
$$

Again substituting $A=87.6$ and $\mu=255$ into the above expressions yields $L_c=14.8$ dB for A-law and 14.9 dB for μ-law. Thus A-law and μ-law quantisers sacrifice signal quality at the top end of input by ~15 dB (and by progressively lower than ~15 dB as you go down the input range) in order to improve the quality of small input signals by 24 dB (for A-law) and by progressively lower as you go up the input range. The end result of this exchange is a lower but consistent SQNR across the entire input range that is more satisfying in communication services than the situation that obtains in uniform quantisation where SQNR is very high at the top end and poor at the bottom end. Note that a piecewise linear approximation of the two laws leads to a lower $\Delta_{\max}$ than the one used above and hence to a companding penalty of 12 dB, which is smaller than the L_c given by Eq. (5.35).

Worked Example 5.5: Quantised Outputs

A sinusoidal signal $x(t) = 5\cos(2\pi \times 250t + \pi/6)$ V, sampled at 8 kHz starting at $t = 0$, fully loads an eight-bit μ-law quantiser ($\mu = 255$). Determine the first 10 quantised output values and calculate SQNR based on the first 10 samples.

The quantiser is described as eight-bit, so it has $N = 2^8 = 256$ quantisation intervals. Sampling frequency $f_s = 8$ kHz, so sampling interval $T_s = 1/f_s = 125$ μs. The steps to follow in solving this problem are shown in Fig. 5.6 and explained below.

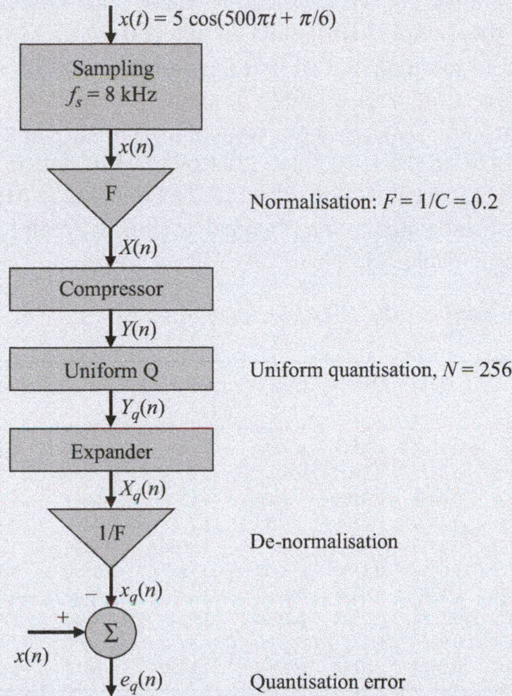

Fig. 5.6 Quantisation of a sinusoidal signal (Worked Example 5.5)

(i) Obtain the values of $x(t)$ at $t = 0, T_s, 2T_s, 3T_s, \ldots, 9T_s$. These are the first 10 samples, denoted by $x(n)$, $n = 0, 1, 2, \ldots, 9$, in the third column of Table 5.1. For example, $x(0)$ is obtained by substituting $t = 0$ into the expression for $x(t)$, so $x(0) = 5\cos(2\pi \times 250 \times 0 + \pi/6) = 5\cos(\pi/6) = 4.3301$; $x(1)$ is obtained by evaluating $x(t)$ at $t = T_s = 125$ μs, so $x(1) = 5\cos(2\pi \times 250 \times 125 \times 10^{-6} + \pi/6) = 5\cos(0.7199) = 3.7592$; $x(2)$ is obtained by evaluating $x(t)$ at $t = 2T_s = 250$ μs; and so on.

(ii) Since the given compressor function has range ± 1, normalise the input signal (of amplitude $C = 5$ V) to this range by multiplying each sample $x(n)$ by the factor $F = 1/C = 0.2$. The result is the normalised input denoted by $X(n)$ in the fourth column of Table 5.1. For example, $X(0) = 0.2x(0) = 0.866$. Retain only the absolute value $|X(n)|$. The sign of $x(n)$ will be assigned to the quantised output at the end.

(iii) Determine compressed output $|Y(n)|$ by using $|X(n)|$ as input to the μ-law function of Eq. (5.25). These results are listed in column 5 of Table 5.1. For example

$$|Y(0)| = \frac{\ln(1 + 255|X(0)|)}{\ln(1 + 255)} = 0.9742$$

(iv) Uniformly quantise $|Y(n)|$ to obtain the normalised quantised output $|Y_q(n)|$ by noting that $|Y(n)|$ falls in one of the $N/2$ uniform intervals of size $\Delta = 2/N = 2/256 = 7.8125 \times 10^{-3}$ that covers the positive quantiser range (0, 1). Numbering these intervals from $j = 0$ to $N/2 - 1$, the sample $|Y(n)|$ falls in interval $j = \lfloor |Y(n)|/\Delta \rfloor$ and is quantised to the midpoint of that interval which is $j\Delta + \Delta/2$, where $\lfloor z \rfloor$ represents the largest integer less than or equal to z, subject to a maximum of $N/2 - 1$. For example, 0.9742 falls in interval $j = \lfloor 0.9742/7.8125 \times 10^{-3} \rfloor = \lfloor 124.69 \rfloor = 124$ and is quantised to $124 \times 7.8125 \times 10^{-3} + 7.8125 \times 10^{-3}/2 = 0.9727$. All the normalised quantised outputs are obtained in this way and are listed in column 6 of Table 5.1.

Table 5.1 Tabulated results (Worked Example 5.5)

n	$t = nT_s$ (μs)	$x(n)$ (V)	$\|X(n)\|$	$\|Y(n)\|$	$\|Y_q(n)\|$	$y_q(n)$ (V)	$X_q(n)$	$x_q(n)$(V)	$e_q(n)$ (mV)
0	0	4.3301	0.8660	0.9742	0.9727	4.8633	0.8588	4.2938	36.3374
1	125	3.7592	0.7518	0.9488	0.9492	4.7461	0.7536	3.7681	−8.8999
2	250	3.0438	0.6088	0.9109	0.9102	4.5508	0.6061	3.0304	13.3824
3	375	2.2114	0.4423	0.8538	0.8555	4.2773	0.4465	2.2326	−21.1326
4	500	1.2941	0.2588	0.7583	0.7617	3.8086	0.2639	1.3195	−25.4535
5	625	0.3270	0.0654	0.5180	0.5195	2.5977	0.0660	0.3300	−2.9880
6	750	−0.6526	0.1305	0.6374	0.6367	−3.1836	−0.1300	−0.6500	−2.6605
7	875	−1.6072	0.3214	0.7968	0.7930	−3.9648	−0.3146	−1.5729	−34.2706
8	1000	−2.5000	0.5000	0.8757	0.8789	−4.3945	−0.5090	−2.5452	45.1536
9	1125	−3.2967	0.6593	0.9253	0.9258	−4.6289	−0.6613	−3.3065	9.7472

(v) Obtain the actual quantised output $y_q(n)$ by (a) denormalising $|Y_q(n)|$ through division by the normalising factor F, and (b) assigning to the result the sign of the input sample $x(n)$. For example, the first sample yields output $y_q(0) = |Y_q(0)|/F = 0.9727/0.2 = 4.8633$ V, and the last sample yields output $y_q(9) = -|Y_q(9)|/F = -0.9258/0.2 = -4.6289$ V. These are the desired quantised outputs and are listed in column 7 of Table 5.1.

(vi) The normalised quantised output $Y_q(n)$ is usually digitally encoded as discussed in Section 5.5. When recovered from its digital format, $Y_q(n)$ must be expanded to remove the distortion introduced by the compressor and yield $X_q(n)$. The μ-law expander is derived by making x the subject of Eq. (5.25). After swapping variables to retain the conventional use of x for input and y for output, this gives the *μ-law expander*

$$y = \frac{1}{\mu}\left[(1+\mu)^{|x|} - 1\right]\mathrm{sgn}(x), \quad -1 \le x \le 1 \tag{5.36}$$

Now substitute $Y_q(n)$ for x in the above equation to obtain the expanded values $X_q(n)$. For example, the last sample yields

$$X_q(9) = \frac{1}{255}\left[(1+255)^{0.9258} - 1\right]\mathrm{sgn}(x(9)) = -0.6613$$

where we have used the fact that $Y_q(n)$ has the same sign as $x(n)$. The full set of values of $X_q(n)$ are in column 8, with the denormalised values $x_q(n) = X_q(n)/F$ in column 9 of Table 5.1.

(vii) Calculate the quantisation error $e_q(n) = x(n) - x_q(n)$ as listed in column 10, being careful to minimise rounding errors. Note that although Table 5.1 lists values rounded to four decimal places, the calculations were performed to greater accuracy.

(viii) The quantisation noise power MSQE, signal power P_s, and finally SQNR are computed as follows using values from relevant columns of Table 5.1:

$$\mathrm{MSQE} = \frac{1}{10}\sum_{n=0}^{9} e_q^2(n) = 5.9975 \times 10^{-4}\ \mathrm{W}$$

$$P_s = \frac{1}{10}\sum_{n=0}^{9} x^2(n) = 6.8946\ \mathrm{W}$$

$$\mathrm{SQNR} = \frac{P_s}{\mathrm{MSQE}} = 1.1496 \times 10^4 = 40.6\ \mathrm{dB}$$

Exercise: Compare the above SQNR with that which would be obtainable using A-law by solving the above problem for the case of an A-law quantiser with $A = 87.6$, given that the *A-law expander* is

$$y = \begin{cases} \frac{1}{A}\exp[(1+\ln A)|x| - 1]\mathrm{sgn}(x), & |x| \ge \frac{1}{1+\ln A} \\ \frac{1+\ln A}{A}|x|\mathrm{sgn}(x), & |x| \le \frac{1}{1+\ln A} \end{cases} \tag{5.37}$$

5.5 PCM

ITU-T [1] defines pulse code modulation (PCM) as

> *a process in which a signal is sampled, and each sample is quantised independently of other samples and converted by encoding into a digital signal.*

The recommended nominal sampling rate is 8000 samples per second (i.e. 8 kHz) with a tolerance of 50 parts per million (ppm), each sample being then quantised using eight bits, which yields a digital signal at a nominal bit rate of 64 kb/s. We illustrated these steps of the PCM process in Fig. 1.6 and have discussed in detail the steps of filtering and sampling (Chapter 4) and quantisation (Sections 5.3 and 5.4). The stage is now set for a discussion of the final step of encoding based on standardised log-quantisers, which leads to digital signals referred to as A-law PCM and μ-law PCM. It is worth noting here again that this description of a base band digital signal as PCM is a regrettable historical misnomer for reasons given in Section 4.7 in the context of the so-called *pulse amplitude modulation*. A more appropriate terminology would be to use ADC in place of PCM so that we have A-law ADC, μ-law ADC, linear ADC (instead of linear PCM), differential ADC (instead of DPCM) and so on. However, current literature seems wedded to PCM without much hope of divorce, so the terminology will be sparingly used in this book for the sole purpose of maintaining continuity with the large body of literature on the subject. But it must be made absolutely clear that PCM is simply a source coding technique by which an analogue signal is presented in digital format as a bit stream without loss of any significant information. Contrary to what the term PCM misleadingly suggests, modulation is not at all involved in this source coding process but takes place elsewhere further down the signal processing chain of a communication system, as Fig. 1.13 makes clear.

The log-quantisation process discussed in the previous section is based on a continuous variation of quantisation step size Δx with input x, from a maximum Δ_{max} at $x = 1$ (normalised) to a minimum Δ_{min} at $x = 0$, each quantisation interval having a unique step size. Practical implementations however use a piecewise linear approximation of the A-law and μ-law compression curves. In what follows we introduce this approximation, discuss the encoding rule (i.e. procedure for assigning a unique set of 8 bits to each of the 256 quantised levels) and evaluate the SQNR of the digital signal.

5.5.1 A-law and μ-law PCM

The quantisation that produces A-law PCM is specified through piecewise linear approximation of the A-law characteristic of Eq. (5.24). This approximation is shown in Fig. 5.7 where the dashed-curve is Eq. (5.24) and the solid lines labelled s_0 to s_7 are the approximations used. Only the positive range of the characteristic is shown. The negative range is similar except for a negative sign, as earlier discussed. Note that the input is normalised to $\pm C$, where $C = 4096$ in order to render all step sizes as integers. The output is divided into 256 equal intervals, 128 of which are in the positive range as shown numbered along the y-axis of Fig. 5.7, with 16 intervals per segment s_0 to s_7. The piecewise linear approximation for A-law PCM actually comprises 13 line segments, namely one line passing through the origin made up of s_0 and s_1 which are collinear, and six distinct lines s_2 to s_7 above and below this one.

The linear approximation means that step sizes are held constant within each segment, and only change from segment to segment, rather than from interval to interval. The maximum step size Δ_{max} is used within segment s_7 and is obtained by dividing the range of s_7 (2048, 4096) by the number of intervals (=16) into which s_7 is divided. Thus

$$\Delta_{max} = \frac{4096 - 2048}{16} = 128$$

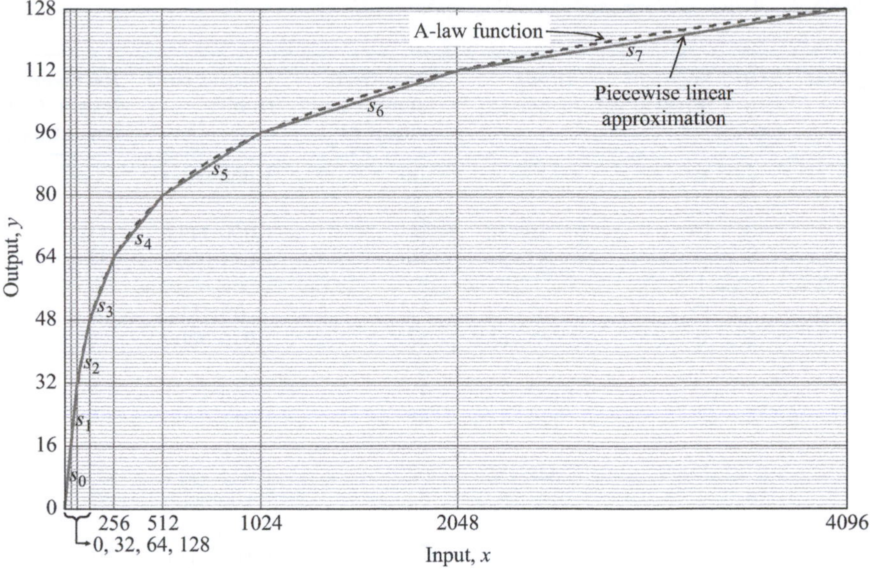

Fig. 5.7 Piecewise linear approximation of A-law compression function

The next step size Δ_6 used in segment s_6 is similarly obtained by dividing the range of s_6 (1024, 2048) by 16:

$$\Delta_6 = \frac{2048 - 1024}{16} = 64$$

and so on until the minimum step size Δ_{min} is reached in segments s_0 and s_1, given by

$$\Delta_{min} = \frac{64 - 32}{16} = \frac{32 - 0}{16} = 2$$

Observe that if the entire range (–4096, 4096) were uniformly quantised using step size Δ_{max}, the number of uniform intervals would be

$$N = \frac{2 \times 4096}{128} = 64$$

which requires $k = \log_2(64) = 6$ bits per sample. But if Δ_{min} were used over the entire input range, the number of uniform intervals would be

$$N = \frac{2 \times 4096}{2} = 4096$$

which requires $k = \log_2(4096) = 12$ bits per sample. Thus in A-law PCM the largest input samples are coarsely quantised at a resolution of six bits/sample, whereas the smallest input samples are finely quantised at a resolution of 12 bits/sample. This means that, compared to eight-bit linear ADC, A-law PCM delivers a four-bit improvement at the bottom end, which, since each extra bit yields a 6 dB increase in SQNR, corresponds to a *companding gain* of 24 dB as discussed in Section 5.4.5. Notice however that this improvement is achieved at the expense of a two-bit shortage for the top end, which corresponds to a *companding penalty* of 12 dB as also earlier discussed.

Table 5.2 A-law coding and decoding specifications

Segment (s)	Step size (Δ)	Segment interval (v)	Input range (X)	Quantised (compressor) output (Y_q)	Output code (8 bit)	Receiver (expander) output (X_q)
0	2	0	0–2	0	1 000 0000	1
		↓	↓	↓	↓	↓
		15	30–32	15	1 000 1111	31
1	2	0	32–34	16	1 001 0000	33
		↓	↓	↓	↓	↓
		15	62–64	31	1 001 1111	63
2	4	0	64–68	32	1 010 0000	66
		↓	↓	↓	↓	↓
		15	124–128	47	1 010 1111	126
3	8	0	128–136	48	1 011 0000	132
		↓	↓	↓	↓	↓
		15	248–256	63	1 011 1111	252
4	16	0	256–272	64	1 100 0000	264
		↓	↓	↓	↓	↓
		15	496–512	79	1 100 1111	504
5	32	0	512–544	80	1 101 0000	528
		↓	↓	↓	↓	↓
		15	992–1024	95	1 101 1111	1008
6	64	0	1024–1088	96	1 110 0000	1056
		↓	↓	↓	↓	↓
		15	1984–2048	111	1 110 1111	2016
7	128	0	2048–2176	112	1 111 0000	2112
		↓	↓	↓	↓	↓
		15	3968–4096	127	1 111 1111	4032

Table 5.2 provides a detailed specification of A-law coding and decoding. The segments s are numbered 0 to 7 in the first column, and have step sizes Δ given in column 2. Each segment is divided into 16 intervals v numbered from 0 to 15 in column 3, with the input range X of each of these intervals given in column 4. All X values in one interval are mapped to a single (quantised) output Y_q given in column 5. For example, the first interval of segment $s=5$ starts at $X=512$ and ends at $X=544$ and produces a single output $Y_q=80$; the second interval of segment $s=5$ starts at $X=544$ and ends at $X=576$ and produces a single output $Y_q=81$; and so on as shown in Fig. 5.8, the size of each interval being the step size of the segment. Column 6 gives the encoded output, which is an eight-bit binary number $b_7b_6b_5b_4b_3b_2b_1b_0$ assigned as follows:

- The most significant bit (MSB) b_7 is set to 1 to indicate a positive input sample and to 0 for a negative input sample.
- The next 3 bits $b_6b_5b_4$ is the binary number equivalent of the segment s (=0 to 7) within which the input sample X falls.
- The last 4 bits $b_3b_2b_1b_0$ is the binary number equivalent of the interval v (=0 to 15) of segment s in which the input sample X falls.

For example, from Fig. 5.8, a positive input $X=789$ lies in interval $v=8$ of segment $s=5$ and so is encoded as 11011000. Finally, at the receiver the PCM codewords of column 6 are converted into the quantised outputs of column 7, each of which is simply the middle of the corresponding input range X in column 4. This final step at the receiver corresponds to passing Y_q (column 5) through an expander to obtain X_q (column 7), as earlier illustrated in Fig. 5.6.

μ-Law PCM is based on quantisation derived from a 15 segment approximation of the μ-law characteristic of Eq. (5.25) as shown in Fig. 5.9. Note that there are eight segments s_0 to s_7 in the positive range and the same number in the negative range (not shown), but s_0 straddles the origin and continues as a single line

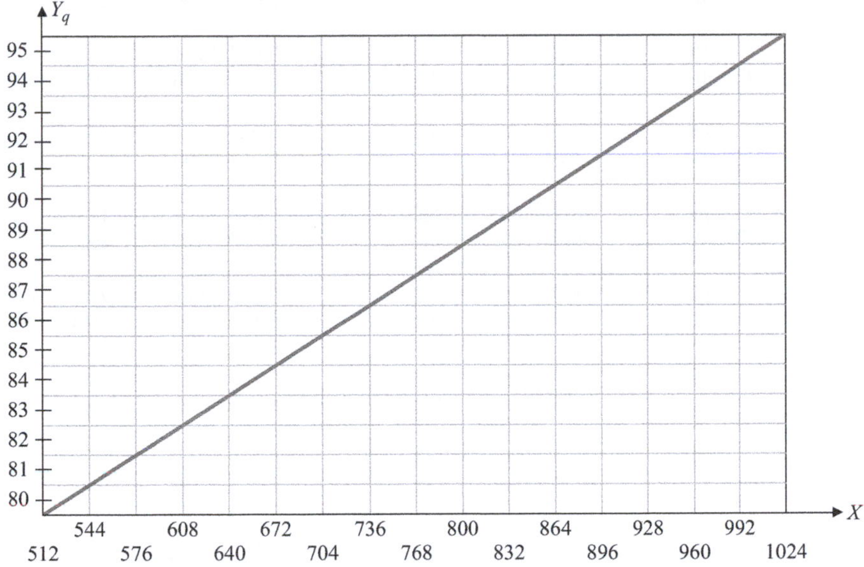

Fig. 5.8 Detailed view of A-law segment s = 5

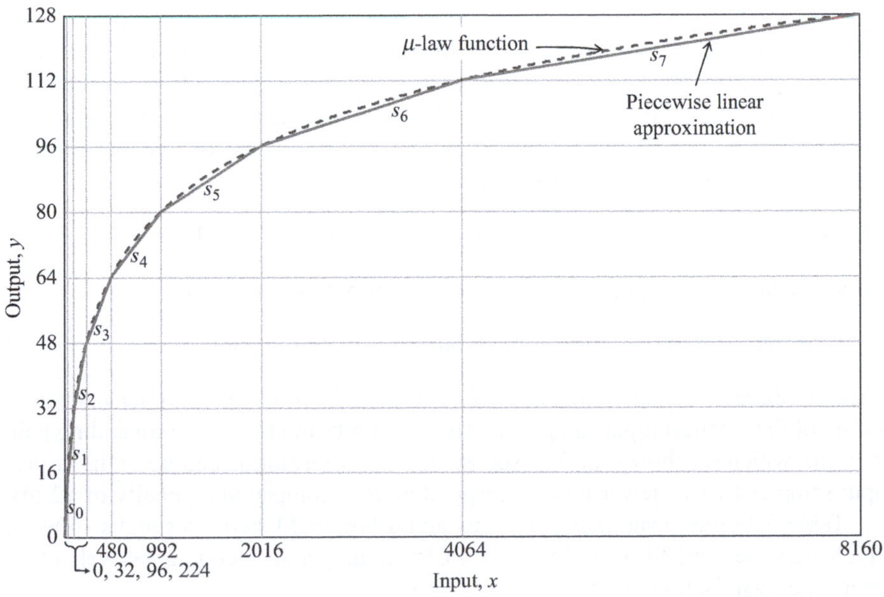

Fig. 5.9 Piecewise linear approximation of μ-law compression function

Table 5.3 μ-Law coding and decoding specifications

Segment (s)	Step size (Δ)	Segment interval (v)	Input range (X)	Quantised (compressor) output (Y_q)	Output code (8 bit)	Receiver (expander) output (X_q)
0	2	0	0–2	0	1 000 0000	1
		↓	↓	↓	↓	↓
		15	30–32	15	1 000 1111	31
1	4	0	32–36	16	1 001 0000	34
		↓	↓	↓	↓	↓
		15	92–96	31	1 001 1111	94
2	8	0	96–104	32	1 010 0000	100
		↓	↓	↓	↓	↓
		15	216–224	47	1 010 1111	220
3	16	0	224–240	48	1 011 0000	232
		↓	↓	↓	↓	↓
		15	464–480	63	1 011 1111	472
4	32	0	480–512	64	1 100 0000	496
		↓	↓	↓	↓	↓
		15	960–992	79	1 100 1111	976
5	64	0	992–1056	80	1 101 0000	1024
		↓	↓	↓	↓	↓
		15	1952–2016	95	1 101 1111	1984
6	128	0	2016–2144	96	1 110 0000	2080
		↓	↓	↓	↓	↓
		15	3936–4064	111	1 110 1111	4000
7	256	0	4064–4320	112	1 111 0000	4192
		↓	↓	↓	↓	↓
		15	7904–8160	127	1 111 1111	8032

into the negative range, giving a total of 15 distinct segments. To render all step sizes as integers, the input is normalised to $\pm C$, where $C = 8160$. Table 5.3 gives detailed specification of μ-law coding and decoding, and is interpreted in the same way as Table 5.2. We see that μ-law PCM uses step sizes that start at a minimum value $\Delta_{\text{min}} = 2$ in the lowest segment s_0 and progressively double in each subsequent segment to a maximum $\Delta_{\text{max}} = 256$ in the top segment s_7. This means — see explanation under A-law — that μ-law PCM is equivalent to a coarse uniform quantisation of the largest input samples using

$$k = \log_2(2C/\Delta_{\text{max}}) = \log_2(2 \times 8160/256) = 6 \text{ bits/sample}$$

and a fine uniform quantisation of the smallest input samples using

$$k = \log_2(2C/\Delta_{\text{min}}) = \log_2(2 \times 8160/2) = 13 \text{ bits/sample}$$

Thus, compared to eight-bit linear ADC, μ-law PCM provides a higher resolution coding of the smallest input samples by five extra bits and hence a companding gain of 30 dB, which is achieved at the expense of a lower resolution coding of the largest input samples by two fewer bits, corresponding to a companding penalty of 12 dB.

Table 5.4 gives a summary of A-law and μ-law PCM step sizes and segments, which may be applied as follows for PCM coding and decoding as well as to determine quantisation error.

Given samples $x(n)$ of an analogue input signal of peak absolute value V_p which fully loads the quantiser, the first step before using Table 5.4 is to normalise

Table 5.4 Summary of A-law and μ-law segments and step sizes

Segment (*s*)	**A-law**			**μ-Law**		
	Step size (Δ)	**Input range**		**Step size (Δ)**	**Input range**	
		X_{min}	X_{max}		X_{min}	X_{max}
0	2	0	32	2	0	32
1	2	32	64	4	32	96
2	4	64	128	8	96	224
3	8	128	256	16	224	480
4	16	256	512	32	480	992
5	32	512	1024	64	992	2016
6	64	1024	2048	128	2016	4064
7	128	2048	4096	256	4064	8160

these inputs to the range $\pm C$ of the table, where $C = 4096$ for A-law and $C = 8160$ for μ-law. That is, scale $x(n)$ to $X(n)$ by multiplying $|x(n)|$ by a scale factor F, where

$$X(n) = |x(n)|F$$
$$F = \begin{cases} 4096/V_p, & \text{A-law} \\ 8160/V_p, & \mu\text{-law} \end{cases} \tag{5.38}$$

To determine the PCM code to which the sample $x(n)$ is converted:

- Look at Table 5.4 (using the LHS of the table for A-law and the RHS for μ-law) and read the segment s in which $X(n)$ calculated above in Eq. (5.38) lies. For example, for A-law, $X(n) = 600$ is in segment $s = 6$, which has step size $\Delta = 32$ and lower limit $X_{min} = 512$, and for μ-law $X(n) = 600$ is in segment $s = 4$, which has lower limit $X_{min} = 480$.
- Next, determine interval v (numbered 0 to 15) of segment s in which $X(n)$ lies:

$$v = \left\lfloor \frac{X(n) - X_{\min}}{\Delta} \right\rfloor \tag{5.39}$$

where $\lfloor z \rfloor$ denotes the integer part of the number z, Δ is the step size and X_{min} the lower limit of segment s. In the above A-law example with $X(n) = 600$, $s = 5$, $\Delta = 32$ and $X_{min} = 512$, we obtain $v = \lfloor 2.75 \rfloor = 2$.
- Finally, the 8-bit PCM codeword for the sample $x(n)$ is obtained by setting the MSB to signify the sign of $x(n)$ as previously discussed, and setting the next three bits to the binary equivalent of s and the last four bits to the binary equivalent of v (using leading zeroes as necessary in both cases). Thus in the above example, the A-law PCM codeword is 1 101 0010.

The sample value $x_q(n)$ to which a PCM codeword $b_7b_6b_5b_4b_3b_2b_1b_0$ is decoded is obtained through the following steps:

$$\begin{aligned} s &= 4b_6 + 2b_5 + b_4 \\ v &= 8b_3 + 4b_2 + 2b_1 + b_0 \\ X_q(n) &= X_{\min} + (v + \tfrac{1}{2})\Delta \\ x_q(n) &= \frac{X_q(n)}{F}(-1)^{b_7+1} \end{aligned} \tag{5.40}$$

where $X_{\min}$ and Δ are respectively the lower limit and step size (read from Table 5.4) of the segment s determined in the first line, F is the scale factor defined in Eq. (5.38) and the last line sets the sign of $x_q(n)$ based on the MSB b_7 and de-normalises to correctly place $x_q(n)$ within the analogue signal range $\pm V_p$. An alternative way to determine the decoded sample $x_q(n)$ from s and v without having to read Table 5.4 is provided by the following formula:

$$x_q(n) = \frac{(-1)^{b_7+1}}{F} \times \begin{cases} 2^{s-1}(2v+1), & \text{A-law}, \quad s=0 \\ 2^{s-1}(2v+33), & \text{A-law}, \quad s=1,2,3,\ldots,7 \\ 2^{s}(2v+33)-32, & \mu\text{-law} \end{cases} \tag{5.41}$$

For example, assuming $V_p = 1$, then $F = 4096$ for A-law and $F = 8160$ for μ-law. Using Eqs. (5.40) and (5.41), an A-law PCM codeword 0 001 1011 would be decoded to the value $x_q(n)$ as follows:

$$\begin{aligned} s &= 4\times 0 + 2\times 0 + 1 = 1 \\ v &= 8\times 1 + 4\times 0 + 2\times 1 + 1 = 11 \\ x_q(n) &= \frac{(-1)^{0+1}}{4096} 2^{1-1}(2\times 11 + 33) = -\frac{55}{4096} = -0.0134 \end{aligned}$$

And a μ-law codeword 1 110 0101 is decoded to the value $x_q(n)$ as follows:

$$\begin{aligned} s &= 4\times 1 + 2\times 1 + 0 = 6 \\ v &= 8\times 0 + 4\times 1 + 2\times 0 + 1 = 5 \\ x_q(n) &= \frac{(-1)^{1+1}}{8160}\left[2^6(2\times 5 + 33) - 32\right] = \frac{64\times 43 - 32}{8160} = 0.3333 \end{aligned}$$

Finally, given N_s samples of an analogue signal $x(n)$, $n = 0, 1, \ldots, N_s - 1$, the quantised values $x_q(n)$ are obtained as discussed above, and hence the SQNR of the PCM signal is estimated as

$$\text{SQNR} = 10\log_{10}\left(\frac{\sum_{n=0}^{N_s-1}\left[x(n) - x_q(n)\right]^2}{\sum_{n=0}^{N_s-1} x^2(n)}\right) \text{dB} \tag{5.42}$$

5.5.2 *SQNR of A-law and μ-law PCM*

We know from Eq. (5.19) that the SQNR of an ideal log-PCM (i.e. one that uses an entirely logarithmic compressor curve) is constant across the entire input range. For an 8-bit quantiser (for which $N = 2^8 = 256$) this constant SQNR is

$$\text{SQNR} = 10\log_{10}\left(\frac{3N^2}{K^2}\right) = \begin{cases} 38.2\ \text{dB}, & \text{for } K = 1 + \ln(A), \quad A = 87.6 \\ 38.1\ \text{dB}, & \text{for } K = \ln(1+\mu), \quad \mu = 255 \end{cases}$$

However, in implementing A-law and μ-law PCM, piecewise linear approximation of the logarithmic curve was employed. It is of interest to assess the impact of this approximation on SQNR. Is the SQNR of A-law and μ-law PCM still constant at ~38 dB across the entire input range? We will carry out this evaluation for an input signal of peak value V_p that has uniform distribution, which means that its sample is equally likely to be located anywhere in the interval ($-V_p$, V_p) and that its peak-to-rms ratio is $R = \sqrt{3}$. The normalised power of this signal is

$$P_s = V_{\text{rms}}^2 = (V_p/R)^2 = V_p^2/3 \tag{5.43}$$

Quantisation noise power is given by MSQE in Eq. (5.6), but here the step size varies from segment to segment, so we obtain MSQE by summing the product of $\Delta_j^2/12$ and the probability P_j that the sample of the input signal lies in the jth segment pair (i.e. including both positive and negative ranges). If $X_{s\,\min} \leq V_p \leq X_{s\,\max}$, where $X_{s\,\min}$ and $X_{s\,\max}$ are the lower and upper limits of segment s (i.e. the peak value of the input signal lies in segment s) then $P_j = 16\Delta_j/V_p$ for segments $j = 0, 1, \ldots, s-1$, and $P_j = (V_p - X_{s\,\min})/V_p$ for $j = s$, so that

$$\begin{aligned}\text{MSQE} &= \frac{1}{12}\sum_{j=0}^{s-1}\Delta_j^2\left(\frac{16\Delta_j}{V_p}\right) + \frac{\Delta_s^2}{12}\left(\frac{V_p - X_{s\,\min}}{V_p}\right)\\ &= \frac{4}{3V_p}\sum_{j=0}^{s-1}\Delta_j^3 + \frac{\Delta_s^2}{12}\left(\frac{V_p - X_{s\,\min}}{V_p}\right)\end{aligned} \tag{5.44}$$

SQNR is the ratio between Eqs. (5.43) and (5.44) expressed in dB:

$$\text{SQNR} = 30\log_{10}(V_p) - 10\log_{10}\left[\frac{\Delta_s^2}{4}(V_p - X_{s\,\min}) + 4\sum_{j=0}^{s-1}\Delta_j^3\right] \text{dB} \tag{5.45}$$

This is an important result that applies to both A-law and μ-law PCM and to any quantiser range $\pm C$. Note however that Tables 5.2–5.4 and Figs. 5.7–5.9 are based on $C = 4096$ for A-law and $C = 8160$ for μ-law. If a different quantiser range is to be used, say $\pm D$, then all step sizes and signal ranges in these figures and tables must be reduced by the factor C/D in order to maintain compressor shape. Let us take two examples to explain how Eq. (5.45) is applied.

Consider A-law PCM with input signal having 0 dB peak relative to quantiser input limit C. That is, the signal fully loads the quantiser and $V_p = C = 4096$. Thus, V_p lies in segment $s = 7$ with step size $\Delta_s = 128$ and lower limit $\Delta_{s\,\min} = 2048$. The SQNR of the PCM representation of this signal is obtained from Eq. (5.45) as

$$\begin{aligned}\text{SQNR} &= 30\log(4096)\\ &\quad -10\log\left[\frac{128^2(4096-2048)}{4} + 4(2^2+2^2+4^2+8^2+16^2+32^2+64^2)\right]\\ &= 108.37 - 69.82 = 38.55\text{ dB}\end{aligned}$$

Next, consider a smaller input signal having peak value 15 dB below quantiser input limit. We apply Eq. (5.45) with V_p 15 dB below 4096, which means

$$V_p = 4096/10^{15/20} = 728.38$$

From Table 5.4 (A-law) we see that V_p lies in segment $s = 5$ with step size $\Delta_s = 32$ and lower limit $X_{s\,\min} = 512$. Substituting into Eq. (5.45) yields

$$\begin{aligned}\text{SQNR} &= 30\log(728.38) - 10\log\left[\frac{32^2(728.38-512)}{4} + 4(2^2+2^2+4^2+8^2+16^2)\right]\\ &= 85.87 - 48.70 = 37.17\text{ dB}\end{aligned}$$

We can therefore see the effect of the piecewise linear approximation on the SQNR of A-law PCM, which slightly exceeds 38 dB at the top end of input signals and is a little lower for smaller input signals. Fig. 5.10 shows a plot of SQNR (dB) of A-law and μ-law PCM versus peak input level V_p in dB relative to quantiser limit C, calculated as discussed above using Eq. (5.45). A peak input level of 0 dB corresponds to an input signal that fully loads the quantiser so that its peak $V_p = C$, whereas a peak input level of −60 dB corresponds to a very small input signal having peak value $V_p = C/1000$. The SQNR of k-bit linear ADC for $k = 8$, 12, 13, calculated using Eq. (5.10), is also shown for comparison. Fig. 5.10 confirms some of the observations made earlier. For example:

- The performance, in terms of SQNR, of (eight bit) A-law PCM coincides with that of a 12-bit linear ADC for small inputs ($V_p \leq -36$ dB).
- The SQNR of μ-law PCM equals that of a 13-bit linear ADC for small inputs ($V_p \leq -48$ dB).
- The vertical gap between the SQNR curves for eight-bit linear ADC and A-law PCM in the small input region ($V_p \leq -36$ dB) is 24 dB. This is the A-law companding gain.
- The vertical gap between the SQNR curves for eight-bit linear ADC and μ-law PCM in the small input region ($V_p \leq -48$ dB) is 30 dB. To obtain this you will need to extend the SQNR curve for μ-law in the small input region (which coincides with the linear graph for 13-bit ADC) backwards until you reach $V_p = -38$ dB where the eight-bit linear graph stops. The vertical gap between the two graphs at this point is 30 dB. This is the μ-law companding gain.
- The SQNR of A-law PCM remains within the range (37, 38.6 dB) over a 30 dB range of peak input levels and degrades by only ~2 dB at peak input $V_p = -36$ dB. The peaks in the slight fluctuation in SQNR in this region coincide with segment boundaries. In comparison, the SQNR of eight-bit linear ADC is an unacceptable 12 dB at $V_p = -36$ dB. We see that due to the

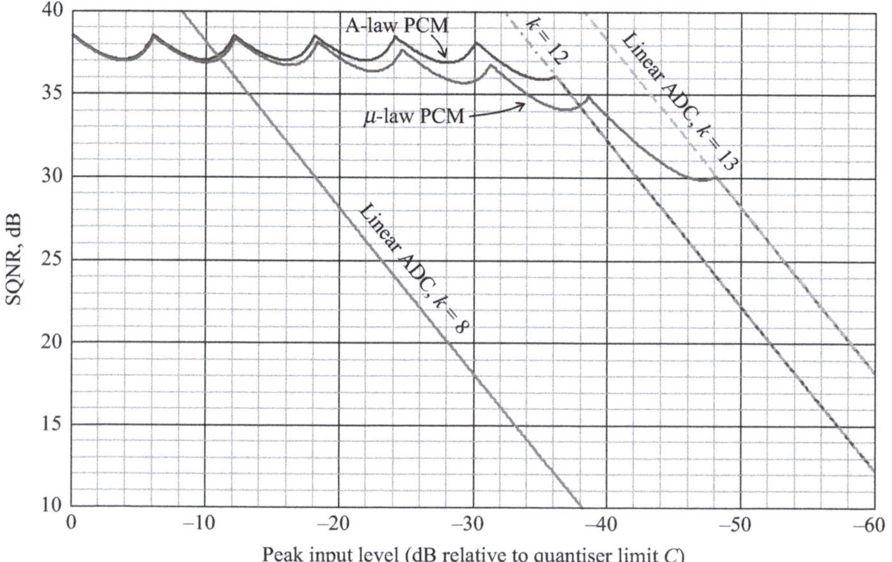

Fig. 5.10 SQNR of A-law PCM, μ-law PCM and k-bit linear ADC (for k = 8, 12, 13)

piecewise linear approximation employed we have not quite kept SQNR exactly constant across the entire range of input levels, but we have succeeded in keeping the variation small to within about ±1 dB over a 36 dB range of input levels. Note that below $V_p = -36$ dB, which is $V_p = 64$ in the ±4096 normalised range, the signal is confined entirely within segments $s = 0$ and $s = 1$ where the step size is constant at $\Delta = 2$. Thus the signal sees a linear ADC and hence SQNR decreases linearly with peak input level in this region.

- The SQNR of A-law PCM is slightly higher than that of μ-law PCM for inputs > -38 dB, but at small inputs < -48 dB μ-law PCM achieves a 6 dB better SQNR than A-law due to its use of a 13-bit/sample coding resolution for small samples compared to A-law's 12 bits/sample.

5.6 Lossy compression

Speech quality is often classified using a measure called *mean opinion score* (MOS), which is the average of the subjective judgement passed by a large group of listeners on the quality of the speech using a scale from 1 to 5, where 1 is bad, 2 is poor, 3 is fair, 4 is good and 5 is excellent. Alternatively, the scale of 1 to 5 gives a degradation category rating (DCR) that indicates the level of the perceived disturbance in the speech as very annoying (1), annoying (2), slightly annoying (3), audible but not annoying (4), and inaudible (5). An MOS of 4 and above indicates speech of *near transparent quality*, also described as *toll quality*. A score between 3.5 and 4 specifies speech of *communication quality*; a score between 3.0 and 3.5 is described as *professional quality*; and a score below 3.0 corresponds to speech of *synthetic quality*.

PCM at 64 kb/s is the toll quality benchmark for *telephone speech* – the name for an audio signal restricted to a maximum frequency component of 3400 Hz. Other audio signals include *wide band speech*, which has a maximum frequency component of 7 kHz, and *wide band audio* (also known as *high fidelity audio*) of bandwidth that covers the entire audible range of frequencies up to a maximum of 20 kHz. Using 64 kb/s PCM for telephone speech however places a high demand on transmission bandwidth, storage capacity and battery life which can be difficult to satisfy in applications such as mobile wireless communication where these resources are usually in very short supply. A wide variety of lossy compression techniques have been developed for application to audio and (both still and moving) image signals. These techniques allow telephone speech for example to be reduced from a bit rate of 64 kb/s to various standardised lower rates such as 32, 16, 4 and 2.4 kb/s. The following trade-offs (illustrated in Fig. 5.11) are involved in the lossy compression of speech in particular.

- *Quality*: As compression ratio increases, leading to lower bit rate and hence lower transmission bandwidth requirement, perceived quality generally decreases. The decline in quality however depends on the processing algorithm. In fact there are some codecs that produce compressed or lower-bit-rate speech having comparable quality to less compressed or higher-bit-rate speech.
- *Processing delay*: This is the time lag between the instants at which the speech signal enters and leaves the coder, and consists of a buffering delay τ_b plus computational delay τ_c. Most compression algorithms work at a time on blocks or segments of the speech signal of duration $\tau_b \geq 20$ ms. The first speech

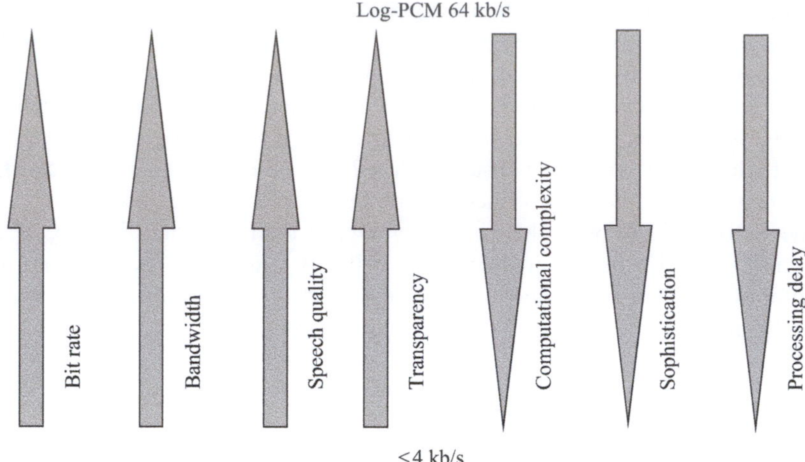

Fig. 5.11 Trade-offs in speech processing

sample in a segment will be τ_b ms old before commencement of computations of duration τ_c needed by the algorithm to produce the codeword representation of the speech segment. The maximum processing delay $\tau_b + \tau_c$ at a coder tends to increase with codec complexity and there will be a further propagation delay τ between transmitter and receiver as well as additional processing delay at the receiver before the speech segment can be delivered to the information sink at destination. For real-time or interactive communication it is vital that total delay is kept at a tolerable level and this requirement can sometimes be a critical factor in delaying the adoption of a given algorithm until advances in DSP technology make sufficiently fast codecs available.

- *Channel errors*: The effect of bit errors tends to be more drastic as bit rate decreases since information is highly compressed. For example, an error in a bit that indicates whether a segment represents voiced or unvoiced speech will have a significant impact on the interpretation and rendering of that entire segment at the decoder.
- *Transparency*: Highly speech-specific codecs might not pass non-speech signals and this lack of transparency generally increases as bit rate decreases.
- *Algorithmic complexity*: This generally increases as more processing is introduced to further increase data compression and hence reduce bit rate. The result is more expensive hardware as well as higher power consumption and hence a shorter battery life for portable communication devices.

Lossy data compression is a wide subject, a detailed treatment of which is beyond the scope of a general digital communications book such as this one. We will therefore limit our interest to the lossy compression of telephone speech and briefly highlight the main types, including waveform coders, vocoders and hybrid coders.

5.6.1 Waveform coders

Waveform coders produce a bit stream that conveys a representation of the original signal waveform. At the receiver, the decoder attempts to reproduce the original waveform from the coded bit stream, but there will always be some added noise arising from the error introduced by quantisation at the encoder. If the original signal is speech and this noise is practically inaudible, the MOS of the reconstructed speech will be between ~4.5 and 5.0 and the coder is said to produce toll quality speech. At the transmitter a waveform coder performs the following processing steps on the original signal:

- *Filtering*: The original signal is first passed through an anti-alias filter, designed as discussed in Section 4.8.
- *Sampling*: The filtered signal $g(t)$ is then sampled at a suitable rate, as discussed in Section 4.4. In this Section, we will refer to the sample of $g(t)$ taken at time $t = nT_s$ as the current sample, denoted simply as $g(n)$. Thus the pth previous sample, taken at an earlier time $t = nT_s - pT_s$ will be denoted by $g(n - p)$.
- *Quantisation*: A k-bit quantiser is employed to map each input $\varepsilon(n)$ to a quantised output $\varepsilon_q(n)$, which is selected from 2^k quantisation levels on the basis of whichever level is the closest to $\varepsilon(n)$. This introduces an inevitable quantisation error $e_q(n)$, where

$$\varepsilon_q(n) = \varepsilon(n) + e_q(n) \tag{5.46}$$

  The quantisation process was discussed in detail in Sections 5.2–5.4. The PCM discussed in Section 5.5 is a special case of waveform coders for which quantiser input $\varepsilon(n)$ is the sampled signal $g(n)$ and the quantiser is non-uniform, employing $k = 8$ bits to code each sample $g(n)$ taken at 8000 samples/s, which results in the standard 64 kb/s bit rate. In that case the linear prediction discussed next is not required. Alternatively, the quantiser might be a *differential quantiser* (introduced in Section 5.3.1) in which case quantiser input $\varepsilon(n)$ is the difference between $g(n)$ and a linear prediction $\hat{g}(n)$.
- *Linear prediction*: A weighted sum of the last p samples, from $g(n - 1)$ to $g(n - p)$, is computed as a prediction of what the current sample $g(n)$ should be. The computation is a linear combination of samples and so the process is referred to as *linear prediction*. The prediction $\hat{g}(n)$ is obtained as follows:

$$\hat{g}(n) = \sum_{j=1}^{p} w_j g_q(n-j) = \sum_{j=1}^{p} w_j\left[g(n-j) + e_q(n-j)\right] \tag{5.47}$$

  where w_j are the predictor (also called filter) coefficients or weights, $e_q(n)$ is the quantisation error introduced into the output $\varepsilon_q(n)$ of the quantiser whose input $\varepsilon(n)$ is the difference between the current sample $g(n)$ and the result of Eq. (5.47). That is

$$\text{Quantiser input, } \varepsilon(n) = g(n) - \hat{g}(n) \tag{5.48}$$

  Notice that the predictor input is

$$g_q(n) = g(n) + e_q(n) \tag{5.49}$$

  and not $g(n)$ which could cause an accumulation of quantisation errors.
- *Encoding*: The quantiser output $\varepsilon_q(n)$ is coded as a bit sequence by an encoder. A discussion of how the encoder maps each of the 2^k levels of the quantiser into a binary codeword was presented in Section 5.5.1 for A-law and μ-law PCM.

Alternatively, the quantiser can be treated as a discrete source that emits 2^k levels or symbols to which codewords can be assigned according to the source coding methods discussed in the next chapter. Whatever the encoding process employed, the resulting bit stream conveys a digital representation of the original waveform g(t) to the receiver. Since k is typically kept small to reduce coding bits and the recovered signal is never an identical copy of g(t) due to the quantisation errors $e_q(n)$ introduced, the encoder bit stream output is described as a *lossy compression* of g(t), or (if g(t) is a speech signal) as *low bit rate speech coding*.

A waveform decoder is required at the receiver which performs the following functions:

- *Decoding*: The decoder receives the coded bit stream and converts it into a sequence of quantised values $\varepsilon_q(n)$, using a procedure dictated by the encoding process at the transmitter.
- *Linear prediction*: The receiver employs an exactly identical prediction algorithm – Eq. (5.47) – as used at the transmitter to predict the current sample from the last p samples. Barring any transmission error, then the sum of the predictor output $\hat{g}(n)$ and decoder output $\varepsilon_q(n)$ gives the current sample $g(n)$ correct to within the quantisation error introduced at the transmitter. This is the case since using Eqs. (5.46), (5.48) and (5.49)

$$\begin{aligned} \hat{g}(n) + \varepsilon_q(n) &= \hat{g}(n) + \varepsilon(n) + e_q(n) \\ &= g(n) + e_q(n) \\ &= g_q(n) \end{aligned} \tag{5.50}$$

- *Reconstruction filter*: The sequence of quantised samples $g_q(n)$ is passed through a reconstruction filter that smooths it to produce the reconstructed analogue waveform $g_r(t)$. The design and operation of the reconstruction filter was discussed in Chapter 4. A subjective measure of the quality of the reconstructed signal $g_r(t)$, giving an indication of how close it is to the original waveform $g(t)$ and how free it is from quantisation noise, is provided by a mean opinion score (MOS) as earlier discussed, whereas an objective measure of the quality of $g_r(t)$ is given by the SQNR discussed in Section 5.5.2.

Fig. 5.12 shows block diagrams of the above operations at the transmitter and receiver. The linear predictor is a finite impulse response (FIR) filter of order p, a block diagram of which is shown in Fig. 5.13. Optimum values of the filter coefficients w_j, $j = 1, 2, 3, \ldots, p$ are determined to minimise the mean square value $\overline{\varepsilon^2(n)}$ of prediction error in a zero-mean stationary segment of the signal g(t) spanning N samples. Equating to zero each of the p partial derivatives of $\overline{\varepsilon^2(n)}$ taken with respect to w_j leads to the Wiener–Hopf equations for the coefficients expressed in matrix form as

$$\begin{bmatrix} R_g(0) & R_g(1) & \cdots & R_g(p-1) \\ R_g(1) & R_g(0) & \cdots & R_g(p-2) \\ \vdots & \vdots & \ddots & \vdots \\ R_g(p-1) & R_g(p-2) & \cdots & R_g(0) \end{bmatrix} \begin{bmatrix} w_1 \\ w_2 \\ \vdots \\ w_p \end{bmatrix} = \begin{bmatrix} R_g(1) \\ R_g(2) \\ \vdots \\ R_g(p) \end{bmatrix} \tag{5.51}$$

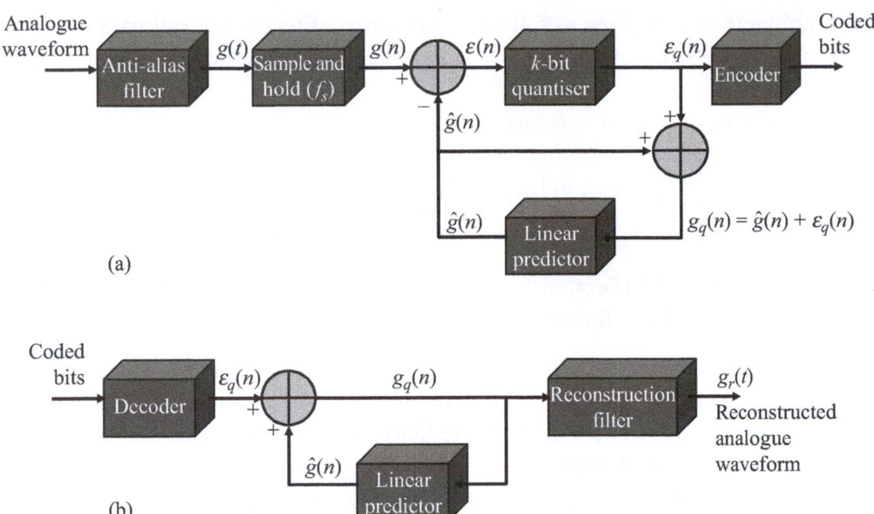

Fig. 5.12 Waveform coder: (a) encoding operation at transmitter; (b) decoding operation at receiver

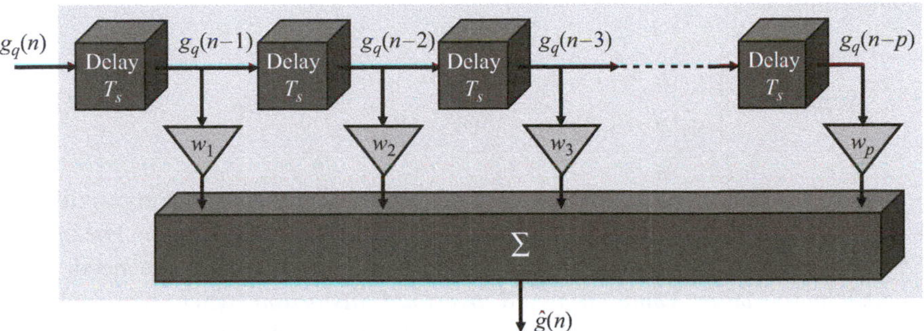

Fig. 5.13 Linear predictor realised as a finite impulse response (FIR) filter of order p

where

$$R_g(m) = \frac{1}{N}\sum_{n=0}^{N-1-m} g_q(n)g_q(n+m) \tag{5.52}$$

is the autocorrelation of $g_q(n)$ computed over the N samples of the signal segment. It gives an indication of how two samples of $g_q(n)$ (separated by m sampling intervals) are related, a value of zero meaning that the two samples are not in any way related. Although the p-by-p autocorrelation matrix of Eq. (5.51) can be inverted to yield the filter coefficients, the autocorrelation values of the signal sequence $g_q(n)$ are not always known. In such situations the *least mean square* (LMS) *algorithm* is an excellent way to iteratively compute the coefficients starting from an arbitrary set of values: The estimate $\hat{w}_{j,n+1}$ of the jth filter coefficient at the next sampling instant is given as follows [2] in terms of its current estimate $\hat{w}_{j,n}$, the jth past quantised signal sample $g_q(n-j)$ and the current prediction error

$$\begin{aligned}\hat{w}_{j,n+1} &= \hat{w}_{j,n} + \mu g_q(n-j)\varepsilon(n) \\ &= \hat{w}_{j,n} + \mu g_q(n-j)\left[g(n) - \sum_{j=1}^{p}\hat{w}_{j,n}g_q(n-j)\right]\end{aligned} \tag{5.53}$$

In addition to PCM (Section 5.5), other special cases and standards of waveform coders include the following:

- *Delta modulation (DM)*: When in Figs. 5.12 and 5.13, $k=1, p=1$ and $w_1=1$, we have the special case of delta modulation (DM). The prediction is simply the previous sample and the linear predictor simplifies to a one-sample delay. In order to keep the difference $\varepsilon(n)$ between adjacent samples small, sampling rate f_s is chosen higher than required by the sampling theorem. Values of $f_s = 16$ kHz and 32 kHz have been used for telephone speech. SQNR increases with sampling frequency by 9 dB per octave [3]. The quantiser output is Δ when the current sample $g(n)$ exceeds the previous sample $g(n-1)$ and $-\Delta$ otherwise, which the encoder codes using binary 1 for Δ and binary 0 for $-\Delta$. At the receiver the quantised sequence $g_q(n)$ is recovered using the following algorithm:

$$g_q(n) = \begin{cases} g_q(n-1)+\Delta, & b(n)=1 \\ g_q(n-1)-\Delta, & b(n)=0 \end{cases} \tag{5.54}$$

  where $b(n)$ denotes the current bit, and Δ is a fixed step size that approximates the amount by which the signal changed from the previous sample. The choice of this step size is a compromise between on the one hand the need for a sufficiently large value to avoid *overload distortion* when the analogue waveform $g(t)$ changes so rapidly that adding or subtracting a step size is insufficient to make the current sample catch up with the previous, and on the other hand the need for a sufficiently small value to minimise *granular noise* during a passage when $g(t)$ changes very slowly so that the reconstructed signal hunts about the slowly changing level of $g(t)$. One solution would be to use a small fixed step size to minimise granular noise and combine that with a very high sampling frequency to minimise overload distortion since this ensures that samples are taken before $g(t)$ has had the time to change by much. This ensures that the DM system retains one of its great features, namely circuit simplicity, but it increases bandwidth requirement towards that of standard PCM. An alternative solution, which increases circuit complexity but keeps bit rate at 32 kb/s or lower (thereby saving bandwidth when compared to PCM) is to adopt a variable step size in what is then referred to as *adaptive delta modulation* (ADM). The step size is increased when necessary to avoid overload distortion and reduced when necessary to minimise granular noise. Encoder and decoder both start with a small step size having an agreed value and then follow the same algorithm for selecting the step size $\Delta(n)$ of the current interval. A simple algorithm might be the following which increases the step size by a factor of λ (>1) when the current bit is the same as the previous and reduces the step size by the same factor if the two bits are different:

$$\Delta(n) = \begin{cases} \Delta(n-1)\lambda, & b(n) \oplus b(n-1) = 0 \\ \Delta(n-1)/\lambda, & \text{Otherwise} \end{cases} \tag{5.55}$$

Further discussion of delta modulation can be found in [3].

- *Adaptive differential PCM (ADPCM)*: If in Fig. 5.12 sampling rate $f_s = 8$ kHz then the use of $k = 2$, 3, 4 and 5 bits/sample at the quantiser leads to coding of the analogue signal at respective bit rates of 16, 24, 32 and 40 kb/s. The ITU-T G.726 standard [4] is the most widely used ADPCM implementation. It uses a combination of one second order and one sixth order linear predictor whose coefficients are constantly updated based on prediction error and reconstructed signal. Adaptive non-uniform quantisation is employed in which the spacing of quantiser levels is adaptively scaled as necessary to maintain SQNR at approximately the same level for different signals. The speech quality of 32 kb/s ADPCM matches that of 64 kb/s PCM. A detailed specification of ADPCM can be found in [4].

Worked Example 5.6: SQNR of Delta Modulation

We wish to determine the SQNR of a delta modulation (DM) signal that conveys telephone speech at 16 kb/s.

Since bit rate $R_b = kf_s$, and $k = 1$ in a DM coder, it follows that sampling frequency $f_s = R_b = 16$ kb/s. The maximum frequency component of telephone speech is $f_m = 3.4$ kHz. To minimise granular noise while avoiding overload distortion, we must choose step size $\Delta_{\min}$ equal to the maximum change in voltage level in one sampling interval ($T_s = 1/f_s$) experienced by the highest frequency component of telephone speech. In the worst case scenario this frequency component also has the maximum amplitude A_m, and its sinusoidal expression is $v_m(t) = A_m \sin(2\pi f_m t)$. Thus,

$$\begin{aligned} \Delta_{\min} &= T_s \frac{dv_m(t)}{dt}\Big|_{\max} \\ &= T_s 2\pi f_m A_m \cos(2\pi f_m t)\big|_{\max} \\ &= 2\pi f_m A_m / f_s \end{aligned} \tag{5.56}$$

In DM, a positive difference signal ε is quantised to step size Δ and a negative difference signal $-\varepsilon$ is quantised to $-\Delta$, which gives quantisation error $|\varepsilon - \Delta|$ and hence (assuming the difference ε is uniformly distributed in the range $-\Delta$ to $+\Delta$) a mean square quantisation error (MSQE) given by

$$\begin{aligned} \text{MSQE} &= \frac{1}{\Delta} \int_{\varepsilon=0}^{\Delta} (\varepsilon - \Delta)^2 d\varepsilon \\ &= \Delta^2/3 \end{aligned} \tag{5.57}$$

Using this expression and the step size specified in Eq. (5.56), and making the reasonable assumption that this quantisation noise power is spread equally over the frequency range $f=0$ to f_s so that at the receiver only a fraction B/f_s of it will pass through the reconstruction filter whose bandwidth $B=f_m$, we obtain

$$\begin{aligned}
\text{SQNR} &= \frac{\text{Signal power}}{\text{Effective quantisation noise power}} \\
&= \frac{A_m^2/2}{(f_m/f_s)(\Delta^2/3)} = \frac{A_m^2/2}{(f_m/f_s)(2\pi f_m A_m/f_s)^2/3} \\
&= \frac{3}{8\pi^2}(f_s/f_m)^3 \\
&= 30\log_{10}\left(\frac{f_s}{f_m}\right) - 14.2 \text{ (dB)} \qquad (5.58)
\end{aligned}$$

Eq. (5.58) is an important result for the SQNR of DM systems. It is remarkable that (unlike PCM) the SQNR of DM does not depend on input signal amplitude. It depends only on sampling frequency f_s (since we really don't have any control over f_m which is fixed), improving by 9 dB per octave increase in f_s.

Substituting $f_s=16000$ and $f_m=3400$ yields SQNR = 5.98 dB, which is quite poor. To improve SQNR we could increase f_s to gain a 9 dB improvement each time fs is doubled. Thus a 32 kb/s DM has SQNR = 15 dB and a 64 kb/s DM has SQNR = 24 dB and so on. Alternatively, we note that the reason for poor SQNR is because we chose a large step size (Eq. (5.56)) which was the minimum required to avoid overload distortion. To improve SQNR without increasing sampling frequency (and hence bit rate) we must resort to ADM to realise a 6 dB improvement in SQNR each time step size Δ is halved. Thus, if we could design a 16 kb/s ADM system that operates mostly with a step size that is a factor of 16 smaller than what is stipulated in Eq. (5.56) it would deliver speech quality at SQNR $\sim$30 dB which is quite good.

5.6.2 Vocoder

Unlike waveform coders, a vocoder (short for vocal tract coder) does not code the speech waveform, but it codes the mechanism that produces the speech signal to enable reproduction of a synthetic quality of the speech at the receiver. This mechanism involves the flow of air (that has been forced from the lungs by muscular action) being modified in the vocal tract to produce different sounds. The vocal tract can be modelled as a tube of non-uniform cross-section. Vowel sounds are produced by resonance of an open vocal tract under various shapes, whereas consonants are produced by turbulence generated in the vocal tract at various points of constriction. A vocoder explicitly models the vocal tract as a filter, partitions the sequence of speech samples $g(n)$ into short segments and extracts (and codes) from each segment the parameters necessary to emulate the air input into the vocal tract and the modifications imposed by the vocal tract to convert that input into output sound for that segment. At the receiver a close version of the original speech signal in each segment is reproduced by taking an appropriate input signal (e.g. white noise) and passing it through a filter (designed as dictated by the coded vocal tract

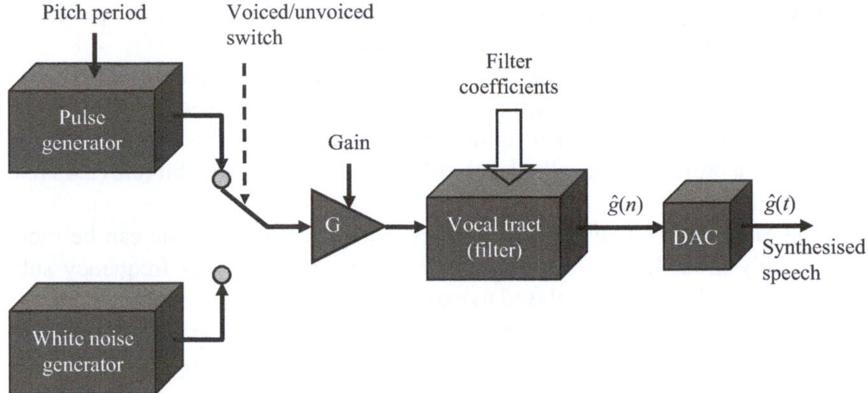

Fig. 5.14 Synthetic speech production in a linear predictive coding (LPC) vocoder

parameters) to give an output that mimics the sound in that segment. Fig. 5.14 shows a block diagram of this process for the simple case of linear predictive coding (LPC) of the vocal tract.

The main source of degradation in reproduced speech quality in vocoders is not quantisation error (although there is a little bit of that) but the use of a model that does not accurately represent the speech production process. For example, each segment of speech (of duration say 20 ms) may be classified either as voiced or unvoiced and reproduced at the receiver by exciting the filter with a pulse train or white noise respectively. In practice each speech frame will be partially voiced and unvoiced, so this binary classification into voiced or unvoiced gives a synthetic quality (of MOS ~2.2) to vocoder speech. Vocoders do however give good intelligibility at very low bit rates, and therefore are widely used in applications that require intelligibility without a serious need for the recognition of speaker identity or emotion. They however involve complex processing and introduce delays >20 ms.

One simple vocoder example is the LPC-10 developed by the US Department of Defence in 1984 and adopted as a federal standard (FS-1015). The vocal tract is modelled using a linear predictor and the speech signal is partitioned into 22.5 ms segments or frames from which the following parameters are extracted: (a) Voiced/unvoiced flag (one bit); (b) gain factor to model the energy of the frame (five bits); (c) pitch period between 51.3 and 400 Hz (six bits); (d) linear predictor coefficients w_1, w_2, w_3 and w_4 quantised using five bits per coefficient (20 bits), and a further 21 bits devoted to error protection if the frame is classified as unvoiced, or, if the frame is voiced, to code a further six coefficients, namely w_5, w_6, w_7, w_8 (four bits each) and w_9 (three bits) and w_{10} (two bits). This sums to 53 bits per frame to which one bit, alternating between 0 and 1, is appended for synchronisation at the receiver, giving a total of 54 bits per 22.5 ms, or a bit rate of 2.4 kb/s.

Improvements to the LPC-10 vocoder approach include

- Using *vector quantisation* (VQ), rather than scalar quantisation, for the filter coefficients: VQ entails coding each set of coefficients by an index to the closest in a list of standard entries known as a *codebook*. Using this method and processing three frames together to exploit their correlations, Wang and

Jay Kuo [5] reported producing synthetic speech of comparable quality to LPC-10 but at a much lower bit rate of 800 b/s.

- Using *mixed excitation linear prediction* (MELP): The input signal used to reproduce each speech segment is a mixture of periodic pulses and white noise (rather than one or the other as in LPC-10). This improves the synthesised speech to a professional quality level of MOS ~3.0 at the same bit rate (2.4 kb/s) as LPC-10.
- Using *Sub-band coding*: This recognises that each speech frame can be more accurately modelled by looking at its content within various frequency sub-bands. So each frame is analysed to extract characteristic parameters in various sub-bands. These parameters are then coded and used at the receiver to reproduce the speech signal. The sub-band analysis and coding can be done in various ways leading to different implementations, such as *multi-band linear predictive* (MB-LPC) [6] and *improved multi-band excited* (IMBE) vocoders that produce professional quality speech (MOS ~3.2) at respective bit rates of 2.4 and 4.15 kb/s. More generally, a fast Fourier transform (FFT) analysis can be carried out on each frame to extract the sets of amplitudes and phases of constituent harmonics. At the receiver the synthesised speech for each segment is obtained as the sum of outputs produced using each set of coded parameters.

5.6.3 Hybrid coder

The general idea of hybrid coders is to employ waveform coding to code the difference between the original speech signal and a synthesised version of the signal. By combining the advantages of waveform coders and vocoders in this way, good reproduction of speech at low bit rates can be achieved. Hybrid coders use a technique called *analysis-by-synthesis*. At the transmitter various combinations of the model parameters are used to synthesise the speech as would be done at the receiver, and the combination yielding the best perceptual match to the original speech signal is selected. Hybrid codecs are therefore quite complex and introduce processing delays that are at least of the same order as vocoders. There are numerous designs of hybrid coders, the distinction being in excitation signal pattern, analysis procedure and type of information transmitted to the receiver. One example is the ITU-T G.729 standard [7] which achieves toll quality speech at 8 kb/s using a technique described as *conjugate-structure algebraic-code-excited linear prediction* (CS-ACELP). Speech signal is processed in 10 ms frames that are analysed to extract various parameters coded using a total of 80 bits, resulting in the 8 kb/s bit rate. The excitation signal at the receiver is constructed using a weighted sum of contributions from an adaptive and a fixed codebook.

5.7 Summary

Transmission of voice and other signals including video and a wide variety of sensor signals that are analogue by nature is now an all-digital process in modern communication networks. These signals must therefore be converted at source into a sequence of numbers or bit stream that contains all the significant information of the original analogue signal. In Chapter 4, we established the minimum rate at which such signals must be sampled in order to capture their information content and permit reconstruction without distortion. We followed that up in this chapter by studying in detail the final digitisation steps, involving quantisation and PCM, that represent the signal samples as a sequence of numbers and ensure that the noise introduced, while unavoidable, is completely under the system designer's control. We examined the intricacies of the concept of quantisation and studied the interplay of design parameters in PCM. Our main focus was on strategies for achieving the two main goals of minimum bandwidth and consistent and acceptable SQNR of the digital signal. In the end we made use of a piecewise linear approximation to a practically non-feasible logarithmic signal compression function and achieved a companding gain of 24 dB that allowed a saving of four bits per sample when compared to uniform quantisation of a similar low-amplitude signal.

PCM speech is a worldwide standard digital signal that conveys near transparent quality voice at 64 kb/s having an SQNR of around 38 dB, which drops by only 2.5 dB over a 36 dB reduction in analogue input signal amplitude. In some applications however, such as mobile telephony, this bit rate is excessive and measures must be found for bit rate reduction through data compression. We discussed lossy compression of telephone speech, also known as low bit rate speech coding, and highlighted techniques such as ADPCM and CS-ACELP that support toll quality voice at only 32 kb/s and 8 kb/s respectively, as well as vocoders that produce synthetic quality speech at bit rates from ~4 kb/s down to ~800 b/s. Lossless compression of non-voice data is the subject of the next chapter.

5.8 References

[1] ITU-T Recommendation G.701. (1993). Vocabulary of digital transmission and multiplexing, and pulse code modulation (PCM) terms. Geneva: International Telecommunications Union

[2] Haykin, S. (2013). *Digital communications systems*. Hoboken: Wiley, p. 301

[3] Otung, I. (2001). *Communication engineering principles*. Basingstoke: Palgrave Macmillan, pp. 359–364

[4] ITU-T Recommendation G.726. (1990). *40, 32, 24, 16 kbit/s adaptive differential pulse code modulation (ADPCM)*. Geneva: International Telecommunications Union

[5] Wang, X., Jay Kuo, C.-C. (1998). 'An 800 bps VQ-based LPC voice coder', *Journal of the Acoustical Society of America*, 103(5), p. 2778

[6] Yeldener, S., Kondoz, A., Evans, B. (1994). 'Multiband linear predictive coding at very low bit rates', *IEE Proceedings – Vision Image and Signal Processing*, 141(5), pp. 289–296

[7] ITU-T Recommendation G.729. (2012). Coding of speech at 8 kbit/s using conjugate-structure algebraic-code-excited linear prediction (CS-ACELP). Geneva: International Telecommunications Union

Chapter 6

Source coding and lossless data compression

In Chapter Six

- ✓ Overview of data compression: An introduction to the different types of data compression and their features.
- ✓ Information content and entropy: If it surprised you when it occurred, then it did inform you, but if you were certain that it would happen before it actually happened then it had no information value. On this basis, we present a detailed discussion of the information content of events focusing on outcomes or characters emitted by a discrete source, and we introduce the concept of entropy to provide a quantitative measure of the average level of surprise conveyed by each outcome of such a source. This lays the foundation for our study of the efficient coding of discrete sources.
- ✓ Entropy coding methods: A detailed and easy-to-follow discussion of the leading lossless data compression techniques, namely Huffman, Lempel-Ziv and arithmetic coding. We introduce the concept of code space and employ graphs, diagrams and worked examples to help bring complete clarity to the discussion.

6.1 Introduction

A large number of data compression techniques have been developed to improve *source coding efficiency* by reducing the average number of bits per second (i.e. bit rate) used to code information emitted by a source. The extent of the reduction is given by the data *compression ratio,* which is the ratio between the number of bits in the original and compressed data or the ratio between the average bit rates of the source and compressed bit streams.

This chapter treats in some detail the subject of lossless data compression, the aim of which is to minimise the number of bits needed to exactly represent a given source message. To understand how this is done and the challenges and constraints that we face in compressing a message, we must delve a little into information theory to introduce the concepts of information content of a character drawn at random from a discrete source and the entropy of such a source. We then discuss three methods of lossless compression generally described as *entropy coding* since they aspire to match average codeword length to the entropy of the source. The methods discussed include Huffman coding, Lempel-Ziv coding and arithmetic coding. In discussing arithmetic coding and the design of prefix codes we find it illuminating to introduce the concept of *code space* and discuss its relationship with codeword length in analogy with the inverse relationship between pulse bandwidth and pulse duration. We use the term *character* to refer to an element or member of the alphabet of a discrete memoryless random source. Note that other literatures on the subject do use the term *symbol*, but we reserve this term to refer to a transmitted energy signal (in Chapter 7 and beyond).

6.2 Classes and features of data compression

Data compression may be *symmetrical* or *asymmetrical*, and *lossless* or *lossy*. In symmetrical data compression the requirements of the compression and decompression algorithms are approximately equal in hardware, software and time, whereas asymmetrical compression are based on algorithms that are significantly more complex and more demanding in hardware, software and time for one of the two operations of compression and decompression. If compression can be done offline whereas it is required to carry out decompression in real-time using inexpensive devices then asymmetric compression that has a more complex and expensive compression process is the preferred option. A good example of such an application is data storage on mass distribution media. A more complex decompression algorithm is advantageous in applications where the aim is to achieve a speedy and more compact backup of data files that will be seldom accessed. However, symmetric compression is best suited for applications involving interactive storage and retrieval (i.e. recording and playback) of data on the same device such as a PC.

Lossless data compression reduces the data size without introducing any errors or loss of information detail. Also called *data compaction*, it is applicable to discrete sources such as text and pixel values of display screens, and works by identifying frequent patterns (or groups of characters) in the data and representing them using shorter codewords (i.e. fewer numbers). Compression ratio varies depending on the sequence of characters from the source and this produces a variable bit rate (VBR) signal that necessitates buffering if transmission is through a constant bit rate (CBR) channel. Lossy data compression, discussed in Section 5.6, achieves much higher compression ratios than its lossless counterpart by removing insignificant information detail as part of its compression measures. Obviously not applicable to text where every character is significant, lossy compression works on speech, audio, and still or moving image. It takes advantage of the limited sensitivity of human audio-visual perception and saves bits that would otherwise be used to represent largely imperceptible details or changes within the original information.

6.3 Information content and entropy

Consider a discrete information source that produces characters drawn from a finite alphabet X consisting of M characters x_j, $j = 1, 2, 3, \ldots, M$. An example of such an alphabet is the ASCII character set listed in Table A.4, ignoring for the moment the codeword assigned to each character (e.g. character B is assigned the codeword 1000010, which to base 10 is the number 66). The source can be treated as a discrete random variable X that produces outcome $X = x_j$ with probability p_j in each signalling interval T_s, where $0 \le p_j \le 1$. Since it is certain (i.e. probability $= 1$) that the output of the source will be one of the M characters it follows that

$$\sum_{j=1}^{M} p_j = 1, \quad 0 \le p_j \le 1 \tag{6.1}$$

where we have made the tacit assumption that the outcome in one interval is not influenced by the outcome in any other interval, in other words the source is *memoryless* so that outcomes in successive intervals are statistically independent.

If a particular outcome is certain, for example if the source is certain to emit a particular character, say x_4 in each interval, then $p_4 = 1$, and $p_j = 0$ for all $j \neq 4$. In this situation there is no uncertainty to resolve by observing the output of the source. We know exactly what is going to happen (i.e. x_4 will occur) so we are not 'surprised' and we gain no 'information' when it occurs. If on the other hand x_4 is very unlikely to be emitted by the source (i.e. $p_4 \ll 1$) then we are highly uncertain that it will occur and very surprised when it does. That is, we gain a lot of information when a highly unlikely outcome occurs because that occurrence resolves the high level of uncertainty that we had beforehand. It makes sense therefore to define the *amount of information* $I(x_j)$ gained after observing outcome x_j, which occurs with probability p_j as

$$I(x_j) = \log(1/p_j) \tag{6.2}$$

The base of the logarithm in Eq. (6.2) is usually chosen as 2, and the resulting information measure is then in a unit of *bit* (which is short for *bi*nary digi*t*). So we gain no ($\equiv$ zero bits of) information when an outcome that is certain to occur occurs, we gain one bit of information if one of two equiprobable outcomes occurs, we gain two bits of information if an outcome with probability ¼ (i.e. one in four chance) occurs and so on. Note that $I(x_j) \geq 0$, indicating that an observation can produce a gain in information or no change in information, but never a loss of information. Fig. 6.1a illustrates this relationship over a range of outcome probabilities.

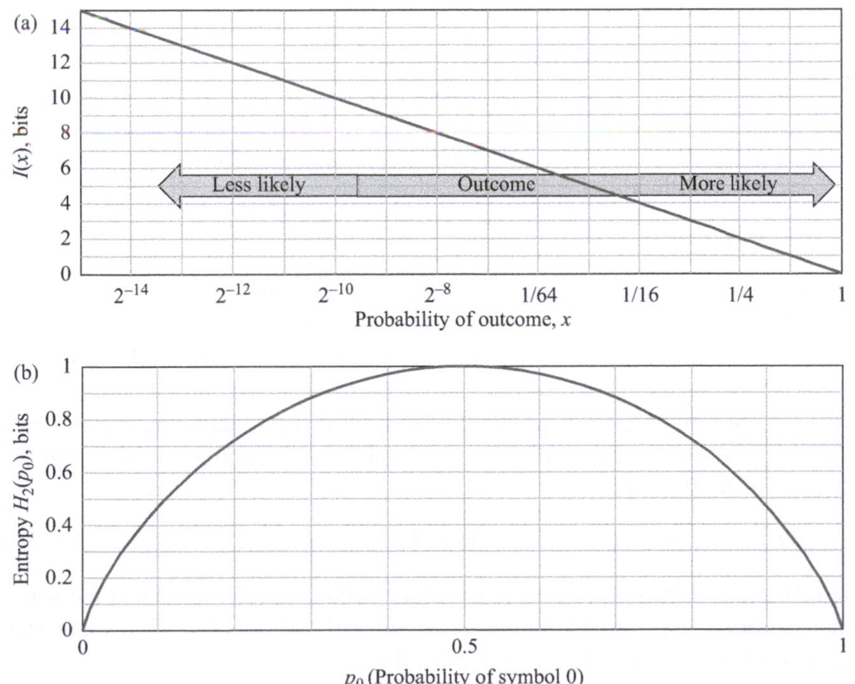

Fig. 6.1 (a) Information I(x) gained when random outcome occurs; (b) entropy of binary source

The amount of information produced by the source (represented by the discrete random variable X) during each observation interval depends on the character emitted during that interval. We may characterise a source X of alphabet size M by a parameter $H(X)$ known as *entropy*, which is the average information content per character emitted by the source, obtained by averaging $I(x_j)$ over all j:

$$H(X) = \sum_{j=1}^{M} p_j I(x_j) = \sum_{j=1}^{M} p_j \log_2\left(\frac{1}{p_j}\right) \quad \text{(bit)} \tag{6.3}$$

This entropy is minimum when there is no uncertainty before observing X, in other words when one of the outcomes x_k is certain, so that $p_k = 1$ and $p_j = 0$ for all $j \neq k$. Thus

$$\begin{aligned} H(X)_{\text{min}} &= p_k \log_2(1/p_k) + \sum_{j\neq k}^{M} p_j \log_2(1/p_j) \\ &= 1 \times \log_2(1/1) + \lim_{\varepsilon\to 0} \sum_{j\neq k}^{M} \varepsilon \log_2(1/\varepsilon) \\ &= 0 + (M-1) \lim_{\varepsilon\to 0} [\varepsilon \log_2(1/\varepsilon)] = 0 + (M-1) \times 0 = 0 \end{aligned} \tag{6.4}$$

$H(X)$ is maximum when there is maximum uncertainty before observing X, which is the case if there is no character that is more likely than any other, in other words when all M characters are equally likely with probability $p_j = 1/M, j = 1, 2, \ldots, M$. Thus

$$\begin{aligned} H(X)_{\text{max}} &= \sum_{j=1}^{M} \frac{1}{M} \log_2\left(\frac{1}{1/M}\right) = \frac{1}{M} \sum_{j=1}^{M} \log_2(M) \\ &= \log_2(M) \end{aligned} \tag{6.5}$$

The entropy of a discrete source X that produces characters drawn from an alphabet of size M is therefore a non-negative quantity

$$0 \leq H(X) \leq \log_2 M$$

The case $M = 2$ is a special source, a binary source represented by the so-called *Bernoulli random variable*, that emits only two characters x_1 and x_2, which may be described variously as outcomes 'failure' and 'success', 'tail' and 'head', 'hit' and 'miss', or binary 0 and binary 1, etc., with respective probabilities p_0 and $p_1 = 1 - p_0$. Substituting these probabilities into Eq. (6.3) yields the entropy $H(B)$ of the binary source (or Bernoulli random variable) as a function of the probability p_0 that character 0 occurs as

$$H(B) = -p_0 \log_2 p_0 - (1 - p_0) \log_2(1 - p_0) \quad \text{(bit)} \tag{6.6}$$

Note that if $p_0 = 1$ it means that character 0 is certain, which means that the source cannot 'surprise' us, so its entropy is zero. Similarly, if $p_0 = 0$ it means that character 1 is certain and entropy is again zero. Thus

$$H(B)|_{p_0=0} = H(B)|_{p_0=1} = 0$$

For values of p_0 in between, the source has an element of surprise and its entropy increases up to a maximum of one bit ($\equiv \log_2 M$) when 0 and 1 are equally likely. Fig. 6.1b gives a plot of this entropy function.

Worked Example 6.1: Entropy

A discrete memoryless source has source alphabet of size $M = 4$ consisting of characters x_1, x_2, x_3, x_4.

(a) Determine the entropy of a source X that emits the above characters with respective probabilities 1/2, 1/4, 1/8, 1/8.

(b) What is the entropy of a source X' in which the four characters are equiprobable?

(c) Determine the entropy of an *extended source*, denoted by X^2 in which each outcome is a block of two successive characters from X. Determine also the entropy of extended source X'^2.

(a) Using the given probabilities in Eq. (6.3) yields

$$\begin{aligned} H(X) &= \frac{1}{2}\log_2\left(\frac{1}{1/2}\right) + \frac{1}{4}\log_2\left(\frac{1}{1/4}\right) + \frac{1}{8}\log_2\left(\frac{1}{1/8}\right) + \frac{1}{8}\log_2\left(\frac{1}{1/8}\right) \\ &= \frac{1}{2}\times 1 + \frac{1}{4}\times 2 + \frac{1}{8}\times 3 + \frac{1}{8}\times 3 \\ &= \frac{7}{4} = 1.75 \text{ bits} \end{aligned}$$

(b) The four characters have equal probability, which means that $p_1 = p_2 = p_3 = p_4 = 1/4$. Under this condition the entropy is maximum and is given by Eq. (6.5) as

$$\begin{aligned} H(X') &= \log_2 M = \log_2 4 \\ &= 2 \text{ bits} \end{aligned}$$

(c) The solution is presented in tabular form in Table 6.1 where column 1 lists all possible compound characters s_1 to s_{16} in the extended source,

Table 6.1 Solution to Worked Example 6.1c

Character pair	**Probability** $P(X^2 = s_k) \equiv p_k$	**Probability** $P(X'^2 = s_k) \equiv p'_k$	**Weighted info** $p_k I(s_k)$	**Weighted info** $p'_k I'(s_k)$
$s_1 = x_1x_1$	1/4	1/16	1/2	1/4
$s_2 = x_1x_2$	1/8	1/16	3/8	1/4
$s_3 = x_1x_3$	1/16	1/16	1/4	1/4
$s_4 = x_1x_4$	1/16	1/16	1/4	1/4
$s_5 = x_2x_1$	1/8	1/16	3/8	1/4
$s_6 = x_2x_2$	1/16	1/16	1/4	1/4
$s_7 = x_2x_3$	1/32	1/16	5/32	1/4
$s_8 = x_2x_4$	1/32	1/16	5/32	1/4
$s_9 = x_3x_1$	1/16	1/16	1/4	1/4
$s_{10} = x_3x_2$	1/32	1/16	5/32	1/4
$s_{11} = x_3x_3$	1/64	1/16	3/32	1/4
$s_{12} = x_3x_4$	1/64	1/16	3/32	1/4
$s_{13} = x_4x_1$	1/16	1/16	1/4	1/4
$s_{14} = x_4x_2$	1/32	1/16	5/32	1/4
$s_{15} = x_4x_3$	1/64	1/16	3/32	1/4
$s_{16} = x_4x_4$	1/64	1/16	3/32	1/4
Entropy: $H(X^2) = \sum p_k I(s_k)$; $H(X'^2) = \sum p'_k I'(s_k)$			7/2 bits	4 bits

each being a pair of characters from the original source. Columns 2 and 3 list the probabilities of the compound characters s_1 to s_{16} in the extended sources X^2 and X'^2, respectively, obtained as the product of the probabilities of the two characters that form each compound character, being careful to use probabilities of source X for column 2 and probabilities of source X' for column 3. Each row of the last two columns gives the term inside the summation sign in Eq. (6.3). This is the *weighted information* of each character. The sum of these entries gives the required entropy of the extended sources in the last row of the table. It can be seen that the entropies of the extended sources are double those of the original sources in (a) and (b). This is not a coincidence because, as a general rule, the entropy of an extended source X^n whose characters consist of blocks of n successive characters taken from source X is

$$H(X^n) = nH(X) \tag{6.7}$$

6.4 Prefix and non-prefix variable-length codes

We are now well placed to consider the task of coding a memoryless discrete source X, which entails selecting *uniquely decodable* binary codewords to represent the source characters. A coding scheme is uniquely decodable if it is possible for the decoder to parse any received sequence of codewords unambiguously into source characters. For example, the coding scheme $x_1 = 1$, $x_2 = 10$, $x_3 = 11$ and $x_4 = 01$ is not uniquely decodable because the received sequence 1101 could be $x_1x_1x_4$ or x_3x_4 or $x_1x_2x_1$. A special *punctuation mark*, referred to simply as *marker*, would need to be inserted between codewords to make this scheme uniquely decodable. This is not a solution without cost however because the marker is an extra overhead that reduces efficiency. On the other hand, you may wish to verify that this other scheme $x_1 = 1$, $x_2 = 01$, $x_3 = 001$ and $x_4 = 000$ is uniquely decodable. Table A.4 shows the ASCII source encoding scheme in which a unique fixed-length seven-bit codeword is assigned to each character. The advantage of fixed-length codewords is that, if a unique codeword is assigned to each source character then the coding scheme is guaranteed to be uniquely decodable since the receiver, once synchronised with the transmitter to correctly identify the starting bit, will map successive blocks of bits of the same length to the correct character according to the coding table.

Table 6.2a–c specifies three examples of uniquely decodable codes, comprising two variable-length codes (a) and (b) and one fixed-length code (c). Code (a) is an example of an important class of codes called *prefix code*, whereas code (b) represents a *non-prefix code*. A prefix code is often defined as a coding scheme in which *no codeword is the prefix (i.e. initial part) of another codeword.* Note that this definition includes fixed-length codes such as (c) as a prefix code. A narrower definition of a prefix code is that it is *a variable-length code in which no valid codeword is the prefix of another valid codeword*. The decoding process for uniquely decodable codes can be modelled by a *decision tree* (actually an upside down tree) as shown in Fig. 6.2a for code (c). The tree consists of nodes

Table 6.2 Code examples: (a) prefix code, (b) uniquely decodable non-prefix code, (c) three-bit fixed-length code, (d)(i) non-uniquely decodable non-prefix code, (d)(ii) two-bit fixed-length code

Variable-length code				Fixed-length code	
Prefix code		Non-prefix code			
Character	Codeword	Character	Codeword	Character	Codeword
x_1	1	x_1	1	x_1	000
x_2	01	x_2	10	x_2	001
x_3	001	x_3	100	x_3	010
x_4	000	x_4	1000	x_4	011
				x_5	100
				x_6	101
				x_7	110
				x_8	111
(a)		(b)		(c)	

Character, x_j	Probability, p_j	Codeword	
		Code d(i)	Code d(ii)
x_1	1/2	1	00
x_2	1/4	10	01
x_3	1/8	11	10
x_4	1/8	01	11

(d)

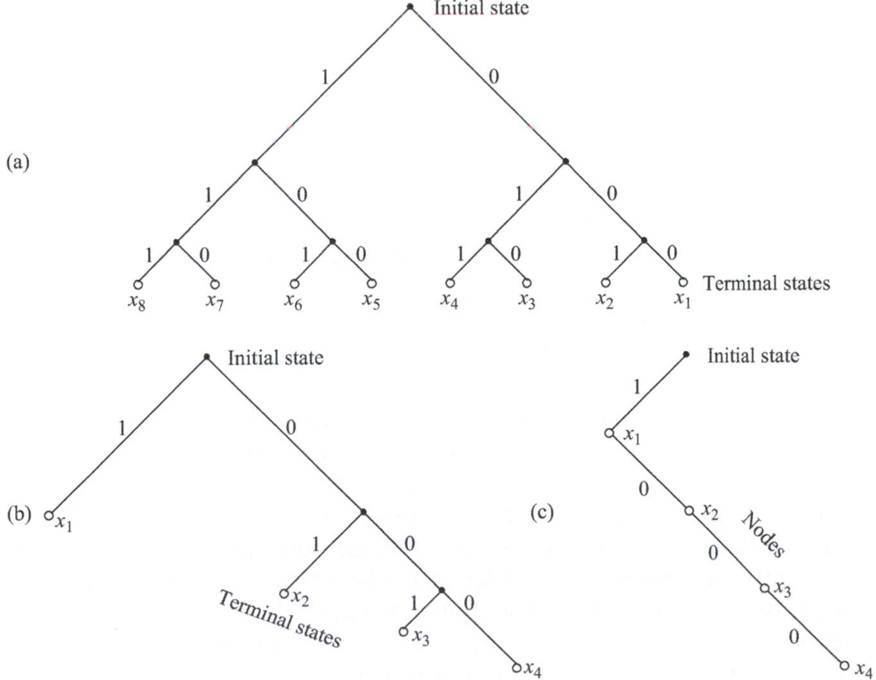

Fig. 6.2 Decision tree for codes in Table 6.2: (a) three-bit fixed-length code; (b) prefix code of maximum length three bits; (c) non-prefix code

and branches. Every node, with the exception of the terminal nodes, has two branches. The decoder always starts at the initial state (starting node) and is moved through the tree by the incoming bit sequence, a binary 0 moving it through the right branch to the next node and a binary 1 moving it through the left branch. Once the decoder reaches a terminal state, it produces the character for that state (indicated on the decision tree) and immediately resets to the initial state, ready to start all over again to decode the next sequence of bits. Thus the sequence of bits 101001 would cause the decoder to take the following action starting at the initial state: Move left-right-left to reach terminal x_6, emit x_6 and return to the initial state. Then move right-right-left to terminal x_2 and emit x_2. In this way the decoded sequence is x_6x_2.

The decoder of a k-bit fixed-length code requires k steps (from the initial state) through the decision tree to get to any one of the 2^k terminal states before it can make a decision. However, in prefix codes of maximum length k bits the decoder can make a decision in fewer than k steps, depending on the incoming sequence. For this reason, prefix codes are sometimes referred to as *instantaneous codes* or *self-punctuating codes*, since the end of a codeword is instantly recognisable once the last bit of the codeword is received and we don't need to wait to check any subsequent bits or to receive a punctuation mark to be sure. The decision tree for code (a) of Table 6.2 is shown in Fig. 6.2b. The decoder operation is as earlier described for Fig. 6.2a. That is, the decoder starts at the initial state and takes one step to the next node for each input bit received — the step taken being through the left branch if the input is binary 1 and through the right branch if binary 0 — until it reaches a terminal state, at which point it emits the character for that state and resets to the initial state to start over. Can you determine what the received sequence 101001 would be decoded to in Fig. 6.2b? The answer is given below.

Code (b) in Table 6.2 is an example of the use of a *marker* (in this case binary 1) to make a non-prefix code uniquely decodable. The decision tree for this code is shown in Fig. 6.2c. This is really not much of a tree because there is only one outgoing branch at each node. The decoding process works as follows. The decoder, starting at the initial state, takes one step to the next node (there's no directional option here) for each bit received. It carries on in this way until the received bit is a binary 1 at which point it emits the character for the current node and moves to node x_1. If no more bit is forthcoming (i.e. end of received sequence), the decoder emits the character of the current node and terminates. Notice that the nodes x_1, x_2 and x_3 are not terminal states. For example, when it reaches node x_1 the decoder does not know whether it will emit character x_1 or move on to node x_2. A punctuation mark, which is binary 1 in this case, is always awaited and relied upon to know when the end of a codeword has been reached. So this is clearly not an instantaneous code. Note also that, unlike the decision trees shown in Fig. 6.2a,b, this one does not reset to the initial state, so the decoder is only ever in the initial state at the start of the decoding process (i.e. at the first received bit). Furthermore, this first bit cannot be a binary 0 and the bit received while the decoder is at node x_4 also cannot be binary 0. The latter means that a sequence of four consecutive 0s is forbidden. Thus this non-prefix code has some inbuilt error detection capability. Can you determine how 101001 would be decoded in this tree (Fig. 6.2c)? The answer is $x_2x_3x_1$. And the answer to the previous question (for Fig. 6.2b) is $x_1x_2x_3$.

6.4.1 Source coding theorem and code efficiency

How efficient are the codes discussed above? Do they minimise the number of bits needed to code a given message (i.e. sequence of characters) from the source? Fixed-length binary codewords are efficient when the source alphabet contains M equiprobable characters and M is a power of 2, but this is rarely the case. For example, an English language text will in general contain more of some characters than others, with on average around 12.6% being the character *e* compared to only around 0.12% being the character *q*. If we assign shorter codewords to the more frequent (i.e. more probable) characters and longer codewords to the less frequent ones then a message can on average be coded using fewer bits thus achieving data compression. An example of this type of *variable-length* coding scheme is the Morse code shown in Table A.1. Characters are coded using dots and dashes, where a dot is a voltage pulse of duration one timing unit (1TU) whereas a dash is a voltage pulse of thrice the duration (i.e. 3TU). The space between a dash or dot and another dash or dot in the same character is 1TU, the space between characters is 3TU and the space between words is 7TU. Thus, according to Table A.1, character E is represented by a short codeword of duration 4TU, whereas the less frequent character Q is represented by a longer codeword of duration 16TU. It is worth noting that Morse code is a non-prefix code. For example, the codeword for character E is a prefix of the codewords for many other characters including A, whose codeword is in turn a prefix of other codewords, and so on. A *marker* is essential to ensure unique decodability, and this is provided by inserting a silence (i.e. *no-pulse*) of duration 3TU after each codeword.

If a variable-length binary coding scheme is applied to code our discrete memoryless source X by assigning a codeword of length l_j bits to character x_j which occurs with probability p_j, where $j = 1, 2, \ldots, M$, then the average codeword length $\bar{l}$ of the source encoder will be

$$\bar{l} = \sum_{j=1}^{M} p_j l_j \quad \text{(bits)} \tag{6.8}$$

On average a message of length N characters will be represented using $N\bar{l}$ bits so it is desirable to make $\bar{l}$ as small as possible in order to achieve a higher compression ratio. If there is to be no loss of information then the best that we can do (i.e. the lowest possible $\bar{l}$, denoted by $\bar{l}_{\min}$) is when each character x_j is represented by a codeword having length l_j bits equal to the information content $I(x_j)$ of outcome x_j. Substituting $I(x_j)$ for l_j into Eq. (6.8) and noting that the resulting expression is the definition of entropy given in Eq. (6.3) yields

$$\bar{l}_{\min} = \sum_{j=1}^{M} p_j I(x_j) = H(X)$$

This is a stipulation of the *source coding theorem* which places a bound on how small $\bar{l}$ can be and hence puts a fundamental limit on data compression. The theorem states that *the average codeword length of a source encoder can be no smaller than the entropy of the source.* That is

$$\bar{l} \geq H(X), \qquad \text{Shannon's source coding theorem} \tag{6.9}$$

The *coding efficiency* of a source encoder gives a measure of how close the average codeword length of the encoder approaches to the source entropy, and is defined by

$$\eta_c = \frac{H(X)}{\bar{l}} \tag{6.10}$$

6.4.2 The code space concept

A further constraint on the codeword lengths l_j of characters x_j, $j = 1, 2, \ldots, M$, is given by the *Kraft inequality* which states:

> *A necessary condition for a binary coding scheme to be uniquely decodable is that the codeword lengths must satisfy*

$$\sum_{j=1}^{M} 2^{-l_j} \leq 1, \qquad \text{Kraft inequality} \tag{6.11}$$

The Kraft inequality can be interpreted as follows to guide the selection of codewords. There is a maximum code space = 1 (RHS of Eq. (6.11)) within which to fit the M codewords representing the M characters of a source alphabet. A codeword of length l occupies an amount 2^{-l} of this space, so the shorter the codeword (measured by its length in bits) the larger the code space it occupies. This has an analogy with the inverse relationship between time and frequency where we found that the shorter the pulse duration the larger is its bandwidth. So if for example you pick the one-bit codeword 0, which has code space $2^{-1} = ½$, then the amount of code space remaining is ½ within which to fit the remaining $M - 1$ codewords if you wish to ensure unique decodability (or, to borrow our bandwidth analogy, if you wish to avoid interference). Just as shorter pulses are desirable because they enable transmission at a higher bit rate, so also are shorter codewords since they allow higher data compression. But while we may increase bandwidth to accommodate shorter and shorter pulses, the maximum code space (= 1) is fixed. To continue our example, having picked codeword 0, if you then select codeword 1, the code space is filled up. Any further codeword selected, whatever its bit combination, will lie in the code space already occupied by these two one-bit codewords and hence will be 'interfered' with. To avoid this 'interference' (i.e. to ensure unique decodability) you must use longer codewords, which occupy smaller code spaces, in order to assemble an optimum set of codewords that covers all the characters in the alphabet without exceeding the total code space available.

Fig. 6.3 gives a pictorial representation of code space on the vertical axis versus codeword length on the horizontal axis, for all binary codewords of length from $l = 1$ to $l = 5$. This is an important diagram, which is interpreted and applied as follows:

- In fixed-length codes (e.g. ASCII code in Table A.4 or code (c) in Table 6.2), codewords are selected from only one column of Fig. 6.3. Columns to the left fill up the code space using fewer codewords. The more characters there are in the source alphabet, the more to the right one must go to get a sufficient number of codewords without exceeding total code space. For example, ASCII codes are selected from column 7 (not shown in Fig. 6.3) where up to 128 codewords can be fitted into the code space.

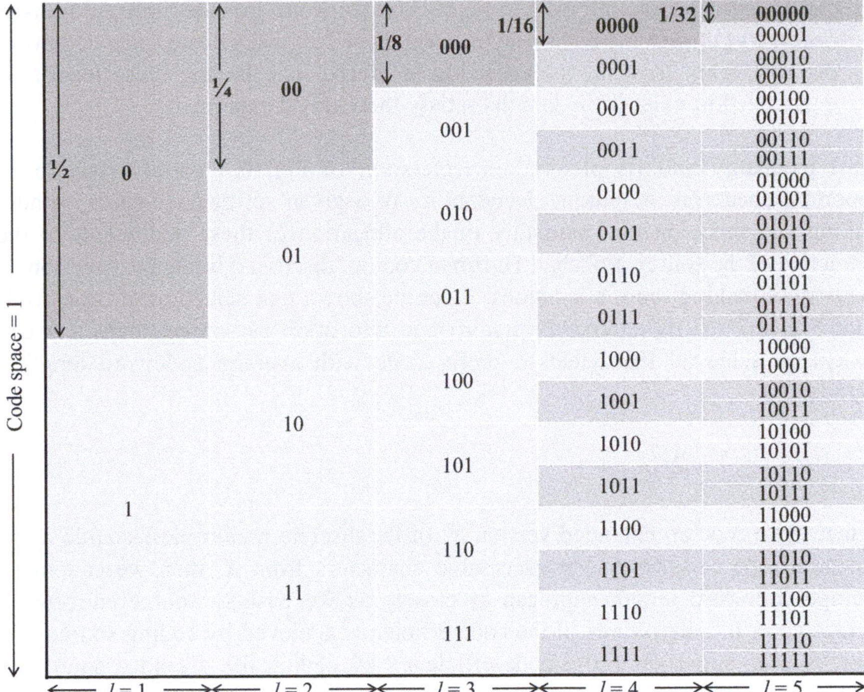

Fig. 6.3 Code space of binary codewords versus codeword length l, from l = 1 to l = 5

- Variable-length codes (e.g. code (b) of Table 6.2) use a pick and mix approach, selecting codewords from different columns and using the columns to the left as much as possible for the more probable characters, subject to the constraint that total code space (= 1) is not exceeded. For example, the codeword selection for code (b) was: 1 (code space = ½), 10 (code space = ¼), 100 (code space = 1/8) and 1000 (code space = 1/16), giving a code space usage of 15/16, which means that there is spare room (= 1/16) enough for an extra four-bit codeword or multiple longer codewords.
- Prefix codes impose a strict order on code space occupancy in addition to the total code space constraint discussed above. Code space is available to only one codeword per row in Fig. 6.3. Once a codeword has been selected then there is no code space available for any other codeword on the same row. Think of codeword 0 in the top row of the first column of Fig. 6.3 as straddling the top two rows of column 2 and the top four rows of column 3 and so on. You should be able to see why there is this extra constraint in code space allocation for prefix codes by observing that all codewords along one row in Fig. 6.3 share a common prefix. So it has to be forbidden to have more than two codewords in a row since the codeword to the left is a prefix of the one to the right. Thus, for example, if codeword 0 is used then the top half of the code space is occupied, which means that the remaining codewords can only be drawn from the bottom half.
- Given any set of codeword lengths $\{l_1, l_2, l_3, \ldots, l_M\}$, if their aggregate code space usage does not exceed the total code space available (= 1), then we

can always find enough rows in Fig. 6.3 (extended to include more columns as necessary) to provide codewords having these lengths, selected one codeword per row. That is, a uniquely decodable prefix code having these lengths is guaranteed to exist if the lengths satisfy the Kraft inequality.

Kraft inequality and its pictorial interpretation in Fig. 6.3 merely provide an important check on the unique decodability of a given set of codewords without making any input on the optimality of the allocation of these codewords to the characters of the source alphabet. Huffman coding, discussed in the next section, is one way of making these allocations. It can be shown that selection of codewords from Fig. 6.3 (or its extended version) and allocation to source characters can always be achieved that yields a prefix code with average codeword length $\bar{l}$ bounded by

$$H(X) \le \bar{l} < H(X) + 1 \tag{6.12}$$

In fact if we code an extended version X^n of the discrete memoryless source X, by allocating one codeword to n successive characters from X, then we can make average codeword length approach as closely as we wish to source entropy by increasing n. In other words, if the code efficiency achieved by coding source X is poor, we can approach 100% code efficiency by coding the extended source X^n instead, for n sufficiently large. However, source codec complexity and buffering delay increase with n, since there will be M^n unique codewords requiring a more cumbersome decision tree and (in a real-time service) the encoder must wait for n signalling intervals of the source to accumulate the n characters needed to produce a single codeword.

Worked Example 6.2: Uniquely Decodable Codes

Tables 6.2a, b, d(i) and d(ii) specify four different codes for the characters x_1, x_2, x_3, x_4, which occur with respective probabilities 1/2, 1/4, 1/8, 1/8.

(a) What is the average codeword length of each code and how does this compare with the entropy of the source?
(b) What is the efficiency of each code?
(c) Do these codes satisfy the Kraft inequality?

(a) The entropy of this source was obtained in Worked Example 6.1a as $H(X) = 1.75$ bits. The respective codeword lengths for characters x_1, x_2, x_3, x_4 are

$$\{l_1, l_2, l_3, l_4\} = \begin{cases} \{1, 2, 3, 3\}, & \text{Code} \quad \text{(a)} \\ \{1, 2, 3, 4\}, & \text{Code} \quad \text{(b)} \\ \{1, 2, 2, 2\}, & \text{Code} \quad \text{(d)(i)} \\ \{2, 2, 2, 2\}, & \text{Code} \quad \text{(d)(ii)} \end{cases}$$

Combining these lengths with their respective probabilities {1/2, 1/4, 1/8, 1/8} according to Eq. (6.8) yields the average codeword length of each code as

$$\bar{l} = \begin{cases} \frac{1}{2}\times 1 + \frac{1}{4}\times 2 + \frac{1}{8}\times 3 + \frac{1}{8}\times 3 = 1.75 \text{ bits}, & \text{Code (a)} \\ \frac{1}{2}\times 1 + \frac{1}{4}\times 2 + \frac{1}{8}\times 3 + \frac{1}{8}\times 4 = 1.875 \text{ bits}, & \text{Code (b)} \\ \frac{1}{2}\times 1 + \frac{1}{4}\times 2 + \frac{1}{8}\times 2 + \frac{1}{8}\times 2 = 1.5 \text{ bits}, & \text{Code (d)(i)} \\ \frac{1}{2}\times 2 + \frac{1}{4}\times 2 + \frac{1}{8}\times 2 + \frac{1}{8}\times 2 = 2 \text{ bits}, & \text{Code (d)(ii)} \end{cases}$$

We see that code (d)(i) has an average codeword length of 1.5 bits, which is lower than source entropy 1.75 bits. But a closer inspection reveals that this code is not uniquely decodable and would cause a loss of information if used. The other three codes are all uniquely decodable. Of these, code (a) has average codeword length exactly equal to source entropy, which is the best we can do in terms of lossless data compression, achieved by using a prefix code in which each character x_j of the alphabet is represented by a codeword of length $l_j = \log_2(1/p_j) \equiv$ information content of outcome x_j. This is possible when source character probabilities are all negative integer powers of 2. Code (b) also implements data compression by using shorter codewords for the more probable characters. However, its average codeword length exceeds source entropy primarily because of the overhead incurred by using a marker (binary 1) in the coding scheme to demarcate codeword boundaries (as earlier discussed). Finally, code (d)(ii) uses fixed-length codewords for each character irrespective of their probability of occurrence. So this represents uncompressed data (similar to ASCII code), and unsurprisingly the codeword length of two bits exceeds source entropy.

(b) Code efficiency is obtained straightforwardly using Eq. (6.10) and the $\bar{l}$ values computed above:

$$\eta_c = \begin{cases} \frac{1.75}{1.75} = 100\%, & \text{Code (a)} \\ \frac{1.75}{1.875} = 93.33\%, & \text{Code (b)} \\ \frac{1.75}{1.5} = 116.7\%, & \text{Code (d)(i)} \\ \frac{1.75}{2} = 87.5\%, & \text{Code (d)(ii)} \end{cases}$$

The main points about these four codes and their performance were highlighted in (a) above. The only additional comment to make here is that if in future you are ever tempted to celebrate your invention of a new source code that has an efficiency higher than 100%, such as code (d)(i), you will do well to check whether the code is uniquely decodable.

(c) In connection with the Kraft inequality of Eq. (6.11)

$$\sum_{j=1}^{M} 2^{-l_j} = \begin{cases} 2^{-1} + 2^{-2} + 2^{-3} + 2^{-3} = 1, & \text{Code (a)} \\ 2^{-1} + 2^{-2} + 2^{-3} + 2^{-4} = 0.9375, & \text{Code (b)} \\ 2^{-1} + 2^{-2} + 2^{-2} + 2^{-2} = 1.25, & \text{Code (d)(i)} \\ 2^{-2} + 2^{-2} + 2^{-2} + 2^{-2} = 1, & \text{Code (d)(ii)} \end{cases}$$

We see that all the codes satisfy the inequality (having a summation ≤ 1) with the exception of code (d)(i), which therefore cannot be uniquely decodable. The problem with code (d)(i) is that it uses too many short codewords occupying large code spaces so that its total code space usage exceeds the maximum available, and this, to borrow our bandwidth analogy one final time, causes mutual interference.

6.5 Huffman coding

David Albert Huffman, while still a PhD student at Massachusetts Institute of Technology (MIT), developed a simple algorithm in 1952 for finding an optimal prefix code. The algorithm provides a systematic and elegant way of constructing a variable-length prefix code, called Huffman code after its inventor, whose average codeword length comes to within one bit of the fundamental limit set by the source entropy. It assigns longer codewords to the least probable characters and shorter codewords to the most probable characters of a source alphabet. The encoder must know the source statistics in advance or it must go through the entire message (prior to character by character encoding) to count the characters in the message and hence compute their probabilities of occurrence. Given a known source alphabet and its statistics consisting of M characters and their probabilities of occurrence, the algorithm constructs a *Huffman tree* starting from the leaves (the full list of characters) and building down, reducing the character count by one in each step until after $M - 1$ steps the final two branches converge on the trunk.

6.5.1 Encoding process

A graphical procedure for constructing the Huffman code is quite simple and is as follows:

1. Form M nodes along a row — a column would do just fine but we use rows here simply because the final structure looks more like an upright tree — and label them with the M characters and their respective probabilities. The algorithm works whether or not the characters are arranged according to their probability, although arranging the M characters on this first row from left to right in descending order of probability will produce a Huffman tree with fewer crisscrossing branches.
2. Form a new row of nodes below the current one by replacing the two least probable nodes in the current row with one new node having probability equal to the sum of the probabilities of the two nodes being combined and replaced. If multiple pairs of nodes qualify as the least probable, then choose any pair. That is, if there is a least probable node and two or more nodes having the same

next lowest probability, then combine the least probable node with any one of the nodes having the next lowest probability; and if there are three or more nodes having the same lowest probability, then combine any two of them. Represent this action diagrammatically by drawing one line downwards from each of the two nodes being replaced to meet at the new node on the new row below. Label the right branch with binary 0 and the left branch with binary 1. Complete the new row by translating any remaining nodes from the current row vertically down into the new row, marking each translation with an adjoining line. Label all nodes in the new row with their probabilities.

3. Repeat step 2 until the final row is formed having just one node labelled with probability 1. The diagram constructed by this process is a *Huffman tree* and this final node is the *trunk*.
4. The Huffman code of each character at the top of the tree is given by the bit sequence encountered while walking upwards along the tree starting from the trunk until you arrive at that character.

The above procedure involves an arbitrary decision in step 2 in the event that candidate nodes have the same probability. Also the choice of which label, binary 0 or 1, to place on the branches is arbitrary but must be consistent. That is, if you choose to label the left branch with binary 0 (and the right with binary 1) or vice versa then you must do that consistently through the entire tree structure. The result of this arbitrariness is that the codeword set generated will not be bit-by-bit identical across all independent implementations. However, the average codeword length will be the same for all codeword sets generated by the procedure on the same source alphabet and statistics. Furthermore, sorting each row before combining the least probable two nodes has no effect on either average codeword length or mean compression ratio, but may in some circumstances (depending on source statistics) reduce the variance of codeword lengths, which helps reduce short-term variation in compression ratio. To include sorting in the above algorithm, then always arrange the characters left to right in order of decreasing probability, making sure to place a new node as high up the order as possible by placing it ahead of any existing nodes that have the same probability as the new node.

Worked Example 6.3: Huffman Coding

Determine the Huffman codes for the following discrete memoryless sources, using the Huffman algorithm. Determine average codeword length, variance of codeword lengths and code efficiency of each code. How does the efficiency of the constructed Huffman code compare with that of a fixed-length code?

(a) A source that produces characters x_1, x_2, x_3, x_4, x_5 and x_6 with respective probabilities 0.3, 0.2, 0.25, 0.1, 0.1 and 0.05.

(b) A source that produces characters x_1, x_2, x_3 and x_4 with respective probabilities ½, ¼, 1/8, 1/8.

(a) The Huffman tree is sketched in Fig. 6.4 following the steps outlined above. In row 1, six nodes are labelled with the given characters and their respective probabilities. The last two nodes in this row have the

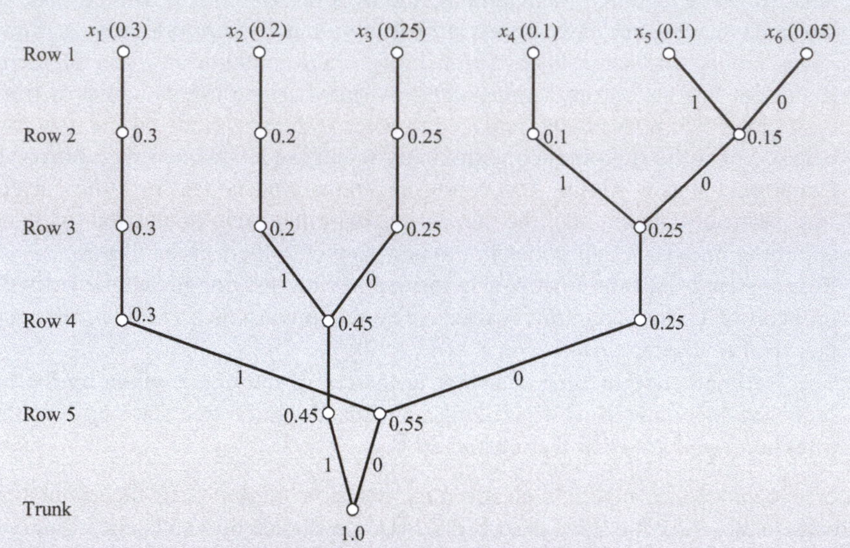

Fig. 6.4 Huffman tree for Worked Example 6.3a

least probabilities and are therefore combined to form a new node in row 2 with probability = 0.1 + 0.05 = 0.15. The merging is indicated on the tree by two branches converging from the two combined nodes in row 1 onto the new node in row 2. The right and left branches are labelled with binary 0 and 1, respectively. The other nodes in row 1 are translated (unchanged) down into row 2 as shown. Note that the decision in this row 1 to combine x_5 and x_6, rather than x_4 and x_6 is arbitrary. It will result in x_5 being represented using a longer codeword than x_4, but they will simply be swapping codewords, which makes no difference to average codeword length since x_4 and x_5 are equiprobable. This procedure is continued through all rows down to the trunk as shown in Fig. 6.4. Now starting from the trunk and going up along the path to x_6 we encounter the binary labels sequence 0000, which is therefore the codeword for character x_6. Similarly, the complete set of codewords are read as

$$x_1 = 01;\ x_2 = 11;\ x_3 = 10;\ x_4 = 001;\ x_5 = 0001;\ x_6 = 0000$$

(i) Source entropy is

$$\begin{aligned} H(X) &= 0.3\log_2\left(\frac{1}{0.3}\right) + 0.2\log_2\left(\frac{1}{0.2}\right) + 0.25\log_2\left(\frac{1}{0.25}\right) \\ &\quad + 0.1\log_2\left(\frac{1}{0.1}\right) + 0.1\log_2\left(\frac{1}{0.1}\right) + 0.05\log_2\left(\frac{1}{0.05}\right) \\ &= 0.5211 + 0.4644 + 0.5 + 0.3322 + 0.3322 + 0.2161 \\ &= 2.366 \text{ bits} \end{aligned}$$

(ii) Combining the codeword lengths {2, 2, 2, 3, 4, 4} with the probabilities of the respective characters represented, we obtain average codeword length as

$$\bar{l} = 0.3 \times 2 + 0.2 \times 2 + 0.25 \times 2 + 0.1 \times 3 + 0.1 \times 4 + 0.05 \times 4$$
$$= 2.4 \text{ bits}$$

(iii) The variance of codeword lengths is defined by

$$\begin{aligned}\sigma^2 &= \sum_{j=1}^{N} p_j (l_j - \bar{l})^2 \\ &= 0.3(2-2.4)^2 + 0.2(2-2.4)^2 + 0.25(2-2.4)^2 \\ &\quad + 0.1(3-2.4)^2 + 0.1(4-2.4)^2 + 0.05(4-2.4)^2 \\ &= 0.54 \text{ square bits}\end{aligned}$$

(iv) Code efficiency of the Huffman code is

$$\eta_c = \frac{H(X)}{\bar{l}} = \frac{2.366}{2.4} = 98.6\%$$

(v) A fixed-length code needs three bits to cover the six characters, so $\bar{l} = 3$ bits, and code efficiency of a fixed-length code is thus

$$\eta_c = \frac{2.366}{3} = 78.9\%$$

The efficiency of the Huffman code is higher than that of the fixed-length code as expected.

(b) This source was analysed in Worked Example 6.2 with the codeword set shown in Table 6.2a and we calculated source entropy $H(X) = 1.75$ bits, average codeword length $\bar{l} = 1.75$ bits and code efficiency $\eta_c = 100\%$. Here we show how the prefix codes given in Table 6.2a were constructed for this source based on its statistics. Fig. 6.5 shows the Huffman tree for this problem constructed as earlier discussed. From this we read the codewords as $x_1 = 1$; $x_2 = 01$; $x_3 = 001$; $x_4 = 000$.

The variance of codeword lengths for this source is

$$\begin{aligned}\sigma^2 &= \sum_{j=1}^{N} p_j (l_j - \bar{l})^2 \\ &= 0.5(1-1.75)^2 + 0.25(2-1.75)^2 + 0.125(3-1.75)^2 \\ &\quad + 0.125(3-1.75)^2 \\ &= 0.6875 \text{ square bits}\end{aligned}$$

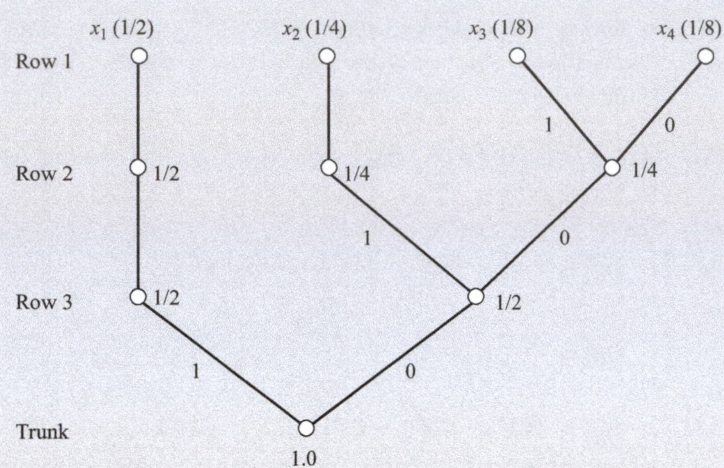

Fig. 6.5 Huffman tree for Worked Example 6.3b

A fixed-length code needs two bits to cover the four characters, so $\bar{l} = 2$ bits and code efficiency of a fixed-length code is thus

$$\eta_c = \frac{1.75}{2} = 87.5\%$$

The efficiency of the Huffman code is again higher than that of the fixed-length code as expected.

It is tempting to think that Huffman codes are always highly efficient seeing as the lowest Huffman code efficiency in the above worked example is 98.6%. Eq. (6.12) indicates that Huffman codes are guaranteed to have average codeword lengths $\bar{l}$ that come to within one bit of the fundamental limit — which is the source entropy $H(X)$. So in the worst case situation (depending on source statistics) $\bar{l}$ is just under $H(X) + 1$ and code efficiency

$$\begin{aligned} \eta_{c\ \min} &= \frac{H(X)}{H(X)+1} \\ &\rightarrow 100\%, \text{ for large } H(X) \\ &\rightarrow 0\%, \text{ for small } H(X) \end{aligned} \qquad (6.13)$$

Thus, as illustrated in the next worked example, Huffman codes have the potential to be very inefficient when the algorithm is applied to low-entropy sources. The solution in such situations is to find a way of coding using less than one bit/character on average, and this is accomplished by coding blocks of n source characters at a time using an average codeword length $\bar{l} < n$, as discussed next.

Worked Example 6.4: Improving the Efficiency of Huffman Codes

A discrete memoryless source emits two characters *a* and *b* with respective probabilities 0.9 and 0.1. We will code blocks of characters from this source, for block size $n = 1, 2, 3, 4$ in order to demonstrate how low-entropy sources can be more efficiently coded using Huffman codes.

The solution to this problem is laid out in Table 6.3. For $n = 1$, the source characters are individually coded. There are only two characters *a* and *b* occurring with respective probabilities 0.9 and 0.1, so the only possible Huffman codewords are 0 and 1, respectively (or vice versa), which means that average codeword length $\bar{l} = 1$ bit. Source entropy is calculated as usual:

$$H(X) = 0.9 \times \log_2(1/0.9) + 0.1 \times \log_2(1/0.1) = 0.469 \text{ bits}$$

We see that the fundamental limit on average codeword length is 0.469 bits, but since Huffman code assigns one codeword to each character of the source it is not possible to go below a length of one bit for any codeword. Thus, the minimum average codeword length of any Huffman code is $\bar{l} = 1$ bit, achievable only in binary sources such as the one at hand. Code efficiency is thus $\eta_c = H(X)/\bar{l} = 46.9\%$, which is quite poor.

Instead of individually coding the characters, let us code blocks of characters by assigning one unique codeword to each block of $n = 2$, 3 and 4 successive characters from the source. This yields extended sources X^2, X^3 and X^4 with the compound characters and probabilities listed in Table 6.3.

Table 6.3 Worked Example 6.4: Improving Huffman code efficiency by coding blocks of symbols

Symbol	Probability	Codeword	Symbol	Probability	Codeword
	X			X^4	
a	0.9	1	*aaaa*	0.6561	1
b	0.1	0	*aaab*	0.0729	0000
$H(X) = 0.469$; $\bar{l} = 1$; $\eta_c = 46.9\%$			*aaba*	0.0729	011
	X^2		*aabb*	8.1×10^{-3}	0001100
aa	0.81	1	*abaa*	0.0729	010
ab	0.09	01	*abab*	8.1×10^{-3}	0001011
ba	0.09	001	*abba*	8.1×10^{-3}	0001010
bb	0.01	000	*abbb*	9×10^{-4}	0001101000
$H(X) = 0.938$; $\bar{l} = 1.29$; $\eta_c = 72.7\%$			*baaa*	0.0729	001
	X^3		*baab*	8.1×10^{-3}	00010001
aaa	0.729	1	*baba*	8.1×10^{-3}	00010000
aab	0.081	000	*babb*	9×10^{-4}	000110111
aba	0.081	001	*bbaa*	8.1×10^{-3}	000111
abb	0.009	01100	*bbab*	9×10^{-4}	000110110
baa	0.081	010	*bbba*	9×10^{-4}	000110111
bab	0.009	01101	*bbbb*	10^{-4}	0001101001
bba	0.009	01110	$H(X) = 1.876$; $\bar{l} = 1.9702$; $\eta_c = 95.2\%$		
bbb	0.001	01111			
$H(X) = 1.407$; $\bar{l} = 1.598$; $\eta_c = 88.0\%$					

Note that the probability of occurrence of a given block is the product of the individual probabilities of constituent characters in the block. For example, the probability of occurrence of block *aab* in extended source X^3 is $0.9 \times 0.9 \times 0.1 = 0.081$. The Huffman codewords of each extended source are listed in the table, derived as earlier discussed in connection with Figs. 6.4 and 6.5. From these, the average codeword lengths and code efficiencies are calculated as also earlier learnt (see previous two worked examples). These results are summarised in the table for each source, from which it can be seen that code efficiency increases dramatically from 46.9% at $n = 1$ to 72.7% at $n = 2$, and steadily towards 100% as n is increased further. It is important to see why there is such a dramatic improvement. At $n = 4$ for example, block *aaaa* is coded using one bit whereas *bbbb* is coded using 10 bits. So we are effectively using 0.25 bits to code the more probable character *a* and 2.5 bits to code the less probable character *b*, which leads to the average codeword length given in Table 6.3 of $\bar{l} = 1.9702$ bits per 4 characters, or 0.4925 bits/character, which is closer to the source entropy of 0.469 bits/character, hence the improved coding efficiency.

6.5.2 Decoding process

We have learnt how the Huffman code table (e.g. Table 6.2a) is created, which the encoder uses as a look-up table to generate the transmitted bit stream that conveys the source message. How will a receiver decode this bit sequence? The receiver must receive a copy of the coding table used by the transmitter, and this in effect means that the receiver has the Huffman tree (e.g. Fig. 6.5) from which the entries in the coding table were read. With only very minor changes, this tree becomes the decision tree (Fig. 6.2b) for the decoder. To see this, consider for example Fig. 6.5 and turn it upside down and rename the trunk as initial state, and the nodes of row 1 as the terminal states of the corresponding characters, and drop all probability labels and intermediate nodes that are not junctions. The result is Fig. 6.2b. This transformation from Huffman tree at transmitter to decision tree at decoder is illustrated in Fig. 6.6. The decision tree is used by the decoder as described in the previous Section: Starting at the initial state the decoder takes one step to the next node for each input bit received – the step taken being through the left branch if the input is binary 1 and through the right branch if binary 0. When a terminal state is reached it means that a codeword has been stepped through, so the decoder produces the character for that state as its output and resets to the initial state to start over with decoding the next codeword.

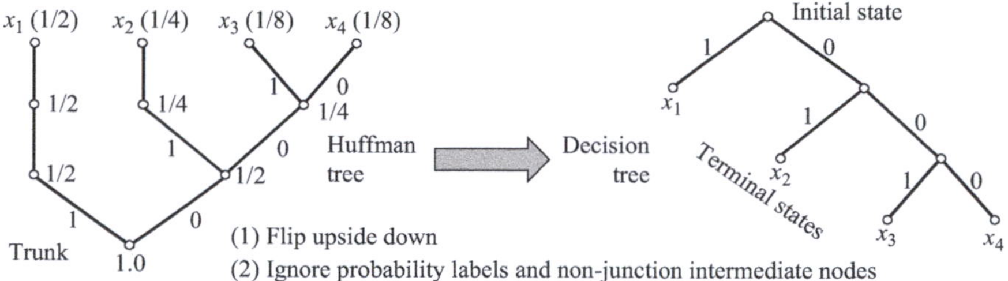

Fig. 6.6 From Huffman tree at transmitter to decision tree at receiver

6.5.3 *Weaknesses of Huffman coding*

Although the Huffman algorithm is an elegant method of source encoding that produces codeword lengths that on average come close to within one bit of the fundamental (entropy) limit, and in some special cases does allow messages or files to be represented using the smallest number of bits theoretically possible without loss of information, the algorithm does have the following significant disadvantages.

- The encoder must know the source statistics in advance and this requires two passes through the entire message or file to be compressed, first to count all the characters in the message in order to compute their individual probabilities, and then to code the message using the code table constructed after the first pass.
- The code table used by the encoder must be saved (and transmitted) together with the coded message since it is essential for decoding. The number of bits spent in this way to convey the code table is a significant overhead that squanders some of the gains of compression, especially in short messages. An attempt to avoid this overhead by using a standard code table for coding and decoding can be counterproductive. Table 6.4 shows the probability distribution of letters in English text based on (i) analysis of all letters in the words listed in the main entries of the Concise Oxford Dictionary (11th edition revised, 2004) [1], and (ii) analysis of over 3 billion letters in 9481 English works [2]. Note that the second compilation gives a better measure of actual usage, since the first includes each English word only once in the analysis irrespective of its frequency of usage. This table can be expanded to give a complete table of probabilities that includes special characters (e.g. the space character between words, punctuation marks, mathematical characters, etc.), which can be used to construct a standard Huffman code table. However, the probability of characters varies significantly not only with language, but also with subject matter and author style. For example, there are more mathematical

*Table 6.4 Probability of letters in English text, based on (i) all main entries in Oxford Concise English Dictionary (*11*th edition,* 2004*) [2]; (ii)* 3,104,375,038 *letters in* 9481 *English works [3]*

Letter	Probability (%)		Letter	Probability (%)	
	(i)	(ii)		(i)	(ii)
A	8.4966	8.000395	N	6.6544	6.903785
B	2.0720	1.535701	O	7.1635	7.591270
C	4.5388	2.575785	P	3.1671	1.795742
D	3.3844	4.317924	Q	0.1962	0.117571
E	11.1607	12.575645	R	7.5809	5.959034
F	1.8121	2.350463	S	5.7351	6.340880
G	2.4705	1.982677	T	6.9509	9.085226
H	3.0034	6.236609	U	3.6308	2.841783
I	7.5448	6.920007	V	1.0074	0.981717
J	0.1965	0.145188	W	1.2899	2.224893
K	1.1016	0.739906	X	0.2902	0.179556
L	5.4893	4.057231	Y	1.7779	1.900888
M	3.0129	2.560994	Z	0.2722	0.079130
Total = 100%					

characters in this textbook than in *Chinua Achebe's Things Fall Apart*, although both are English texts. The price paid for coding a message using the wrong probability distribution is a longer coded message that in some cases might even exceed the length of the uncompressed message (i.e. the message coded using fixed-length codewords). To take an extreme example, if a Huffman code table based on the probability distribution of letters in Table 6.4 were used to compress a message made up predominantly of the letter 'Q', the compressed file would be larger than before compression. So there is really no reliable alternative to using a Huffman code table built on the message at hand.

- The algorithm assumes a static character probability and a memoryless source neither of which are strictly true. In the building of words and messages in every language, characters in successive intervals exhibit some dependence, which means that the source has memory. For example, in an English text, if the current letter is a 'q' then the probability of the next letter being a 'u' increases dramatically from 2.84% on average for a 'u' to over 99%; and if two consecutive characters are the same letter then the probability of the next character being that same letter changes to ~0% irrespective of the letter. You can think of many more examples that lead to the conclusion that the probability of occurrence of a character in a given location (≡ signalling interval) within a message is a context-dependent value that is often very different from the global probability values in Table 6.4. Performance could be improved by extending Table 6.4 to include all *bigrams* (two-letter combinations) and *trigrams* and perhaps even *quadrigrams*. In this way, for example, rather than coding the letters 'q' and 'u' separately, we would code them together as 'qu' using the same codeword length as would be used for 'q' alone, thus saving bits. However, this approach leads to a very complex codec with a hugely unwieldy code table containing some entries that will be rarely encountered in most messages. Furthermore, even if successfully implemented in spite of its complexity, the block coding approach for block sizes $n = 1$ to 4 bits would represent only a partial solution since the context-dependence effect on probability extends beyond four adjacent characters. For example, given the sequence of characters 'oratory', it is clear (assuming an English language text) that the next character cannot be a letter. So the context here, as in countless other instances, changes the probability of occurrence of any letter to zero, the only options being a punctuation or space character whose probabilities would therefore dramatically increase in the context.
- Finally, Huffman's approach of assigning a unique codeword to each source character means that each character is represented by an integer number of bits (irrespective of its information content) and this leads to sub-optimal performance (i.e. poor coding efficiency, as shown in Eq. (6.13)) when the algorithm is applied to low-entropy sources. Huffman codes are optimal only if all character probabilities are integer powers of ½, which is rarely the case in practice. Recall that source entropy is the fundamental limit on the average number of bits needed to represent (without loss of information) each character emitted by the source. Since Huffman's absolute minimum is one bit per codeword, it means that code efficiency (≡ ratio between source entropy and average number of bits per codeword) will tend towards zero as source entropy decreases below one bit. If we code blocks of source characters, rather than individual source characters (see Worked Example 6.4), we can raise efficiency towards 100%, but this solution is at the expense of increased codec complexity.

6.6 Lempel-Ziv coding

The Lempel-Ziv algorithm is an adaptive lossless data compression technique developed by Abraham Lempel and Jacob Ziv and published in 1977 and 1978 [3, 4]. In 1984, Terry Welch [5] suggested a simpler logic for the algorithm to allow cheaper implementation and faster execution, and he called this variant of the algorithm Lempel-Ziv-Welch (LZW). Fundamentally, Lempel-Ziv is a universal compression algorithm that treats source data whatever their type as a sequence of bits, rather than a string of known characters. It makes only one pass through the source data and dispenses with the need for prior knowledge of the underlying statistics of the source, including its characters and their occurrence probability. Compression ratio and hence the compressed message size is variable, depending very much on message type, content and size. At start-up the algorithm is inefficient and actually leads to an expansion as code lengths exceed the bit patterns that they replace until patterns are encountered that repeat earlier patterns. Very short messages therefore do not compress well, and very long messages consisting of sections with significantly different bit patterns will also experience a decline in compression efficiency. Data types (such as floating point arrays) that look much like a random sequence of 1s and 0s will compress poorly (compression ratio $\sim$1), whereas those with repeated character patterns and structured blank spaces (e.g. computer program source codes such as COBOL files) will have good compression ratios between 2 and 6.

The UNIX shell compression program *compress* is an implementation of LZW, whereas *gzip* is a free data compression software based on a combination of LZ77 (the 1977 Lempel-Ziv algorithm) and Huffman coding. Created under the GNU project, *gzip* achieves better compression than both LZW and Huffman coding. There are several variants of the Lempel-Ziv algorithm but the fundamental idea is to parse the source data into a dictionary containing blocks of characters or patterns, each new entry being the longest sequence of characters not previously encountered. Thus, if an entry in the dictionary is not a single character encountered for the first time, then it has some earlier entry as its prefix. If we denote the nth entry in the dictionary as P_nK_n, then P_n is its prefix which is a pattern of characters already in the dictionary, and K_n is the next character that appends to make P_nK_n the longest sequence not previously encountered. The compressed (coded) output is produced by coding P_nK_n using a fixed length code made up of the address of P_n in the dictionary followed by K_n. If however a new entry is a single character encountered in the message for the first time, then there is no prefix P_n and the address portion is set to zero. There is a saving in the number of bits required to represent P_nK_n (and hence data compression) when eventually, beyond the early part of the input file, most patterns P_n are several characters in length and therefore much longer than the pointer. It is important to note that a character here refers to a convenient grouping of bits, such as a six-bit, eight-bit or nine-bit character, etc. or even one-bit, and not necessarily the ASCII character size.

6.6.1 Run-length encoding (RLE)

It should be clear from the above description that the Lempel-Ziv algorithm is a generalisation of *run-length encoding* (RLE), a lossless data compression scheme in which a pattern is more narrowly defined as a consecutive sequence (called *run*) of the same character. There are different variants of RLE depending on how the pattern is coded. In one implementation each pattern is coded using three fields, namely an escape code (such as the number 255) followed by the repeating character and then by the length of the run. An escape code in the input data is represented using two consecutive escape codes. As an example, given the following input message consisting of a string of eight-bit characters (listed below as comma-separated integers for easy reading), the RLE output code would be as shown:

Input Message: 5, 9, 4, 5, 9, 4, 211, 211, 211, 211, 211, 211, 211, 255, 48, 48, 48, 48, 48

RLE Code: 5, 9, 4, 5, 9, 4, 255, 211, 7, 255, 255, 255, 48, 5

RLE-recognised patterns in the input message are underlined above. The limitation of RLE is obvious even in this simple example in that the algorithm does not detect the repetition of the sequence 5, 9, 4 within the message, a situation that the Lempel-Ziv algorithm would efficiently handle. RLE is therefore ineffective for compressing text, where, with the exception of the space character, a run-length of 3 or more is extremely rare. However, RLE is excellent for application in facsimile (fax) transmission where the picture elements (called pixels and comprising 1728 pixels/line and 1188 lines/page (normal resolution standard) or 2376 lines/page (high-resolution standard)) are quantised into only two levels, either black (B) or white (W). A fax message, obtained for example by scanning a page of this textbook, will therefore contain long runs of B or W depending on the line. RLE can capture this run much faster than Lempel-Ziv, which by its design can only gradually build up a profile of patterns in a given message.

6.6.2 Lempel-Ziv encoding process

Table 6.5 illustrates the Lempel-Ziv encoding process. For simplicity we choose a source with a four-character alphabet *a, b, c*, ◇, where ◇ signifies EOF (end of file).

Table 6.5 Lempel-Ziv encoding procedure: 40 *input characters have been represented using* 15 *codewords*

Input data: ***c a c a a a a a c b b c c b c a b a a c c b b a a c c c b c b b b b a a c c c*****◇**

Dictionary address	**Dictionary entry**		**Coded output**	
	Pattern	**Fixed-length format**	**Codeword**	**Binary format**
1	*c*	<0, *c*>	<0, *c*>	000010
2	*a*	<0, *a*>	<0, *a*>	000000
3	*c a*	<1, *a*>	<1, *a*>	000100
4	*a a*	<2, *a*>	<2, *a*>	001000
5	*a a c*	<4, *c*>	<4, *c*>	010010
6	*b*	<0, *b*>	<0, *b*>	000001
7	*b c*	<6, *c*>	<6, *c*>	011010
8	*c b*	<1, *b*>	<1, *b*>	000101
9	*c a b*	<3, *b*>	<3, *b*>	001101
10	*a a c c*	<5, *c*>	<5, *c*>	010110
11	*b b*	<6, *b*>	<6, *b*>	011001
12	*a a c c c*	<10, *c*>	<10, *c*>	101010
13	*b c b*	<7, *b*>	<7, *b*>	011101
14	*b b b*	<11, *b*>	<11, *b*>	101101
15	*a a c c c* ◇	<12, ◇>	<12, ◇>	110011

The input data example comprises 40 characters listed in the first row of the table. The encoder makes a single pass through the input data (i.e. sequence of characters to be compressed) during which it constructs a dictionary (columns 1 to 3) and generates the coded output (columns 4 and 5). The dictionary entries are numbered 1, 2, 3,... in column 1 and this serves as the address m of the entry made in the dictionary by the encoder in sequential order starting at $m=1$. The sub-sequence P_nK_n parsed by the encoder at each step is the *longest sequence of characters not previously encountered in the input data*. We underline this sub-sequence in row 1 and also list it in column 2. The encoder writes P_nK_n into the dictionary and then presents it in the fixed-length format <Address of P_n, K_n> as the coded output of the current step. For the data sequence given, the first sub-sequence P_1K_1 is the single character 'c', so $K_1=c$ and there is no prefix P_1 (which we refer to as $P_1=$ Null, and assign dictionary address 0). The first dictionary entry is thus $K_1=c$, and the first coder output is <Address of P_1, K_1> ≡ <0, c>. In the second step ($n=2$), the encoder parses the sub-sequence $P_2K_2=$ 'a', which is also a single character with no prefix, so the second dictionary entry at $m=2$ is a, and the second coded output is <0, a>. In the next step $n=3$, the sub-sequence is $P_3K_3=$ '$c\ a$', which is a string with a prefix $P_3=c$ already in the dictionary at address $m=1$. The encoder therefore writes the third entry '$c\ a$' in the dictionary and outputs <1, a> as the third codeword. You are encouraged to work through Table 6.5 to see that the encoder proceeds in this way (writing an entry in the dictionary and producing a coded output at every step) until at step $n=15$, the sub-sequence is $P_{15}K_{15}=$ '$a\ a\ c\ c\ c\ \lozenge$', which has prefix $P_{15}=a\ a\ c\ c\ c$ located at address 12 in the dictionary, and has next character $K_{15}=\lozenge$ that signifies EOF. Thus the encoder produces the final codeword <12, ◊> for the input data. It should be noted that the dictionary is a local code book used by the encoder but not transmitted or saved with the coded message. There is enough information in the transmitted codewords (column 5) for the receiver to decode the message on the fly while at the same time reconstructing an exact copy of the dictionary.

So where is the compression in all these? In this illustration the input data chosen led to 15 sub-sequences and hence 15 dictionary entries involving 4 characters each of which may be identified using two bits ($a \equiv 00$, $b \equiv 01$, $c \equiv 10$, $\lozenge \equiv 11$). Thus, including address 0 for null prefix, four bits are needed to code dictionary addresses (from $0 \equiv 0000$ to $15 \equiv 1111$) and two bits for next character K_n, which leads to the six-bit codewords listed in the final column of Table 6.5, where the first four bits specify the dictionary address of the prefix and the last two-bits specify the next character. It can be seen that in the early part of the data, there is no compression at all. In fact there is an expansion, since the first sub-sequence is a two-bit character but is coded using six bits, the second is also a two-bit character coded with six bits, the third is a four-bit string coded using six bits; and so on. However, by the 15th sub-sequence we find that a 12-bit string is coded using six bits, which represents compression by a factor of 2. Overall, we went from an 80-bit input to a 90-bit output, which represents a small expansion. But this was a short 40-character message that terminated before we could begin to reap the rewards of our early investment in coding short prefixes at a loss.

The Lempel-Ziv algorithm is a smart and adaptive method of reducing the redundancies (in the form of repeated patterns) that are inherent in an information-bearing bit stream (i.e. data file). The encoder uses a compact format to memorise every new pattern that it finds in the file so that if, as is likely due to redundancy, that pattern reappears at a subsequent point in the file then an index is inserted with which a decoder can recall the pattern. For there to be any savings in the number of bits needed to represent the data file, the memorised pattern must be longer than the index used to reference it. Since a new pattern is longer than its immediate prefix by only one character, it takes long input sequences (i.e. large data files) to progressively increase the lengths of memorised patterns in order to maximise bit savings. However, this benefit cannot grow indefinitely since index length (in bits) increases logarithmically as the number of patterns memorised: If you have M patterns, you need $\log_2 M$ bits to index them all.

If we divide an arbitrarily large file into multiple sections of equal length and find that the dictionaries constructed on each section contain the same set of patterns, the data source is said to be *ergodic*. This means that the redundancy statistics of the data remain the same across all sufficiently large sections of the file as its size grows indefinitely. Thus character patterns can be continually re-applied and extended in their length as one goes deeper into the file, which improves coding efficiency. The Lempel-Ziv algorithm therefore asymptotically approaches 100% coding efficiency. It is remarkable that (on the condition of ergodicity) this asymptotic approach towards 100% coding efficiency is an inherent feature of the algorithm that does not require special measures which would increase codec complexity as in the case of Huffman coding.

To assess the performance of the algorithm in the example of Table 6.5, we note that the source emits four characters with probabilities 0.325, 0.275, 0.375 and 0.025 (based on a count of the number of each character in the input sequence of 40 characters). Thus source entropy is

$$\begin{aligned} H(X) &= 0.325 \log_2(1/0.325) + 0.275 \log_2(1/0.275) \\ &\quad + 0.375 \log_2(1/0.375) + 0.025 \log_2(1/0.025) \\ &= 1.7029 \text{ bits} \end{aligned}$$

which means that the fundamental limit on the coding of this source is 1.7029 bits per character on average. Fixed-length coding (i.e. uncompressed input data) has two bits per character, and thus an efficiency of 1.7029/2 = 85.1%, whereas the Lempel-Ziv code uses 90 bits for 40 characters, or 2.25 bits/character, which gives a code efficiency of 1.7029/2.25 = 75.7%. It can be seen that for this short 40 character message, the coding efficiency of the algorithm is far from the limiting value of 100%. However, as message length increases indefinitely the efficiency would increase asymptotically towards 100%, allowing us to approach towards the fundamental limit of 1.7029 bits/character (on average) for coding information from this particular four-character source.

Just for comparison, we note that Huffman coding would have represented this source more efficiently by using the codewords 000 for ◊, 001 for *b*, 01 for *a*, and 1 for *c*, which gives an average codeword length of 1.925 bits/character and hence an efficiency of 88.5%. The problem however is that the occurrence probabilities of the characters would have had to be known in advance to inform the selection of Huffman's variable length codewords. It is important to note the difference between the two parameters, namely coding efficiency and compression ratio, which we have used here to assess code performance. Both compare the number of bits used by the code to represent the message against some standard, which for coding efficiency is the fundamental minimum number of bits stipulated by Shannon's source-coding theorem, whereas for compression ratio the comparison is against the uncompressed number of bits resulting from each unique character in the message being assigned a unique fixed-length binary codeword. So in the example of Table 6.5, the respective coding efficiencies for Huffman, uncompressed and Lempel-Ziv techniques are 88.5%, 85.1% and 75.7%, and the respective compression ratios are 80/77 = 1.039, 80/80 = 1.0 and 80/90 = 0.889.

The address field of the Lempel-Ziv codeword <Address of P_n, K_n> is usually of a specified length k bits. To be able to use k bits to number every entry in the dictionary, its size (and hence the maximum number of sub-sequences that can be saved) is limited to

$$N_s = 2^k - 1$$

reserving address 0 for the null prefix. When compressing a sufficiently long input data file the dictionary will therefore eventually become full, and this can be dealt with in one of two ways:

- No further sub-sequences are added to the dictionary. Coding continues as usual using the N_s prefixes that are already in the dictionary, but this means that there can be no further increase in the number of characters per sub-sequence and therefore the asymptotic improvement in coding efficiency is halted.
- New sub-sequences continue to be added to the dictionary at the addresses of little-used previous sub-sequences that are therefore replaced. The replacement is carried out in a deterministic way that is followed at encoder and decoder so that both will make exactly the same decisions. This approach constantly updates the dictionary without increasing its size, which allows the encoder to better handle non-ergodic data streams by adapting to changing character patterns.

6.6.3 Lempel-Ziv decoding process

The decoding process is as straightforward as encoding. Just as the encoder creates a dictionary and coded outputs on the fly, the decoder similarly creates an identical dictionary as well as the decoded message. The decoder receives only the succession of codewords (e.g. column 4 of Table 6.5, which is the more readable non-binary format for column 5). First it initialises its local dictionary address pointer to $m = 1$, and then reads the codewords <Address of P_n, K_n> in succession starting at $n = 1$. For each codeword the decoder performs the following operations: (1) It obtains the sub-sequence P_n from the specified address in its local dictionary and then outputs the string P_nK_n as the message represented by the codeword. If the address is zero (as it must be for the first codeword read) then P_n = Null. (2) It writes the entry P_nK_n into its local dictionary at address m and increments m by one.

If the dictionary becomes full, the decoder will no longer increment m after each codeword but will follow exactly the same procedure as used at the transmitter, which is either to not write any further entries into the dictionary or to set the value of m to replace older entries according to a common deterministic algorithm. In this way the decoder's dictionary will be exactly identical to the encoder's at the same point in the message, which ensures that the content placed by the encoder at any location in its own dictionary will be the same content retrieved by the decoder from its local dictionary created on the fly.

6.6.4 Strengths and drawbacks of Lempel-Ziv

The Lempel-Ziv algorithm is a universal lossless data compression technique that asymptotically approaches 100% coding efficiency in long messages. The encoder needs no prior knowledge of source alphabet or statistics and the decoder avoids the overhead of requiring a copy of the codebook (i.e. dictionary) used by the encoder. The decoder has sufficient information in its own local dictionary created on the fly to make an immediate decision on the message represented by the current codeword, so decoding delay is negligible. Furthermore, the coding and decoding algorithms are simple and fast to implement. However, the coding efficiency of short messages is poor, and this also applies to long messages composed of short sections with significantly different character patterns. Moreover, a message compressed using the Lempel-Ziv algorithm is totally intolerant to error. If there is an error in even a single codeword (in the compressed message) that error may propagate through the recovered message as the prefix in the errored codeword is reused repeatedly at several points in the recovered message. As an example, consider that if in Table 6.5 there was an error in any of the last two bits of the second codeword <0, *a*> that error would be carried forward into codewords 4, 5, 10, 12, 15 and beyond, leading to a garbled message. It is therefore essential to protect compressed messages by using a cyclic redundancy check computed on the entire set of Lempel-Ziv codewords that represents the original message. Decoding is abandoned and the message is declared corrupted and unusable if any error is detected. Fortunately, errors are extremely rare in modern storage devices such as computer hard disks where compressed files are stored.

6.7 Arithmetic coding

We have studied Huffman and Lempel-Ziv codes in detail in the last two sections. Huffman code views a message as a sequence of characters and produces a variable-length codeword for each character in the message, coding efficiency being achieved by using fewer bits (i.e. shorter codewords) to represent the more frequent characters. Lempel-Ziv code on the other hand treats a message as a collection of repeatable patterns of characters of variable length and assigns a fixed-length codeword to each pattern, coding efficiency being realised by representing a long pattern of characters using a fixed number of fewer bits. *Arithmetic coding*, the subject of this section, is however the state of the art in lossless data compression.

Arithmetic coding views a message as one string of interdependent characters, and produces a *single codeword for the entire message*. Yes the codeword is long, but it is very efficient because, unlike Huffman code which assigns bits to characters based on their fixed message-global probability, arithmetic code has the ability to assign bits to the next character in the message based on its variable context-dependent probability at that point in the message. More specifically, Huffman dedicates the same number of bits to code a particular character such as ‘u’ based on its estimate of the probability of the character obtained by counting all characters in the message. We call this the *message-global probability*. So if you went through an English message and asked Huffman what was the probability of the letter ‘u’ at each location, it would always answer with the same value, say 3%, and therefore would feel justified to use the same number of bits, say k_3, to code each instance of a ‘u’ in the message. The exciting innovation that arithmetic coding brings is the means to decodably assign different numbers of bits to code ‘u’ at different locations in the same message in accordance with the variation in the probability of a 'u' at different points in the message.

To comment further on the above example, it is well known that in English text the letter 'q' is almost always followed by the letter 'u'. So when the current character in a message is the letter 'q', there is little surprise when the next character is the letter 'u'. That is, this 'u' contains little information so coding it with k_3 bits as Huffman does is clearly inefficient and a waste of bits. However, at some other location such as after letter 'i', the character 'u' would be more surprising and hence contain more information thus justifying its representation using k_3 bits or a little more. The arithmetic coding algorithm makes it possible to represent each character in a message using a number of bits that equals the information content of that character at that location in the message, thereby achieving 100% coding efficiency. Any shortcoming in coding efficiency is due entirely to errors in the answers given at each point in the message to the crucial question: What is the character that can occur next and with what probability? In other words, given a source model that accurately predicts the context-dependent distribution of probabilities for the next character, the arithmetic coding technique delivers optimal coding of the message. In what follows we explain how the arithmetic codeword is computed at the encoder and interpreted at the decoder in the simple case of a fixed source model, deriving the codeword first in decimal form and then as a binary sequence. We then discuss how a simple adaptive source model might be incorporated for improved efficiency.

6.7.1 Arithmetic encoding process

Let the message to be coded be of length N characters emitted by a source with an alphabet of size M characters, $\{x_1, x_2, x_3, \ldots, x_M\}$, where x_j is the jth character, and the last character x_M is dedicated for signalling EOF. The source emits one character in each signalling interval numbered from $n = 1$ for the first character to $n = N$ for the last character in the message. Let the source probability distribution in the nth signalling interval be the set of numbers

$$\{p_{1n}, p_{2n}, p_{3n}, \ldots, p_{Mn}\}, \quad n = 1, 2, 3, \ldots, N$$

where $0 < p_{jn} < 1$ is the probability of occurrence of the jth character x_j in the nth signalling interval. The simplest source probabilistic model is a *fixed model* describing a discrete *memoryless* source in which outcomes in successive intervals are statistically independent and the probability distribution is static so that

$$p_{j1} = p_{j2} = p_{j3} = \cdots = p_{jN} = p_j, \quad j = 1, 2, 3, \ldots, M$$

That is, character x_j has a constant occurrence probability p_j in every signalling interval. In reality most sources (e.g. English text) will be best described by an *adaptive model* in which p_{jn} is a context-dependent probability conditioned by what characters were emitted in previous signalling intervals up to $n - 1$. Arithmetic coding is however not tied to any particular source model. The tasks of source modelling and source coding are clearly separated, the only requirement being that encoder and decoder must use the same source model. Whatever the source, if the probabilistic model is accurate, arithmetic coding enables coding of the source at its theoretical entropy bound. We are concerned here only with the source coding process.

The encoder steps through the message starting at the first character (in signalling interval $n = 1$) until the Nth character (in interval $n = N$), and progressively constructs a single codeword to uniquely represent the entire message. The codeword is completed only after the Nth character has been read but encoder output bits are released on the fly as they are ready. In building up the sequence of bits that upon completion constitutes the codeword, the encoder assigns fewer coding bits, which may even be fractional, to the more probable characters — since their occurrence springs less surprise and therefore gives less information — and assigns more coding bits to the less probable characters that contain more information. Recalling our code space discussion following Eq. (6.11), this means that the amount of code space allocated to each character is proportional to its probability.

The encoder therefore starts by partitioning the entire code space range [0, 1) into M intervals of respective sizes $p_{11}, p_{21}, p_{31}, \ldots, p_{M1}$, where the first code space interval $[0,\ p_{11})$ belongs to character x_1, the second code space interval $[p_{11},\ p_{11}+p_{21})$ belongs to character x_2, and so on as shown in Fig. 6.7a. Note that $[a,\ b)$ is used to denote the half-open interval $a \le z < b$. The encoder then reads the first character (emitted by source in the first signalling interval $n=1$) and selects the code space interval allocated to this character. Let the first emitted character be the j_1^{th} character x_{j_1}. Then the selected code space interval at $n=1$ is $[a_1,\ b_1)$, where

$$a_1 = \sum_{i=1}^{j_1-1} p_{i1}; \quad b_1 = \sum_{i=1}^{j_1} p_{i1} \tag{6.14}$$

The bits allocated to the next character (emitted by source in interval $n=2$) will be appended to the codeword so far constructed, thereby increasing its length and therefore narrowing the occupied code space — recall the inverse relationship between code space and codeword length. If the second character is very probable, only a small number of bits will be devoted to it, and so there will be only a small increase in codeword length and hence only a small reduction in occupied code space. The encoder therefore partitions the currently occupied code space $[a_1,\ b_1)$ into M intervals in proportion to $\{p_{12}, p_{22}, p_{32}, \ldots, p_{M2}\}$, the predicted probabilities for interval $n=2$, and then selects the sub-interval belonging to the character actually emitted. So if the 2nd emitted character is the j_2^{th} character x_{j_2} then the occupied code space interval after reading the first two characters $x_{j_1}x_{j_2}$ is $[a_2,\ b_2)$ given by

$$a_2 = a_1 + S_1\sum_{i=1}^{j_2-1} p_{i2}; \quad b_2 = a_1 + S_1\sum_{i=1}^{j_2} p_{i2}; \quad S_1 = b_1 - a_1 \tag{6.15}$$

where S_1 is the previous code space range. The encoder continues in this manner, as illustrated in Fig. 6.7 up to $n=3$, taking the following action at each step n: (1) Partitioning the previous code space range according to the probability distribution for the current interval provided by the source model, and (2) selecting the partition that belongs to the character actually emitted by the source in the current interval. The encoder reaches the final (EOF) character at step $n=N$, at which point the final occupied code space interval $[a_N,\ b_N)$ is given by

$$a_N = a_{N-1} + S_{N-1}\sum_{i=1}^{M-1} p_{iN}; \quad b_N = a_{N-1} + S_{N-1}; \quad S_{N-1} = b_{N-1} - a_{N-1} \tag{6.16}$$

So how is this code space interval $[a_N,\ b_N)$ a codeword? To see this, take a moment to look again at the code space of binary codewords shown in Fig. 6.3. Notice for example that the codeword 01 occupies the code space interval [1/4, 1/2) of size 0.25 in column $l=2$, and that the codeword 10011 occupies code space interval [19/32, 20/32) of size 0.03125 in column $l=5$. Thus given a code space interval such as the one at hand $[a_N,\ b_N)$, we can write down the corresponding codeword.

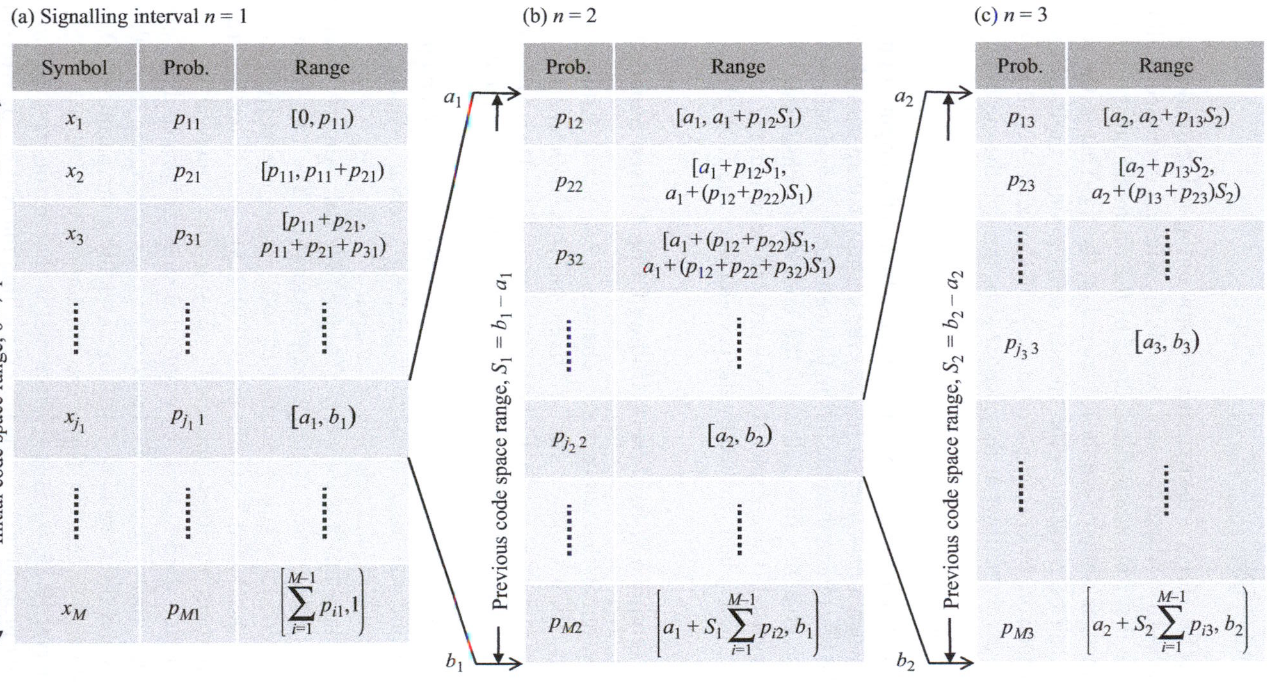

Fig. 6.7 Partitioning of code space into intervals equal to symbol probability

The interval size $S_N = b_N - a_N$ determines codeword length l — the smaller the codespace interval the longer the codeword — whereas the start of the interval a_N determines the sequence of bits that constitutes the codeword as follows:

$$\begin{aligned} l &= \lfloor -\log_2 S_N \rfloor; \quad S_N = b_N - a_N \\ \text{Codeword} &= \lfloor 2^l a_N \rfloor|_{\text{base 2 (to } l \text{ digits)}} \end{aligned} \tag{6.17}$$

The first line of Eq. (6.17) is the rightmost column of Fig. 6.3 (extended with as many columns as necessary) whose row size exceeds S_N, whereas in the second line we are shifting out the first l most significant digits of the fractional binary number a_N. In practice the encoder can produce this codeword on the fly. That is, the early part of the codeword can be released as the encoder steps through the message until the sequence of codeword bits is completed at the EOF character. Recall that code space interval progressively narrows as the encoder steps through the message:

$$[0,\ 1) \rightarrow [a_1,\ b_1) \rightarrow [a_2,\ b_2) \rightarrow [a_3,\ b_3) \rightarrow [a_4,\ b_4) \rightarrow \cdots$$

Thus the encoder is moving deeper into Fig. 6.3 from left to right. As it does so, it looks out for when a new interval $[a_n, b_n)$ fits entirely within a single row. If that happens in the kth column then the bits in that cell (i.e. row and column) of Fig. 6.3 unambiguously give the first k bits of the codeword, which can therefore be immediately released. The completed codeword will be of length l bits given by the first line of Eq. (6.17) and reached by the encoder after the EOF character.

Worked Example 6.5: Arithmetic Coding Using a Fixed Source Model

Consider a source that emits three characters x_1, x_2, x_3 with respective probabilities 0.5, 0.4 and 0.1 that are fixed and context-independent, where the last character x_3 is the EOF character. We wish to determine the arithmetic code for the message string $x_1x_2x_1x_1x_2x_3$ from this source, and compare with Huffman and Lempel-Ziv's.

Number of source characters $M = 3$, message length $N = 6$ and predicted probabilities are the same in every signalling interval, namely $p_1 = 0.5$, $p_2 = 0.4$, $p_3 = 0.1$ for characters x_1, x_2 and x_3.

The coding steps are shown in Fig. 6.8. We partition the initial code space [0, 1) in proportion to the given probabilities. The first character (at signalling interval $n = 1$) is x_1, so the first code space interval [0, 0.5), which belongs to x_1, is selected. Next, the interval [0, 0.5) is partitioned according to the character probabilities applicable to signalling interval $n = 2$. Use Eq. (6.15) if you need to, or simply note that since the probabilities are 0.5, 0.4 and 0.1, it means that the first half of [0, 0.5) is assigned to x_1, the next $4/10^{\text{th}}$ to x_2 and the remaining $1/10^{\text{th}}$ to x_3, yielding the partitions [0, 0.25), [0.25, 0.45) and [0.45, 0.5). Since the character emitted at $n = 2$ is x_2, the second interval [0.25, 0.45) is selected and is then partitioned according to the character probabilities for the next signalling interval $n = 3$. We continue in this way, obtaining the results shown in Fig. 6.8 until after coding the last character x_3, the final occupied code space interval is

$$[a_N,\ b_N) = [0.293,\ 0.295)$$

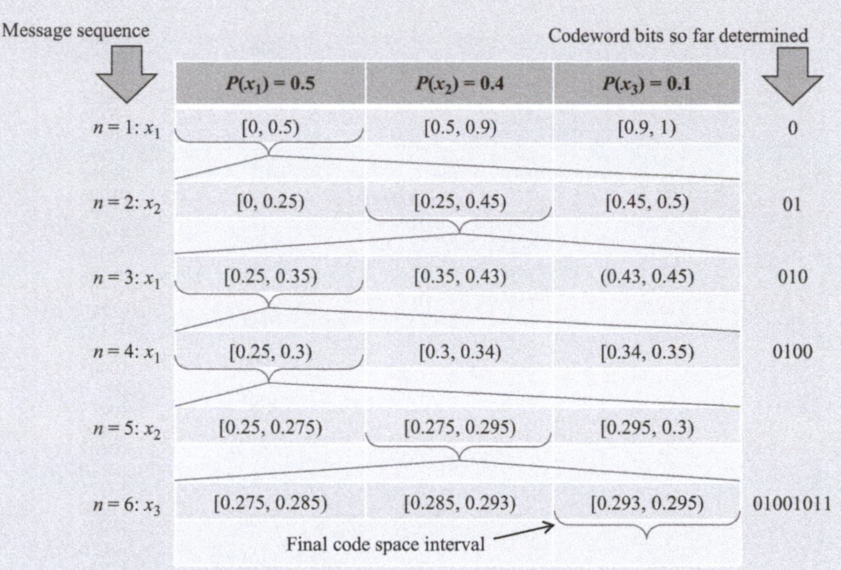

Fig. 6.8 Arithmetic encoding steps for Worked Example 6.5

Using Eq. (6.17) we obtain the codeword length and codeword as

$$l = \lfloor -\log_2(0.295 - 0.293) \rfloor = \lfloor -\log_2(0.02) \rfloor = \lfloor 8.9658 \rfloor = 8$$

$$\text{Codeword} = \lfloor 2^8 \times 0.293 \rfloor|_{\text{base 2 (to } l \text{ digits)}} = 75|_2 = 01001011$$

To further clarify the arithmetic encoding process and demonstrate how the encoder is able to release codeword bits on the fly, consider Fig. 6.9 where we superimpose this solution on a code space diagram. Double-headed arrows demarcate code space intervals labelled with the source character sequence that would take the encoder into that interval. The solid bold arrow indicates the interval selected in response to the character actually emitted by the source. Notice that having observed x_1 at $n=1$, the encoder is in a code space range [0, 0.5) that lies entirely in the first row of column 1, so the encoder knows that the first bit of the codeword has to be a 0. At $n=2$, the code space range [0.25, 0.45) lies entirely in the second row of column 2 (which covers the code space range 0.25 → 0.5 and belongs to code prefix 01), so the encoder knows that the first two bits of the codeword have to be 01. The bits that have been so determined can be released as the encoder advances through the message sequence. This is shown on the RHS of Fig. 6.8. Going further, the observation of character x_1 at $n=3$ causes the encoder to select code space range [0.25, 0.35) which lies entirely in the third row of column 3, thus determining the first three bits of the codeword as 010. Similarly, the occurrence of character x_1 at $n=4$ causes the range [0.25, 0.3) to be selected which lies within row 5 of column 4, thus determining the first four (prefix) bits of the codeword as 0100. To complete this coding process, the code space diagram is continued in Fig. 6.10 for prefix 0100 up to column $l=9$.

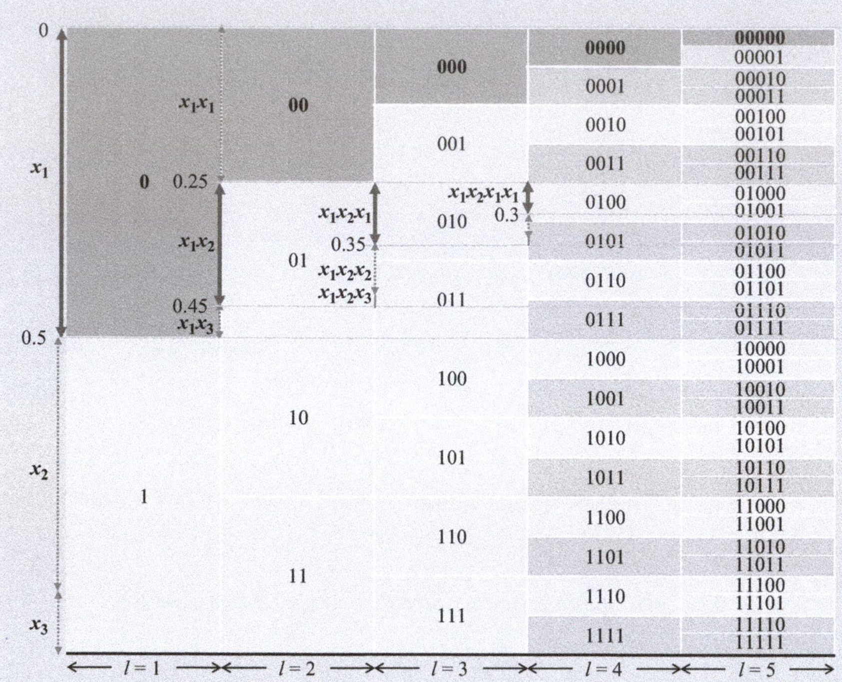

Fig. 6.9 An arithmetic codeword can be viewed as a trail through code space

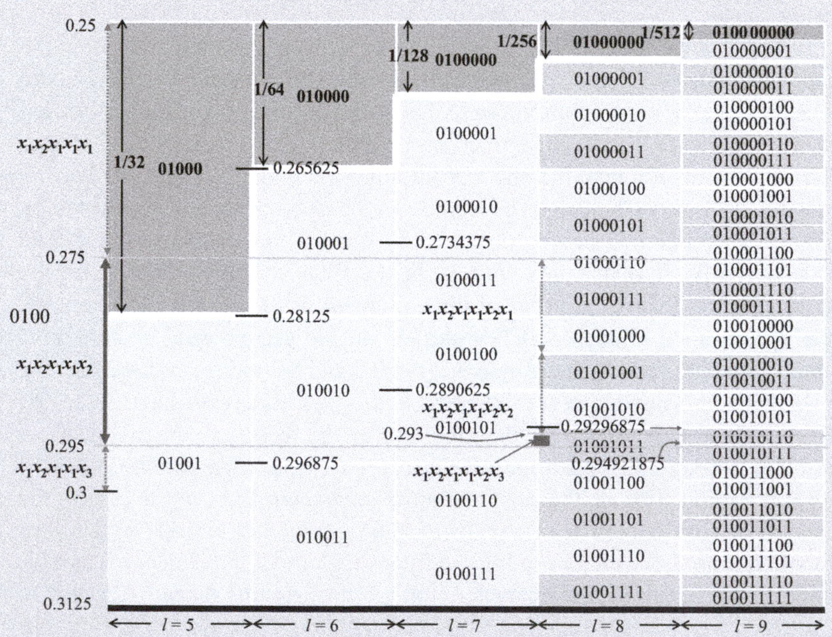

Fig. 6.10 Code space for codewords with prefix 0100 up to length $l = 9$ for Worked Example 6.5

At $n = 5$, the emitted character x_2 causes the encoder to select code space range [0.275, 0.295) which straddles two rows (01000 and 01001) of column 5, which means that the encoder cannot at this point tell what the next bit of the codeword will be. Finally, at $n = 6$ the emitted character is the EOF character x_3 and this causes the encoder to choose its terminating code space range [0.293, 0.295) of size 0.002. The rightmost column within which this range can fit into a single row is column 8 where codewords occupy a code space range of $1/256 = 0.00390625$. So the codeword for this message sequence is of length eight bits and is given by the codeword of the row in column 8 that contains the lower limit (i.e. 0.293) of the terminating code space range. Thus codeword is 01001011.

The arithmetic codeword we have computed is eight bits long. Huffman code would assign codeword 1 to character x_1 and codewords 01 and 00 to characters x_2 and x_3, respectively, so that the six-character message would be represented using nine bits, which is slightly longer than the arithmetic code. Lempel-Ziv would process this message string into four codewords $<0, x_1>$, $<0, x_2>$, $<1, x_1>$ and $<2, x_3>$ each requiring four bits (since two bits are needed to index a dictionary that has four entries, and two bits are needed for the three characters), which gives a total of 16 bits. As earlier discussed, Lempel-Ziv cannot be competitive when used on such a short message.

6.7.2 Arithmetic decoding

The codeword generated by an arithmetic encoder is actually a trail through code space from left to right. For example, the trail of the codeword 01001011 of the previous worked example is signposted by the solid double-headed arrows in Figs. 6.9 and 6.10. If an arithmetic decoder can unambiguously determine the trail created by a given codeword it will successfully decode the message represented. This task is made possible by the decoder using exactly the same source probabilistic model as the encoder, including an identical ordering of characters. The decoder computes the character probabilities for each signalling interval n using exactly the same model as the encoder, and partitions the current code space accordingly so that it knows the code space boundaries for each character. By examining the coded binary sequence the decoder is able to tell which code space partition the encoder selected and hence which source character the encoder observed. In this way the decoder is able to regenerate the original source character sequence or message.

Let us consider the decoding of the codeword 01001011 from the previous worked example. Using the first set of character probabilities (for $n = 1$), the decoder computes the code space intervals for characters x_1, x_2 and x_3 as shown on the LHS of the first column in Fig. 6.9. It then reads the first bit of the codeword and deduces that bit 0 implies that the encoder selected the interval for character x_1. Thus the decoder generates x_1 as the message character at $n = 1$. It then creates the second set of partitions (on LHS of column 2 of Fig. 6.9), and seeing that the first two bits of the codeword is 01 it knows that the encoder selected the second interval of this partition which can only mean that the character sequence was x_1x_2. Working in this way the decoder correctly deduces message sequence $x_1x_2x_1x_1$ from the first four bits 0100 of the codeword. The next codeword bit (following 0100) is binary 1, which makes it certain that it was not character x_1 at $n=5$ (since this would have produced 0 as the next bit — see partitions on LHS of the first column of Fig. 6.10). So was it x_2 or x_3 (EOF) at $n=5$? By examining a few more bits of the codeword, the decoder is able to ascertain that the codeword could only be produced by x_2 at $n=5$ (leading to the partition shown on the LHS of column 4 in Fig. 6.10) followed by x_3 at $n=6$, which by convention signals end of message.

6.7.3 Arithmetic precision issue

It would appear that arithmetic coding requires performing arithmetic with infinite precision in order to reach the final code space interval (equivalent to deriving the binary codeword) that represents a long message sequence. This would however only be the case if in constructing the codeword, by progressively appending bits as discussed above, we had no option but to penetrate deeper and deeper into code space, where interval sizes decrease with codeword length l as 2^{-l}. Notice though that the code space diagram expands to the right in exactly the same manner, irrespective of column (i.e. depth): The next column to the right is the result of partitioning each row of the current column into two and appending bits 0 and 1, respectively to the codeword of the current row to form the codewords of the upper and lower partitions. It is not surprising that if the prefix bits 0100 are dropped from all codewords in Fig. 6.10 we obtain Fig. 6.9 except for a scale factor in code space interval — since Fig. 6.9 spans the interval [0, 1) of size = 1, whereas Fig. 6.10 spans the interval [0.25, 0.3125) of size 2^{-4}.

Therefore the need for infinite precision arithmetic can be avoided if every time after determining prefix bits we return to the start of the code space diagram ($l = 1$) to continue the search for subsequent codeword bits. In other words, we penetrate into the code space diagram only as far as necessary to establish the identity of the next few bits and then return to the starting point in column one. This is accomplished through repeated *normalisation* whereby the code space interval is scaled by a factor of 2^k for every k codeword bits determined. For example, having determined the first four bits of our coding example as 0100, we scale the range [0.25, 0.3125) by a factor of 16 to [0, 1) so that the interval [0.25, 0.3) becomes [0, 0.8). We then continue with coding in the usual way by partitioning this magnified interval in proportion to character probabilities and selecting the partition belonging to the character actually observed. It is important to understand when and how to carry out the normalisation discussed here. For example, if at $n=1$, the emitted character causes interval [0.2, 0.4) to be selected, the first codeword bit can be immediately determined as 0, and the interval doubled to [0.4, 0.8) for $n=2$. That is straightforward! If the selected interval at $n=1$ is [0.55, 0.85), the first codeword bit is 1 (since this interval fits entirely within code space range [0.5, 1) that belongs to codeword 1) so the interval is doubled to [0.1, 0.7) for $n=2$. This needs care to do it correctly by not only doubling the size of the interval from 0.3 to 0.6, but also translating to ensure that the gaps to code space boundaries at either end of the original interval are also doubled from 0.05 and 0.15 to 0.1 and 0.3. However, if at $n=1$ the selected interval is [0.45, 0.68), it is not possible to tell the identity of the next codeword bit since this interval spans 0.5, which is the code space boundary between bit 0 and bit 1. In this case we must proceed deeper into code space without scaling.

6.7.4 *Adaptive model*

The example we presented on arithmetic coding and decoding made use of a fixed source model in which the distribution of character probabilities is the same in all signalling intervals. However, arithmetic coding can work just as readily with adaptive context-dependent probabilistic models. In the absence of a detailed knowledge of source statistics (except that it has an alphabet size of M characters), the simplest way to achieve adaptivity is to begin (at signalling interval $n = 1$) with each character x_j having equal probability and then to continually update the probability of each character in subsequent signalling intervals to reflect the number of occurrences of each character in the message so far. Denoting the probability of occurrence of character x_j in signalling interval n by p_{jn}, we can specify the probabilistic model as follows:

$$p_{jn} = \begin{cases} \dfrac{1}{M}, & n = 1 \\ \dfrac{N_j + 1}{M + n - 1}, & n = 2, 3, 4, \ldots, N \end{cases} ; \quad j = 1, 2, 3, \ldots, M \tag{6.18}$$

where N is the message length (in number of characters), and N_j is the cumulative total number of occurrences of character x_j in all signalling intervals up to $n - 1$. For large N, the probability values for rare characters will tend to zero in the above equation as n increases. It is however impossible to code a character if it has a predicted probability of zero in the interval in which it is by chance encountered, since no code space partition would have been allocated for that character. In using Eq. (6.18) it will therefore be necessary to set a limit on the minimum value of probability to prevent it being rounded to zero as a result of limitations in the precision to which arithmetic can be performed on a computer. When it becomes necessary to enforce such a probability limit for any character in any signalling interval, it must be borne in mind that this artificially inflates the probability of that character merely for computational convenience, and therefore the probability values of all other characters in that signalling interval must be proportionately reduced as required to ensure that

$$\sum_{j=1}^{M} p_{jn} = 1, \quad \text{for all } n$$

6.7.5 *Weaknesses of arithmetic coding*

Arithmetic coding will deliver optimal coding of a message if supplied with an accurate model of source statistics, whatever the source entropy. For low-entropy sources, Huffman coding can match this performance only by resorting to an unwieldy block coding approach using a sufficiently large block size n, whereas Lempel-Ziv coding will asymptotically approach optimality only in sufficiently long messages produced by an ergodic source. Arithmetic coding appears to be the ultimate state-of-the-art in entropy coding, suggesting that future research should be focused on improving its algorithmic computational speed and developing more sophisticated and accurate source models. Coding of a general English text at 2.2 bits/character on average using arithmetic coding in conjunction with dynamic

Markov modelling has been reported [6], which is a compression ratio of 3.2 compared to a seven-bit ASCII representation. However, arithmetic coding is computationally more demanding than both Huffman and Lempel-Ziv, a lot of manipulations being required at every signalling interval especially for updating source statistics. The arithmetic codeword is also highly sensitive to bit errors and must be protected using a cyclic redundancy check. Furthermore, as elaborated below, an EOF character is essential for message termination, which can add up to two extra bits to the codeword.

Although the encoder computes the final code space range $[a_N, b_N)$, it is actually only a single number within this range that is sent to the decoder as the codeword. See for example the conversion of a_N to a binary codeword in Eq. (6.17). Without a dedicated end of message character, the decoder would not know when to stop decoding. To see why this is so, consider a simple example of a source that emits characters x_1 and x_2 with probabilities 0.6 and 0.4. Following the encoding procedure that we have learnt, message x_1 produces final code space range $[a_N, b_N) = [0, 0.6)$; message x_1x_1 produces $[a_N, b_N) = [0, 0.36)$; $x_1x_1x_1$ produces $[a_N, b_N) = [0, 0.216)$; $x_1x_1x_1x_1$ produces $[a_N, b_N) = [0, 0.1296)$; and so on. If the decoder received $[a_N, b_N)$ there would be no ambiguity regarding the message length, but that would needlessly double the number of codeword bits (to cover both a_N and b_N). So the decoder receives a_N only, which is the same for all the listed sequences and therefore it cannot tell whether the message ends with a single or multiple x_1. Consider introducing an EOF character $\Diamond$ to the alphabet $\{x_1, x_2, \Diamond\}$ with probabilities $\{0.5, 0.4, 0.1\}$. The message x_1 would then be issued as $x_1\Diamond$ and assigned codespace range $[a_N, b_N) = [0.45, 0.5)$; and message x_1x_1 would be issued as $x_1x_1\Diamond$ and assigned code space range $[a_N, b_N) = [0.225, 0.25)$, which correspond to different codewords a_N. Thus the ambiguity is resolved by introducing an EOF character. Clearly, in all but the shortest messages, this message termination overhead is insignificant.

6.8 Summary

The last three chapters have been about the important subject of source coding, which in digital communications deals with how to efficiently represent information from a source as a sequence of integer numbers (usually expressed in binary, hence a bit stream). If the source is analogue then the coding is *intrinsically lossy* because it is only through quantisation that a continuous-value waveform can be constrained to discrete values that can then be mapped onto integer numbers. As a starting point, the design of efficient coders for analogue sources exploits the limitations of human audio-visual perception by eliminating *insignificant details* from the source information on the basis that they are humanly undetectable. For example, a 30 kHz audio component from a musical instrument will not be heard by a human. In practice this means that an analogue waveform can be replaced by a digital signal that is a *filtered*, *sampled* and *quantised* version of the original but conveys information that appears identical in every respect to the original.

Further savings in the number of coding bits can be made by reducing inherent redundancy in the resulting digital signal. Viewed in the time domain this means that we recognise that the current sample contains contributions from past (previously coded) samples. So rather than code the signal samples as though they were independent, we exploit this knowledge to code the current sample more efficiently using fewer bits. Viewed in the frequency domain, signal redundancy means that information content is not evenly spread across the signal bandwidth, rather there are pockets of concentration that deserve more coding bits and swathes of insignificant content that can be ignored altogether. Beyond the exploitation of source redundancy and human perception limitations, any further savings in coding bits *must* sacrifice perceived signal quality, for example by using coarser approximations that introduce perceptible distortions.

These analogue source coding ideas and concepts have been explored in detail in this book, starting in Chapter 4 where we specified filtering and sampling rate requirements, and culminating in Chapter 5 where we encountered toll quality telephone speech at bit rates from 64 kb/s through 32 kb/s down to 8 kb/s, and poorer quality speech at 2.4 kb/s and even 800 b/s.

The efficient coding of discrete sources was the subject of Chapter 6. Like analogue sources, a discrete source contains inherent redundancy. However, unlike analogue sources, it does not produce insignificant details. Thus, approximations — fine or coarse — are not applicable. An example of a discrete source is the keyboard for English text. The message 'BILL JILL' does contain some redundancy but no insignificant detail. Any change or dropping of even one character changes the message in a significant way. For example, 'KILL JILL' and 'ILL JILL' are both noticeably different from the original. Therefore to preserve the integrity of information from discrete sources, the adopted coding method must be *lossless* in that it should be possible to recover an exact copy of the original message from the coded sequence of bits.

Furthermore, unlike analogue sources where quality can be subjectively traded for a gradual reduction in coding bits, a discrete source does have an objective fundamental limit on the minimum number of coding bits required to represent a message. This is given by the product of the entropy of the source (expressed in bits) and the length of the message (in number of characters), and allows us to define source coding efficiency by reference to this limit.

In this chapter we presented an introduction to information theory covering the definition of information content and the calculation of source entropy. We then studied in detail the design of prefix and non-prefix codewords and then devoted the rest of the chapter to a discussion of the three entropy encoding approaches, namely Huffman, Lempel-Ziv and arithmetic coding. Huffman does poorly (using the yardstick of coding efficiency) when applied to low-entropy sources, whereas Lempel-Ziv asymptotically approaches 100% coding efficiency for any source if the message is sufficiently long. Arithmetic coding provides a vehicle for optimal coding of a discrete source at 100% efficiency, but its concept is not intuitive like Huffman coding, and its tutorials are often poorly presented. I hope that our use of the concept of code space in explaining the encoding and decoding processes has brought complete clarity to the subject.

Now that we have obtained an efficient coding of information as a bit stream, how do we design voltage pulses to represent these information bits in a physical circuit and to convey them from a transmitter at the source to a receiver at a different location? Line coding is the subject of the next chapter.

6.9 References

[1] Oxford Dictionaries. (no date). Available at http://oxforddictionaries.com/words/what-is-the-frequency-of-the-letters-of-the-alphabet-in-english (Accessed: 5 August 2013)

[2] Relative frequencies of letters. (no date). Available at http://www.cryptograms.org/letter-frequencies.php (Accessed: 5 August 2013)

[3] Ziv, J., Lempel, A. (1977). 'A universal algorithm for sequential data compression', *IEEE Transactions on Information Theory*, IT-23(3), 337–343

[4] Ziv, J., Lempel, A. (1978). 'Compression of individual sequences via variable-rate coding', *IEEE Transactions on Information Theory*, IT-24(5), 530–536

[5] Welch, T. (1984). 'A technique for high-performance data compression', *Computer*, 17(6), 8–19

[6] Witten, I. H., Radford, M. N., Cleary, J. G. (1987). 'Arithmetic coding for data compression', *Communications of the ACM*, 30(6), 520–540

Chapter 7
Line codes

In Chapter Seven

- ✓ Classes of line codes: We discuss various classes of line codes, viz. level versus transition, instantaneous versus memory, return to zero versus non-return to zero, unipolar versus bipolar, and binary, ternary, quaternary and *M*-ary. Code radix, efficiency and redundancy are also introduced, and their impact on link bit rate and repeater spacing are explored.
- ✓ Autocorrelation of coded waveforms: We learn how to apply the rules of a given line code to obtain statistical information that enables the calculation of the autocorrelation function of a coded waveform representing a random bit stream.
- ✓ Power spectral density (PSD) of coded waveforms: It is not only the bandwidth of a line code that is critical to the transmission of a coded waveform, the spectral shape matters just as much. Both parameters are captured in the PSD, which we learn how to evaluate.
- ✓ Basic line codes: We introduce a wide variety of basic line codes that produce output pulses by coding the input bit stream one bit at a time. We discuss their PSD and BER and assess their suitability for various applications.
- ✓ Advanced line codes: We learn the operation of sophisticated line codes including block codes and run-length-limited codes and the benefits they bring especially to the design of long-distance links. Various terminologies (DS, RDS, DSV, etc.) are introduced and applied.

7.1 Introduction

This chapter deals in detail with the issues taken into account in selecting suitable base band voltage waveforms (called *line codes*) to represent bit streams in a physical circuit. We examine the most common line codes and explore in some detail their waveform characteristics as well as power spectral density (PSD). This gives a sound basis for the selection of the most suitable code for a given digital communication application.

We first develop the skills required to determine the bandwidth of line codes and to evaluate their spectral characteristics. The autocorrelation function of energy and power signals is defined and it is shown that this function forms a Fourier transform (FT) pair with the energy or PSD of the signal. Starting with a random binary waveform and then moving on to a general line code waveform we learn how to derive autocorrelation functions and hence power spectral densities using statistical information based on the rules of a given line code. We then introduce and define a large selection of line codes and assess their suitability based on the characteristics of (1) spectral content, (2) DC content, (3) timing content, (4) bandwidth, (5) transparency, (6) error detection and performance monitoring, (7) code efficiency, (8) error probability and (9) codec complexity. You may wish to refer to Section 1.3.4 now to refresh your understanding of these characteristics. The definitions of code radix, code efficiency and code redundancy given in that discussion are explored further in the next section with an emphasis on their practical significance and impact on the bit rate and repeater spacing of transmission systems.

7.2 Classes and efficiency of line codes

Line codes are base band waveforms used to electrically represent the output bit stream of a source coder, such as the PCM bit stream of the previous section. A wide range of line codes have been developed over the years some of which have been adopted in various digital transmission applications. Line codes are classified in various ways. A line code may be a *level code* in which information is conveyed in the voltage level, a *differential code* (also called *transition code*) where information is conveyed in a *change in level*, or a *phase code* where information is conveyed in the phase of the symbol sent. A line code may also be an *instantaneous code* where the current output depends only on the current input or it may have *memory* in which case the output depends on current as well as some past input. A line code is described as *return to zero* (RZ) if it uses fractional-width pulses in which the voltage level is non-zero only for a fraction (typically first half) of the bit interval; but is said to be *nonreturn to zero* (NRZ) if it uses full-width pulses in which a specified voltage level is maintained for the entire bit interval. A line code is said to be *unipolar* if its pulses have a single polarity, either all positive or all negative; but is described as *bipolar* if two polarities are used (i.e. both positive and negative pulses are involved). Finally, a line code may be classified according to the number of significant voltage levels that it uses for signalling. The minimum number of levels is two (e.g. 0 and $+A$, or 0 and $-A$, or $+A$ and $-A$) and codes that use this minimum are described as *binary line codes*; those that use three levels are called *ternary line codes*; and those that use four levels are *quaternary line codes*. In general, those that use M voltage levels are *M-ary line codes*. Most line codes are either binary or ternary.

It is worth noting that some literature classify line codes as unipolar and polar. However, we emphasise here that all (non-zero) line codes have polarity and hence are 'polar'. So the question is usually whether the code has a single polarity and therefore is *unipolar* or has two polarities, making it *bipolar*. Furthermore, polarity should not be confused with the number of significant voltage levels in the code, as this is covered by the M-ary classification. For example, a line code that employs three voltage levels $+A$, 0 and $-A$ for signalling is a *ternary bipolar code*; if it uses levels $+2A$, A and 0 (or $-2A$, $-A$ and 0) it is a *ternary unipolar code*; if it uses levels $+A$ and $-A$ then it is a *binary bipolar code*; but if it uses levels 0 and $-A$ (or 0 and $+A$) it a *binary unipolar code*.

The efficiency η_c and redundancy $\Re$ of a line code is defined as follows:

$$\begin{aligned}\eta_c &= \frac{\text{Used information per code symbol}}{\text{Potential information per code symbol}} = \frac{\text{Bits per code symbol}}{\log_2(\text{Code radix})}\\ \Re &= \frac{\text{Unused information per code symbol}}{\text{Used information per code symbol}} = \frac{1}{\eta_c} - 1\end{aligned} \tag{7.1}$$

where code radix is the number of *significant* voltage levels (or more generally number of states) used as symbols, and the simplification for $\Re$ follows from the fact that unused information is the difference between potential and used information. Note the importance of the term 'significant' in the definition of code radix. To understand what makes a voltage level significant, consider that a code that uses a positive half-width pulse of amplitude A for binary 1 and a negative half-width pulse for binary 0 has only two significant levels $-A$ and $+A$, even though each pulse contains a zero level. The number of significant levels is exactly as would be the case if the pulses were full-width. A half-width pulse is a strategy aimed at improving *timing content*. It does not in any way change *potential information content*. However, a line code that alternately uses a positive half-width pulse and negative half-width pulse for binary 1 and a 'no pulse' for binary 0 has 3 significant voltage levels $-A$, 0 and A. Again, this number is exactly as it would be if the pulses were full-width.

The importance of code efficiency η_c is that it sets the maximum possible bit rate $R_{b\max}$ in a transmission system of bandwidth B as

$$R_{b\max} = 2B\eta_c \log_2(\text{Code radix}) \tag{7.2}$$

Thus the higher the code efficiency, the higher the information transmission rate possible in a given channel. Conversely, by making $2B$ the subject of the above equation and noting that this corresponds to the maximum symbol rate $R_{s\max} = 2B$, we see that higher code efficiencies allow the use of lower symbol rates to support a given bit rate. A lower symbol rate implies a larger symbol period with lower bandwidth, and hence smaller distortion in band limited transmission systems. So high-efficiency codes are preferred on long-distance transmission links to reduce symbol distortion and therefore increase the link length traversed before distortion becomes excessive. This increases the minimum distance between repeaters, which lowers overall system cost.

7.3 Power spectral density of line codes

The bit stream represented by a line code is random and so is the line code waveform. This means that it is not possible to take the FT of a line code to obtain a measure of its bandwidth. We get around this difficulty by determining bandwidth through the PSD of the line code, which we obtain by using statistical information about the line code to calculate its *autocorrelation function* and then applying the relationship between autocorrelation and PSD as discussed below.

7.3.1 Autocorrelation and spectral density

Consider the convolution of an energy signal $g(t)$ with a time-reversed version of itself $g(-t)$. From the definition of convolution given in Eq. (2.8) and the explanation in Worked Example 2.1, this operation is the area under the product of the first function $g(t)$ and the time-reversed and shifted (to the right by τ) version of the second function $g(-t)$. These two actions on $g(-t)$ have the following effect: First, time reversal changes it to $g(t)$, and then the shift changes it to $g(t-\tau)$. The resulting area will obviously depend on τ and is denoted by $R_g(\tau)$. Thus

$$g(t) * g(-t) = R_g(\tau) = \int_{-\infty}^{\infty} g(t)g(t-\tau)dt \tag{7.3}$$

Eq. (7.3) defines the *autocorrelation function* $R_g(\tau)$ of the energy signal $g(t)$ and gives a measure (in units of energy) of the match and common area between the signal and a delayed version of itself. Note that $R_g(\tau)$ is not a time-dependent waveform, but a quantity that depends only on a *scanning* or *searching parameter* τ. Clearly, at $\tau = 0$ we are comparing $g(t)$ to itself and the match will be greatest. The autocorrelation function $R_g(\tau)$ should therefore be maximum at $\tau = 0$.

Let $G(f)$ denote the FT of $g(t)$, which means that $g(t)$ is equivalently the inverse Fourier transform (IFT) of $G(f)$, a twin relationship represented as $g(t) \rightleftharpoons G(f)$. Then, from the time-reversal property stated in Eq. (C.12) of Appendix C, $g(-t) \rightleftharpoons G(-f)$. Furthermore, the convolution of $g(t)$ and $g(-t)$ corresponds in the frequency domain to a multiplication of their respective FT, so that

$$g(t) * g(-t) = R_g(\tau) \rightleftharpoons G(f)G(-f) = G(f)G^*(f) = |G(f)|^2$$

where we have used the fact that, by definition of FT (see Eq. (C.6)), $G(-f)$ is the complex conjugate of $G(f)$, denoted by $G^*(f)$. But from Eq. (2.22) the square of the magnitude of $G(f)$ is the energy spectral density of $g(t)$, denoted by $\Psi_g(f)$. Therefore

$$R_g(\tau) \rightleftharpoons \Psi_g(f) \tag{7.4}$$

We may obtain a similar relation for a power signal, i.e. that the power spectral density $S_g(f)$ of a power signal $g(t)$ is the FT of its autocorrelation function, provided we define autocorrelation function for a power signal in a way that is consistent with the relationship between energy and power as follows:

$$R_g(\tau) = \begin{cases} \displaystyle\int_{-\infty}^{\infty} g(t)g(t-\tau)dt, & \text{Energy signal} \\ \displaystyle\frac{1}{T}\int_{-T/2}^{T/2} g(t)g(t-\tau)dt, & \text{Periodic power signal} \\ \displaystyle\lim_{T\to\infty}\frac{1}{T}\int_{-T/2}^{T/2} g(t)g(t-\tau)dt, & \text{Nonperiodic power signal} \end{cases} \tag{7.5}$$

The autocorrelation function of a real-valued signal has the following important properties:

- $R_g(\tau)$ has even symmetry about $\tau = 0$. That is, $R_g(\tau) = R_g(-\tau)$. When comparing $g(t)$ with a delayed version of itself, what matters is the amount of shift, not the direction. Thus

$$R_g(\tau) = \int_{-\infty}^{\infty} g(t)g(t-\tau)dt = \int_{-\infty}^{\infty} g(t)g(t+\tau)dt \tag{7.6}$$

- $R_g(\tau)$ has a maximum value at $\tau = 0$ equal to the energy E of the energy signal or power P of the power signal:

$$R_g(\tau)_{\max} = R_g(0) = \begin{cases} E, & \text{Energy signal} \\ P, & \text{Power signal} \end{cases} \tag{7.7}$$

- Autocorrelation and spectral density form an FT pair:

$$R_g(\tau) \rightleftharpoons \begin{cases} \Psi_g(f), & \text{Energy signal} \\ S_g(f), & \text{Power signal} \end{cases} \tag{7.8}$$

This last property makes $R_g(\tau)$ particularly useful in digital communications, providing a means of determining the spectral content of signals that are not Fourier transformable. In the next two sections, we obtain the PSD of a random power signal by computing its autocorrelation function from a statistical description of the signal and then taking the FT of $R_g(\tau)$.

7.3.2 PSD of random binary waveforms

Fig. 7.1 shows a random binary waveform $g(t)$ in which a rectangular pulse of duration T_s is sent with amplitude $+A$ for binary 1 and $-A$ for binary 0, and binary 1s and 0s are equally likely to occur. A version $g(t-\tau)$ of the waveform delayed by τ is also shown in dotted outline. This waveform is a nonperiodic and random power signal, so to obtain its PSD we first determine $R_g(\tau)$ using the third line of Eq. (7.5) and the statistical information provided. To do this we span N bits and an interval $T = NT_s$ and let $N \rightarrow \infty$ to satisfy Eq. (7.5). When we shift $g(t)$ by amount $|\tau|$ — the direction being immaterial — we see that if $|\tau| \leq T_s$, the kth bit (denoted by b_k) in $g(t)$ overlaps the kth bit in $g(t-\tau)$ over a time width $T_s - |\tau|$. This overlap region is shaded in Fig. 7.1 for the first few bits. In each overlap region the product $g(t)g(t-\tau)$ is $(A) \times (A) = A^2$ for $b_k = 1$, and $(-A) \times (-A) = A^2$ for $b_k = 0$, and the area is $A^2(T_s - |\tau|)$. Thus the total area under $g(t)g(t-\tau)$ contributed by an overlap between corresponding bits in $g(t)$ and $g(t-\tau)$ is

$$NA^2(T_s - |\tau|)$$

Notice that the shift by $\tau \leq T_s$ also creates an overlap between b_k in $g(t-\tau)$ and b_{k+1} in $g(t)$ over an interval τ. In this region the product $g(t)g(t-\tau)$ will be A^2 if $b_k = b_{k+1}$, and will be $-A^2$ if $b_k \neq b_{k+1}$. Since these two bits are equally likely to be same or different, the total contribution to the area from the overlap of adjacent bits is zero. Furthermore, any shift by $\tau > T_s$ will only create overlaps between non-corresponding bits b_k in $g(t-\tau)$ and b_{k+n} in $g(t)$, where $n = 1, 2, 3, \ldots$, and this also

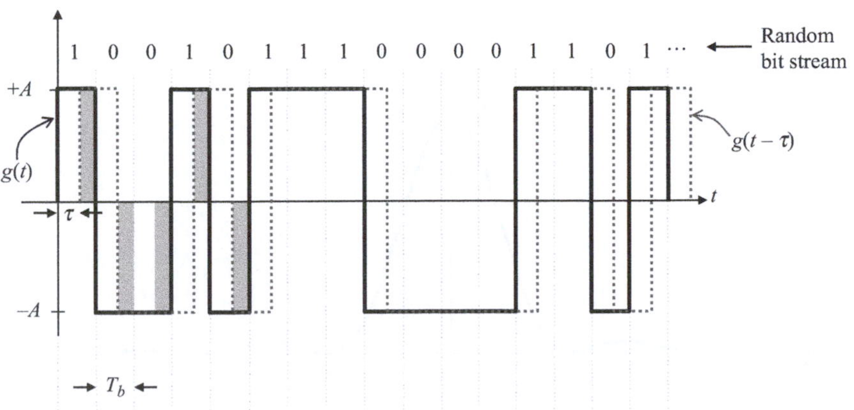

Fig. 7.1 Random binary waveform

makes zero total contribution to the area since b_k and b_{k+n} are equally likely to be the same ($+A$ and $+A$, or $-A$ and $-A$) or different ($+A$ and $-A$, or $-A$ and $+A$). Thus

$$
\begin{aligned}
R_g(\tau) &= \lim_{T\to\infty} \frac{1}{T} \int_{-T/2}^{T/2} g(t)g(t-\tau)dt \\
&= \begin{cases} \lim\limits_{N\to\infty} \dfrac{1}{NT_s} NA^2(T_s - |\tau|), & |\tau| \le T_s \\ 0, & \text{otherwise} \end{cases} \\
&= \begin{cases} A^2(1 - |\tau|/T_s), & |\tau| \le T_s \\ 0, & \text{otherwise} \end{cases} \\
&= A^2 \operatorname{trian}\left(\frac{\tau}{2T_s}\right)
\end{aligned}
$$

Thus the autocorrelation function of a bipolar random binary waveform of amplitude A and bit duration T_s is a triangular pulse having duration $2T_s$ and amplitude A^2. The FT of a triangular pulse is listed in Appendix C, from which we obtain the FT of $R_g(\tau)$ and hence the PSD of $g(t)$ as

$$S_g(f) = A^2 T_s \operatorname{sinc}^2(fT_s) \tag{7.9}$$

This PSD is shown in Fig. 7.2 along with the autocorrelation function. We see that the null bandwidth of a random binary waveform is $1/T_s$, which is the symbol rate (and in this binary system also the bit rate) of the transmission.

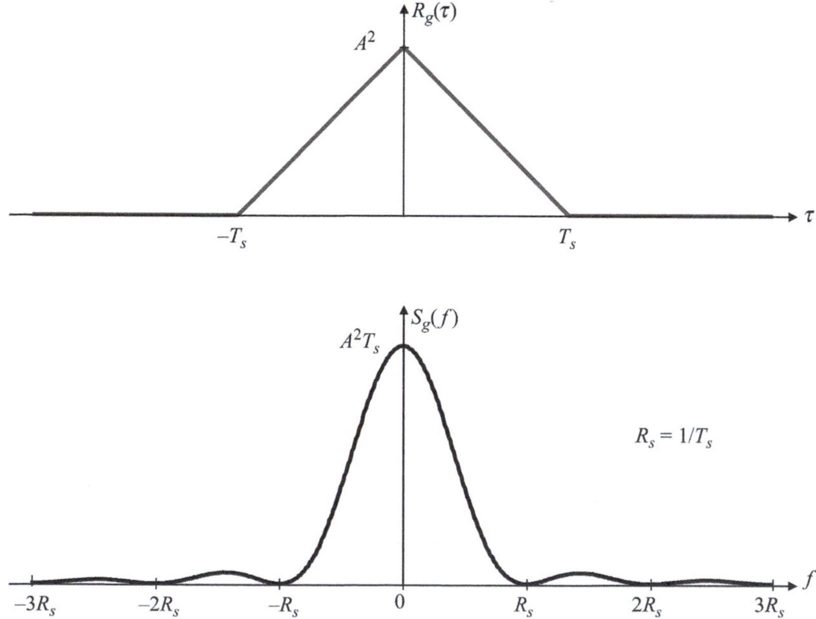

Fig. 7.2 Autocorrelation function and PSD of random binary waveform of amplitude A, rectangular pulse width T_s, and pulse rate $R_s = 1/T_s$

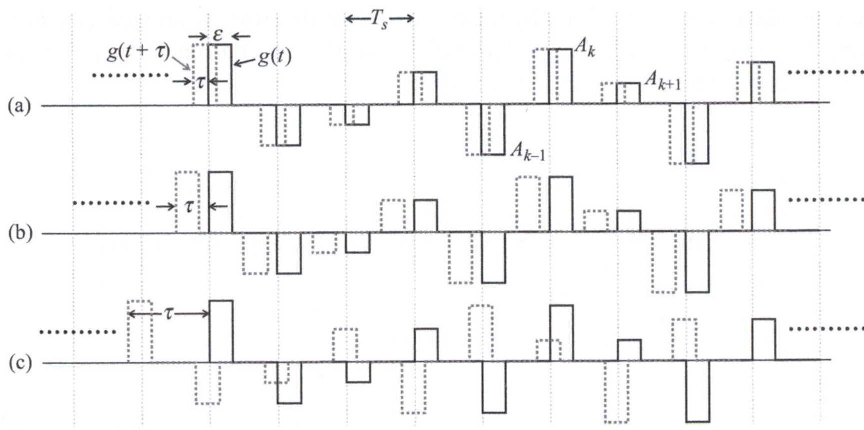

Fig. 7.3 Random rectangular pulse train g(t): (a) g(t) and $g(t+\tau)$ for $\tau < \varepsilon$; (b) g(t) and $g(t+\tau)$ for $\varepsilon < \tau < T_s - \varepsilon$; (c) g(t) and $g(t+\tau)$ for $T_s - \varepsilon < \tau < T_s + \varepsilon$

7.3.3 PSD of general line code waveforms

Consider a waveform $g(t)$ that is a random rectangular pulse train consisting of pulses of width ε spaced T_s seconds apart, with the kth pulse having amplitude A_k. Fig. 7.3 shows $g(t)$ along with $g(t+\tau)$ for three different ranges of shift τ. It can be seen that there is overlap between the pulses of $g(t)$ and $g(t+\tau)$ only for shifts τ of magnitude in the regions

$$|\tau - nT_s| < \varepsilon, \quad n = 0, \pm 1, \pm 2, \pm 3, \ldots$$

Outside these regions, e.g. Fig. 7.3b where $\varepsilon < |\tau| < T_s - \varepsilon$, the pulses do not overlap and hence $R_g(\tau) = 0$. In the regions of overlap stated above, the kth pulse of $g(t)$ overlaps with the $(k+n)^{\text{th}}$ pulse of $g(t+\tau)$ over an interval $\varepsilon - |\tau - nT_s|$. The area of the product of the two pulses is thus $A_k A_{k+n}[\varepsilon - |\tau - nT_s|]$, and the total area of $g(t)g(t+\tau)$, which is $R_g(\tau)$, is obtained by summing these contributions over all N pulses of $g(t)$ — spanning an interval $T = NT_s$ — in the limit $N \to \infty$:

$$\begin{aligned} R_g(\tau) &= \lim_{N\to\infty} \frac{1}{NT_s} \sum_{k=1}^{N} A_k A_{k+n}[\varepsilon - |\tau - nT_s|] \\ &= \frac{\varepsilon}{T_s}\left[1 - \frac{|\tau - nT_s|}{\varepsilon}\right] \lim_{N\to\infty} \frac{1}{N} \sum_{k=1}^{N} A_k A_{k+n} \\ &= \frac{\varepsilon}{T_s}\left(\overline{A_k A_{k+n}}\right)\left[1 - \frac{|\tau - nT_s|}{\varepsilon}\right], \quad n = 0, \pm 1, \pm 2, \ldots \end{aligned}$$

Thus $R_g(\tau)$ is a sequence of triangular pulses of duration 2ε and amplitude $\Lambda_n = \varepsilon\left(\overline{A_k A_{k+n}}\right)/T_s$ located along the τ-axis at $\tau = nT_s$, $n = 0, \pm 1, \pm 2, \pm 3, \ldots$, where $\overline{A_k A_{k+n}}$ is the average value of the product $A_k A_{k+n}$, and $\Lambda_n = \Lambda_{-n}$ since $R_g(\tau)$ is an even function. We write this as

$$R_g(\tau) = \sum_{n=-\infty}^{\infty} \Lambda_n \operatorname{trian}\left(\frac{\tau - nT_s}{2\varepsilon}\right); \quad \Lambda_n = \varepsilon\left(\overline{A_k A_{k+n}}\right)/T_s$$

We now obtain the PSD of $g(t)$ by taking the FT of the aforementioned equation, using the listing in Section C.2 in Appendix C for the FT of a triangular pulse and the time-shifting property in Eq. (C.8). Thus

$$\begin{aligned}
S_g(f) &= \varepsilon\,\mathrm{sinc}^2(f\varepsilon)\sum_{n=-\infty}^{\infty}\Lambda_n \exp(-j2\pi nfT_s) \\
&= \varepsilon\,\mathrm{sinc}^2(f\varepsilon)\left[\Lambda_0 + \sum_{n=1}^{\infty}\Lambda_n\{\exp(j2\pi nfT_s) + \exp(-j2\pi nfT_s)\}\right] \\
&= \varepsilon\,\mathrm{sinc}^2(f\varepsilon)\left[\Lambda_0 + 2\sum_{n=1}^{\infty}\Lambda_n\cos(2\pi nfT_s)\right]
\end{aligned}$$

These are important results that apply to any line code using rectangular pulse shapes. To summarise

$$\begin{aligned}
R_g(\tau) &= \sum_{n=-\infty}^{\infty}\Lambda_n\,\mathrm{trian}\left(\frac{\tau - nT_s}{2\varepsilon}\right) \\
S_g(f) &= \varepsilon\,\mathrm{sinc}^2(f\varepsilon)\left[\Lambda_0 + 2\sum_{n=1}^{\infty}\Lambda_n\cos(2\pi nfT_s)\right] \\
\Lambda_n &= \varepsilon\left(\overline{A_k A_{k+n}}\right)/T_s
\end{aligned} \tag{7.10}$$

The task of evaluating the PSD of any line code in order to assess its spectral content (such as null bandwidth, DC component, bandwidth that contains 99% of power, etc.) therefore simplifies to the following steps:

1. Determine the width ε of the pulse used by the line code to represent each symbol in terms of the pulse spacing T_s, where $1/T_s$ is the symbol rate. The width ε might encompass more than one voltage level or pulse, say M levels. If these are treated as a single composite pulse of duration ε, then $\varepsilon/T_s = 1 - \Delta/T_s$, where Δ is the gap between the end of one composite pulse and the start of the next one, and T_s is the spacing between adjacent composite pulses. The pulse rate in this case is $1/T_s$. However, if the levels are treated as M separate pulses then $\varepsilon/T_s = 1$, and the pulse rate is $1/\varepsilon$, where ε is the width of the narrowest level or pulse.
2. Using the coding rule of each line code and information on the probability of each bit or grouping of bits, determine the mean square value of pulse amplitudes $\overline{A_k^2}$, and the average $\overline{A_k A_{k+n}}$ of the product of two pulses in separate intervals k and $k+n$, and multiply these by ε/T_s to obtain Λ_0 and Λ_n, $n = 1, 2, 3, \ldots$. Some line codes yield Λ_0 and Λ_n in simple analytic forms, but others require a simulation-based approach in which a waveform representing a sufficiently long pseudorandom bit stream is generated according to the line code rules and is analysed to obtain estimates of Λ_0 and Λ_n.
3. The PSD is given by the second line of Eq. (7.10) using the values of ε, T_s, Λ_0 and Λ_n obtained in the first two steps.

   As a simple example of these steps, consider the random binary waveform discussed in the previous Section. This is a basic line code known as *bipolar nonreturn to zero* (BNRZ) that employs full-width pulses, so $\varepsilon = T_s$. And it uses pulse amplitude $+A$ for binary 1 and $-A$ for binary 0. So $\Lambda_0 = A^2\varepsilon/T_s = A^2$. Next, to evaluate Λ_n, we take the kth and $(k+n)^{\text{th}}$ bits, multiply their amplitudes together and take the average of this product over $k=1$ to $N \to \infty$. The only possible bit combinations will be 00, 01, 10 and 11, which yield respective products A^2, $-A^2$, $-A^2$ and A^2. Assuming that binary 1 and binary 0 occur randomly with equal probability, these di-bits occur with equal probability ¼ yielding an average

$$\Lambda_n = \left[\frac{1}{4}(A^2) + \frac{1}{4}(-A^2) + \frac{1}{4}(-A^2) + \frac{1}{4}(A^2)\right]\frac{\varepsilon}{T_s} = 0, \quad n = 1, 2, 3, \ldots$$

Substituting $\varepsilon = T_s$, $\Lambda_0 = A^2$ and $\Lambda_n = 0$ into Eq. (7.10) yields

$$S_g(f) = A^2 T_s \operatorname{sinc}^2(fT_s)$$

which you will recognise as Eq. (7.9). The next worked example further explains how the coding rule and statistical information are utilised in calculating Λ_n.

Worked Example 7.1: PSD of RZ and AMI Line Codes

The return to zero (RZ) line code represents binary 1 using a half-width rectangular pulse of amplitude A, and binary 0 with no pulse, whereas the alternate mark inversion (AMI) line code (see Fig. 1.17) differs only in using alternate-polarity pulses for binary 1. We wish to determine the PSD of each line code, assuming that binary 1s and 0s are equally likely.

The task here is straightforward: Determine ε (in terms of T_s) as well as Λ_0 and Λ_n for each line code and then substitute these into Eq. (7.10) to obtain the required PSD.

First, both line codes employ half-width pulses, so

$$\varepsilon = T_s/2; \quad \Rightarrow \frac{\varepsilon}{T_s} = \frac{1}{2}$$

Next, RZ has pulse amplitude A_k restricted to two values, namely 0 for binary 0 and A for binary 1, whereas in AMI A_k has three values, namely 0 for binary 0, and $\pm A$ alternating for binary 1. To obtain Λ_0, note that this requires $\overline{A_kA_k}$ that involves only one bit, the kth bit which is either binary 0 or binary 1 with equal probability ½. The product $A_kA_k = 0$ for binary 0, and $A_kA_k = A^2$ for binary 1. Thus for both codes

$$\begin{aligned}\Lambda_0 &= \frac{\varepsilon}{T_s}\left(\overline{A_kA_k}\right) = \frac{1}{2}\left(0 \times \frac{1}{2} + A^2 \times \frac{1}{2}\right) \\ &= A^2/4\end{aligned}$$

The next step is to evaluate Λ_n, $n \geq 1$. Consider first Λ_1 that requires $\overline{A_kA_{k+1}}$ and involves 4 combinations of two adjacent bits 00, 01, 10 and 11. Each combination occurs with probability ¼, but the product $A_kA_{k+1} = 0$ for the first three (since binary 0 is represented by amplitude 0), and for the last combination $A_kA_{k+1} = A^2$ for RZ, and $-A^2$ for AMI in view of the alternating polarity rule. Thus

$$\Lambda_1 = \frac{\varepsilon}{T_s}\left(\overline{A_kA_{k+1}}\right)$$

$$= \begin{cases} \frac{1}{2}\left(0 \times \frac{1}{4} + 0 \times \frac{1}{4} + 0 \times \frac{1}{4} + A^2 \times \frac{1}{4}\right) = A^2/8, & \text{RZ} \\ \frac{1}{2}\left(0 \times \frac{1}{4} + 0 \times \frac{1}{4} + 0 \times \frac{1}{4} - A^2 \times \frac{1}{4}\right) = -A^2/8, & \text{AMI} \end{cases}$$

Next, Λ_2 requires $\overline{A_kA_{k+2}}$ that involves two bits separated by one bit position. That is, it involves the first and last bits of the eight equally likely three-bit combinations 000, 001, 010, 011, 100, 110, 101, 111. There is a non-zero contribution in only the last two combinations, where the product is A^2 for RZ, yielding

$$\overline{A_kA_{k+2}} = A^2 \times \frac{1}{8} + A^2 \times \frac{1}{8} = A^2/4 \quad \text{for RZ}$$

For AMI a little more thought is required to respect its coding rule. $A_kA_{k+2} = -A^2$ for the combination 101, whereas $A_kA_{k+2} = A^2$ for the combination 111 (in view of the alternating polarity rule), so that $\overline{A_kA_{k+2}} = 0$ for AMI. In fact, for RZ A_kA_{k+n} will always involve 4 equally likely two-bit combinations only one of which contributes product A^2, so that $\overline{A_kA_{k+n}} = A^2/4$ in RZ for all $n > 1$. But in AMI, A_kA_{k+n} will have a non-zero contribution from the $(n+1)$-bit combination 1xxx..x1, where xxx..x denote $n-1$ bits, each of which is equally likely to be 0 or 1, which means that half of the 2^{n-1} combinations contain an even number of 1s resulting in $A_kA_{k+n} = -A^2$, and the other half contain an odd number of 1s resulting in $A_kA_{k+n} = A^2$. The average $\overline{A_kA_{k+n}}$ is thus zero for $n \geq 2$ in AMI. The results to use in Eq. (7.10) are therefore

$$\Lambda_0 = A^2/4; \qquad \varepsilon = T_s/2, \qquad \text{RZ and AMI}$$

$$\Lambda_n = A^2/8, \qquad n = 1, 2, 3, \ldots \qquad \text{RZ}$$

$$\Lambda_n = \begin{cases} -A^2/8, & n = 1 \qquad \text{AMI} \\ 0, & n = 2, 3, 4, \ldots \end{cases}$$

This yields the PSD of AMI and RZ line codes as

$$S_{AMI}(f) = \frac{T_s}{2}\text{sinc}^2\left(f\frac{T_s}{2}\right)\left[\frac{A^2}{4} + 2\left(-\frac{A^2}{8}\right)\cos(2\pi fT_s)\right]$$

$$= \frac{A^2T_s}{8}\text{sinc}^2(fT_s/2)[1 - \cos(2\pi fT_s)]$$

$$= \frac{A^2T_s}{4}\text{sinc}^2(fT_s/2)\,\sin^2(\pi fT_s)$$

$$S_{RZ}(f) = \frac{T_s}{2}\text{sinc}^2\left(f\frac{T_s}{2}\right)\left[\frac{A^2}{4} + \frac{A^2}{4}\sum_{n=1}^{\infty}\cos(2\pi nfT_s)\right]$$

$$= \frac{A^2T_s}{8}\text{sinc}^2\left(f\frac{T_s}{2}\right)\left[1 + \sum_{n=1}^{\infty}\cos(2\pi nfT_s)\right]$$

$$= \frac{A^2T_s}{16}\text{sinc}^2(fT_s/2)\left[1 + \frac{1}{T_s}\sum_{n=-\infty}^{\infty}\delta(f - n/T_s)\right]$$

$$= \frac{A^2T_s}{16}\text{sinc}^2(fT_s/2) + \frac{A^2}{16}\sum_{n=-\infty}^{\infty}\text{sinc}^2(n/2)\delta(f - n/T_s)$$

where, in the second line of $S_{RZ}(f)$ above, we used the identity

$$\sum_{n=1}^{\infty}\cos(2\pi nfT_s) = \frac{1}{2}\left[\frac{1}{T_s}\sum_{n=-\infty}^{\infty}\delta(f - n/T_s) - 1\right] \tag{7.11}$$

To summarise, the PSD of AMI and RZ line codes are

$$\begin{aligned} S_{AMI}(f) &= \frac{A^2T_s}{4}\text{sinc}^2(fT_s/2)\sin^2(\pi fT_s) \\ S_{RZ}(f) &= \frac{A^2T_s}{16}\text{sinc}^2(fT_s/2) + \frac{A^2}{16}\sum_{n=-\infty}^{\infty}\text{sinc}^2(n/2)\delta(f - n/T_s) \end{aligned} \tag{7.12}$$

Fig. 7.4a,b shows the RZ and AMI power spectral densities from which we observe the following about the spectral content of the two codes:

- The PSD of RZ has an impulse (represented by an arrow) of weight $A^2/16$ at the origin, an indication that the code contains an undesirable DC offset with DC power equal to $A^2/16$. AMI on the other hand has no DC content seeing that its PSD is zero at $f=0$.
- RZ contains impulses at odd integer multiples of the bit rate R_s. This makes clock extraction easier with RZ than with AMI since the impulse at R_s can be filtered and used to provide precise bit timing information to synchronise the receiver with the transmitter. However, in addition to the DC offset noted above, the continuous portion of the PSD of RZ is non-zero at $f=0$. When passed through an AC-coupled transmission system (which blocks DC and also behaves as a high pass filter) this causes the spectrum and hence line code waveform to be distorted, thus increasing the likelihood of bit error.
- The null bandwidth of RZ is $2R_s$ which is double that of BNRZ shown in Fig. 7.2. It should come as no surprise that there is indeed a price to pay for the extra clock content in RZ when compared to BNRZ.

Following a similar method to that used in the previous worked example, analytic forms for the parameters ε and Λ_n for some of the line codes defined in the next Section have been obtained and substituted into Eq. (7.10) to give PSD, which are listed in Eq. (7.13). The presence of an impulse at $f=0$ is indicative of an

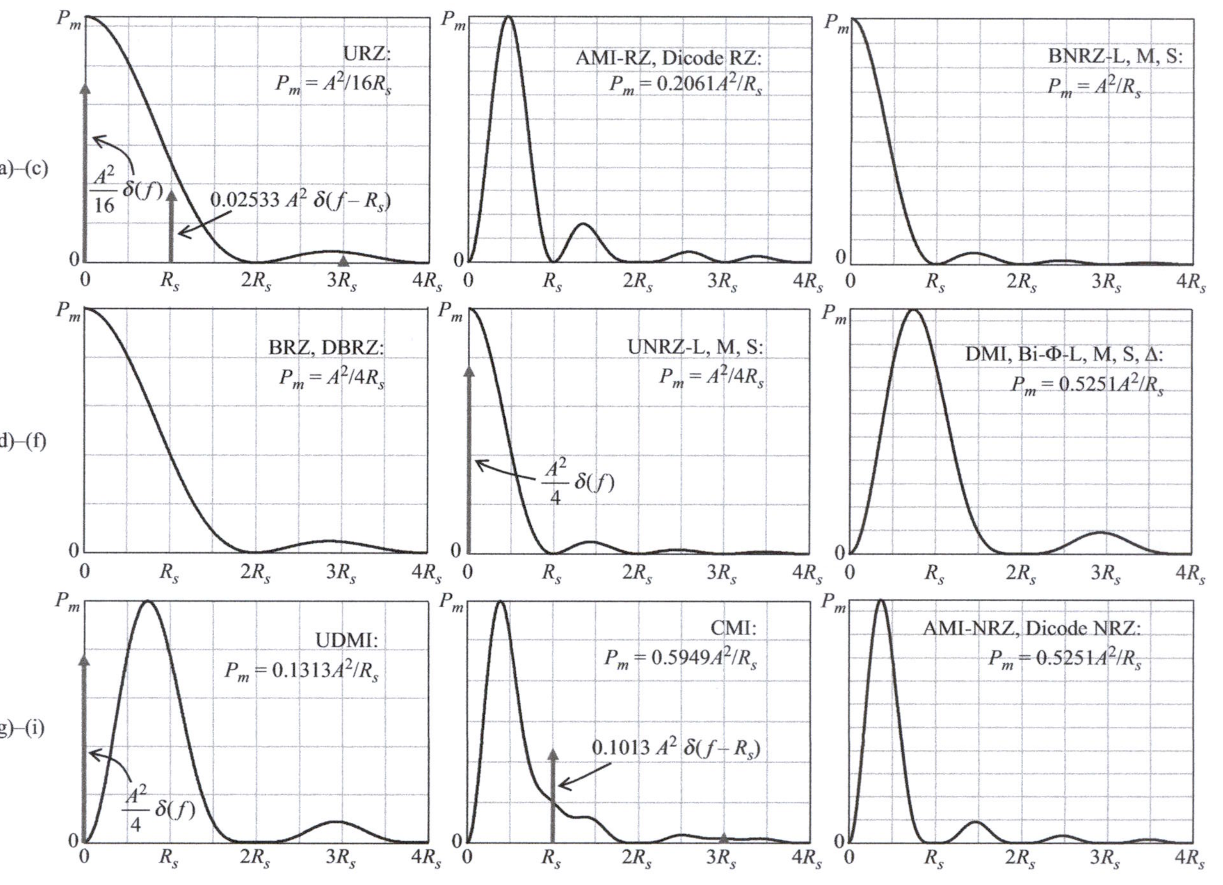

Fig. 7.4 PSD of various line codes of amplitude A and bit rate R_s

undesirable DC offset having power equal to the weight of the impulse. The PSD of line codes provide complete information about spectral content that can be crucial in selecting between two codes that are comparable in other aspects such as clock content, error detection or complexity. The PSD of many line codes, including some not listed in Eq. (7.13), are plotted in Figs. 7.4 and 7.5.

$$
S(f)=\begin{cases}
A^2T_s\,\mathrm{sinc}^2(fT_s), & \text{BNRZ-L,M,S}\\
\frac{1}{4}A^2T_s\,\mathrm{sinc}^2(fT_s)+\frac{1}{4}A^2\delta(f), & \text{UNRZ-L,M,S}\\
\frac{1}{4}A^2T_s\,\mathrm{sinc}^2(fT_s/2), & \text{BRZ, DBRZ}\\
\frac{A^2T_s}{16}\mathrm{sinc}^2\left(\frac{fT_s}{2}\right)+\frac{A^2}{16}\sum_{n=-\infty}^{\infty}\mathrm{sinc}^2\left(\frac{n}{2}\right)\delta\left(f-\frac{n}{T_s}\right), & \text{URZ}\\
\frac{1}{4}A^2T_s\,\mathrm{sinc}^2(fT_s/2)\sin^2(\pi fT_s), & \text{AMI-RZ, Dicode RZ}\\
A^2T_s\,\mathrm{sinc}^2(fT_s)\sin^2(\pi fT_s), & \text{AMI-NRZ, Dicode NRZ}\\
A^2T_s\,\mathrm{sinc}^2(fT_s/2)\sin^2(\pi fT_s/2), & \text{Bi-}\Phi\text{-(L, M, S, }\Delta\text{), DMI}\\
\frac{1}{4}A^2T_s\,\mathrm{sinc}^2(fT_s/2)\sin^2(\pi fT_s/2)+\frac{1}{4}A^2\delta(f), & \text{UDMI}\\
\frac{A^2T_s}{8}\mathrm{sinc}^2\left(\frac{fT_s}{2}\right)[3-2\cos(2\pi fT_s)-\cos(3\pi fT_s)] & \text{CMI}\\
\quad+\frac{A^2}{\pi^2}\sum_{n=-\infty}^{\infty}\delta(f-[2n+1]R_s)/(2n+1)^2 &
\end{cases}
\tag{7.13}
$$

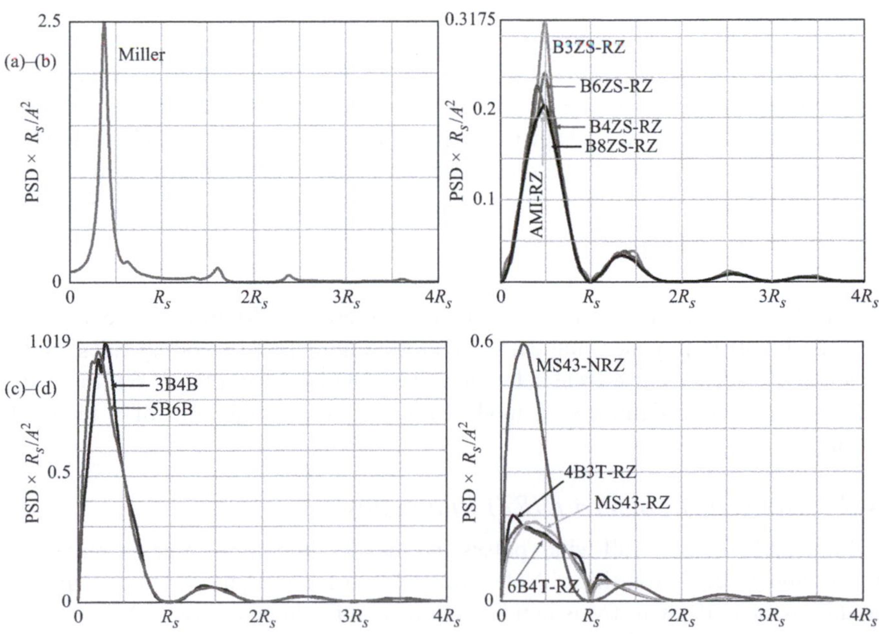

Fig. 7.5 PSD of advanced line codes of amplitude A and bit rate R_s

Table 7.1 Comparison of bandwidth and spectral content of various line codes

	Null bandwidth B_n	Power in B_n	90% Bandwidth	95% Bandwidth	99% Bandwidth
BNRZ-L, S, M	R_s	90.28%	$0.849R_s$	$2.073R_s$	$10.286R_s$
UNRZ-L, S, M	R_s	95.14%	$0.536R_s$	$0.849R_s$	$5.214R_s$
BRZ, DBRZ	$2R_s$	90.28%	$1.698R_s$	$4.146R_s$	$20.573R_s$
URZ	$2R_s$	90.41%	$1.579R_s$	$3.775R_s$	$20.749R_s$
AMI-RZ, Dicode RZ	R_s	79.6%	$1.7R_s$	$4.3R_s$	$20.5R_s$
AMI-NRZ, Dicode NRZ	R_s	85.57%	$1.524R_s$	$3.219R_s$	$15.375R_s$
Bi-Φ-L, S, M, Δ and DMI	$2R_s$	85.57%	$3.048R_s$	$6.439R_s$	$30.751R_s$
CMI	$2R_s$	87.91%	$2.938R_s$	$5R_s$	$25R_s$
UDMI	$2R_s$	92.785%	$1.27R_s$	$3.048R_s$	$15.08R_s$
Miller	$2R_s$	92.505%	$1.581R_s$	$3.052R_s$	$15.313R_s$
B3ZS-RZ	R_s	79.6%	$1.7R_s$	$4.3R_s$	$20.5R_s$
B4ZS-RZ	R_s	79.6%	$1.7R_s$	$4.3R_s$	$20.5R_s$
B6ZS-RZ	R_s	79.6%	$1.7R_s$	$4.3R_s$	$20.5R_s$
B8ZS-RZ	R_s	79.6%	$1.7R_s$	$4.3R_s$	$20.5R_s$
3B4B	R_s	87.7%	$1.4R_s$	$2.5R_s$	$12.7R_s$
5B6B	R_s	88.6%	$1.3R_s$	$2.4R_s$	$11.7R_s$
4B3T-RZ	R_s	78.0%	$1.7R_s$	$4.2R_s$	$20.6R_s$
MS43-RZ	R_s	78.4%	$1.7R_s$	$4.2R_s$	$20.5R_s$
MS43-NRZ	R_s	88.3%	$1.4R_s$	$2.5R_s$	$12.4R_s$
6B4T-RZ	R_s	77.9%	$1.7R_s$	$4.2R_s$	$20.6R_s$

Table 7.1 compares the spectral characteristics of 31 line codes. The null bandwidth of each line code is given in column 2 as a multiple of symbol rate R_S. The third column gives the amount of power contained inside the null bandwidth as a percentage of total waveform power, and the last three columns give the bandwidth that contains 90%, 95% and 99% of total power. The results are for rectangular shaped pulses and represent the spectral content of each line code prior to filtering which is usually employed to shape the pulses in order to reduce spectral amplitudes outside the null bandwidth. We discuss filtering for pulse shaping in Chapter 8.

7.4 Basic line codes

Fig. 7.6 shows a wide selection of basic coded waveforms. We introduce below the line code involved in generating each waveform and define its coding rules. Note that some of the codes discussed below have been omitted from Fig. 7.6 if the code is the result of a straightforward modification of the coding rule of a featured waveform.

7.4.1 Nonreturn to zero (NRZ) line codes

NRZ line codes employ full-width pulses and may be unipolar or bipolar, binary or ternary, and level or transition. Note that NRZ line codes are all *instantaneous* (having no memory). Our focus in this section is on the binary class of NRZ codes. Ternary NRZ codes will be mentioned in the next section along with their RZ counterparts.

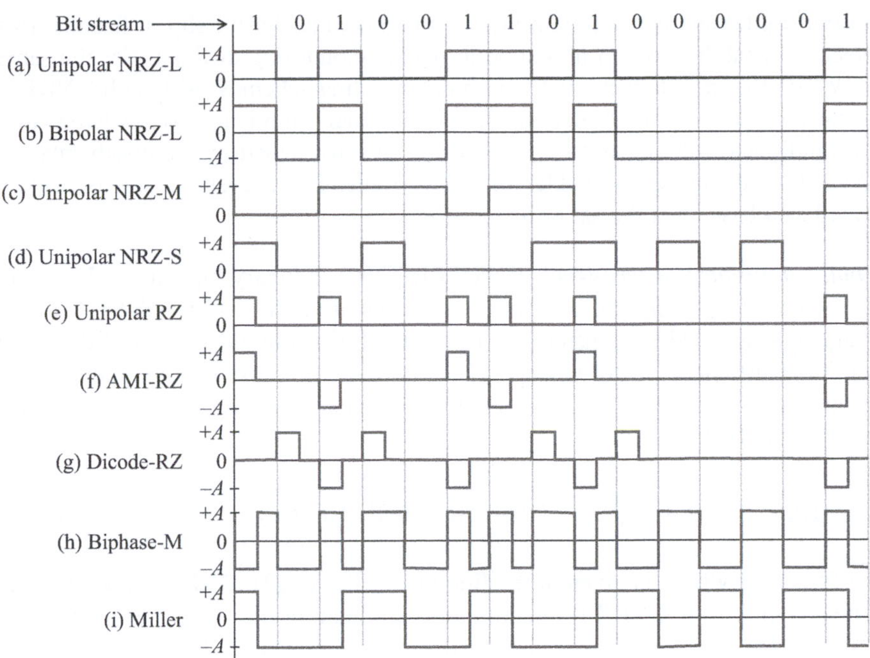

Fig. 7.6 Waveforms of various line codes

***Unipolar nonreturn to zero-level (UNRZ-L)*:** This is a binary unipolar level code in which binary 1 is signalled by transmitting a pulse of height $+A$ and duration equal to the bit interval T_s, whereas binary 0 is signalled by no pulse, which can be viewed as a pulse of height 0 V, for the entire bit interval. Since the scheme basically involves switching on a pulse to signal binary 1 and switching it off to signal binary 0, UNRZ-L is described as *on-off* signalling. Assuming equally likely 1s and 0s, the PSD of this code is given in the second line of Eq. (7.13) and the probability of bit error (called bit error rate (BER) and derived in Chapter 10) in a matched receiver and an additive white Gaussian noise (AWGN) channel is

$$\text{BER} = \frac{1}{2}\text{erfc}\left(\sqrt{\frac{E_b}{2N_o}}\right), \quad (\text{UNRZ-L, URZ, Miller, Bi-}\Phi\text{-M, S}) \tag{7.14}$$

where E_b is the average energy per bit, N_o is the noise power per unit bandwidth and erfc is the complementary error function defined in Appendix D.1. In a system of transmission bandwidth B and bit rate R_b, the ratio E_b/N_o is related to bandwidth efficiency η and the ratio between carrier power and noise power, called carrier to noise ratio (C/N), by

$$\frac{C}{N} = \eta\frac{E_b}{N_o}; \quad \eta = \frac{R_b}{B} \tag{7.15}$$

The UNRZ-L line code has many significant disadvantages: It contains a DC offset that conveys no information and therefore represents a waste of power. Its PSD is maximum at $f = 0$, which will cause significant waveform distortion in AC-coupled transmission systems that block DC. It does not have guaranteed clock content since a sequence of all-1s or all-0s produces a waveform void of transitions that if sustained for long enough will result in loss of synchronisation between receiver and transmitter and hence timing jitter with a potential for bit slip. It has no error detection capability, which makes performance monitoring difficult. For the same E_b/N_o, it also has a higher BER than bipolar coding schemes. Although UNRZ-L has the advantages of 100% coding efficiency, codec simplicity and low bandwidth (with null bandwidth equal to R_s), these are far outweighed by the disadvantages. UNRZ-L is therefore unsuitable for use in transmission systems and is usually treated as a basic data format or *uncoded data*.

***Bipolar nonreturn to zero-level (BNRZ-L)*:** This is a binary bipolar level code in which binary 1 is signalled by transmitting a positive full-width pulse ($+A$), and binary 0 is signalled by transmitting a negative full-width pulse ($-A$) of the same amplitude and energy as for binary 1. Thus BNRZ-L employs two antipodal pulses $-A$ and $+A$, which improves bit error performance, i.e. lowers BER (when compared to UNRZ-L) to

$$\text{BER} = \frac{1}{2}\text{erfc}\left(\sqrt{\frac{E_b}{N_o}}\right), \quad (\text{BNRZ-L, BRZ, Bi-}\Phi\text{-L}) \tag{7.16}$$

The PSD of BNRZ-L is given in the first line of Eq. (7.13). BNRZ-L code has the same advantages as UNRZ-L in addition to eliminating the DC offset problem of UNRZ-L and achieving a lower BER. However, it retains all the other significant drawbacks of UNRZ-L and these also make it unsuitable for digital transmission applications. Furthermore, BNRZ-L is susceptible to *polarity inversion* whereby a level inversion in transit causes all symbols to be incorrectly interpreted at the receiver. This susceptibility may be overcome by using transition coding or by preceding the information bits in level codes by a known bit sequence that enables the receiver to detect the presence or otherwise of polarity inversion and hence to correctly interpret the information bits that follow.

***UNRZ-M, UNRZ-S*:** Unipolar nonreturn to zero-mark (UNRZ-M) and unipolar nonreturn to zero-space (UNRZ-S) are binary unipolar transition codes that convey information through a change in voltage level, rather than in the level itself. UNRZ-M represents a binary 1 (called *mark* in the days of telegraphy) by a transition in voltage level between 0 and $+A$, and binary 0 (formerly called *space*) by the absence of a transition at the start of the bit interval. Note that any voltage level initiated at the start of a bit interval is maintained for at least the duration of the bit interval, making this an NRZ scheme. Furthermore, the output 'change' or 'no change' signal is determined exclusively by the current input bit, making this an instantaneous code with no memory. UNRZ-S works on a similar principle but represents binary 0 ($\equiv$ space) with a transition at the start of the bit interval and a binary 1 with no transition. Both schemes have probability of error given by

$$\text{BER} = \text{erfc}\left(\sqrt{\frac{E_b}{2N_o}}\right), \quad (\text{UNRZ-M, UNRZ-S}) \tag{7.17}$$

and the same power spectral density as UNRZ-L given by the second line of Eq. (7.13). Susceptibility to polarity inversion has been eliminated and bandwidth is low, but both codes still have significant disadvantages, namely (1) DC offset, (2) maximum PSD at $f=0$, (3) BER that is double that of UNRZ-L and (4) poor clock content since a run of 0s can cause the UNRZ-M waveform to go for an unlimited period without the crucial transitions in voltage level relied upon by the receiver to maintain synchronisation with the transmitter. Similarly, a run of 1s can cause the UNRZ-S waveform to be devoid of transition for an unlimited period.

***BNRZ-M, BNRZ-S*:** Bipolar nonreturn to zero-mark (BNRZ-M) and bipolar nonreturn to zero-space (BNRZ-S) are binary bipolar transition codes that work on the same principle as UNRZ-M and UNRZ-S except that the voltage levels between which transitions are made are $-A$ and $+A$. Thus BNRZ-M represents binary 1 by a transition between $-A$ and $+A$ at the start of the bit interval, and binary 0 by the absence of a transition, whereas BNRZ-S represents binary 0 by a transition and binary 1 by the absence of a transition. They improve on their unipolar counterparts by eliminating DC offset and reducing BER to

$$\text{BER} = \text{erfc}\left(\sqrt{\frac{E_b}{N_o}}\right), \quad (\text{BNRZ-M, BNRZ-S, DBRZ, Bi-}\Phi\text{-}\Delta) \tag{7.18}$$

However, the other significant deficiencies remain, especially poor clock content and significant spectral content around $f = 0$, and BER, although lower than for UNRZ-S and UNRZ-M, is double that of BNRZ-L. The PSD is given in the first line of Eq. (7.13) and plotted in Fig. 7.4c, from which it can be seen that (in common with all NRZ codes) null bandwidth is R_s, which is low.

Fig. 7.7 shows the arrangement for transition coding and decoding of a binary sequence m_k. An XOR gate combines the current input m_k with the previous output d_{k-1} to produce the current output d_k, which yields NRZ-M when represented using separate voltage levels for binary 1 and binary 0. If an XNOR gate is used instead, then the result is NRZ-S. At the receiver, NRZ-M decoding is achieved by using an XOR gate to combine the current received bit with the previous received bit, whereas NRZ-S decoding uses an XNOR gate instead. Table 7.2 shows the output sequence d_k produced by the arrangements in Fig. 7.7 in response to an input sequence $m_k = 1000101110$. A comment column is provided to draw attention to

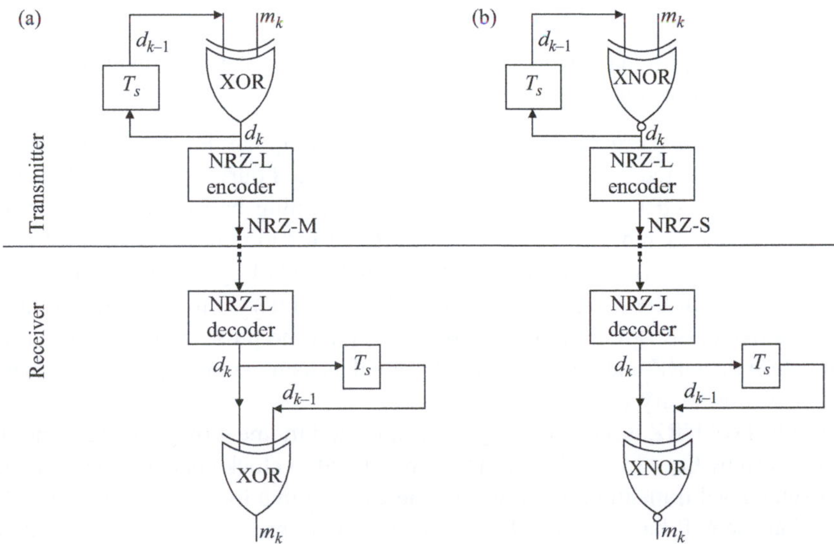

Fig. 7.7 Differential coding of binary data m_k showing (a) NRZ-M coding and decoding using XOR gate; (b) NRZ-S coding and decoding using XNOR gate

Table 7.2 Table of differential coding. At the start ($k = 0$) there is no previous output and d_{k-1} is chosen (arbitrarily) to be 0 or 1. Which one you choose makes no difference to the coding rule (i.e. which bit causes a change or no change in output)

t	k	Input m_k	Previous Output, d_{k-1}	Output, $d_k = m_k \oplus d_{k-1}$	Comment	d_{k-1}	Output, $d_k = \overline{m_k \oplus d_{k-1}}$	Comment
0	0	1	**0 (chosen)**	1	Change	**0**	0	No change
T_s	1	0	1	1	No change	0	1	Change
$2T_s$	2	0	1	1	No change	1	0	Change
$3T_s$	3	0	1	1	No change	0	1	Change
$4T_s$	4	1	1	0	Change	1	1	No change
$5T_s$	5	0	0	0	No change	1	0	Change
$6T_s$	6	1	0	1	Change	0	0	No change
$7T_s$	7	1	1	0	Change	0	0	No change
$8T_s$	8	1	0	1	Change	0	0	No change
$9T_s$	9	0	1	1	No change	0	1	Change
			⟵ XOR differential coder: ⟶			⟵ XNOR differential coder ⟶		

output transitions. It is clear that the XOR arrangement does correctly implement the NRZ-M coding rule of a change (i.e. transition) in output in response to a binary 1 input and no change in output when the input is binary 0. Similarly, the output of the XNOR arrangement changes when the input is binary 0 and stays the same when the input is binary 1. At the start of coding ($k = 0$) there is no previous output, so d_{-1} is pre-selected as 0 in the table. You may wish to verify that the relevant coding rule would still be correctly implemented by choosing $d_{-1} = 1$.

7.4.2 Return to zero (RZ) line codes

RZ line codes employ half-width pulses and may be unipolar or bipolar, binary or ternary, and level or transition.

***Unipolar return to zero (URZ)*:** This is a binary unipolar level code in which binary 1 is represented by a positive pulse of amplitude A during the first half of the bit interval followed by no pulse (0 V) in the remaining half of the bit interval, whereas binary 0 is represented as no pulse (0 V) for the entire bit interval. For brevity we will specify the shape of a transmitted symbol by using the characters '+', '−' and '0' in the corresponding parts of the bit interval to respectively indicate a positive pulse $+A$, a negative pulse $-A$, and a 'no pulse' in that position. In this notation the URZ coding rule is simply stated as using symbol (+, 0) for binary 1 and (0, 0) for binary 0.

The PSD of URZ is shown in Fig. 7.4a and listed in line 4 of Eq. (7.13), and its BER is given by Eq. (7.14). There is an improvement in clock content (compared to NRZ) since level transitions are now guaranteed even in a long run of 1s. Fig. 7.4a shows that the null bandwidth of RZ is $2R_s$, so this improvement in clock content does come at a price. In general the bandwidth of codes that employ half-width pulses is double that of codes that use full-width pulses, in view of the inverse relationship between time (namely pulse duration) and frequency (i.e. pulse bandwidth). RZ is however significantly disadvantaged by having a DC offset as well as a large spectral content around $f = 0$. Furthermore, a long run of 0s still leaves the coded waveform without level transition and it has no error detection, which does not help with performance monitoring.

***Bipolar return to zero (BRZ)*:** This is a binary bipolar level code in which a positive half-width pulse (+, 0) is used to represent binary 1 and a negative half-width pulse (−, 0) is used for binary 0. The PSD of this code is shown in Fig. 7.4d and listed in line 3 of Eq. (7.13). Since the coding involves two antipodal pulses, the BER is the same as for BNRZ and is given in Eq. (7.16). BRZ is an improvement on URZ in terms of a lower BER, no DC offset, and guaranteed clock content whatever the bit sequence. However, it is susceptible to polarity inversion and lacks error detection capability. Its spectral content is also, like URZ, maximum around $f = 0$.

***Alternate mark inversion (AMI-RZ, AMI-NRZ)*:** AMI-RZ is a ternary bipolar level code that employs alternate polarity pulses to represent binary 1 and no pulse to indicate binary 0. An example is shown in Fig. 7.6f where it can be seen that pulses (+, 0) and (−, 0) are used alternatingly to represent binary 1. Two distinct voltage levels are used to represent binary 1 and one distinct level to represent binary 0, which means that AMI-RZ has a code radix of 3 and is therefore a *ternary* code. The consequence is that code efficiency η_c is less than 100%, the exact value being obtained from Eq. (7.1) by noting that here a code of radix 3 conveys one bit per symbol sent, so that

$$\eta_{c,AMI\text{-}RZ} = \frac{1}{\log_2(3)} = 63.1\%$$

and redundancy

$$\Re_{AMI\text{-}RZ} = \frac{1}{\eta_c} - 1 = 58.5\%$$

This redundancy delivers an important benefit, namely error detection. Any *violation* of the alternating polarity rule is definitely an indication of error and this can be relied upon for performance monitoring on low-error-rate links. At high error rates however certain combinations of multiple errors (e.g. two consecutive 1s being received as 0s) will go undetected. The BER of AMI-RZ coded waveform is given by

$$\begin{aligned} \text{BER} &= \frac{3}{4}\text{erfc}\left(\sqrt{\frac{E_b}{2N_o}}\right) + \frac{1}{4}\text{erfc}\left(\sqrt{\frac{2E_b}{N_o}}\right) \quad \text{(AMI-RZ, AMI-NRZ)} \\ &\approx \frac{3}{4}\text{erfc}\left(\sqrt{\frac{E_b}{2N_o}}\right) \end{aligned} \tag{7.19}$$

The PSD of AMI-RZ is listed in line 5 of Eq. (7.13) and plotted in Fig. 7.4b. Due to the alternating polarity rule, there is no build-up of a DC offset — an improvement on URZ, and the PSD is in fact zero at the origin which minimises distortion of the coded waveform in AC-coupled transmission systems. The null bandwidth of AMI-RZ is R_s, which at first gives the impression that it has the same low bandwidth as NRZ codes. However, on closer inspection of spectral content (Table 7.1) we see that 20.4% of AMI-RZ power lies in the frequency range above R_s, compared to only 4.86% of power in UNRZ codes. Spectral occupancy can be improved somewhat by using full-width pulses, leading to *AMI-NRZ* whose PSD is plotted in Fig. 7.4i and listed in line 6 of Eq. (7.13). The null bandwidth of AMI-NRZ is also R_s, but Table 7.1 shows that only 14.4% of power is contained at frequencies $f > R_s$, which is a reduction of 6% on AMI-RZ.

The significant drawback of AMI line codes whereby a run of 0s results in a waveform without transition over a significant period can be mitigated by scrambling the bit stream prior to coding in order to reduce the likelihood of a long run of 0s. With this remedy AMI line codes have been adopted in transmission systems particularly due to its apt spectral shape, performance monitoring capability, and codec simplicity. Furthermore, timing recovery is straightforward because squaring or full-wave rectifying an AMI-RZ waveform converts it into URZ with a spectral line at the bit rate R_s (see Fig. 7.4a). However, AMI has a higher BER than URZ and its low code efficiency (63%) limits its use to short-distance links, e.g. on coaxial cables connecting communications equipment within one building. It is worth noting that AMI codes have memory: Binary 0s are coded instantaneously but when the current bit is a binary 1 then there is a choice of pulse polarity that depends on the identity of previous bits. This influence of past input is a characteristic of processes or systems with memory.

***Dicode (RZ, NRZ)*:** This is a ternary bipolar level coding scheme with memory. Alternate polarity pulses are employed to represent a change in input bit (which is the case when a binary 1 in the current interval is preceded by a binary 0 in the previous interval, or binary 0 in the current interval is preceded by binary 1 in the previous), whereas no pulse is sent when there is no change in bit (i.e. the previous and current bit intervals are either both 1s or both 0s). See Fig. 7.6g. Dicode therefore has memory since the current output symbol is determined by the current and previous input bits. More specifically, as shown in the circuit arrangement of Fig. 7.8, dicode RZ is the result of AMI coding the bit sequence obtained by XOR summing each bit of an input binary sequence with the previous bit. At the start of this process (at $k = 0$), there is no previous input bit m_{k-1}, so m_{-1} is preselected as either binary 1 or binary 0. If the pulses used are half-width the coding scheme is called dicode RZ, whereas full-width pulses produce dicode NRZ. The PSD of dicode RZ is the same as AMI-RZ, as given in line 5 of Eq. (7.13) and plotted in

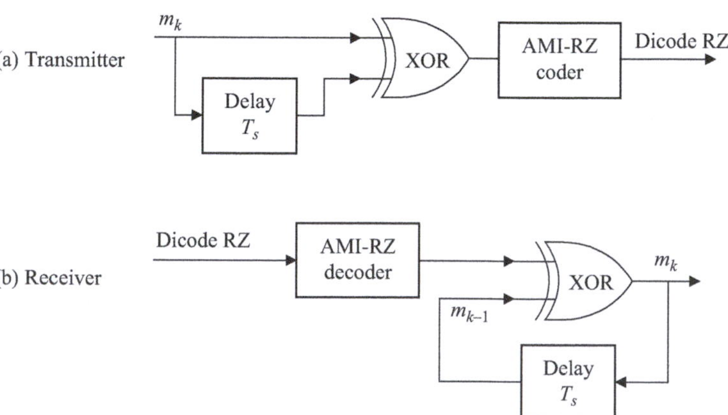

Fig. 7.8 (a) Dicode RZ is produced by differential coding of binary data m_k using an XOR gate followed by AMI-RZ coding; (b) decoding arrangement

Fig. 7.4b, and the PSD of dicode NRZ is the same as AMI-NRZ in Fig. 7.4i and line 6 of Eq. (7.13). The main weaknesses of dicode are higher BER and codec complexity compared to AMI. Dicodes are related to AMI by differential coding and therefore have double the BER of AMI:

$$\text{BER} \approx \frac{3}{2}\text{erfc}\left(\sqrt{\frac{E_b}{2N_o}}\right) \quad \text{(Dicode RZ, Dicode NRZ)} \tag{7.20}$$

***Differential bipolar RZ (DBRZ)*:** This is a binary bipolar transition coding scheme with memory in which a half-width pulse is always sent. The pulse polarity is *changed* to indicate a change in input bit, whereas when there is no bit change this is indicated by there being no change in pulse polarity. Thus DBRZ represents a *change in input state* by *a change in output state*, and *no change in input state* by *no change in output state*. It has all the benefits of BRZ discussed earlier as well as the same PSD shown in Fig. 7.4d, and overcomes the problem of susceptibility to polarity inversion albeit at the expense of an increase in BER given in Eq. (7.18).

7.4.3 Biphase line codes

This is a class of binary bipolar phase codes in which the symbols that represent binary 1 and 0 may contain one or both of the two voltage levels ($-A$, A) that are employed, but are distinguished in 'phase'. All biphase codes are binary and bipolar, but may be phase, level or transition as indicated below in the four types of biphase codes defined.

***Manchester or biphase-L (Bi-Φ-L)*:** This line code employs two symbols of opposite phase to represent binary 1 and 0. Binary 1 is represented by a symbol that consists of a half-width positive pulse of amplitude A in the first half of the bit interval followed by a half-width negative pulse of the same amplitude in the second half of the bit interval. Binary 0 is represented using a symbol of opposite phase, i.e. a negative half-width pulse of amplitude A followed by a positive half-width pulse of the same amplitude. Thus this is a binary bipolar instantaneous phase code. Noting that the two symbols are antipodal, and taking (completely arbitrarily) the symbol for binary 1 as reference, Bi-Φ-L can be viewed as representing binary 1 by a phase of 0° and binary 0 by a phase of 180°, hence its classification as a biphase code.

There is however a valid alternative view of Bi-Φ-L as a transition line code that represents binary 1 by a downward transition in voltage level in the middle of the bit interval and binary 0 by an upward transition in voltage level in the middle of the interval, the transitions at the start of each interval being overheads that signify no information but are included for clock content. A coded waveform example is shown in Fig. 7.9a. The PSD of Bi-Φ-L is shown in Fig. 7.4f with an expression listed in line 7 of Eq. (7.13), where it can be seen that there is no DC offset and no spectral content around $f = 0$. Other good features of the Manchester code are that clock content is guaranteed because there is a mid-interval transition in every bit interval, and error rate is low as given by Eq. (7.16). However, there is no error detection and the code is susceptible to polarity inversion. Nevertheless, the code is used in the Ethernet standard IEEE 802.3 for local area networks (LAN).

***Biphase mark (Bi-Φ-M)*:** This is a binary bipolar instantaneous transition code that represents binary 1 by *two transitions*, one at the start of the bit interval and the second at the middle of the bit interval, and represents binary 0 by only *one transition*

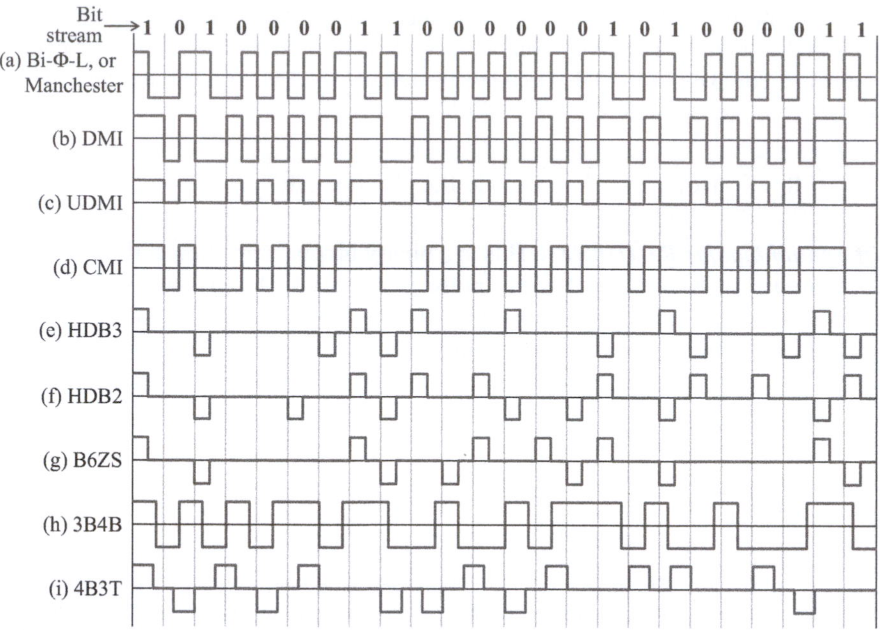

Fig. 7.9 More line code waveforms

at the start of the bit interval. The transition at the start of each bit interval does not add any information about the identity of the bit in that interval and is therefore an overhead that is present for the benefit of clock content. Bit identification is provided only by the mid-interval transition, which is present for binary 1 and absent for binary 0. This code is also called *differential Manchester line code*. Binary 1 is represented either by symbol (+, −) or (−, +), which means a positive half-width pulse of amplitude A occupying the first half of the bit interval and a negative half-width pulse of the same amplitude occupying the second half of the bit interval, or a negative half-width pulse followed by a positive half-width pulse; and binary 0 is represented by either a positive full-width pulse (+, +) or a negative full-width pulse (−, −). Note that the symbols for binary 1 have opposite phases, as do the symbols for binary 0, hence the biphase classification. Fig. 7.6h shows a coded waveform example. Bi-Φ-M has the same PSD as Bi-Φ-L (line 7 of Eq. (7.13) and Fig. 7.4f) and the same guaranteed clock content, and also lacks error detection. Its main improvement on Bi-Φ-L is that it is not affected by polarity inversion, but it has a BER, given by Eq. (7.14), which is larger than that of Bi-Φ-L. It also has a higher codec complexity. Bi-Φ-M is used in the IEEE 802.5 standard for token ring LANs.

***Biphase space (Bi-Φ-S)*:** This has the same characteristics as Bi-Φ-M and works on exactly the same principle, but reverses the roles of mark and space. Thus Bi-Φ-S represents binary 0 by *two transitions*, one at the start and the second at the middle of the bit interval, and binary 1 by only *one transition* at the start of the bit interval. That is, binary 0 is represented either by symbol (+, −) or (−, +), and binary 1 by either (+, +) or (−, −). There are therefore two symbol options for each bit and in each bit interval the choice is guided by the rule that there must be a voltage transition at the start of each bit interval. The type of symbols allocated to binary 0 and binary 1 ensures that there is always an additional transition in the middle of the bit interval for binary 0 and never a mid-interval transition for binary 1.

***Conditioned biphase-L (Bi-Φ-Δ)*:** This represents a *change* in input bit by a downward transition in the middle of the bit interval using symbol (+, −), whereas when there is no change in input bit this is represented by an upward mid-interval transition using symbol (−, +). Thus conditioned biphase-L is produced through differential coding followed by biphase-L. That is, each bit of an input sequence is XORed with the previous bit and the resulting sequence is coded using biphase-L. Conditioned biphase-L (denoted by Bi-Φ-Δ) has the same PSD as Bi-Φ-L (line 7 of Eq. (7.13) and Fig. 7.4f) and a BER given by Eq. (7.18).

***Miller code*:** This line code, inappropriately called delay modulation in some literature, is a binary bipolar transition code with memory that operates according to the following three coding rules: (1) Binary 1 is represented by a mid-interval transition; (2) binary 0 is represented by the absence of a mid-interval transition; and (3) there is a transition at the end of any binary 0 interval that is followed by another binary 0. Fig. 7.6i shows an example of a Miller coded waveform from which it can be seen that the above three rules are implemented by using two symbol options (+, −) and (−, +) for binary 1, and (+, +) and (−, −) for binary 0. The symbol selection made in each interval is dictated entirely by the above rules, and the memory characteristic of the code is due to the fact that the current output is influenced not only by the current bit but by the previous as well. For example, if the current bit is binary 0 then the output symbol used is influenced by the identity of the previous bit, in that if the previous bit was binary 1 the selection is a random choice between (+, +) and (−, −), but if the previous bit was binary 0 the selection must be the other binary 0 symbol not used in the previous interval. Notice that the symbols for binary 1 have opposite phases, as do the symbols for binary 0, hence the biphase classification. The PSD of Miller code has good clock content, and manages to achieve this while retaining a narrow bandwidth and an appropriate spectrum that contains no DC offset and has relatively small low-frequency components, as can be seen in the PSD shown in Fig. 7.5a. The BER of Miller code is given by Eq. (7.14), which is comparable to other less suitable basic line codes. The main drawback of Miller code is its lack of error detection which makes performance monitoring difficult.

7.5 Advanced codes

The line codes discussed below offer flexibility for eliminating DC offset and ensuring adequate clock content whatever the input bit sequence. Most of them also allow routine performance monitoring and have high code efficiency, with the only significant drawback being codec complexity. There are two broad categories of these codes, namely run length limited (RLL) codes that limit the length of a run of voltage levels without transition, and block codes that convert blocks of input bits into coded symbols according to assignments in a code table.

7.5.1 Run length limited (RLL) line codes

RLL codes are designed to address one of the main shortcomings of AMI line codes whereby the coded waveform is devoid of transition for the duration of an unbroken run of 0s in the input bit stream however long that may be. BnZS (bipolar with *n* zeros substitution) is the main type of RLL code, although coded mark inversion (CMI) and differential mode inversion (DMI) are usually also classed as RLL.

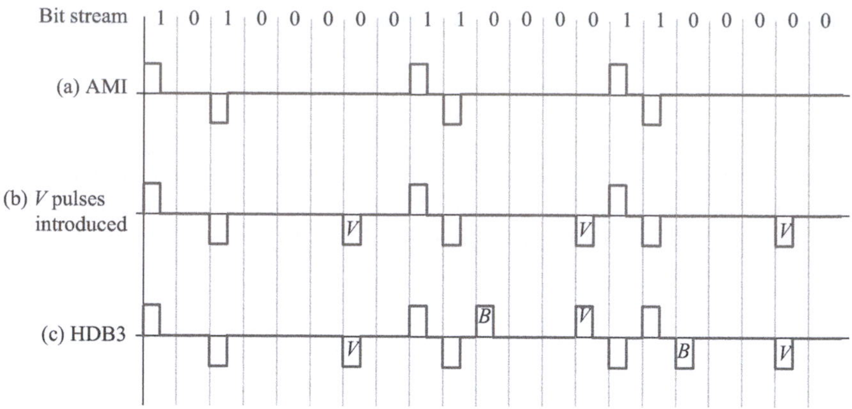

Fig. 7.10 Steps in modifying AMI to produce HDB3

***BnZS (Bipolar with n zeros substitution)*:** BnZS is a class of ternary bipolar level codes based on a modification of AMI to limit the maximum number of no pulses to $n-1$. The input bit stream is coded as usual according to AMI rules except that each pattern of n consecutive 0s is represented by a special pattern of n symbols that includes at least one pulse, known as a *violation (V) pulse*, and if necessary a second pulse known as a *balancing (B) pulse*. As explained below, both pulses are recognisable at the receiver as representing binary 0. Different values of n lead to different codes, such as B3ZS, B4ZS, B6ZS and B8ZS, all of which work on the principle. The case of $n=4$, i.e. B4ZS, is also called *high-density bipolar with 3 zeros maximum* (HDB3). This alternative naming convention may be extended to apply to all BnZS codes so that B3ZS is HDB2, B6ZS is HDB5 and so on. The operation of HDB3 is discussed below in detail and a brief guidance is subsequently given about other commonly used BnZS codes.

***HDB3 (≡ B4ZS)*:** Fig. 7.10 shows the steps involved in modifying an AMI waveform to obtain HDB3 coded waveform. The top waveform (a) is the AMI code of the indicated bit stream. In the second waveform (b) a violation (V) pulse has been introduced in every fourth position of 4 consecutive 0s in order to improve clock content by limiting the maximum number of consecutive 0s to three. The introduced pulse is called a violation pulse since it violates the alternating polarity rule. Each V pulse is deliberately chosen to have the same polarity as the previous pulse; otherwise the receiver would incorrectly interpret the pulse as a binary 1. Notice however that, although this introduction of V pulses now guarantees clock content it has created a new problem: All V pulses have the same polarity and this introduces an undesirable DC offset in the modified waveform. The only way to prevent this build-up of DC is to ensure that a new V pulse does not have the same polarity as the previous V pulse. This is achieved as shown in the third waveform (c) in Fig. 7.10 as follows: Before inserting a new V pulse we check to see if it would have the same polarity as the previous V pulse. If so, we represent the first binary 0 of the four consecutive 0s by a pulse, called *balancing (B) pulse* that obeys the alternating polarity rule, and then insert the new V pulse to violate this B pulse. Note that a B pulse is used only when it is necessary to prevent two successive V pulses having the same polarity, and that a B pulse when used is always placed in the first bit position and a V pulse always in the fourth bit position of a pattern of four consecutive 0s.

In summary, to modify AMI into HDB3, replace four consecutive no pulses 0 0 0 0 by

- 0 0 0 – if preceding pulse is negative and number of pulses since last bipolar violation is odd.
- 0 0 0 + if preceding pulse is positive and number of pulses since last bipolar violation is odd.
- – 0 0 – if preceding pulse is positive and number of pulses since last bipolar violation is even.
- + 0 0 + if preceding pulse is negative and number of pulses since last bipolar violation is even.

where '0' indicates a no pulse, '–' a negative half-width pulse and '+' a positive half-width pulse. Note that when coding a bit stream, a *B* pulse cannot be used before a *V* pulse has been used for the first time because the necessity for a *B* pulse would not yet have arisen.

So how does a decoder correctly interpret a HDB3 waveform? Well, it proceeds as follows:

- Every no pulse is a binary 0.
- Every *V* pulse is a binary 0.
- Every pulse violated after just two no pulses is a *B* pulse and hence a binary 0.
- Any other pulse is a binary 1.

Other BnZS codes work on a similar basis and are obtained by replacing every block of *n* consecutive 0s in AMI code by the patterns specified below for B3ZS, B6ZS and B8ZS.

***B3ZS (≡ HDB2)*:** This limits the maximum run of no pulses to two bit intervals. It achieves this by replacing three consecutive no pulses 0 0 0 by

- 0 0 – if preceding pulse is negative and number of pulses since the last bipolar violation is odd.
- 0 0 + if preceding pulse is positive and number of pulses since the last bipolar violation is odd.
- – 0 – if preceding pulse is positive and number of pulses since the last bipolar violation is even.
- + 0 + if preceding pulse is negative and number of pulses since the last bipolar violation is even.

Note that the first two replacement patterns are 00V (i.e. *no* pulse, *no* pulse, *V* pulse) and the last two are B0V (i.e. *B* pulse, *no* pulse, *V* pulse). As discussed under HDB3, the receiver will recognise a *V* pulse (due to alternating polarity rule violation) and a *B* pulse (which in this case is any pulse violated after just one no pulse) and will decode both to binary 0.

***B6ZS (≡ HDB5)*:** This limits the maximum run of no pulses to five bit intervals and achieves this by replacing six consecutive no pulses 0 0 0 0 0 0 by

- 0 + – 0 – + if the last pulse was positive.
- 0 – + 0 + – if the last pulse was negative.

Note that each of the above is a pattern 0VB0VB that the receiver will recognise and decode as six consecutive binary 0s.

***B8ZS (≡ HDB7)*:** This limits the maximum run of no pulses to seven bit intervals and achieves this by replacing eight consecutive no pulses 0 0 0 0 0 0 0 0 by

- 0 0 0 + – 0 – + if the last pulse was positive.
- 0 0 0 – + 0 + – if the last pulse was negative.

Again each of the above is the pattern 000VB0VB which the receiver will also recognise and decode as eight consecutive binary 0s.

Examples of coded waveforms for HDB3 ≡ B4ZS, HDB2 ≡ B3ZS and B6ZS are shown in Fig. 7.9e to 7.9g. The power spectral densities of these codes are shown in Fig. 7.5b, and can be seen to be very similar to the PSD of AMI. Thus the modifications discussed above have delivered the intended benefit of guaranteed clock content without adversely affecting the spectral characteristics of the waveform. Code efficiency is also still 63%, the same as for AMI, but codec complexity has increased. BnZS codes are used mostly for conveying TDM signals (the T-series in North America and the E-series in Europe) between equipment in the TDM hierarchy. On a final note, although Fig. 7.9e,f feature only half-width pulses, all BnZS codes, like AMI, can be implemented using either half-width or full-width pulses.

***Coded mark inversion (CMI)*:** This is a binary bipolar level code that represents binary 1 (≡ mark) using alternating polarity full-width pulses (+, +) or (−, −) and binary 0 by the symbol (−, +), which comprises a negative half-width pulse in the first half of the bit interval followed by a positive half-width pulse in the second half of the bit interval. The possibility of code violation (e.g. two consecutive full-width pulses of the same polarity) allows error detection and hence performance monitoring. Fig. 7.9d shows a CMI waveform example. CMI has good clock content with a guarantee of one transition in a period no longer than 1.5 bit intervals. The PSD is plotted in Fig. 7.4h and shows that CMI also has good synchronisation properties since it contains a spectral line at the symbol rate R_s which can be readily extracted at the receiver. Furthermore, its low-frequency components are insignificant and it has no DC offset. Its error rate performance is given by

$$\text{BER} = \frac{3}{4}\text{erfc}\left(\sqrt{\frac{E_b}{2N_o}}\right) + \frac{1}{4}\text{erfc}\left(\sqrt{\frac{E_b}{N_o}}\right) \approx \frac{3}{4}\text{erfc}\left(\sqrt{\frac{E_b}{2N_o}}\right) \tag{7.21}$$

which is approximately the same as for AMI and the BnZS series.

CMI however has a lower code efficiency than AMI, which may be determined by noting that in CMI a bit interval is divided into two halves each of which contains two significant levels (+, −). Thus four levels, with potential information content $\log_2 4 = 2$ bits, are used to convey one bit, which gives CMI code efficiency $\eta_c = ½ = 50\%$. It is important to see how this consideration differs from AMI which has a code efficiency of 63%. In AMI, the second half of the bit interval is at a fixed level 0 V for both binary 1 and binary 0 and therefore is unused. That is, the receiver would gain no information whatsoever about the identity of the bit in an AMI interval by looking at the voltage level of the second half of the interval. RZ pulses are used purely for the purpose of improving clock content and carry a bandwidth overhead but do not change information content from what it would be with NRZ pulses. So AMI uses a total of three significant voltage levels (+, 0, −) all in the first half of the bit interval, with a potential information content of $\log_2 3$, to convey one bit of information, and this is what gives AMI the code efficiency $\eta_c = 1/\log_2 3 = 63\%$ determined earlier.

CMI's low efficiency is certainly a drawback, but its codec simplicity and ease of synchronisation make it ideal for high bit rate links operating over short distances, such as the coaxial cable connection conveying the European E4 TDM signal at 140 Mb/s between equipment in one building. A unipolar version of CMI, denoted as *UCMI*, obtained by using a 0 V level in place of the negative level is suitable for application in optical fibre line systems.

Differential mode inversion (DMI): DMI is a binary bipolar level code that represents binary 1 using alternating polarity full-width pulses (+, +) or (−, −), just like CMI. However, DMI represents binary 0 using either symbol (−, +) or (+, −) selected as necessary to ensure that the coded waveform does not stay at one level for longer than one bit interval. For example, if the previous bit was a binary 1 coded using symbol (−, −) and the current bit is a binary 0, then the symbol option (+, −) must be selected. See a waveform example in Fig. 7.9b. The PSD of DMI is plotted in Fig. 7.4f and is unsurprisingly the same as for the biphase codes, with the attractive features of zero DC offset and low spectral content near $f = 0$. It also has good clock content with a guarantee of one or two transitions in a period of one bit interval, but unlike Manchester line code (Bi-Φ-L) it is not susceptible to polarity inversion and it allows performance monitoring. A unipolar version of this code *(UDMI)*, suitable for optical fibre applications, obtained by using 0 V in place of the negative voltage level, is shown in Fig. 7.9c along with its PSD in Fig. 7.4g.

7.5.2 Block line codes

The line codes discussed so far have worked on the basis of mapping one input bit or one input change to one output code symbol. Block coding on the other hand takes n input bits at a time and maps them into an output codeword consisting of m coded symbols as illustrated in Fig. 7.11. If the output symbols are binary the block code is described as a binary block code designated nBmB, where $m > n$. If on the other hand the output symbols are ternary (having three possible voltage levels +, 0, −) the block code is described as a ternary block code designated nBmT.

Since there are 2^n possible input combinations to be assigned codewords from a pool of 2^m codewords in nBmB and 3^m codewords in nBmT, there is a surplus of codewords that gives great flexibility in the choice of codewords to achieve advantageous code characteristics such as good clock content and zero DC offset. For example, in a 3B4B code there are 16 codewords to choose from for 8 input blocks, and codewords such as − + − + with frequent transitions are preferred to codewords such as − + + + with fewer transitions. Furthermore, codewords containing an equal number of positive and negative pulses (e.g. + − + −), referred to as *balanced codewords*, are preferred over *unbalanced codewords* (such as + + + −) that contain an unequal number of positive and negative pulses.

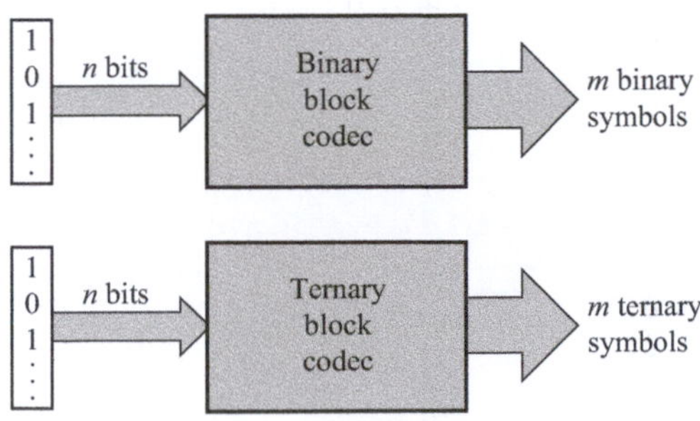

Fig. 7.11 Block coding

Digital sum (DS) is the count of the imbalance between negative and positive pulses in a sequence of symbols. If the count is taken over only one codeword then the DS is called the *disparity* of the codeword. And if we keep a count of the cumulative sum of the disparity of each transmitted code word we obtain an important variable known as *running digital sum* (RDS). There is usually not enough balanced codewords to go round, so unbalanced codewords will have to be assigned to some input blocks. Assignment of unbalanced codewords is done in pairs, so that any input block that is represented by an unbalanced codeword has a positive-disparity codeword along with its negative-disparity counterpart (e.g. $+ + - +$ and $- - + -$) at its disposal to balance out each other. When coding such an input block, the coder uses the positive-disparity option if RDS is negative, and the negative-disparity option if RDS ≥ 0. In this way the RDS will be an integer confined within a small range such as $(-1, +1)$ and the coded waveform will have no DC offset.

Fig. 7.9h,i shows 3B4B and 4B3T coded waveforms based respectively on the coding tables of Tables 7.3 and 7.4. Notice in Fig. 7.9 that the 3B4B waveform

Table 7.3 3B4B coding table

Input	**Output codeword**			**Disparity**
	Negative-disparity option	**Balanced**	**Positive-disparity option**	
000	− + − −		+ − + +	±2
001		− − + +		0
010		− + − +		0
011		− + + −		0
100		+ − − +		0
101		+ − + −		0
110		+ + − −		0
111	− − + −		+ + − +	±2

Table 7.4 4B3T coding table

Input	**Output codeword**			**Disparity**
	Negative-disparity option	**Balanced**	**Positive-disparity option**	
0000		+ 0 −		0
0001		− + 0		0
0010		0 − +		0
0011		+ − 0		0
0100	− − 0		+ + 0	±2
0101	0 − −		0 + +	±2
0110	− 0 −		+ 0 +	±2
0111	− − −		+ + +	±3
1000	− − +		+ + −	±1
1001	+ − −		− + +	±1
1010	− + −		+ − +	±1
1011	− 0 0		+ 0 0	±1
1100	0 − 0		0 + 0	±1
1101	0 0 −		0 0 +	±1
1110		0 + −		0
1111		− 0 +		0

Table 7.5 Coding steps used to obtain the 3B4B waveform of Fig. 7.9h

Input	RDS before	Codeword	RDS after
101	−1 (Pre-selected)	+ − + −	−1
000	−1	+ − + +	+1
011	+1	− + + −	+1
000	+1	− + − −	−1
000	−1	+ − + +	+1
101	+1	+ − + −	+1
000	+1	− + − −	−1
011	−1	− + + −	−1

consists of four binary pulses per interval of three input bits and the 4B3T waveform consists of three ternary RZ pulses per interval of four input bits. The coding procedure for 3B4B is documented in Table 7.5 and works as follows. Column 1 is the bit stream at the top of Fig. 7.9 taken three bits at a time. The initial value of RDS (at the start of coding) is preselected as −1 (see row 2, column 2). The first block of three bits is 101 and this is represented using the codeword + − + −, according to Table 7.3. The RDS is then updated by adding the disparity of this codeword to the 'RDS before' to obtain the 'RDS after', which becomes the 'RDS before' for the next row where the next block of bits is coded. This next block happens to be 000, and because RDS = −1, we select the positive-disparity codeword option + − + + in Table 7.3. This has disparity +2 and so the updated RDS is 'RDS before' +2 = −1 + 2 = +1, which again is transferred to the next row as 'RDS before'. Coding proceeds in this manner until the entire bit stream has been covered, three bits at a time, and yields the sequence of codewords in column 3, which is what is plotted in Fig. 7.9h. As an exercise you may wish to use this procedure and the 4B3T coding table of Table 7.4 to verify the 4B3T waveform plotted in Fig. 7.9i.

In addition to good timing content and zero DC offset, one of the great strengths of block coding is the possibility of achieving high coding efficiencies which allows transmission at a lower symbol rate and hence overall system cost savings through a greater spacing of repeaters on long-distance links. An nBmB binary block code uses m binary symbols (having code radix 2) to convey n bits, whereas an nBmT ternary block code uses m ternary symbols (having code radix 3) to convey n bits. Therefore their code efficiencies are given by

$$\begin{aligned}\eta_{n\mathrm{B}m\mathrm{B}} &= \frac{n}{m}\\ \eta_{n\mathrm{B}m\mathrm{T}} &= \frac{n}{m\log_2 3}\end{aligned} \tag{7.22}$$

As examples, the code efficiencies of 3B4B, 4B3T and 6B4T are respectively 75%, 84.12% and 94.64%. If each of these codes was considered for application in carrying an E4 signal that has bit rate $R_b = 139264$ kb/s, what would be the respective symbol rates? Noting that the codes respectively produce four symbols per three bits, three symbols per four bits and four symbols per six bits, and that the service requirement has 139 264 000 bits per second, it follows that the number of symbols per second (called baud) will be

$$R_s = \begin{cases} 139264000 \dfrac{\text{bits}}{\text{s}} \times \dfrac{4 \text{ symbols}}{3 \text{ bits}} = 185.6853 \text{ Mbaud} & \text{(3B4B)} \\ 139264000 \dfrac{\text{bits}}{\text{s}} \times \dfrac{3 \text{ symbols}}{4 \text{ bits}} = 104.448 \text{ Mbaud} & \text{(4B3T)} \\ 139264000 \dfrac{\text{bits}}{\text{s}} \times \dfrac{4 \text{ symbols}}{6 \text{ bits}} = 92.8427 \text{ Mbaud} & \text{(6B4T)} \end{cases}$$

Notice that the more efficient codes require lower symbol rates to support the same bit rate. It is worth pointing out that for a given m, the efficiency of an nBmB block code is maximum when n takes on its maximum value of $n = m - 1$. Under this condition coding efficiency increases with m, but this increases codec complexity (and hence cost) as well as the maximum length of a run of like pulses, which degrades clock content.

The PSD of various block codes are plotted in Fig. 7.5c,d and show that block codes also have excellent spectral characteristics, including negligible content at $f = 0$ as well as a small null bandwidth equal to symbol rate R_s, which in the case of 3B4B contains as much as 87.7% of total power. The PSD of an MS43 code is also shown in Fig. 7.5d. This code is a variant of the 4B3T code with codeword assignments that lead to a slight reduction in low-frequency content [1]. The biggest drawback of block codes is their complexity which translates to expensive codecs. However, on long-distance links their use can lead to savings in overall system cost since they are more efficient and therefore allow a wider spacing of repeaters. 3B4B and other binary block codes are used on optical fibre transmission lines where binary (on-off) signalling is a requirement. TAT-8 (transatlantic optical fibre) system for example uses 24B1P, which represents *24* input *B*its as 24 Binary symbols plus *1* odd *P*arity symbol and therefore has an efficiency of 24/25 = 96%. The 6B4T line code was specified for long-distance coaxial cable systems at 139264 kb/s which have largely been replaced by optical fibre lines. Performance monitoring is an important function needed to gain knowledge of transmission link quality. It would be wasteful to install a decoder just to check each codeword for coding rule violation to facilitate error detection. Rather, errors can be detected at a repeater without decoding by monitoring the digital sum variation (DSV) of received pulses. An error will eventually cause the DSV to exceed its limits.

7.6 Summary

Line codes for application in digital base band transmission systems are intended to satisfy a number of features some of which contain elements of mutually conflicting requirements. The desirable characteristics include spectral content and shape that are suited to the transmission channel, no DC content, small bandwidth, guaranteed timing content, transparency and bit sequence independence, error detection and performance monitoring capability, high code efficiency, low error probability, and low codec complexity. No one code in the world has all these features so a decision on which code to use must be guided by the priorities of the particular application. The purpose of this chapter was to equip the reader with the tools needed to analyse the performance of line codes and to make informed decisions on their suitability for a desired application.

Following a careful classification of line codes that included clarification of commonly misused terms such as polar codes, we engaged in a detailed autocorrelation and spectral analysis of binary and general random rectangular waveforms and provided a collection of equations, graphs and tables for the evaluation of the spectral content of a wide range of line codes. Given a common line code, you should now hopefully be able to answer these key questions: Does it contain DC? Does it possess relatively small low-frequency components? What is its bandwidth? What percentage of the line code waveform power lies outside its null bandwidth? Would the line code suffer significant distortion in an AC-coupled channel? What symbol rate does it require to support a given bit rate? We also discussed a wide range of basic and advanced line codes with an emphasis on evaluating each of the above features and assessing the suitability of the code for different applications. Block codes were introduced as offering great flexibility in selecting codewords that satisfy the features that are most important to the intended application. This often involves trade-offs. For example, the only way to design a code with error detection capability is by introducing redundancy, and this necessarily sacrifices code efficiency. And to guarantee timing content, the line code waveform needs to contain frequent level transitions, which is achieved by sacrificing bandwidth. The three features of guaranteed clock content, absence of DC content and bit-sequence independence are however fundamental features that must be satisfied by every line code before it can be considered for application in base band transmission systems. AMI could be cited as an exception that does not have a guaranteed clock content — a problem solved in HDB3 and similar codes — but is widely used. However, it must be noted that AMI is always used in conjunction with a scrambler, which may therefore be considered to be an integral part of the codec, in order to improve the clock content of the coded waveform.

All line code waveforms discussed in this chapter have been presented as rectangular or stair case waveforms. However, waveforms having a rectangular shape are never transmitted due to their excessive spectral sidelobes which cause inter-symbol interference when transmitted through band limited channels. In the next chapter we derive filter specifications for use in shaping base band waveforms prior to transmission or (in the case of band pass transmission) before they are applied to modulate a carrier.

7.7 Reference

[1] Xiong, F. (2000). *Digital modulation techniques*. Norwood, MA: Artech House, pp. 17–85

Chapter 8

Transmission through band limited AWGN channels

In Chapter Eight

✓ Bandwidth effect on symbol rate: You will learn why symbol duration should not be smaller than $1/2B$ in a base band channel of bandwidth B.

✓ Intersymbol interference (ISI): We study various methods of eliminating ISI, including Nyquist, raised cosine, square root raised cosine, cosine, sine filtering and duobinary.

✓ Noise effect on bit rate: We examine the limitation placed by noise on the number of bits that can be represented by each transmitted symbol.

✓ Mutual information and channel capacity: An introduction to these information theory concepts with application to the discrete memoryless channel, the binary symmetric channel and the continuous channel.

✓ Channel coding theorem and information capacity law: The most significant results of information theory and their implications on communication system design and trade-offs.

✓ Matched filter: A heuristic derivation of the matched filter and its implementation as a correlation receiver. A mathematical approach is also included for completeness.

✓ Eye diagram: A brief introduction to a useful diagnostic tool for visualising timing error, ISI and noise degradation in a transmission system.

8.1 Introduction

This chapter deals with the constraints set on symbol rate (i.e. how many pules we transmit per second) by the finite bandwidth of the transmission system or channel, and with the constraints set on bit rate and signal power by the noisiness of the transmission channel. We carefully quantify the challenges at hand and then explore practical measures that can be taken at the transmitter to minimise *intersymbol interference* (ISI) and at the receiver to maximise correct detection of incoming symbols in the presence of noise.

Fig. 8.1 illustrates the transmission of a rectangular pulse $x(t)$ (also called symbol) of duration T_s through an ideal low pass LTI channel (i.e. filter — note that every channel or system is a filter of some sort). At symbol duration T_s, the number of symbols transmitted each second, or symbol rate (denoted by R_s) is

$$R_s = \frac{1}{T_s} \qquad \text{(symbols per second} \equiv \text{baud)} \tag{8.1}$$

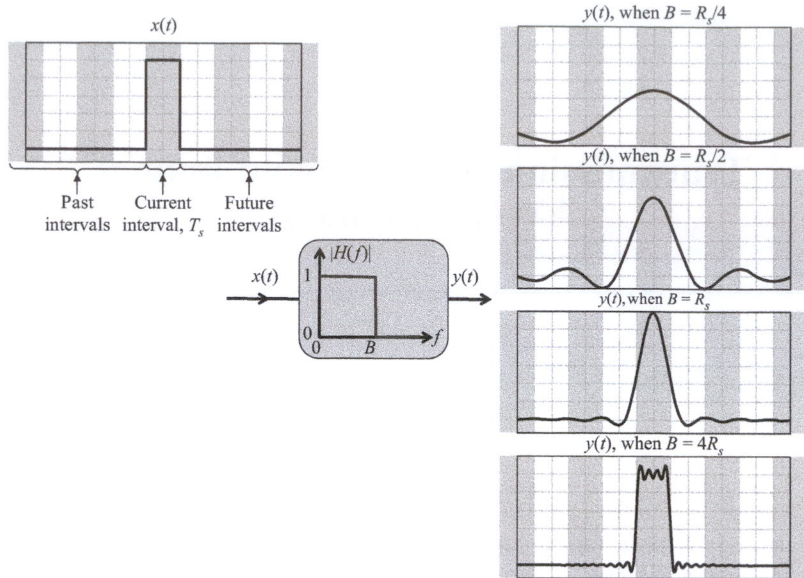

Fig. 8.1 Rectangular pulse x(t) of duration T_s transmitted through ideal low pass LTI channel of bandwidth B. Symbol rate $R_s = 1/T_s$

The output $y(t)$ of the channel is shown for various values of channel bandwidth from $B = R_s/4$ to $B = 4R_s$. It can be seen that pulse spreading occurs whereby the duration of the received pulse exceeds T_s so that it contributes significant energy into adjacent symbol intervals, a phenomenon known as *intersymbol interference* (ISI). If not addressed, ISI will lead to symbol detection errors at the receiver as the true identity of the current symbol is blurred by contributions from previous symbols.

From Fig. 8.1, one obvious solution to ISI would be to increase channel bandwidth B, since pulse spreading reduces as B increases. This behaviour is to be expected from the inverse relationship between time and frequency: The channel narrows the bandwidth of the input pulse $x(t)$ to B at the output, so the duration of the output pulse $y(t)$ broadens beyond T_s in response. The amount of pulse broadening decreases as bandwidth narrowing lessens (i.e. as B becomes larger), and vice versa. However, attempting to solve the ISI problem by increasing channel bandwidth is an expensive solution, which may even be non-feasible if it requires bandwidth to be increased beyond what is physically possible in the transmission medium. In Section 8.2, we explore effective and economic solutions through filtering that controls pulse broadening to ensure that the received pulse has zero value at the decision instants of all adjacent intervals.

Whatever the anti-ISI filtering measure employed, the channel bandwidth must be at least wide enough to pass the fundamental frequency f_0 of the fastest-changing sequence of transmitted pulses, namely $\{A, 0, A, 0, A, 0, \dots\}$ or $\{A, -A, A, -A, A, -A, \dots\}$, etc., where A = pulse amplitude. As can be seen from Fig. 8.2, if the channel has sufficient bandwidth to pass the fundamental sinusoidal waveform then sampling this waveform at the midpoint of each symbol interval enables the receiver to completely identify the pulse sequence. However, if the channel bandwidth is smaller than f_0, the fundamental sinusoidal waveform will be blocked making it impossible for the receiver to detect this fastest changing pulse sequence.

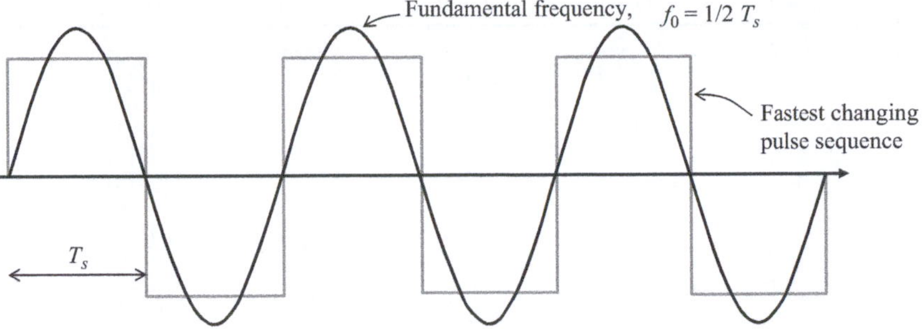

Fig. 8.2 Fundamental frequency of fastest changing pulse sequence

All other pulse sequences will change less slowly and will therefore have a lower fundamental frequency and hence require less bandwidth, so a bandwidth that passes f_0 is adequate for detecting all possible pulse sequences having pulse duration T_s.

For base band transmission this means that the minimum pass band must be from DC to f_0, whereas for band pass transmission obtained by modulating a carrier f_c, it means that minimum pass band must be from $f_c - f_0$ to $f_c + f_0$. Since $f_0 = 1/2T_s = R_s/2$, it follows that a base band system can support transmission at a symbol rate that is at most twice the bandwidth. That is,

$$R_{s\,\max} = \begin{cases} 2B, & \text{Base band system} \\ B, & \text{Band pass system} \end{cases} \tag{8.2}$$

Furthermore, in order to convey bits, there needs to be M unique symbols each of which is used to represent a unique combination of $k = \log_2 M$ bits, the assignment of which is agreed between transmitter and receiver. In general, the transmission system is described as M-ary, but in the special cases $M = 2$, 3 and 4 it is described as *binary*, *ternary* and *quaternary*, respectively. Beyond $M = 3$, the number of unique symbols M is usually chosen to be an integer power of 2 to ensure full utilisation of all possible bit combinations. The task of the receiver is to identify each received symbol and then to locally generate the group of k bits corresponding to that symbol. One or more bit errors occur whenever there is symbol mis-identification. Each transmit symbol may be described as possessing one of M distinct states that differ in amplitude, phase, frequency or a combination of these three parameters. Since each received symbol delivers k bits, the bit rate R_b of an M-ary transmission system is given in terms of symbol rate R_s by

$$R_b = R_s k = R_s \log_2 M \qquad \text{(bits per second)} \tag{8.3}$$

which from Eq. (8.2) has a maximum

$$R_{b\,\max} = 2B \log_2 M \tag{8.4}$$

Can we therefore indefinitely increase the bit rate through a channel of bandwidth B simply by increasing M, as illustrated in Fig. 8.3 for the case where the M symbols differ only in amplitude between 0 and A? In Fig. 8.3, the receiver associates any symbol having amplitude within a shaded interval with the k bits assigned to that interval. It can be seen that as M increases these amplitude intervals become smaller. For a fixed value of A, the symbols would remain distinct as $M \to \infty$ only if the

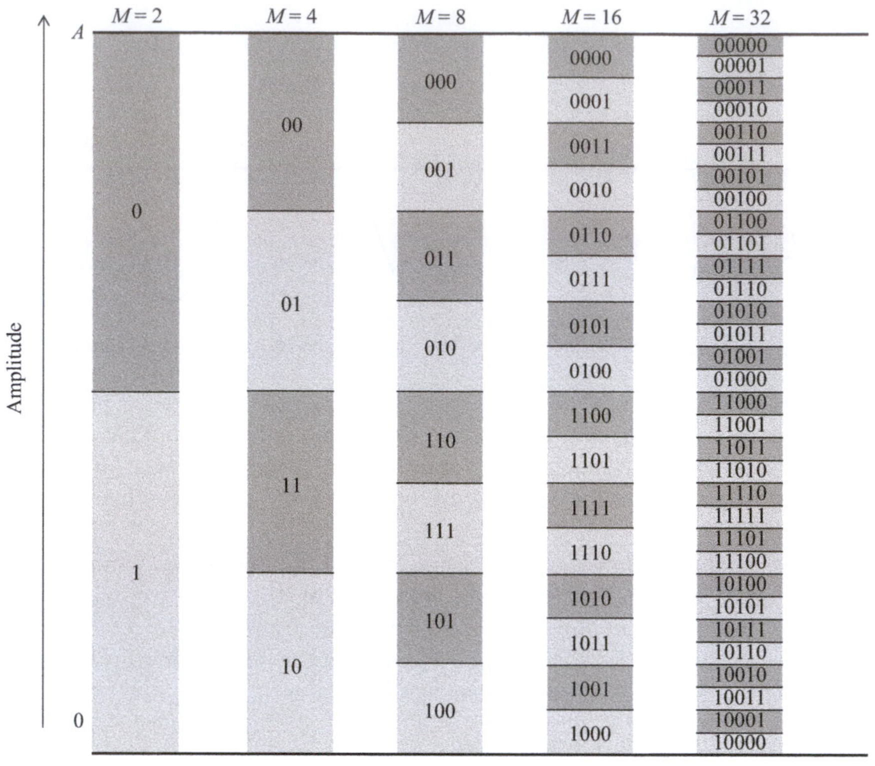

Fig. 8.3 M distinct states represent $\log_2 M$ bits per state as shown for amplitude states. Bits allocated to adjacent states differ in only one bit position, an arrangement called gray coding

channel was *noiseless* and the receiver had the capability to distinguish between infinitesimally close levels. In the presence of noise power P_n, adjacent levels are blurred by noise if they are any closer in rms value than $\sqrt{P_n}$. Combining received signal power P_s and noise power P_n, the full range of rms values of the symbol levels is $\sqrt{P_n + P_s}$, which at a minimum separation $\sqrt{P_n}$ gives a maximum number of distinguishable levels or states in the presence of noise as

$$M = \frac{\sqrt{P_n + P_s}}{\sqrt{P_n}} = \sqrt{1 + P_s/P_n}$$

Substituting this for M into Eq. (8.4) yields the maximum possible bit rate, called *channel capacity* of a noisy channel of bandwidth B (if the symbols are to remain distinct thus making *error-free* operation possible) as

$$\begin{aligned} R_{b\,\max} &= 2B \log_2\left[\sqrt{1 + P_s/P_n}\right] \\ &= B \log_2(1 + P_s/P_n) \qquad \text{(bits per second)} \end{aligned} \tag{8.5}$$

This is an important equation the implications of which we will discuss in Section 8.3 following a discussion of the formal meaning of channel capacity in information theory. It should, however, be clear from this introduction that when transmitting through a band limited noisy channel, ISI places a limit on symbol rate as stated in Eq. (8.2), whereas noise precludes us from indefinitely increasing M while maintaining error-free transmission at a fixed signal power (or maximum amplitude A) and it is this, when signal power is finite, that places a limit on bit rate as stated in Eq. (8.5).

8.2 ISI and bandwidth constraint on symbol rate

Fig. 8.4 shows a base band transmission system in which a base band signal $x(t)$ produced by a symbol generator (in this case a line coder) at point (a) in a transmitter is detected at point (b) in a receiver. The base band signal consists of a sequence of pulses spaced T_s seconds apart, which means that the symbol rate is $R_s = 1/T_s$. The worst case situation for pulse spreading occurs when the pulses have very narrow widths so that the nth pulse can be approximated by an impulse of weight b_n which carries the identity of the nth bit. For example, $b_n = 1$ for binary 1 and $b_n = -1$ for binary 0. Note that the discussion that follows is equally applicable to M-ary transmission, where b_n can take on values drawn from a set of M amplitudes. Thus,

$$x(t) = \sum_n b_n \delta(t - nT_s)$$

Due to the finite bandwidth of the transmission system from symbol generator output at point (a) to detection point at (b), the base band signal $y(t)$ arriving at point (b) will be a sequence of spread pulses. Normalising the transmission path to unit gain and zero delay, we may write:

$$y(t) = \sum_n b_n h(t - nT_s)$$

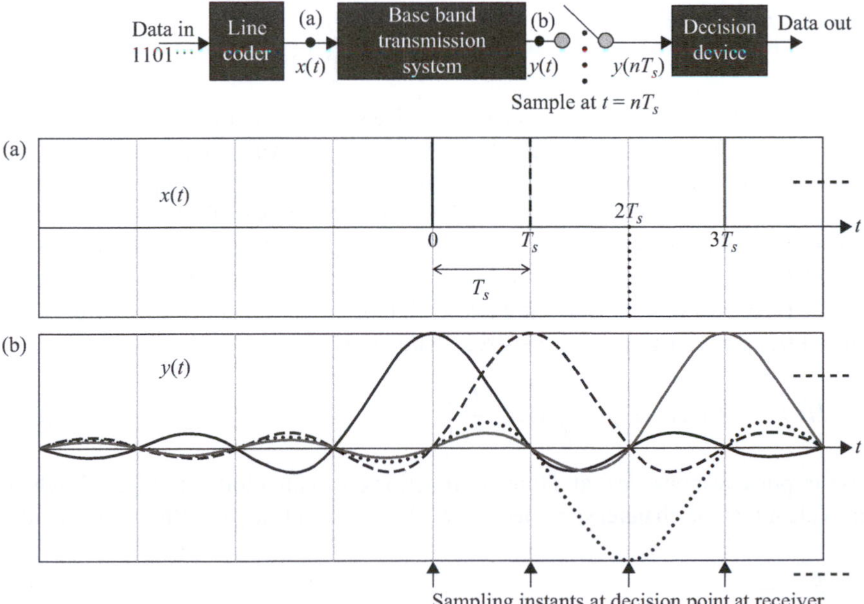

Fig. 8.4 Zero-ISI pulse spreading. The sequence of narrow pulses at (a) is received as a sequence of sinc pulses at (b)

Since $h(t)$ is the output pulse in response to an input impulse $\delta(t)$, it represents the *impulse response* of the entire transmission channel from point (a) to point (b).

At the detection point, $y(t)$ is sampled at intervals of T_s in synchrony with the transmitter and the sample value $y(nT_s)$ is passed to a decision device which compares it to one threshold level (in the case of binary transmission) or multiple levels (for $M > 2$) to reach a decision as to which bit(s) were transmitted in that interval. Thus to avoid ISI it is necessary and sufficient for the nth pulse $h(t - nT_s)$ at the detection point to have (normalised) unit value at its own sampling instant $t = nT_s$, and a value of zero at all other sampling instants $\ldots, (n-2)T_s, (n-1)T_s, (n+1)T_s, (n+2)T_s, \ldots$. This means that we require that at the sampling instant $t = mT_s$, the contribution $b_n h(mT_s - nT_s)$ from $h(t - nT_s)$ to the sample output $y(mT_s)$ should be zero in all cases except $m = n$ when the contribution equals b_n. Stated mathematically,

$$h(mT_s - nT_s) = \begin{cases} 1 & \text{for } m = n \\ 0 & \text{otherwise} \end{cases} \tag{8.6}$$

So long as there is correct timing at the receiver, the values and hence waveform of $h(t - nT_s)$ between sampling instants are immaterial. Eq. (8.6) is the *Nyquist criterion for zero ISI* which allows a number of solutions as discussed below.

8.2.1 Nyquist filtering

A sinc impulse response

$$h(t) = \text{sinc}(t/T_s) \tag{8.7}$$

satisfies Eq. (8.6) since

$$\text{sinc}\left(\frac{mT_s - nT_s}{T_s}\right) = \text{sinc}(m - n) = \begin{cases} 1 & \text{for } m = n \\ 0 & \text{otherwise} \end{cases} \tag{8.8}$$

This is illustrated in Fig. 8.4 for a binary impulse sequence in (a) corresponding to a bit sequence 1101 .... It can be seen that among the received spread pulses arriving at (b) only the current pulse has a non-zero value at each sampling instant. This means that if the impulse response of the transmission path from symbol output point at transmitter to symbol sampling point at receiver is as given by Eq. (8.7), then ISI-free operation is guaranteed in a properly synchronised receiver. Taking the FT of Eq. (8.7) — consult the table of FT pairs in Appendix C — yields the required transfer function of an ISI-free channel as

$$H(f) = T_s \text{rect}(fT_s) = \frac{1}{R_s}\text{rect}\left(\frac{f}{R_s}\right) \tag{8.9}$$

These impulse response $h(t)$ and transfer function $H(f)$ are plotted in Fig. 8.5, which makes clear that the channel specified by $H(f)$ is an ideal low pass filter of bandwidth

$$B = \frac{R_s}{2} \tag{8.10}$$

known as the *Nyquist bandwidth* for ISI-free transmission at symbol rate R_s. Such a channel is called the *ideal Nyquist channel* and allows transmission at a symbol

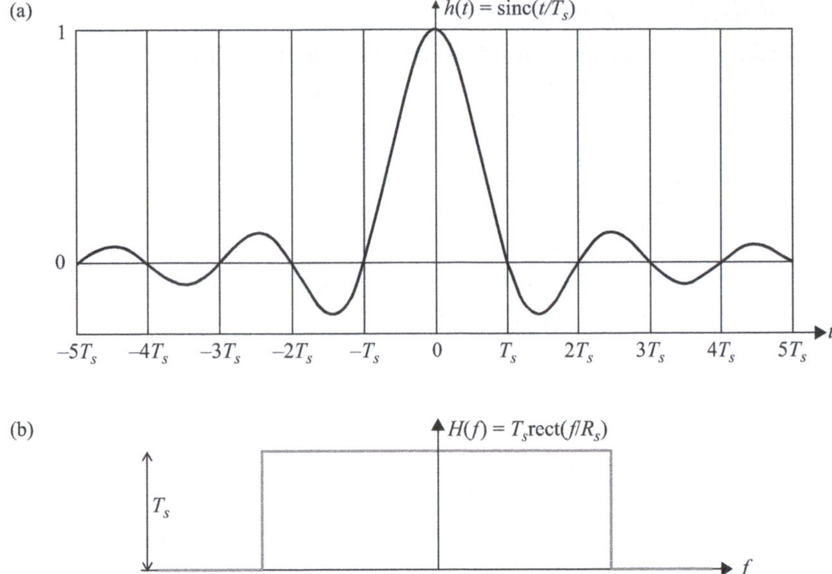

Fig. 8.5 (a) Impulse response h(t), and (b) transfer function H(f) of ideal Nyquist channel for zero-ISI

rate, called *Nyquist rate*, which is twice the available bandwidth. This is the very best that we can do without incurring ISI. That is,

$$R_{s\,\max} = 2B$$

which you may recognise as Eq. (8.2) in the introduction. The pass band gain of $1/R_s$ is not significant to the scheme of ISI elimination and may be normalised to unit gain. It is important to emphasise that Eq. (8.9) does not specify a single filter for the transmission system; rather it lays down the filtering characteristic which must be satisfied by the *entire* transmission path from symbol output point at the transmitter to the sampling point at the receiver. Typically this path will consist of several filters, in which case Eq. (8.9) specifies what the product of their transfer functions must be for ISI-free operation.

There are, however, two practical problems with this solution to ISI, namely

- It requires an overall filtering action that produces constant gain up to frequency $f = R_s/2$, and infinite attenuation beyond this cut-off frequency. As discussed in Section 2.2 and in Worked Example 4.4, such a sharp transition cannot be achieved in real time because it requires non-causality whereby future inputs contribute to current output. So our ideal scenario of ISI-free transmission at a symbol rate that is twice the channel bandwidth is practically unrealisable.
- The envelope of the sinc pulse $\text{sinc}(t/T_s)$ decays very slowly with t as $1/|t|$, which leaves significant residual energy in adjacent intervals, a contribution that is only avoided if we sample at precisely the right timings $t = 0$, T_s, $2T_s$, $3T_s, \ldots$. Thus the Nyquist channel imposes a very stringent requirement on timing accuracy. Any timing error leads to significant contribution to each sampled output from previous and future (yes, future!) sinc pulses at each mistimed sampling instant.

8.2.2 Raised cosine filtering

We may address the problem of residual energy highlighted earlier by introducing a factor which forces a more rapid decay of the impulse response envelope. Now consider the set of impulse response functions:

$$h(t) = \left[\frac{\cos(\pi \alpha t/T_s)}{1 - 4(\alpha t/T_s)^2}\right] \mathrm{sinc}(t/T_s) \tag{8.11}$$

where $0 \le \alpha \le 1$ is a dimensionless parameter. These all satisfy the Nyquist criterion for zero ISI stated in Eq. (8.6) since

$$\begin{aligned} h(mT_s - nT_s) &= \frac{\cos\left(\pi\alpha \dfrac{mT_s - nT_s}{T_s}\right)}{1 - 4\alpha^2\left(\dfrac{mT_s - nT_s}{T_s}\right)^2} \mathrm{sinc}\left(\frac{mT_s - nT_s}{T_s}\right) \\ &= \frac{\cos[(m-n)\pi\alpha]}{1 - 4\alpha^2(m-n)^2} \mathrm{sinc}(m-n) = \begin{cases} 1 & \text{for } m = n \\ 0 & \text{otherwise} \end{cases} \end{aligned}$$

This can be readily seen in Fig. 8.6a, where the impulse responses plotted for three values of the parameter $\alpha = 0$, 0.5, 1 have zero crossings at all non-zero integer multiples of T_s. Notice that when $\alpha = 0$, Eq. (8.11) reduces to Eq. (8.7), the sinc impulse response and the rate of decay of the tails of the function is left unchanged. At the other end of the scale, $\alpha = 1$ gives the fastest rate of decay of the tails of the

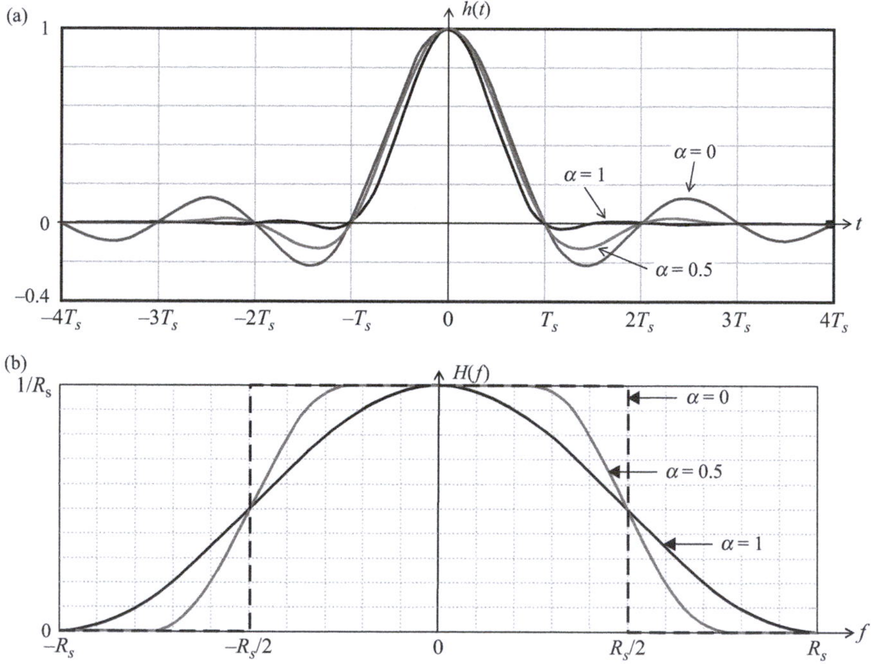

Fig. 8.6 (a) Impulse response h(t), and (b) transfer function H(f) of raised cosine filter for zero-ISI at various values of parameter α

impulse response function so that it contains negligible energy outside its main lobe, and therefore contributes negligible ISI at slightly mistimed sampling instants. By now we have come to expect that this gradual narrowing of the duration of the impulse function as α is increased from 0 to 1 must come at a price of bandwidth broadening.

Taking the FT of Eq. (8.11) yields the corresponding transfer function $H(f)$ plotted in Fig. 8.6b. The analytic form of this transfer function is

$$H(f) = \frac{1}{R_s} \times \begin{cases} 1, & |f| \le f_1 \\ \frac{1}{2}\left[1 + \cos\left(\pi \frac{|f| - f_1}{f_2 - f_1}\right)\right], & f_1 \le |f| \le f_2 \\ 0, & |f| \ge f_2 \end{cases} \tag{8.12}$$
$$f_1 = (1-\alpha)R_s/2; \;\; f_2 = (1+\alpha)R_s/2; \;\; 0 \le \alpha \le 1$$

For $\alpha > 0$, this spectrum represents a low pass filter having a constant gain portion up to $f = f_1$, followed by a portion in which the gain decreases gradually until it reaches zero at $f = f_2$. The decrease in gain with frequency in the interval $f_1 \le f \le f_2$ follows a cosine function to which the number 1 is added to limit the result to non-negative values. This filter is therefore called a *raised cosine filter*. Its null bandwidth was obtained in Worked Example 2.5 as

$$B = \begin{cases} (1+\alpha)\frac{R_s}{2}, & \text{Base band} \\ (1+\alpha)R_s, & \text{Band pass} \end{cases} \tag{8.13}$$

and it can be seen that at $\alpha = 1$, the (base band) raised cosine filter bandwidth $B = R_s$, which is double the bandwidth of the unrealisable Nyquist channel, so that as expected the reduction in the tails of the filter's impulse response (to make the filter more realisable and improve its tolerance to timing error) is achieved at the expense of an increase in channel bandwidth.

Other noteworthy features of the raised cosine filter include:

- The parameter α is the *roll-off* factor that controls the size of the frequency interval $f_2 - f_1 = \alpha R_s$ over which the filter transitions from maximum normalised unit gain in the pass band to zero gain in the stop band. $\alpha = 0$ corresponds to the special case of the Nyquist filter having no transition band at all, whereas $\alpha = 1$ gives the *full-cosine roll-off filter* having the most gradual roll-off or transition that starts with unit normalised gain at $f = 0$ and reaches zero gain at $f = R_s$.
- The raised cosine filter gain response exhibits *antisymmetry* about the vertical at the Nyquist bandwidth $R_s/2$. This means that starting at $f = R_s/2$ the gain response increases by the same amount when one moves to the left along the frequency axis by say Δf as it decreases when one moves by Δf to the right.
- Strictly speaking, the raised cosine filter is unrealisable because it exhibits zero gain or infinite attenuation at $f = f_2$ and beyond. However, its gradual roll-off makes the raised cosine filter characteristic easier to approximate than the ideal Nyquist filter using a realisable *tapped delay line* (also known as *finite-duration impulse response* (FIR)) filter. The required number of taps and hence filter complexity increases as α decreases.

- Putting $\alpha = 1$ in Eq. (8.11) and simplifying yields the impulse response of the full-cosine roll-off filter as

$$h(t) = \frac{\sin(2\pi t/T_s)}{2\pi t/T_s\left[1 - 4(t/T_s)^2\right]} \tag{8.14}$$

This has zero crossings at $2\pi t/T_s = n\pi,\ n = \ldots,\ \pm 3,\ \pm 2,\ 2,\ 3, \ldots$ or

$$t = n\frac{T_s}{2}, \quad n = \ldots,\ \pm 3,\ \pm 2,\ 2,\ 3, \ldots$$

So that beyond the main lobe this pulse has double the zero-crossing rate of the sinc pulse, a feature which is very useful in extracting timing information on T_s for synchronisation.

8.2.3 *Square root raised cosine filtering*

A practical arrangement for zero-ISI transmission is as shown in Fig. 8.7a. We assume that equalisation has been employed to eliminate channel distortion as illustrated in Fig. 1.22 and discussed in Section 2.7, so the channel is shown as distortionless. Fig. 8.7a is actually applicable to both base band and band pass transmission systems, except that in a band pass system the line coder and line decoder blocks would be replaced by a modulator and a demodulator, respectively. The transmit and receive filters serve different fundamental roles, the former being required for shaping the pulses generated by the line coder or modulator in order to reduce ISI, whereas the latter helps to reduce the impact of noise on symbol detection.

Assuming (as before) for convenience that the line coder produces a weighted impulse in each signalling interval, then the output of the transmit filter in each interval is the impulse response $h_x(t)$ also proportionately weighted. This pulse passes through the distortionless channel and arrives along with additive white Gaussian noise $w(t)$ at the input of the receive filter. We will show in Section 8.4 that optimum detection of $h_x(t)$ in the presence of $w(t)$ is achieved when the receive

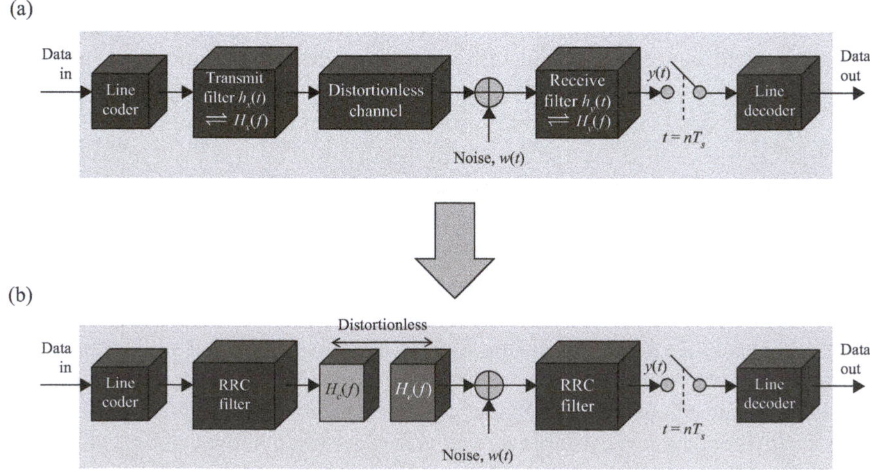

Fig. 8.7 Arrangement for zero-ISI transmission system

filter is a *matched filter* having an impulse response that is a time-reversed and delayed version of $h_x(t)$. That is,

$$h_y(t) = h_x(T_s - t) \tag{8.15}$$

Taking the FT of both sides, noting that time-reversing a real signal has the effect of complex-conjugating its Fourier transform, and that a delay of T_s in the time domain corresponds to a factor $\exp(-j2\pi fT_s)$ in the frequency domain, yields the gain response of the receive filter as

$$\begin{aligned} |H_y(f)| &= |H_x^*(f)\exp(-j2\pi fT_s)| = |H_x^*(f)||\exp(-j2\pi fT_s)| \\ &= |H_x(f)| \end{aligned} \tag{8.16}$$

Normalising the transmission path to unit gain and zero delay and ignoring $w(t)$ — since our focus in this section is solely on ISI, the output pulse $y(t)$ at the sampling point of the receiver in response to an impulse $\delta(t)$ generated by the line coder is

$$\begin{aligned} y(t) &= \delta(t) * h_x(t) * h_y(t) \\ &= h_x(t) * h_x(T_s - t) \equiv h(t) \end{aligned} \tag{8.17}$$

where we have used the fact — see Eq. (2.44) — that convolving a signal with $\delta(t)$ leaves the signal unchanged, and that for zero ISI, $y(t)$ must be the raised cosine filter impulse response $h(t)$ given in Eq. (8.11) with transfer function $H(f)$ given in Eq. (8.12). Taking the FT of Eq. (8.17) and retaining the magnitude yields

$$\begin{aligned} |H(f)| &= |FT[h_x(t) * h_x(T_s - t)]| \\ &= |H_x(f)H_x^*(f)\exp(-j2\pi fT_s)| \\ &= |H_x(f)|^2 \end{aligned} \tag{8.18}$$

Thus, in view of Eq. (8.16) we may state that

$$\begin{aligned} |H_x(f)| = |H_y(f)| &= \sqrt{|H(f)|} \\ &\equiv \sqrt{\text{Raised cosine filter gain response}} \\ &\equiv |H_{RRC}(f)| \end{aligned} \tag{8.19}$$

That is, the requirements of zero ISI and optimum detection in the presence of white noise may be jointly satisfied by using a pair of identical *square root raised cosine (RRC) filters*, one at the transmitter to shape the output pulses of the line coder and the other as the matched filter at the receiver. Fig. 8.7b shows the transmission block diagram based on this remarkable result. Using Eq. (8.12) in (8.19) yields the transfer function $H_{RRC}(f)$ of the RRC filter as

$$H_{RRC}(f) = \frac{1}{\sqrt{R_s}} \times \begin{cases} 1, & |f| \le f_1 \\ \cos\left(\dfrac{\pi}{2}\dfrac{|f| - f_1}{f_2 - f_1}\right), & f_1 \le |f| \le f_2 \\ 0, & |f| \ge f_2 \end{cases} \tag{8.20}$$

$$f_1 = (1 - \alpha)R_s/2;\; f_2 = (1 + \alpha)R_s/2;\; 0 \le \alpha \le 1$$

The inverse FT of this expression gives the impulse response $h_{RRC}(t)$ as follows:

$$h_{RRC}(t) = \frac{1/\sqrt{T_s}}{1 - (4\alpha t/T_s)^2}\left[(1-\alpha)\text{sinc}\left((1-\alpha)\frac{t}{T_s}\right) + \frac{4\alpha}{\pi}\cos\left(\pi(1+\alpha)\frac{t}{T_s}\right)\right] \tag{8.21}$$

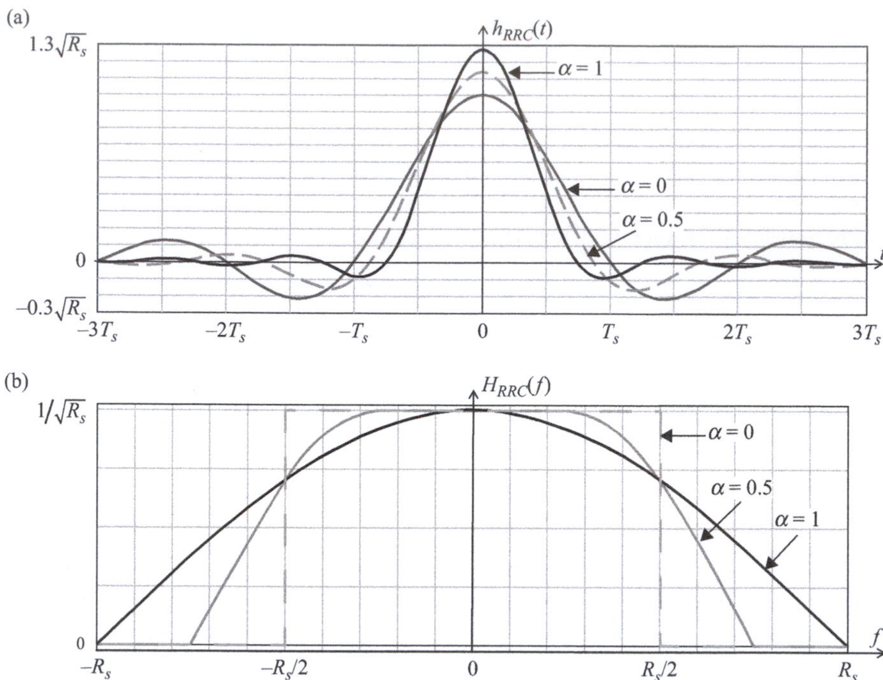

Fig. 8.8 (a) Impulse response $h_{RRC}(t)$, and (b) transfer function $H_{RRC}(f)$ of root raised cosine filter at various values of parameter α

These RRC filter impulse response and transfer function are plotted in Fig. 8.8, and, like the raised cosine type, may be closely approximated using a tapped delay line. Notice that except for the case $\alpha = 0$, the zero crossings of the RRC impulse response are not at integer multiples of T_s. Therefore, a single RRC filter cannot eliminate ISI although it has exactly the same bandwidth as its raised cosine counterpart. RRC filters must be used in pairs as shown in the block diagram of Fig. 8.7b so that after passing through *both* filters a transmitted pulse will arrive at the sampling point of the receiver having been filtered by the equivalent of one raised cosine filter causing it to have zero crossings at non-zero integer multiples of the sampling interval T_s, which therefore averts ISI.

A few words are in order on how the channel equaliser $H_e(f)$ shown in Fig. 8.7b might influence the specification of the transmit and receive filters. Channel attenuation tends to increase with frequency, so the equaliser would need to have a gain that increases with frequency in order to compensate. If that is the case, then the noise reaching the receive filter, having passed through the equaliser, will no longer be white but will be 'coloured', having an amplitude spectrum that increases with frequency. Under this condition, a matched filter is one that attenuates the higher frequency components more drastically in order to reduce noise, maximise SNR and optimise symbol detection. The RRC receive filter must therefore be modified accordingly. However, this will also attenuate the desired pulse energy at these frequencies, so the RRC transmit filter has to be modified to proportionately boost the high-frequency components of the pulse in preparation for their increased attenuation in the receive filter. In this way the combination of transmit and receive filters still yields a raised cosine filter and hence ISI-free pulses at the decision point, and there is also optimum performance in the presence of noise, but the transmit and receive filters are no longer identical RRC filters as specified in Eq. (8.20).

It is worth noting that the well-known scheme of pre-emphasis and de-emphasis employed in analogue FM (frequency modulation) transmission is based on the aforementioned principle. Noise at the output of the frequency discriminator is coloured, increasing as the square of frequency. So a low pass filter (called *de-emphasis filter*) is placed at the output of the FM demodulator to attenuate high-frequency components more drastically in order to reduce noise. In preparation for this 'controlled distortion' of the message signal, a high pass filter (called *pre-emphasis filter*) is used at the transmit end to proportionately boost the high-frequency components of the message. For a comprehensive discussion of FM and other analogue modulation techniques, see reference [1].

8.2.4 Duobinary signalling

Note from the outset that this is an entirely different technique that should not be confused with the raised cosine filtering technique discussed above. A raised cosine filter averts ISI by operating on each pulse individually and independently to shape it into having zero-crossings at non-zero integer multiples of sampling interval at the detection point in the receiver. The duobinary signalling technique on the other hand averts ISI at the receiver by combining two or more adjacent pulses at the transmitter, thereby introducing a controlled amount of ISI. To borrow from medicine, the technique protects against the ISI disease by vaccinating each transmitted pulse with a controlled amount of ISI.

The technique of zero-ISI transmission discussed here is variously described in the literature as *partial response signalling* or *correlative coding*, and was first introduced by Adam Lender in 1963 [2] as *duobinary signalling*, the prefix *duo* indicating a doubling in bit rate (or operation at half the bandwidth) when compared to binary transmission using a raised cosine filter of roll-off $\alpha = 1$. Our discussion will assume all filtering operation to be based at the transmitter, but actual systems share filtering between the transmitter and the receiver [3]. In this case a procedure similar to our presentation on the square root raised cosine filter is followed to derive filter specifications for the transmitter and receiver necessary to achieve the required overall filter characteristic. Different extensions and modifications to the basic duobinary signalling scheme are possible, all based on the same principle of introducing a controlled ISI into the transmitted pulses in order to gain various advantages [4]. We will discuss in detail the basic duobinary implementation that leads to the *cosine filter*, introduce a variant called *modified duobinary* that leads to the *sine filter*, comment briefly on *polybinary signalling* and discuss the trade-off involved in this class of zero-ISI techniques.

8.2.4.1 Cosine filter

Consider the arrangement shown in Fig. 8.9a. We wish to determine the impulse response $h(t)$ of the filter comprising the highlighted portion of the block diagram. The filtering process involves summing the pulses in the current and previous signalling intervals and then passing this combined pulse through an ideal low pass filter (LPF). But don't worry about the ideal LPF because we don't intend to build it. What we will build is the overall highlighted filter having impulse response $h(t)$ and transfer function $H(f)$ to be determined by making the input an impulse $\delta(t)$ as shown.

(a)

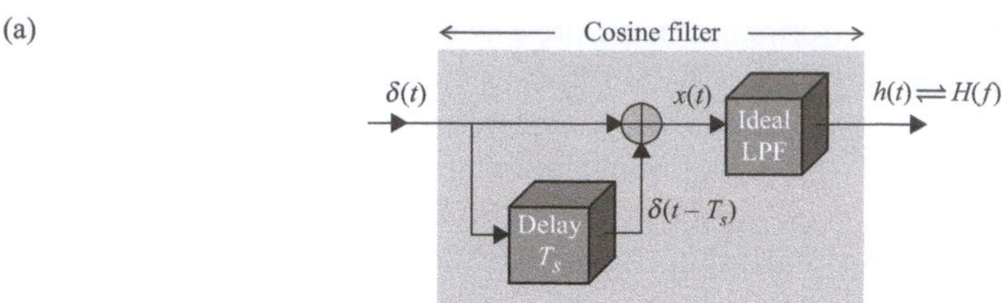

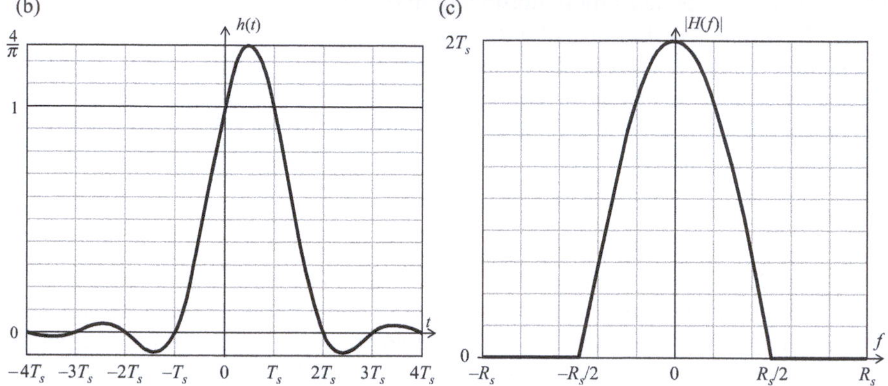

Fig. 8.9 The cosine filter: (a) block diagram; (b) impulse response; (c) transfer function

The input to the ideal LPF is

$$x(t) = \delta(t) + \delta(t - T_s)$$

Taking the FT of this equation gives the signal spectrum at the input of the ideal LPF as

$$X(f) = 1 + \exp(-j2\pi f T_s)$$

Multiplying this by the transfer function of the ideal LPF (which is $T_s\text{rect}(fT_s)$ — see Eq. (8.9) and Fig. 8.5b) yields the spectrum at the output of the ideal LPF as

$$\begin{aligned} H(f) &= [1 + \exp(-j2\pi f T_s)]T_s\text{rect}(fT_s) \\ &= T_s\text{rect}(fT_s) + T_s\text{rect}(fT_s)\exp(-j2\pi f T_s) \end{aligned} \tag{8.22}$$

Noting that $H(f)$ consists of one rectangular function and another that is scaled by the factor $\exp(-j2\pi fT_s)$, which corresponds to a delay T_s in the time domain, we see that the inverse FT, which gives $h(t)$, will comprise an impulse function and a delayed impulse function. That is,

$$h(t) = \text{sinc}\left(\frac{t}{T_s}\right) + \text{sinc}\left(\frac{t - T_s}{T_s}\right) \tag{8.23}$$

We may manipulate Eq. (8.22) into the following more familiar form for the gain response of the filter

$$\begin{aligned} |H(f)| &= \left|\left[e^{j\pi fT_s} + e^{-j\pi fT_s}\right]e^{-j\pi fT_s}T_s\text{rect}(fT_s)\right| \\ &= \left|2T_s\cos(\pi fT_s)\text{rect}(fT_s)e^{-j\pi fT_s}\right| \\ &= 2T_s\cos(\pi fT_s)\text{rect}(fT_s) \\ &= \begin{cases} \dfrac{2}{R_s}\cos(\pi f/R_s), & -\dfrac{R_s}{2} \le f \le \dfrac{R_s}{2} \\ 0, & \text{Otherwise} \end{cases} \end{aligned} \tag{8.24}$$

These impulse response and gain response functions are plotted in Fig. 8.9b,c. The gain of the filter varies according to a cosine function and therefore the filter is known as a *cosine filter*. It is remarkable that it has exactly the Nyquist bandwidth of $R_s/2$, with zero gain above this cut-off frequency. However, unlike the Nyquist filter, the cosine filter does not have a sharp transition from pass band to stop band so it may be more readily synthesised using a tapped delay line to achieve a very high but non-infinite attenuation beyond $f = R_s/2$. It can also be seen that while the raised cosine filter impulse response (see Fig. 8.6a) is non-zero only at one sampling point $nT_s = 0$, the cosine filter impulse response has non-zero value $h(t) = 1$ at two sampling points $t = nT_s$ for $n = 0, 1$. This shows that energy in one input pulse has been spread over two signalling intervals, which is indicative of the controlled ISI introduced. Furthermore, unlike the Nyquist filter, the cosine filter impulse response decays rapidly beyond its main lobe, making the transmission system more tolerant to slight errors in the setting of sampling instants.

We will discuss the use of a cosine filter to facilitate zero-ISI transmission based on the block diagram shown in Fig. 8.10. The precoder section in the transmitter, which performs modulo-2 summing of the current bit and the previous output to produce the current output, is essential in every application of the cosine filter in order to prevent error propagation at the receiver whereby an error in one bit corrupts future decoding decisions. Cosine filters can be used with any M-ary symbol generator, but here we will employ a binary bipolar line coder, which produces an impulse of weight $b_n = +1$ for binary 1 and weight $b_n = -1$ for binary 0. To understand the operation of the system we will examine how the message sequence $m_n = 100101$ is handled, where $m_1 = 1$, $m_2 = 0$, $m_3 = 0$, and so on, are the message bits in the first, second, third signalling intervals, etc. The processing of this bit sequence by the transmission system is shown in the table of Fig. 8.11, where the first column is the signalling interval n; the second column is the message sequence m_n; the third column is the previous output d_{n-1} of the precoder; the fourth column is the current output d_n of the precoder, which is also the line coder input; the fifth column is the line coder output and cosine filter input b_n; $y(t)$ is the waveform at the output of the cosine filter (plotted in the graph of Fig. 8.11), which is sampled at $t = nT_s$ to obtain the sample y_n listed in the seventh column; and the last column is the recovered message sequence $\hat{m}_n$. For comparison, we also show the waveform $z(t)$ that would reach the sampling point if the message sequence was transmitted through a raised cosine filter of roll-factor $\alpha = 1$ instead, first by applying it directly to the line coder — a raised cosine filter does not require precoding — and then passing the line coder output through the raised cosine filter.

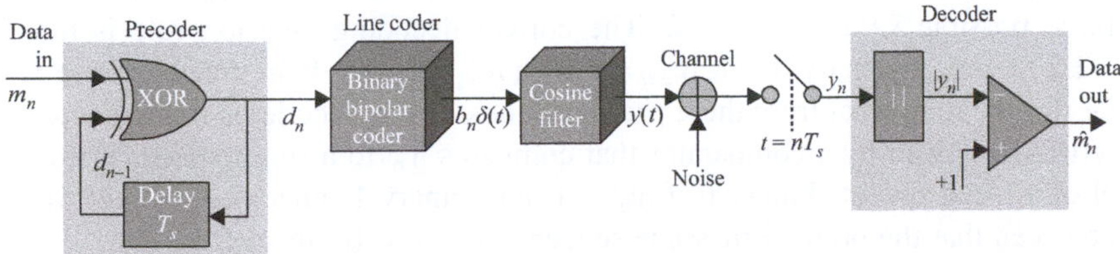

Fig. 8.10 Block diagram of zero-ISI transmission system using a cosine filter. The pre-coder section is essential and must have the structure shown, but the line coder and corresponding decoder can be a design other than the ones shown here

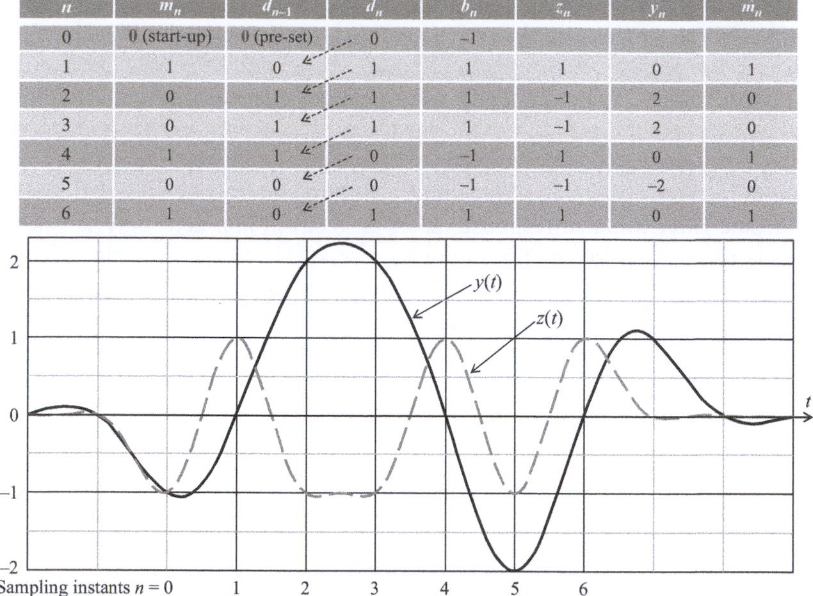

n	m_n	d_{n-1}	d_n	b_n	z_n	y_n	$\hat{m}_n$
0	0 (start-up)	0 (pre-set)	0	–1			
1	1	0	1	1	1	0	1
2	0	1	1	1	–1	2	0
3	0	1	1	1	–1	2	0
4	1	1	0	–1	1	0	1
5	0	0	0	–1	–1	–2	0
6	1	0	1	1	1	0	1

Fig. 8.11 Operation of block diagram of Fig. 8.10 *for input data* = 100101. *The waveform reaching the sampling point is y(t) for a cosine filter and z(t) for a raised cosine filter*

This waveform $z(t)$ is sampled at $t = nT_s$ to obtain the sequence z_n listed in the sixth column of the table. The decision device that converts z_n into the recovered message sequence is not shown in Fig. 8.11, but is a comparator that decides in favour of binary 1 if $z_n > 0$ and in favour of binary 0 otherwise.

Still on Fig. 8.11, at start-up (signalling interval $n = 0$), there is no previous output bit d_{-1}, so we define a start-up bit $m_0 = 0$, which is not part of the message and combine this with $d_{-1} = 0$ to obtain $d_0 = 0$. The rest of the precoding then proceeds with message bits m_n combined with the previous output (column 3), which is obtained by translating from column 4 as shown by the arrows in the table. The line coder converts d_n to normalised impulses $-\delta(t)$ for $d_n = 0$, and $\delta(t)$ for $d_n = 1$. Thus the impulses have (normalised) weights ± 1 given by b_n in the table. This sequence of impulses is processed by the cosine filter to produce the waveform $y(t)$ at the sampling point. Note that the channel does add noise to $y(t)$, which has been ignored in Fig. 8.11 since we are here primarily concerned with ISI. Sampling $y(t)$ at sampling instants nT_s, $n = 1, 2, \ldots, 6$, yields y_n which has one of three possible values -2, $0, +2$. The correct decoding decision rule is that if $y_n = \pm 2$ then $\hat{m}_n$ = binary 0, but if $y_n = 0$ then $\hat{m}_n$ = binary 1. In practice noise will make y_n vary somewhat from these precise values, so the decoder is implemented as a rectifier followed by a comparator that compares $|y_n|$ to a (normalised) threshold level of $+1$ and outputs binary 0 if $|y_n| > 1$, and binary 1 otherwise. In this way it can be seen that the original message sequence is correctly recovered.

8.2.4.2 Signal power trade-off

We found earlier that we could avert ISI by using a realisable raised cosine filter of roll-off factor $\alpha > 0$. The price we paid then was in using a larger bandwidth

$B=(1+\alpha)R_s/2$ than the minimum that is theoretically possible, namely the Nyquist bandwidth $R_s/2$. However, by using a cosine filter we are able to operate at Nyquist bandwidth using a realisable filter and free from ISI-induced error. There is always a price or trade-off in engineering, so what is the cosine filter trade-off? To understand this trade-off take a look again at Fig. 8.11 and notice that the raised cosine filter converts a binary input ± 1 from a line coder into output $z(t)$ from which the sampled sequence z_n is obtained, having two possible values ± 1 (normalised). In the presence of noise these two values will of course map into two intervals separated by a threshold level chosen to be equal to the mean level (which would be zero in this case, assuming that binary 1 and binary 0 are equally likely in the message sequence m_n). The cosine filter on the other hand converts the same binary input into output waveform $y(t)$ which when sampled yields a ternary sequence y_n having values drawn from the alphabet $\{-2, 0, 2\}$. In other words, the raised cosine filtering technique maintains binary signalling (or whatever M-ary signalling is delivered to it by the line coder used), whereas the cosine filtering technique transforms binary signalling into ternary, or more generally, M-ary signalling into $(2M-1)$-ary. This happens because a cosine filter introduces a controlled ISI between the current and previous pulses, so that if these had binary levels ± 1, the possible outcomes will be adding two positive pulses, or two negative pulses, or two opposite polarity pulses, which leads to the output levels +2, −2 or 0, respectively. Precoding allows each level to be linked to a unique identity of the bit in the current interval thus making instantaneous decoding possible as illustrated in Fig. 8.11.

It was important to explain the transformation of M-ary signalling into $(2M-1)$-ary signalling by a cosine filter before stating the consequence, which is that M-ary signalling is more robust to channel noise than $(2M-1)$-ary signalling at the same signal power. See the illustration in Fig. 8.3 for M amplitude states and note that the states are closer together as M increases, which means that channel noise added to a pulse sent with amplitude in the midpoint of one interval can more readily cause the pulse to be received with amplitude in an adjacent interval, giving rise to bit errors. To maintain the same bit error rate (BER) as M increases, we must increase signal amplitudes (i.e. enlarge the range $0 \rightarrow A$ in Fig. 8.3) and hence signal power in order to maintain a constant spacing between states. Although Fig. 8.3 was drawn for M-ary signalling based on amplitude shift keying, these comments are also applicable to phase shift keying, where the states have equal amplitude but different phases and can be treated as being arranged on a circle centred at the origin, with radius equal to normalised signal amplitude. As the number of states on the circle increases, the only way to maintain a constant distance between the states is to increase the radius of the circle and hence signal amplitude and power.

Cosine filtering therefore eliminates ISI while operating at Nyquist bandwidth but requires more signal power in order to ensure that its ternary transmission (assuming a binary line coder) has the same BER as a zero-ISI binary transmission using a raised cosine filter at larger bandwidth. So this is further evidence of a fact emphasised throughout this book that engineering just does not do free lunch: In building a realisable ISI-free system using a raised cosine filter the price we paid was more bandwidth than the minimum, but in building the system at the minimum bandwidth based on a cosine filter the price we pay is more signal power (for the same transmission quality).

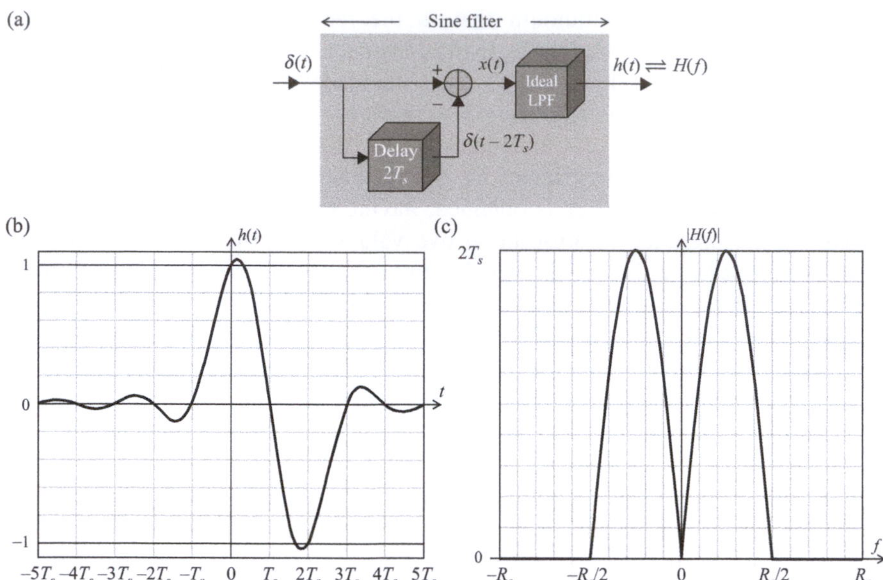

Fig. 8.12 The sine filter: (a) block diagram; (b) impulse response; (c) transfer function

8.2.4.3 Sine filter

Consider a modification to the basic scheme discussed above whereby in Fig. 8.9 we increase the delay to $2T_s$ and subtract (rather than add) this delayed element as shown in Fig. 8.12a. The result is a *sine filter* with impulse response

$$h(t) = \text{sinc}\left(\frac{t}{T_s}\right) - \text{sinc}\left(\frac{t - 2T_s}{T_s}\right) \tag{8.25}$$

and gain response, derived using a similar manipulation as for the cosine filter, given by

$$|H(f)| = \begin{cases} \frac{2}{R_s}|\sin(2\pi f/R_s)|, & -\frac{R_s}{2} \leq f \leq \frac{R_s}{2} \\ 0, & \text{Otherwise} \end{cases} \tag{8.26}$$

These are shown in Fig. 8.12b,c. It can be seen that the sine filter has a spectral null at DC and its output pulse is therefore well suited to capacitor- or transformer-coupled channels that block DC. This sine filter is referred to in the literature as a *modified duobinary filter*. Notice in Fig. 8.12 that the impulse response of the sine filter has non-zero value $h(t) = \pm 1$ at two sampling points $t = nT_s$ for $n = 0$, 2, which indicates that in this variant of the scheme, controlled ISI is introduced between each pulse and its next but one subsequent neighbour. The impulse response of the sine filter also decays rapidly beyond its twin main lobe which minimises ISI in the event of small timing errors at the receiver.

8.2.4.4 Polybinary signalling

Another modification to the basic scheme discussed here leads to the so-called *polybinary signalling* in which the delay element in Fig. 8.9 is replaced by a tapped delay line so that controlled ISI is introduced among multiple adjacent pulses resulting in multilevel signalling and hence further improvement in bandwidth efficiency without ISI, but at the price of a larger signal power requirement. See reference [5] for a discussion of the application of polybinary signalling in optical transmission systems.

8.3 Noise constraint on bit rate

In the previous section, we examined in some detail how the finite bandwidth of the transmission channel causes pulse spreading and necessitates the use of special filters to avert ISI, a measure that constrains symbol rate to a maximum of double the channel bandwidth. At that time we ignored channel noise entirely and treated the channel as effectively noiseless. In this section we wish to quantify the constraints that channel noise will place, not on symbol rate because that is a channel bandwidth issue discussed earlier, but on bit rate. In fact in the absence of channel noise, bit rate can be made indefinitely large even though symbol rate is constrained by channel bandwidth.

8.3.1 Characterisation of a discrete memoryless channel

Consider a *discrete memoryless channel* depicted in Fig. 8.13a whose input is a random variable X that takes on one value in each signalling interval. This value is drawn at random from a finite alphabet $\{x_1, x_2, \ldots, x_M\}$ of size M symbols with respective probabilities $\{p(x_1), p(x_2), \ldots, p(x_M)\}$. In response to each input the channel produces an output that is simply the current input plus a sample of random noise. Note that additive noise is the only difference between output and input

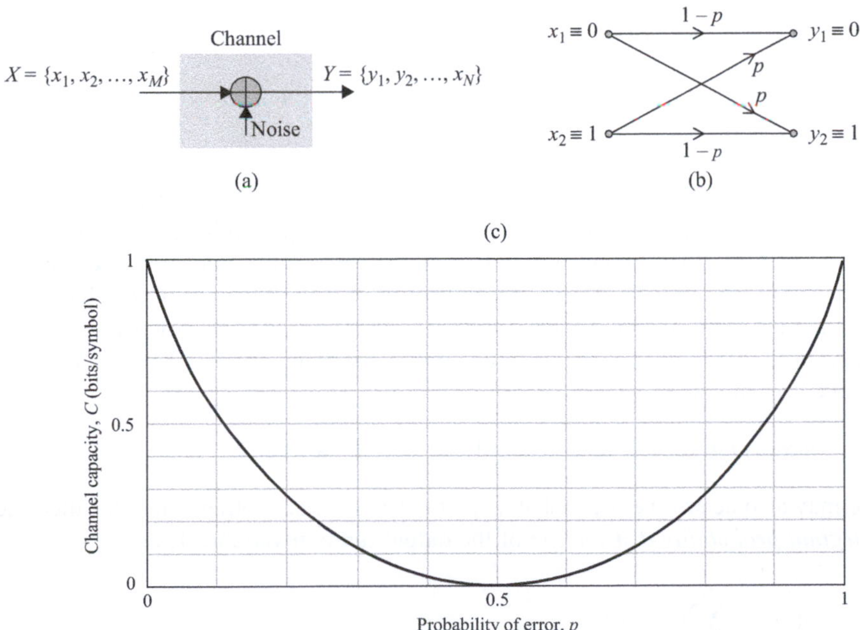

Fig. 8.13 (a) Discrete memoryless channel with input X and output Y; (b) transition probability of a binary symmetric channel (BSC); (c) channel capacity of BSC

considered here, the effect of intersymbol interference (ISI) having been eliminated as discussed in Section 8.2. We partition the output space into N regions and associate a unique output symbol with each region so that the output may be said to be a random variable Y that takes on one value in each signalling interval drawn at random from a finite alphabet $\{y_1, y_2, \ldots, y_N\}$ of size N symbols. Although the type of symbols is completely general, it may help add physical meaning to the discussion if you think of the input as a set of pulse amplitudes x_j, $j = 1, 2, \ldots, M$, and the output space as the real number axis partitioned by $N - 1$ threshold levels into N contiguous and non-overlapping intervals each of which is represented by a quantised value y_k, $k = 1, 2, \ldots, N$.

Due to random noise, we cannot be certain ahead of time which symbol of the output alphabet the channel will produce in response to a given input symbol, but we can characterise the channel by a set of transition probabilities that give the probability of receiving y_k if x_j is sent, denoted by $p(y_k|x_j)$, and read 'probability y_k given x_j', for $j = 1, 2, \ldots, M$ and $k = 1, 2, \ldots, N$. This leads to $M \times N$ transition probabilities which can be conveniently arranged in an $M \times N$ matrix, called the *channel matrix* as follows:

$$\mathbf{P} = \begin{bmatrix} p(y_1|x_1) & p(y_2|x_1) & \cdots & p(y_N|x_1) \\ p(y_1|x_2) & p(y_2|x_2) & \cdots & p(y_N|x_2) \\ \vdots & \vdots & \ddots & \vdots \\ p(y_1|x_M) & p(y_2|x_M) & \cdots & p(y_N|x_M) \end{bmatrix} \tag{8.27}$$

Note that each row of **P** corresponds to a particular input and each column to a particular output. The sum of all elements in each row is equal to one because when a particular input occurs then the output is certain (probability = 1) to be any one of the N output symbols. That is,

$$\sum_{k=1}^{N} p(y_k|x_j) = 1, \quad \text{for all } j \tag{8.28}$$

Given this channel matrix and the respective *prior probabilities* $p(x_j)$, $j = 1, 2, \ldots, M$ with which each input symbol is drawn from the input alphabet, we can obtain the probability that output y_k and input x_j will occur together (i.e. in the same signalling interval). This is the probability that the response of the channel to a particular input x_j will be the particular output y_k and is known as the *joint probability distribution* of the input and output random variables X and Y, denoted by $p(x_j, y_k)$ and given by

$$p(x_j, y_k) = p(y_k|x_j)p(x_j), \quad j = 1, 2, \ldots, M; \quad k = 1, 2, \ldots, N \tag{8.29}$$

We may also determine the probability of receiving each output symbol, called the *marginal probability distribution* of the output random variable Y, as

$$\begin{aligned} p(y_k) &= \sum_{j=1}^{M} p(x_j, y_k) \\ &= \sum_{j=1}^{M} p(y_k|x_j)p(x_j), \quad k = 1, 2, \ldots, N \end{aligned} \tag{8.30}$$

The special case in which the channel matrix is an $M \times M$ identity matrix with ones in the diagonal and zeros elsewhere,

$$\mathbf{P} = I_M = \begin{bmatrix} 1 & 0 & \cdots & 0 \\ 0 & 1 & \cdots & 0 \\ \vdots & \vdots & \ddots & \vdots \\ 0 & 0 & \cdots & 1 \end{bmatrix} \tag{8.31}$$

describes an *error-free channel*, where a given input is always guaranteed to lead to a specific output. That is, $x_1 \to y_1$, $x_2 \to y_2, \ldots, x_M \to y_M$, and for convenience the output may be normalised and treated as identical to the input.

A discrete memoryless channel of particular interest is the *binary symmetric channel* (BSC) which has the following two features:

- $M = N = 2$, meaning that there are two input symbols x_1, x_2, which we associate with binary 0 and binary 1, respectively; and two output symbols y_1, y_2, which we also identify with binary 0 and 1, respectively.
- The probability of receiving a 0 when a 1 is sent is the same as the probability of receiving a 1 when a 0 is sent. This is called the probability of error, denoted by p.

It follows, in view of Eq. (8.28), that the BSC channel matrix is

$$\mathbf{P} = \begin{bmatrix} 1-p & p \\ p & 1-p \end{bmatrix} \tag{8.32}$$

which is more usefully represented in the form of a *transition probability diagram* as shown in Fig. 8.13b, where p is the probability of a *bit flip* and $1-p$ is the probability of receiving the bit that was sent.

8.3.2 Mutual information and channel capacity

The purpose of a telecommunication channel is to convey information X from a source to a destination, which are often geographically separated. The destination therefore has no access to the channel input X and must rely on observing the channel output Y to resolve its uncertainty about X. In a *perfect* channel Y (normalised) is identical to X so that observing Y yields full information $H(X)$ about X, where $H(X)$ is the entropy of X introduced in Chapter 6. In a *useless* channel on the other hand X and Y are independent so that observing Y yields no information about X. In the former, the channel has the capacity to deliver information from source to destination, whereas in the latter it has no such capacity, and in between the conditions of *perfect* and *useless* the channel has a reduced capacity.

As an example, if $p = 0$ in the BSC channel represented by Fig. 8.13b then we can resolve all uncertainty about the channel input by observing the channel output. Under this condition of the channel, a binary 0, for example, observed at the output gives us full information about the input, confirms that the input is certainly binary 0 (since the probability of a bit flip is zero) and therefore resolves all uncertainty. Similarly, if the channel condition is such that $p = 1$, we can also resolve all uncertainty about channel input by observing the channel output.

We know that there is always a bit flip (since $p = 1$) so if we observe a binary 0 at the output we can tell that the input is certainly binary 1. However, for values of p in between, there will still be some unresolved uncertainty about the identity of the channel input even after observing the output. In fact if $p = 0.5$, meaning that an input bit is equally likely to be flipped or not, then observing the output gives us no information about the input and therefore does not resolve any of our prior uncertainty about the input. So when $p = 0$ or $p = 1$, then observing the output resolves all uncertainty and gives us as much information about the input as we would gain by directly observing the input. This amount of information is on average the source entropy $H(X)$, as discussed in Chapter 6. But if $p = 0.5$, then we gain zero information about the input by observing the output. For values of p between 0 and 0.5, or between 0.5 and 1 (non-inclusive), the amount of information gained about the input by observing the output is between 0 and $H(X)$ (non-inclusive).

We may also be interested in resolving our uncertainty about the output Y by observing the input X. By a similar consideration it is clear that if $p = 0$ or $p = 1$, then observing X resolves all uncertainty about Y and gives us the same amount of information about Y as we would gain by directly observing Y. But if $p = 0.5$, then observing the input cannot resolve any of our uncertainty about Y. That is, even if we know that the input is binary 0, the output is still equally likely to be 0 or 1 (since the bit is equally likely to be flipped or not flipped by the channel) and therefore we are still as uncertain about the output after observing the input as we were before observing the input. What is really happening is that the channel parameter p controls the relationship between the input random variable X and the output random variable Y. If $p = 0.5$, then X and Y are independent so that knowing one of them gives zero information about the other. If $p = 0$, then X and Y are exactly identical (with correlation coefficient $\rho = 1$) so that knowing one of them gives full information $H(X) = H(Y)$ about the other. If $p = 1$, then X and Y are exactly opposite (with correlation coefficient $\rho = -1$) so that knowing one again gives full information $H(X) = H(Y)$ about the other. For all other values of p, the random variables X and Y are related so that knowing X gives some information larger than zero but less than $H(Y)$ about Y, and knowing Y gives some information larger than zero but less than $H(X)$ about X. When $p = 0$ the channel is said to be *error-free* or *perfect* or *noiseless*, when $p = 1$ the channel is *error-certain* and therefore *perversely perfect*, and when $p = 0.5$ the channel is *error-crippled* or *uselessly noisy*, and for all other values of p the channel is simply *noisy* or *error-impaired*.

We can generalise and formalise the above thoughts by defining a quantity called the *mutual information* of the discrete random variables X and Y as

$$I(X;Y) = \sum_{k=1}^{N}\sum_{j=1}^{M} p(x_j, y_k)\log_2\left[\frac{p(x_j, y_k)}{p(x_j)p(y_k)}\right] \tag{8.33}$$

which gives a measure (in bits) of how much information is shared by X and Y, i.e. how much information we can gain about X by observing Y. This is also the same amount of information that we can obtain about Y by observing X. Note that since

$$p(x_j, y_k) = p(x_j|y_k)p(y_k) = p(y_k|x_j)p(x_j) \tag{8.34}$$

then Eq. (8.33) may be expressed in the following alternative forms:

$$\begin{aligned}
I(X;Y) &= \sum_{k=1}^{N}\sum_{j=1}^{M} p(x_j, y_k)\log_2\left[\frac{p(x_j|y_k)}{p(x_j)}\right] \\
&= \sum_{k=1}^{N}\sum_{j=1}^{M} p(x_j, y_k)\log_2\left[\frac{p(y_k|x_j)}{p(y_k)}\right] = I(Y;X) \\
&= \sum_{k=1}^{N}\sum_{j=1}^{M} p(y_k|x_j)p(x_j)\log_2\left[\frac{p(y_k|x_j)}{\sum_{j=1}^{M} p(y_k|x_j)p(x_j)}\right]
\end{aligned} \tag{8.35}$$

The second form of Eq. (8.35) emphasises the *symmetry* of mutual information, whereas the last form underlines that $I(X;\ Y)$ depends only on the input probability distribution $p(x_j)$ and the inherent property of the channel encapsulated in the channel matrix **P**, which contains the transition probabilities. In the special case of a BSC, the channel is characterised simply by p, the probability of error or bit-flip.

If X and Y are *independent*, then $p(x_j, y_k) = p(x_j)p(y_k)$ so that the term in square brackets in Eq. (8.33) reduces to one and its logarithm is therefore zero for all j and k, which means that the RHS is zero and hence

$$I(X;Y) = 0, \quad \text{for } X \text{ and } Y \text{ independent} \tag{8.36}$$

If on the other hand X and Y are *identical*, then

$$\begin{aligned}
M &= N, \quad \text{and} \\
p(x_j, y_k) &= \begin{cases} p(x_j) = p(y_k), & \text{for } j = k \\ 0, & \text{for } j \neq k \end{cases}
\end{aligned} \tag{8.37}$$

Using Eq. (8.37) in (8.33) and manipulating yields

$$\begin{aligned}
I(X;Y) &= \left\{\sum_{j=1}^{M} p(x_j, y_1)\log_2\left[\frac{p(x_j, y_1)}{p(x_j)p(y_1)}\right] + \cdots + \sum_{j=1}^{M} p(x_j, y_M)\log_2\left[\frac{p(x_j, y_M)}{p(x_j)p(y_M)}\right]\right\} \\
&= p(x_1, y_1)\log_2\left[\frac{p(x_1, y_1)}{p(x_1)p(y_1)}\right] + \cdots + p(x_M, y_M)\log_2\left[\frac{p(x_M, y_M)}{p(x_M)p(y_M)}\right] \\
&= p(x_1)\log_2\left[\frac{p(x_1)}{p(x_1)p(x_1)}\right] + \cdots + p(x_M)\log_2\left[\frac{p(x_M)}{p(x_M)p(x_M)}\right] \\
&= \sum_{j=1}^{M} p(x_j)\log_2\left[\frac{1}{p(x_j)}\right] \\
&= H(X)
\end{aligned}$$

Proceeding in a similar way but this time retaining $p(y_k)$ rather than $p(x_j)$ when $j = k$ yields equality with $H(Y)$. Thus if X and Y are identical then mutual information equals the entropy of each random variable,

$$I(X;Y) = H(X) = H(Y), \quad \text{for } X \text{ and } Y \text{ identical} \tag{8.38}$$

Since $H(X)$ is the information gained about X by observing X and $I(X;\ Y)$ is the information gained about X by observing Y, it follows that the difference between

$H(X)$ and $I(X; Y)$ is the uncertainty remaining (i.e. the information not yet gained) about X after observing Y, called the *conditional entropy* and denoted by $H(X|Y)$. Thus,

$$H(X|Y) = H(X) - I(X;Y) \tag{8.39}$$

It follows from this definition and Eqs. (8.36) and (8.38) that

$$\left.\begin{aligned} H(X|Y) &= H(X) \\ H(Y|X) &= H(Y) \end{aligned}\right\}, \quad \text{for } X \text{ and } Y \text{ independent} \tag{8.40}$$

$$H(X|Y) = H(Y|X) = 0, \quad \text{for } X \text{ and } Y \text{ indentical} \tag{8.41}$$

In line with the definition and physical meaning of source entropy discussed in Chapter 6, *conditional entropy* $H(X|Y)$ and *joint entropy* $H(X, Y)$ of the random variables X and Y are specified in terms of relevant probability distributions as

$$\begin{aligned} H(X|Y) &= \sum_{k=1}^{N}\sum_{j=1}^{M} p(x_j, y_k)\log_2\left[\frac{p(y_k)}{p(x_j, y_k)}\right] \quad \text{(Conditional entropy)} \\ H(X,Y) &= \sum_{k=1}^{N}\sum_{j=1}^{M} p(x_j, y_k)\log_2\left[\frac{1}{p(x_j, y_k)}\right] \quad \text{(Joint entropy)} \end{aligned} \tag{8.42}$$

Mutual information has the following useful properties.

- *Symmetry:*

  $$I(X;Y) = I(Y;X) \tag{8.43}$$

  which means that observing the output resolves the same amount of uncertainty about the input as sending an input will resolve about the output.
- *Equivalent forms:*

  $$\begin{aligned} I(X;Y) &= H(X) - H(X|Y) && \text{(a)} \\ &= H(Y) - H(Y|X) && \text{(b)} \\ &= H(X) + H(Y) - H(X,Y) && \text{(c)} \\ &= H(X,Y) - H(X|Y) - H(Y|X) && \text{(d)} \end{aligned} \tag{8.44}$$

  The derivation of (c) is left as an exercise, but (d) follows by noting that $I(X; Y) = \text{(a)} + \text{(b)} - \text{(c)}$.
- Finally, mutual information is always nonnegative. That is,

  $$I(X;Y) \geq 0 \tag{8.45}$$

  with the equality holding only when X and Y are independent as proved in Eq. (8.36).

We are now in a position to formally define the capacity of a channel as follows:

> *Channel capacity C is the maximum of the mutual information between the channel input and output in a signalling interval, the maximisation being over all possible input distributions $p(x_j)$, $j = 1, 2, \ldots, M$.*
>
> That is

$$C = \max_{p(x_j)} I(X;Y) \quad \text{(Channel capacity)} \tag{8.46}$$

It should be noted that channel capacity as defined above is in units of information per use of the channel, which is therefore bits per symbol. Thus the rate of use of the channel or symbol rate is constrained by the channel's bandwidth (as discussed in Section 8.2), whereas the number of bits per symbol (called channel capacity) is constrained by the channel's noisiness or probability of error as demonstrated next for a binary symmetric channel.

Worked Example 8.1: Channel Capacity of a BSC

We wish to determine the channel capacity C of a binary symmetric channel (BSC) the inherent property of which is its error probability p, which leads to the transition probability diagram of Fig. 8.13b.

The channel has two inputs 0, 1 and corresponding outputs 0, 1, with transition probabilities $p(1|0)=p(0|1)=p$, and $p(0|0)=p(1|1)=1-p$. Furthermore, the input probability distribution that maximises source entropy also maximises mutual information. From the discussion in Chapter 6 (see Fig. 6.1b) this distribution is $p(0)=p(1)=½$. What remains in the evaluation of C is to substitute these values in the last line of Eq. (8.35) for $I(X;Y)$ bearing in mind that $x_1 \equiv 0$, $x_2 \equiv 1$, $y_1 \equiv 0$, $y_2 \equiv 1$ and $M=N=2$. Thus

$$
\begin{aligned}
C &= p(0|0)p(0)\log_2\{p(0|0)/[p(0|0)p(0)] + p(0|1)p(1)\}\\
&\quad + p(0|1)p(1)\log_2\{p(0|1)/[p(0|0)p(0)] + p(0|1)p(1)\}\\
&\quad + p(1|0)p(0)\log_2\{p(1|0)/[p(1|0)p(0)] + p(1|1)p(1)\}\\
&\quad + p(1|1)p(1)\log_2\{p(1|1)/[p(1|0)p(0)] + p(1|1)p(1)\}\\
&= 0.5(1-p)\log_2\{(1-p)/[0.5(1-p)+0.5p]\}\\
&\quad + 0.5p\log_2\{p/[0.5(1-p)+0.5p]\}\\
&\quad + 0.5(1-p)\log_2\{p/[0.5p+0.5(1-p)]\}\\
&\quad + 0.5(1-p)\log_2\{(1-p)/[0.5p+0.5(1-p)]\}\\
&= (1-p)\log_2\{(1-p)/[0.5(1-p)+0.5p]\}\\
&\quad + p\log_2\{p/[0.5p+0.5(1-p)]\}\\
&= (1-p)\log_2\{(1-p)/[0.5]\} + p\log_2\{p/[0.5]\}\\
&= (1-p)[\log_2(1-p)+1] + p[\log_2(p)+1]\\
&= (1-p)\log_2(1-p) + 1 - p + p\log_2(p) + p\\
&= 1 + p\log_2(p) + (1-p)\log_2(1-p)
\end{aligned}
$$

We see that the BSC channel capacity,

$$C = 1 + p\log_2(p) + (1-p)\log_2(1-p) \tag{8.47}$$

depends only on the probability of channel error p and is plotted in Fig. 8.13c. It reaches a maximum $C=1$ bit per symbol at $p=0$ when the channel is *perfect* or *error-free* and at $p=1$ when the channel is *perversely perfect* or *error-certain*, and is a minimum $C = 0$ at $p = 0.5$ when the channel is *uselessly noisy* or *error-crippled*. For all other values of p the BSC channel is simply *noisy* or *error-impaired* with capacity

$$0 < C < 1 \text{ bit per symbol.}$$

8.3.3 Channel coding theorem and information capacity law

Shannon's channel coding theorem states that

> *There exists a coding scheme which can be used to transmit with a vanishingly small probability of error over a channel of capacity C at a rate not greater than C bits per symbol. For rates greater than C it is not possible by any encoding method to have an arbitrarily small probability of error.*

This theorem is widely regarded as the single most important result of information theory [6]. It states what bit rate is possible in error-free transmission but does not show how to construct the encoding algorithm that delivers such performance. To briefly examine the application of the theorem, consider the block diagram of Fig. 8.14 in which a discrete memoryless source X generates a symbol stream consisting of one symbol every T_s seconds. The source entropy being $H(X)$ means that on average the source generates $H(X)$ bits of information per symbol and hence $H(X)/T_s$ bits per second. An encoder codes these bits, generally by adding redundant bits, which necessitates the use of the channel once every T_c seconds to convey the source bits.

The theorem entails that error-free transmission is possible if

$$\frac{H(X)}{T_s} \leq \frac{C}{T_c} \tag{8.48}$$

If the source is a binary source emitting equiprobable 1s and 0s, then $H(X) = 1$. Noting that

$$r = T_c/T_s \equiv \text{Code rate} \tag{8.49}$$

we see that the theorem simplifies to the condition,

$$r \leq C \tag{8.50}$$

Thus the theorem indicates that there is an encoder with code rate $r \leq C$ that will enable us to have reliable (error-free) communication over a noisy binary

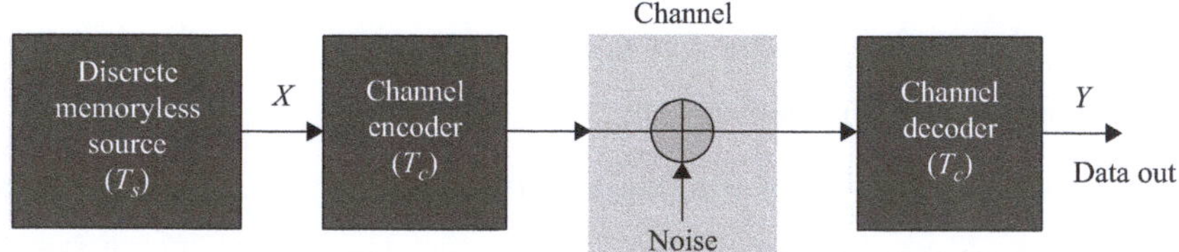

Fig. 8.14 Transmission of a symbol stream generated by a discrete memoryless source through a channel by means of a channel encoder at the transmitter and channel decoder at the receiver

symmetric channel. Fig. 8.13c shows that when a BSC channel has probability of error $p = 0$ or 1 then the channel capacity $C = 1$, and it follows from Eq. (8.50) that we can communicate reliably over a channel in this perfect condition using an encoder with code rate $r = 1$. In other words, no redundancy is necessary in the encoding to ensure reliable transmission. When $p = 0.5$, we see that $C = 0$ and it is impossible to provide reliable transmission since the theorem stipulates that code rate r must be zero. A BSC channel with $p = 0.5$ is error-crippled beyond the possibility of redemption through redundancy. For all other values of p the theorem insists that an encoder does exist with code rate $r \leq C$, which will guarantee reliable error-free communication.

Such an encoder takes k message bits, adds $n - k$ redundant bits through some processing algorithm and outputs n coded bits for a code rate $r = k/n \leq C$. For example, if in Fig. 8.13c $p = 0.2$, then $C = 0.28$. Amazingly, the theorem implies that given this noisy channel which randomly flips one in five of the bits sent through it we can still communicate reliably (with vanishingly small error probability) by finding an (unspecified) encoder with code rate ≤ 0.28. It is well known that a *repetition code* (in which each message bit is sent n times for a code rate $r = 1/n$) can be used to achieve vanishingly small error probabilities as $n \to \infty$ and $r \to 0$. What the theorem, however, indicates is that it is not necessary for r to approach zero and for us to pay the high price of excessively growing redundancy in order to make transmission more and more reliable. Yes, redundancy is essential, the theorem concedes, but this only needs to be at a code rate r that can be as high as the channel capacity C. The theorem categorically rules out any possibility of achieving reliable communication while saving on redundancy to the extent that $r > C$.

What we have done so far has been to derive a formula (Eq. (8.47)) for channel capacity C for the special case of a binary symmetric channel characterised by its error probability p (see Worked Example 8.1 and Fig. 8.13c). We then applied this result in Shannon's channel coding theorem to assess the implication of the theorem on a binary source that transmits equiprobable 1s and 0s through a BSC that randomly flips bits with probability p, and found that the condition for reliable communication laid down by the theorem in this case is that the code rate r of the encoder cannot exceed C.

To extend the application of Shannon's channel coding theorem to a more general communication system we must evaluate Eq. (8.46) for the channel capacity of such a system. This is a rather more involved derivation than Worked Example 8.1, and has been presented in [6] for a *continuous channel* of bandwidth B with average transmitted signal power P that is subject to additive white Gaussian noise (AWGN) having average noise power per unit bandwidth N_o. The derivation yields the capacity of such a channel as

$$C = B\log_2\left(1 + \frac{P}{N_oB}\right) \quad \text{bits per second} \tag{8.51}$$

This is the celebrated *information capacity law*, the most significant equation in information theory, referred to as the *Shannon-Hartley law* in recognition of the work of Claude Shannon [7] building on the early work of Hartley [8]. This equation lays down the rule governing how bandwidth and signal power may be exchanged in the design of a transmission system affected by noise. Shannon's channel coding theorem stipulates that it is possible to transmit information with an arbitrarily small probability of error at a bit rate R_b not exceeding the channel capacity C given by Eq. (8.51). If the bit rate exceeds C then there is no encoding system that can deliver reliable communication through the channel. Assuming transmission at the maximum possible bit rate $R_b = C$, Eq. (8.51) may be expressed in terms of the average signal energy per bit E_b by noting that

$$E_b = \frac{\text{Energy}}{\text{second}} \div \frac{\text{bit}}{\text{second}} = P/R_b \tag{8.52}$$

Thus, making the substitution $P = E_b R_b$ in Eq. (8.51) and manipulating to make E_b/N_o the subject of the equation yields the following form for the information capacity law:

$$\frac{E_b}{N_o} = \frac{2^{R_b/B} - 1}{R_b/B} = \frac{2^{\eta} - 1}{\eta} \tag{8.53}$$

The term $\eta = R_b/B$ is the *bandwidth efficiency* of the system in bits/second/hertz, whereas E_b/N_o is the dimensionless ratio between signal energy per bit and noise power per unit bandwidth, which gives an indication of the *power efficiency* of the system — the smaller the less power used and hence the more power efficient. Eq. (8.53) is plotted in Fig. 8.15 as E_b/N_o in dB versus bandwidth efficiency. We can now itemise the important implications of the Shannon-Hartley law of Eqs. (8.51) and (8.53) combined with Shannon's channel coding theorem stated at the beginning of this section:

- If signal power is fixed or in short supply for whatever reason, you can increase bit rate R_b without sacrificing system reliability (i.e. degrading noise performance or increasing probability of error) by increasing bandwidth B. Bit rate R_b increases proportionately with increase in bandwidth. In practice this exchange is facilitated by staying with the same modulation or line coding

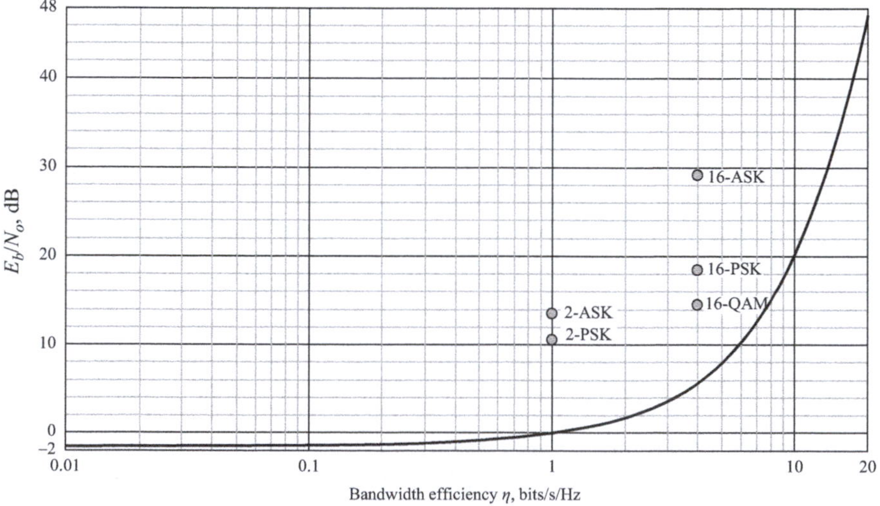

Fig. 8.15 Trade-off between E_b/N_o and bandwidth efficiency stipulated by the Shannon-Hartley information capacity law. Operating points of M-ary systems shown are for $BER = 10^{-6}$. For a system to be reliable its operating point must lie in the region above the curve

scheme which has a fixed number of bits k per symbol, but increasing bandwidth in order to allow symbol rate R_s to be increased without incurring ISI, following the relationship between symbol rate and bandwidth laid down in Eq. (8.13). In this way, bit rate $R_b = kR_s$ is increased.

- If bandwidth is fixed, you can still increase bit rate R_b (without sacrificing system reliability) by increasing signal power. However, the required increase in power is exponential, unlike the linear increase in bandwidth involved in the previous exchange. For example, Eq. (8.51) shows that if transmitting over a channel bandwidth $B = 10$ kHz, the lowest signal-to-noise ratios (SNR $= P/N_oB$) required for reliable communication at bit rates $C = 10$ to 40 kb/s are

$$\text{SNR} = 2^{C/B} - 1$$
$$= \begin{cases} 1, & \text{Bit rate } C = 10\ \text{kb/s} \\ 3, & \text{Bit rate } C = 20\ \text{kb/s} \\ 7, & \text{Bit rate } C = 30\ \text{kb/s} \\ 15, & \text{Bit rate } C = 40\ \text{kb/s} \end{cases}$$

Notice that to increase bit rate by a factor of 4 (from 10 to 40 kb/s) SNR (and hence signal power, assuming noise power is fixed) has to be increased by a factor of 15. In general, if the current reliable communication system operating point is (SNR, R_b, B) and bit rate is to be increased by a factor n without increasing bandwidth, then signal power must be increased exponentially so that the new operating point is

$$\left([\text{SNR} + 1]^n - 1, nR_b, B\right)$$

To facilitate this exchange, we would increase k (the number of bits per symbol) at a fixed symbol rate R_s (since bandwidth is fixed) by moving to higher M-ary modulation schemes which requires an exponential increase in power if transmitted symbols are to continue to be detected at the same bit error rate (BER) as for the lower M-ary scheme.

- There is no lower limit on the transmission bandwidth required to support any given bit rate. For example, Eq. (8.51) indicates that it is possible to reliably communicate at $R_b = 100$ kb/s over a 10 Hz bandwidth by using a signal-to-noise ratio SNR $\geq$ 30,103 dB. And if the bandwidth is reduced to just 1 Hz, then reliable communication at 100 kb/s over this 1 Hz bandwidth is still possible provided SNR $\geq$ 301,030 dB; and so on without limit. The practical constraint is of course obvious: If noise power is ~ -210 dBW/Hz (typical), received signal power would need to be in excess of 300,820 dBW or $10^{30{,}076}$ MW to allow reliable communication at 100 kb/s over a 1 Hz bandwidth. You would need to install a Sun transmitter to get that kind of power, but it would vaporise the entire world on being turned on. So while in theory there is no lower limit on required bandwidth, in practice there is an indirect limit due to constraints on achievable or safe or authorised levels of signal power.
- Although increasing channel bandwidth allows us to reduce the signal power needed for reliable communication, there is a lower limit on signal power below which reliable communication is not possible even at infinite transmission bandwidth. This limit is given in terms of E_b/N_o and is called the

Shannon limit. Noting that bandwidth efficiency $\eta \to 0$ as $B \to \infty$, we evaluate Eq. (8.53) in the limit $\eta \to 0$ to obtain

$$\begin{aligned}
\left.\frac{E_b}{N_o}\right|_{B\to\infty} &= \lim_{\eta\to 0}\frac{2^\eta - 1}{\eta} \\
&= \left.\frac{2^\eta \ln 2}{1}\right|_{\eta=0} \\
&= \ln 2 = 0.69315 \\
&= -1.6\ \text{dB} \qquad \text{(Shannon limit)}
\end{aligned} \tag{8.54}$$

where we made use of L'Hôpital's rule (differentiating the numerator and denominator) in the second line. The channel capacity at this Shannon limit is obtained by evaluating Eq. (8.51) in the limit $B \to \infty$,

$$\begin{aligned}
C_{S\ \text{Lim}} &= \lim_{B\to\infty}\left[B\log_2\left(1+\frac{P}{N_o B}\right)\right] \\
&= \lim_{B\to\infty}\left[B\log_e\left(1+\frac{P}{N_o B}\right)\right]\log_2 e \\
&= \lim_{B\to\infty}\left[B\frac{P}{N_o B}\right]\log_2 e \\
&= \frac{P}{N_o}\log_2 e \qquad \text{bits per second}
\end{aligned} \tag{8.55}$$

- The Shannon-Hartley law provides a benchmark against which all digital transmission systems can be measured. Practical systems will fall short of this performance in the sense that the practical system having bandwidth efficiency η will require a value of E_b/N_o that is higher than what is specified by the curve of Fig. 8.15 in order to operate at an acceptable bit error rate (BER) in an AWGN channel. For example, assuming a raised cosine filter of roll-off factor $\alpha = 0$, the bit rate and bandwidth efficiency of M-ary PSK, ASK and QAM modulation schemes (discussed in Chapter 9) are

$$\begin{aligned}
R_b &= B\log_2 M \\
\eta &= \frac{R_b}{B} = \log_2 M
\end{aligned}$$

We derive the BER versus E_b/N_o curves for these modulation schemes in Chapter 10, but will make use of those results here. If a BER of 10^{-6} is acceptable for the communication service and $M = 16$, we find that these schemes require E_b/N_o values of 29.13, 18.44 and 14.40 dB for 16-ASK, 16-PSK and 16-QAM, respectively. Furthermore, these schemes have bandwidth efficiency $\eta = 4$, and we find from Fig. 8.15 that a benchmark Shannon system operating at bandwidth efficiency $\eta = 4$ would deliver error-free performance while using $E_b/N_o = 5.74$ dB. We therefore conclude that all three modulation schemes are less power-efficient in an AWGN channel than the benchmark, the poorest being the 16-ASK which uses 23.4 dB more power than the ideal system. The best of the three systems is the 16-QAM, but it still uses 8.7 dB more power than the ideal. To approach closer to Shannon's benchmark we would have to deploy redundancy through error control coding, which we discuss in Chapter 11. Used in conjunction with an encoder/decoder, the 16-QAM demodulator might receive a signal at $E_b/N_o = 7.88$ dB which it demodulates, incurring a high BER of 10^{-2}. This high-BER bit stream is then passed through the decoder which employs the in-built redundancy to correct most of the errors thereby reducing the BER to a more acceptable value of say 10^{-6}. In this way, the 16-QAM transmission system would be only 2.14 dB less power efficient than Shannon's benchmark system.

- It is worth emphasising that Shannon's channel coding theorem stipulates that all realisable reliable communication systems affected by AWGN must have operating points that lie in the region above the Shannon-Hartley curve of Fig. 8.15. A system that lies in the region below the curve cannot be reliable. For example, the theorem categorically rules out the possibility of achieving reliable communication in an AWGN channel using uncoded 16-QAM at $E_b/N_o = 4$ dB. To illustrate this point further we have marked the operating points of the three 16-ary systems discussed above as well as 2-PSK and 2-ASK systems all judged on the basis of BER $\leq 10^{-6}$ being the acceptable performance target and therefore constituting a reliable system. The vertical distance on this graph between the operating point of a given system and the Shannon-Hartley curve gives the excess signal power requirement of the system when compared to the Shannon-Hartley benchmark. For example, 2-PSK requires 10.53 dB more power, and is said to be 10.53 dB less power efficient than the ideal system.
- Finally, Fig. 8.15 provides an excellent encapsulation of the trade-off between bandwidth and signal power in reliable communication systems that are subject to noise. Systems with operating points located on the RHS of Fig. 8.15 trade high power for bandwidth efficiency. In other words, these systems spend more power in order to use less bandwidth or in order to send more bits per second in the allocated bandwidth, whereas systems on the LHS trade bandwidth for a saving in signal power. These LHS systems use bandwidth less efficiently in order to use power more efficiently.

8.4 The matched filter

A matched filter is designed to maximise the signal-to-noise ratio at its output for a given input pulse that is contaminated by noise. The situation is as illustrated in Fig. 8.16a in which a pulse $g(t)$ has AWGN $w(t)$ added to it. Our task is to specify a filter which will produce an output $y(t)$ that has maximum signal-to-noise ratio. This ensures that the sample of $y(t)$ taken at $t = T_s$ will contain the smallest possible amount of noise perturbation $\tilde{n}(T_s)$. The matched filter was introduced in Section 1.3.8 (which you may wish to review). In this section, we derive the matched filter's impulse response $h(t)$ and transfer function $H(f)$ required to optimise the detection of the pulse in the presence of AWGN. We will do so first following heuristic arguments before providing a sound mathematical basis and then discussing a practical implementation approach using correlation processing.

8.4.1 Heuristic approach

The transfer function $H(f)$ of the matched filter in Fig. 8.16a may be obtained by making three increasingly prescriptive observations. Once $H(f)$ has been determined the impulse response is obtained by taking the inverse FT:

1. The bandwidth of the filter must be just enough to pass the incoming signal $g(t)$. If filter bandwidth is too wide, noise power is unnecessarily admitted, and if it is too narrow then some signal energy is blocked. Thus, the filter transfer

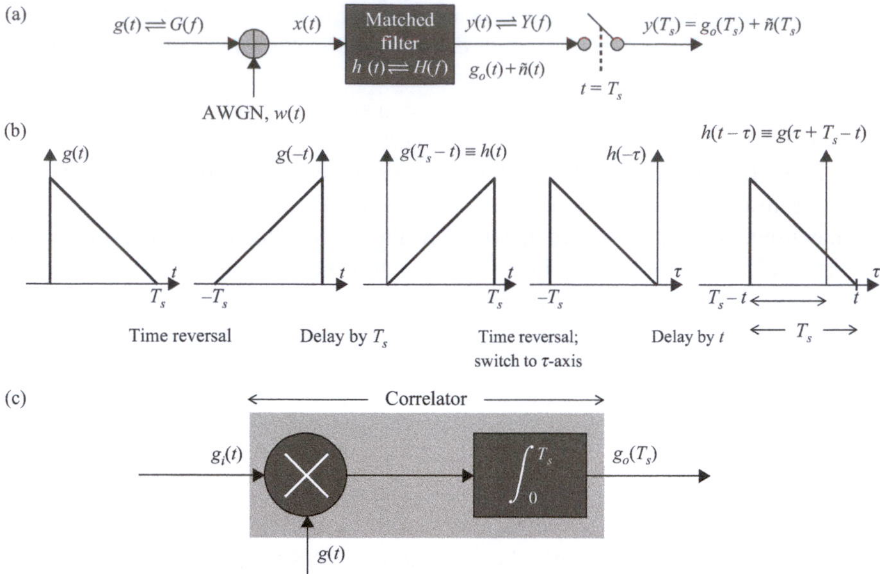

Fig. 8.16 Optimum detection of a signal g(t) in the presence of additive white Gaussian noise (AWGN): (a) the matched filter; (b) graphical illustration of relationship between input pulse g(t) and matched filter impulse response h(t); (c) correlation implementation of matched filter

function $H(f)$ must span exactly the same frequency band as the Fourier transform $G(f)$ of the incoming pulse $g(t)$. What should the spectral shape of $|H(f)|$ be like compared to $|G(f)|$?

2. The gain response $|H(f)|$ of the filter should not necessarily be flat within its pass band. Rather, it should be such that the filter attenuates the white noise significantly at those frequencies where $G(f)$ is small — since these frequencies contribute little to the signal energy. And the filter should boost those frequencies at which $G(f)$ is large in order to maximise the output signal energy. Therefore, the filter should be tailored to the incoming pulse, with a gain response that is small where $G(f)$ is small and large where $G(f)$ is large. In other words, the gain response of the filter should be identical in shape to the amplitude spectrum of the pulse. That is,

$$|H(f)| = K|G(f)| \tag{8.56}$$

where K is some constant.

3. To complete the specification of the filter, its phase response is required. The filter output $y(t)$ may be written as

$$y(t) = g_o(t) + \tilde{n}(t) \tag{8.57}$$

where $g_o(t)$ is the signal component and $\tilde{n}(t)$ is coloured noise — coloured because after white noise passes through a filter, its amplitude spectrum is no longer flat but is stronger at some frequencies than at others. The maximum instantaneous output signal power occurs at the sampling instant $t = T_s$ if every frequency component (i.e. cosine function) in $g_o(t)$ is delayed by the same amount T_s and has zero initial phase so that

$$g_o(t) = A_1 \cos[2\pi f_1(t - T_s)] + A_2 \cos[2\pi f_2(t - T_s)] + A_3 \cos[2\pi f_3(t - T_s)] + \cdots \tag{8.58}$$

which results in the maximum possible signal sample at $t = T_s$ given by

$$g_o(T_s) = A_1 + A_2 + A_3 + \cdots$$

where $A_1, A_2, A_3, \ldots$ are the amplitudes of the sinusoidal components of $g_o(t)$ of respective frequencies $f_1, f_2, f_3, \ldots$. These frequencies are infinitesimally spaced, giving rise to a continuous spectrum $G_o(f)$. Rewriting Eq. (8.58) in the form

$$\begin{aligned} g_o(t) = {} & A_1 \cos(2\pi f_1 t - 2\pi f_1 T_s) \\ & + A_2 \cos(2\pi f_2 t - 2\pi f_2 T_s) \\ & + A_3 \cos(2\pi f_3 t - 2\pi f_3 T_s) \\ & + \cdots \end{aligned}$$

makes it clear that the phase spectrum of $g_o(t)$ is

$$\phi_o(f) = -2\pi f T_s$$

From the discussion in Chapter 2, this output phase response is the sum of the input signal's phase spectrum $\phi_g(f)$ and the filter's phase response $\phi_H(f)$. Thus,

$$\phi_H(f) = -2\pi f T_s - \phi_g(f) \tag{8.59}$$

Eq. (8.56) gives the required filter gain response, and Eq. (8.59) gives its phase response. What remains is for us to combine these two into a single expression for the filter's transfer function $H(f)$ and then take the inverse FT of $H(f)$ to obtain the filter's impulse response $h(t)$:

$$\begin{aligned} H(f) &= K|G(f)|\exp[j\phi_H(f)] \\ &= K|G(f)|\exp\left[-j\phi_g(f)\right]\exp(-j2\pi f T_s) \\ &= KG^*(f)\exp(-j2\pi f T_s) \end{aligned} \tag{8.60}$$

where the asterisk (*) denotes complex conjugation, a process in which every occurrence of the imaginary number j in $G(f) \equiv |G(f)|\exp\left[j\phi_g(f)\right]$ is replaced by $-j$. Noting that complex conjugation of $G(f)$ corresponds to a time reversal of the real signal $g(t)$ to give $g(-t)$, and that multiplying $G^*(f)$ by the exponential term $\exp(-j2\pi f T_s)$ corresponds to delaying $g(-t)$ by T_s (consult Appendix C if in doubt), we see that the inverse FT of Eq. (8.60) yields

$$h(t) = Kg(T_s - t) \tag{8.61}$$

This is an important result which states that if the detection filter is to be matched to an AWGN-corrupted pulse $g(t)$ of duration T_s — matched in the sense of maximising the output signal power when compared to the noise power at the decision instant $t = T_s$ — then the filter must be designed with an impulse response that is (except for a non-zero scale factor K) a time reversed replica of the pulse, delayed by the sampling interval T_s. See Fig 8.16b for a graphical illustration of the process of converting $g(t)$ into $g(T_s - t)$ for any given value of T_s. It may also be useful to refer to Fig. 2.4 and Worked Example 2.1 for more in-depth information.

8.4.2 Mathematical basis

This section presents a mathematically rigorous derivation of the matched filter specification. Although very illuminating, it may be skipped without loss of essential foundation if you are satisfied with the heuristic approach taken above.

The matched filter with transfer function $H(f)$ and impulse response $h(t)$ maximises the instantaneous signal power at its output at $t = T_s$ compared to the average output noise power P_n, where T_s is the sampling instant and signalling interval, and noise is due to AWGN $w(t)$ at the filter input. Eq. (2.29) gives the output noise power, repeated here for convenience as

$$P_n = \frac{N_o}{2}\int_{-\infty}^{\infty} |H(f)|^2 df \tag{8.62}$$

The signal component $g_o(t)$ at the filter output has spectrum

$$G_o(f) = G(f)H(f)$$

and it follows by definition of the inverse FT (see Appendix C) that the signal sample at $t = T_s$ is

$$g_o(T_s) = \int_{-\infty}^{\infty} G_o(f)\exp(j2\pi fT_s)df = \int_{-\infty}^{\infty} H(f)G(f)\exp(j2\pi fT_s)df \tag{8.63}$$

Since signal power at $t = T_s$ is proportional to the square of $g_o(T_s)$, it means that we seek a filter transfer function $H(f)$ that maximises the ratio γ between $|g_o(T_s)|^2$ and P_n, namely

$$\gamma = \frac{\left|\int_{-\infty}^{\infty} H(f)G(f)\exp(j2\pi fT_s)df\right|^2}{\frac{N_o}{2}\int_{-\infty}^{\infty} |H(f)|^2 df} \tag{8.64}$$

Let us invoke *Schwarz's inequality*, which states as follows:

Given two complex functions $H_1(f)$ and $H_2(f)$ in the real variable f satisfying the conditions:

$$\int_{-\infty}^{\infty} |H_1(f)|^2 df < \infty, \quad \int_{-\infty}^{\infty} |H_2(f)|^2 df < \infty$$

then

$$\left|\int_{-\infty}^{\infty} H_1(f)H_2(f)df\right|^2 \le \int_{-\infty}^{\infty} |H_1(f)|^2 df \int_{-\infty}^{\infty} |H_2(f)|^2 df \tag{8.65}$$

The equality holds in this relation if, and only if

$$H_1(f) = KH_2^*(f) \tag{8.66}$$

where K is a constant and the asterisk denotes complex conjugation.

Comparing the LHS of Eq. (8.65) to the numerator of Eq. (8.64) and making the following identifications:

$$H(f) \equiv H_1(f), \quad G(f)\exp(j2\pi f T_s) \equiv H_2(f)$$

we see that the numerator of Eq. (8.64), and hence γ is maximum when Eq. (8.66) is satisfied. That is when,

$$H(f) = KG^*(f)\exp(-j2\pi f T_s) \tag{8.67}$$

This is therefore the transfer function specification for a matched filter that maximises the ratio between signal power and noise power in a sample of its output taken at $t = T_s$ when the input signal is the pulse $g(t)$ corrupted by white noise $w(t)$. To obtain the impulse response of the filter, we apply the definition of the inverse FT to Eq. (8.67) and manipulate as follows:

$$\begin{aligned}
h(t) &= \int_{-\infty}^{\infty} H(f)\exp(j2\pi ft)df \\
&= \int_{-\infty}^{\infty} KG^*(f)\exp(-j2\pi f T_s)\exp(j2\pi ft)df \\
&= \int_{-\infty}^{\infty} KG(-f)\exp[-j2\pi f(T_s - t)]df \\
&= K\int_{-\infty}^{\infty} G(\lambda)\exp[j2\pi\lambda(T_s - t)]d\lambda \\
&= Kg(T_s - t)
\end{aligned} \tag{8.68}$$

In the above, the third line follows from noting that the Fourier transform $G(f)$ of a real signal $g(t)$ satisfies $G^*(f) = G(-f)$; the fourth line from making the substitution $\lambda = -f$; and the last line by definition of the inverse Fourier transform. Thus the impulse response of a matched filter that gives optimum detection of a pulse $g(t)$ in the presence of white noise is simply a time-reversed version of the pulse delayed by the sampling interval T_s. The constant K is a scale factor which may be normalised to unity.

8.4.3 Matched filter as a correlator

A matched filter designed to detect a known pulse $g(t)$ will have the impulse response $h(t)$ given by Eq. (8.61). The output $g_o(t)$ of this filter in response to a received input pulse $g_i(t)$ is obtained by convolving $g_i(t)$ with the impulse response of the filter. Thus,

$$\begin{aligned}
g_o(t) &= g_i(t) * h(t) \\
&= \int_{-\infty}^{\infty} g_i(\tau)h(t-\tau)d\tau \\
&= \int_{0}^{T_s} g_i(\tau)g(T_s - t + \tau)d\tau
\end{aligned} \tag{8.69}$$

where the second line is merely the definition of the convolution integral, and in the last line we made use of the fact that the input pulse is of duration T_s to narrow the range of integration, and we substituted $h(t) \equiv g(T_s - t)$, normalising the constant K to unity. Refer to Fig. 8.16b for a graphical aid in seeing how $h(t-\tau)$ becomes $g(T_s - t + \tau)$. You will notice the series of steps that transforms the known pulse $g(t)$ for which the matched filter is designed, shown in the first graph of Fig. 8.16b, to $h(t-\tau)$ in the last graph. By comparing these two graphs you will be able to see that the last is simply the first plotted on a τ-axis and made to start earlier (i.e. advanced) by $T_s - t$, hence it is $g(\tau + T_s - t)$. Sampling the matched filter output at $t = T_s$ simply means replacing t by T_s on both sides of Eq. (8.69). This yields the result,

$$\begin{aligned} g_o(T_s) &= \int_0^{T_s} g_i(\tau) g(T_s - T_s + \tau) d\tau \\ &= \int_0^{T_s} g_i(\tau) g(\tau) d\tau \end{aligned} \tag{8.70}$$

The RHS of this equation shows that to obtain $g_o(T_s)$ — which is the matched filter's output sampled at $t = T_s$ — we can use an arrangement that multiplies the incoming pulse $g_i(t)$ by the known pulse $g(t)$ and then integrates the product over the pulse interval $0 \rightarrow T_S$. This process is known as the *correlation* of $g_i(t)$ with $g(t)$. A block diagram of this implementation of the matched filter is shown in Fig. 8.16c. Note that in Eqs. (8.69) and (8.70), time t is measured from the start of each pulse interval, hence the range 0 to T_s even though we are in the nth pulse interval, where n can take on any positive integer value. The correlator performs an *integrate-and-dump* operation whereby an accumulator is reset to zero at the start of each pulse interval, and the product of $g_i(t)$ and $g(t)$ is then accumulated for a period T_s at the end of which the accumulator contains the result $g_o(T_s)$ for that interval. This result is passed to a decision device and the accumulator is immediately reset to zero ready to repeat the same computation in the next pulse interval. In the next Chapter we will examine in more detail the use of a bank of correlators for signal detection.

8.5 Eye diagram

The *eye diagram*, also referred to as *eye pattern*, is a useful analysis tool that provides a visual display of the combined effects of ISI, timing error and noise on pulses at the detection point (i.e. matched filter output) of a communication system. It gives an indication of the likelihood of decision errors at the receiver due to the degradation of the incoming symbol. Fig. 8.17a shows a binary bipolar waveform representing the 20 bits listed at the top, one in each symbol interval of duration T_s, and filtered by a raised cosine filter of roll-off factor $\alpha = 1$. If we superimpose all 20 waveforms in the 20 signalling intervals onto one interval spanning $-T_s/2$ to $T_s/2$, we obtain the result displayed in part (b) of the figure, which is called an *eye diagram* because it resembles the human eye. The eye diagram of an actual transmission can easily be displayed on an oscilloscope. The symbols in successive

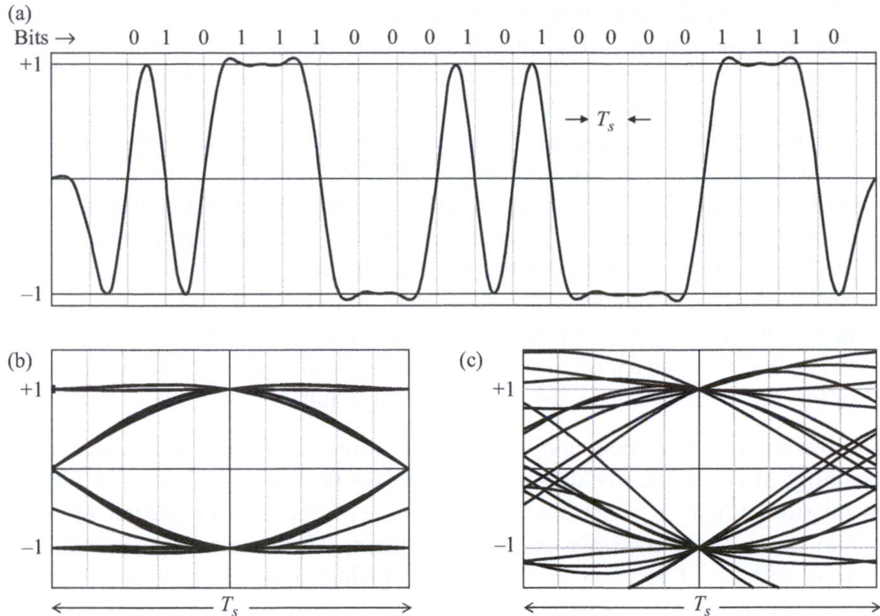

Fig. 8.17 (a) Binary bipolar waveform filtered with RC filter of α = 1; (b) eye-diagram of this waveform; (c) eye diagram of the same line code if filtered with Nyquist filter (α = 0)

intervals will be automatically superimposed on the screen if the oscilloscope is triggered using the receiver's clock signal.

Features of the eye diagram may be interpreted as follows to obtain useful information about system performance:

- The horizontal width of the eye opening gives the interval over which sampling should be performed to obtain the value $y(T_s)$ — discussed in the previous section (see Fig. 8.16a) — needed for a decision. It indicates the timing error that can be tolerated in the sampling instants at the receiver. In Fig. 8.17b, the vertical line that runs through the centre of the eye opening marks the best sampling instant.
- How rapidly the eye opening narrows as you move away from its centre gives an indication of the sensitivity of the transmission system to timing error. To see this, compare the eye diagrams in Fig. 8.17b,c, the latter being the eye diagram of a binary bipolar waveform (not shown) for the same bit stream as in part (a) of the figure but filtered using a Nyquist filter (equivalent to a raised cosine filter with $\alpha = 0$). Notice that the two eye diagrams have equal height at the centre, so both have the same tolerance to noise. However, the opening in (c) closes more rapidly as you move away from the centre, meaning that the Nyquist filtered system is more sensitive to timing error. This issue was explored in detail in Section 8.2. To emphasise again, both systems (b) and (c) have zero ISI at the eye centre, which corresponds to the precise best sampling instant. Away from this centre, both systems have some ISI, indicated by the narrowing of both eyes as you approach the signalling interval boundaries, but system (c) has more ISI than (b), as evidenced by the more rapid closing of the eye in (c).

- The height of the eye opening at its centre gives an indication of the noise margin of the system. A horizontal line that runs through the centre of the eye opening is the decision threshold, in the sense that when the sample $y(T_s)$ has a value above this line the receiver decides in favour of binary 1 (in this binary transmission system), otherwise the decision is in favour of binary 0; or vice versa depending on bit assignment plan. Recall from Fig. 8.16a that $y(T_s) = g_o(T_s) + \tilde{n}(T_s)$, where $g_o(T_s)$ and $\tilde{n}(T_s)$ are respectively the signal and noise components of the sample. A decision error will occur whenever $\tilde{n}(T_s)$ causes the sample to trespass across the decision boundary. The eye diagrams of Fig. 8.17b,c were plotted for the ideal case of a noiseless system. In general, noise will narrow the eye opening across the entire width, and this is illustrated in Fig. 8.18a, which shows the effect of noise on the binary transmission system of Fig. 8.17b. In extreme cases noise might cause full closure of the eye opening.

A narrowing of the eye opening by ISI, noise or timing jitter signifies an increased probability of error. The eye diagram discussed so far has been those of a binary transmission system in which there is just one eye opening. In general the eye diagram of an M-ary transmission system contains $M-1$ vertically stacked eye openings, with $M-1$ decision thresholds each of which appears on the eye diagram as a horizontal line passing through the centre of each eye opening. Fig. 8.18b–d shows the eye diagram of M-ary ASK for $M=4$, and raised cosine filter roll-off $\alpha=1$, 0.5 and 0, respectively. It can be seen from these three diagrams that sensitivity to timing error (explained above) increases as α decreases. The eye diagram is a useful diagnostic tool that provides a qualitative assessment of system performance and is able to highlight the severity or otherwise of timing error, noise, and pulse distortion due to ISI. A quantitative measure of the impact of these defects on

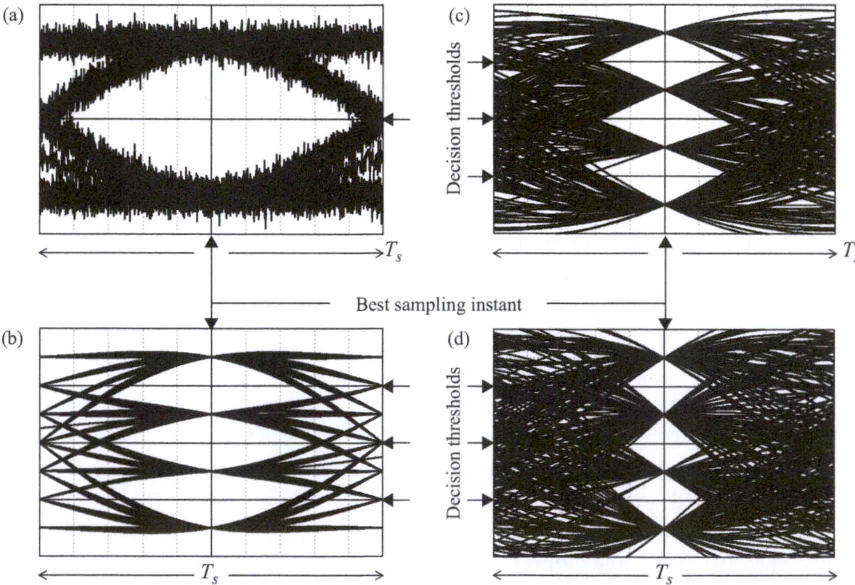

Fig. 8.18 Eye diagrams of (a) noisy binary bipolar system (e.g. BPSK) filtered with RC filter having roll-off factor $\alpha=1$*; (b) 4-ASK,* $\alpha=1$*; (c) 4-ASK,* $\alpha=0.5$*; (d) 4-ASK,* $\alpha=0$

system performance is, however, usually obtained through detailed simulations to evaluate bit error rate, packet error rate, packet delay variations and other performance parameters of interest.

8.6 Summary

We have learnt that in digital communications information is conveyed through transmission of a sequence of pulses or symbols, each of which is an energy signal of finite duration T_s, from a transmitter to a receiver through a channel. By mutual agreement between transmitter and receiver, each unique transmitted pulse represents a unique group of one or more bits. The task of the receiver is to establish the identity of the received pulse in each interval and then to locally generate the agreed group of bits corresponding to the identified pulse. If a pulse is misidentified then the wrong group of bits will inadvertently be generated by the receiver and this gives rise to one or more bit errors in the interval T_s. The pulse detection process involves taking one sample per interval T_s of the waveform arriving at the detection point and passing each sample on to a decision device which compares its value to one or more threshold levels and decides accordingly. There are three main potential causes of detection error:

- The sample is corrupted by contributions from symbols sent in previous intervals. This is called *intersymbol interference* (ISI) and we dealt in detail with the solutions to this problem in Section 8.2. We found to start with that this problem constrains the symbol interval T_s to be no smaller than $1/2B$ in a base band channel, or $1/B$ in a pass band channel, where B is the channel bandwidth. This effectively means that we cannot avoid ISI if we transmit at a rate faster than $2B$ symbols per second in a base band channel or B symbols per second in a pass band channel. Furthermore, we presented solutions to the ISI problem in which filters are used to shape the transmitted pulses to ensure that upon arrival at the detection point they have zero value at adjacent sampling instants, except when controlled ISI was employed in a technique called *duobinary signalling*, which we also discussed.
- The sampling instant is mistimed. If the receiver asks the right question, 'Who are you?' at the wrong time, it might get a misleading answer. Even with well-designed zero-ISI filters in place, contributions from previous symbols are more significant as you move away from the correct sampling instant of the current symbol. An eye diagram demonstrates this effect most vividly with the eye opening narrowing away from its centre. Thus synchronisation between transmitter and receiver is crucial for reliable operation of a digital communication system. The subject of synchronisation was briefly discussed in Section 1.3.7, and in Chapter 7 we examined various line code designs to ensure that the transmitted waveform contains sufficient clock content to aid receiver synchronisation. A more detailed treatment of synchronisation and clock extraction, including the operation of the phase-locked loop (PLL) can be found in [1] and [9].
- The sample is corrupted by noise. This problem constrains transmitted signal power in a manner discussed in detail in Section 8.3 and beautifully encapsulated in the Shannon-Hartley information capacity law. In particular, we need to ensure that signal power is sufficiently large compared to noise power so

that the inevitable addition of random noise in the channel will not cause the sample value taken at the detection point to trespass beyond its decision boundary. This consideration gives rise to the need to employ a signal processing algorithm or filter at the receiver that maximises the signal-to-noise ratio of the pulse at the decision point. The specification of such a filter called the matched filter was the focus of Section 8.4, where we also studied the correlation receiver implementation of this filter.

We also briefly discussed the eye diagram which is a useful analysis tool that provides a visual display of the combined effects of ISI, timing error and noise on pulses at the detection point of a communication system, and facilitates a qualitative assessment of system performance.

8.7 References

[1] Otung, I. (2001). *Communication engineering principles*. Basingstoke: Palgrave Macmillan

[2] Lender, A. (1963). 'The duobinary technique for high-speed data transmission', *Transactions of the American Institute of Electrical Engineers, Part I: Communications and Electronics*, 82(2), pp. 214–218

[3] Newcombe, E. A., Pasupathy, S. (1980). 'Effects of filtering allocation on the performance of a modified duobinary system', *IEEE Transactions on Communications*, COM-28(5), pp. 749–752

[4] Kabal, P., Pasupathy, S. (1975). 'Partial response signaling', *IEEE Transactions on Communications*, COM-23(9), pp. 921–934

[5] Walklin, S., Conradi, J. (1999). 'Multilevel signalling for increasing the reach of 10 Gb/s lightwave systems', *Journal of Lightwave Technology*, 17(11), pp. 2235–2248

[6] Haykin, S. (2014). *Digital communication systems*. Hoboken: Wiley

[7] Shannon, C. E. (1948). 'A mathematical theory of communication', *The Bell System Technical Journal*, 27, pp. 379–423, 623–656

[8] Hartley, R. V. L. (1928). 'Transmission of information', *The Bell System Technical Journal*, 7, pp. 535–563

[9] Sklar, B. (2001) *Digital communications fundamentals and applications*. Upper Saddle River, NJ: Prentice Hall PTR

Chapter 9

Transmitted digital signals

In Chapter Nine

✓ A graphical introduction to the concept of orthogonality of signals, followed by a mathematical description.

✓ Geometric representation of signals in digital communications: This is an important application of orthogonality which culminates in signal space diagrams, a truly invaluable tool for representing digital transmission systems and analysing their performance in the presence of noise and other distortions.

✓ Digital modulation: An introduction to the energy signals (called symbols) employed in base band and modulated digital transmission systems. Digital modulator block diagrams are also discussed.

✓ Complex representation of signals: You will learn how to compute and apply the Hilbert transform and the in-phase and quadrature components of signals, the base band equivalent representation of band pass signals and their canonical and envelope forms, etc.

✓ Coherent detection: You will be able to calculate the correlation of transmitted symbols, and understand the operation of the correlation receiver and the unique noise-bandwidth trade-off that is possible with M-ary FSK.

9.1 Introduction

This chapter deals with the geometric and complex representations of transmitted signals and noise in digital communications, and the correlation of signals. Working carefully through the material presented here will give you a thorough grounding in these vital concepts which have extensive applications in communication systems design and analysis, as highlighted at various points throughout the chapter. Computer simulation has now become an indispensable tool in the design of a communication system and the analysis of its performance. This chapter also aims to equip you with an excellent understanding of how transmitted signals and noise are represented for simulation purposes.

We first explain signal orthogonality and introduce a number of types of signals that are orthogonal to each other, using graphical plots to better illustrate the concept. We then discuss how an arbitrary signal can be represented or realised as a linear combination of N orthogonal signals called basis functions, and note that the Fourier series and the sampling theorem (where a signal is perfectly reconstructed from its samples) are examples of this application. We follow this up with expressions for the signals transmitted in base band and modulated digital communication systems. We note however that in all cases of modulated digital transmission, except M-ary FSK with $M > 2$, the number of component basis functions $N = 2$. This allows us to represent each transmitted signal as a vector in two-dimensional signal space, with one basis function identified along the x-axis (called the real or in-phase axis) and the other lying along the y-axis (called the imaginary or quadrature axis). This leads naturally to a complex representation which we develop for both band limited white noise and band pass signals. We also discuss a measure of the similarity between two signals in terms of their correlation coefficient which is a number in the range $(-1, +1)$, and present important applications of correlation in communication systems, e.g. the correlation receiver and the detection in M-ary FSK systems of extremely weak signals in the presence of noise. The chapter concludes with a discussion of suitable modulator block diagrams for the generation of the M-ary signals discussed.

9.2 Geometric representation of signals

In digital transmission, information is conveyed using a finite set of M distinct symbols $g_0(t), g_1(t), \ldots, g_{M-1}(t)$, where $M = 2$ for binary, and $M > 2$ for multilevel or M-ary transmissions. Each symbol $g_k(t)$ is an energy signal, being of finite duration T_s. We wish to introduce a geometric representation of the set of symbols $\{g_k(t)\}$ which is an excellent way of visualising the corresponding transmitted states $\{S_k\}$ of a system and their energies and closeness to each other. This representation also greatly facilitates the evaluation of the impact of additive white Gaussian noise (AWGN) on the system. But first let us explain the concept of orthogonal signals.

9.2.1 Orthogonality of signals

The mathematical operation of integration is very easily performed in communication systems using a suitably designed low pass filter and yields the area under a function or waveform of a signal. Given two signals $g_1(t)$ and $g_2(t)$, if their product $g_1(t)g_2(t)$ has zero area over an interval of duration say T_s whereas the areas of $g_1^2(t)$ and $g_2^2(t)$ are both non-zero over the same interval, then the two signals are said to be orthogonal over the interval T_s. The principle of orthogonality finds extensive applications in communication systems as we will learn in this chapter. For example, if a binary transmission system sends bit 1 using signal $g_1(t)$ and bit 0 using $g_2(t)$, then we can determine at the receiver which bit was sent during each interval as follows. Multiply the incoming signal by $g_1(t)$ and compute the area of the resulting product. If this area is zero — in practice smaller than a set threshold — we conclude that the incoming signal is $g_2(t)$ and hence that bit 0 was sent. But if this area is significant (i.e. larger than the set threshold), then it is concluded that the incoming signal is $g_1(t)$ and hence that bit 1 was sent.

Fig. 9.1 illustrates that two sinusoidal signals $g_1(t)$ and $g_2(t)$ having a phase difference of 90° are orthogonal over an interval T_s containing an integer number of cycles, where

$$g_1(t) = A_c \sin(2\pi f_c t); \quad g_2(t) = A_c \cos(2\pi f_c t)$$

Note in this figure that when we multiply each signal by itself, the area of the result is non-zero and equals the energy of each signal in the interval T_s. But when we multiply one signal by the other, the area of the product waveform is zero since there are clearly equal amounts of negative and positive regions. This feature can

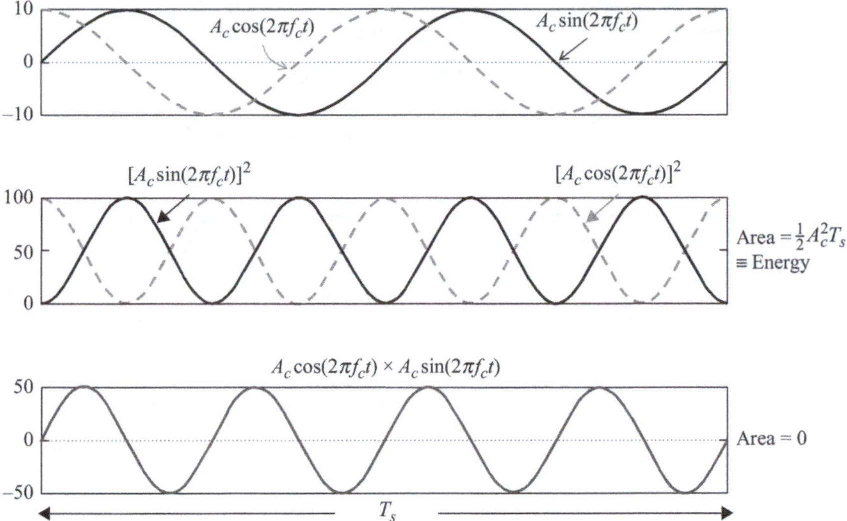

Fig. 9.1 Orthogonality of sine and cosine waveforms of the same frequency f_c

be applied to improve the efficiency of bandwidth utilisation by transmitting two bit streams or message signals simultaneously at the same carrier frequency f_c. One bit stream is sent through the 'in-phase' channel after being modulated onto the 'cosine carrier' $g_2(t)$; whereas the other bit stream is borne on the 'sine carrier' $g_1(t)$ through the quadrature channel. At the receiver there are two processing branches, namely the in-phase branch which first multiplies the incoming signal by $g_2(t)$ before computing the area, and the quadrature branch where the incoming signal is multiplied by $g_1(t)$. It is easy to see that by virtue of orthogonality the signal transmitted in one channel does not make any contribution to the area computed in the other channel and therefore the two bit streams can be separately recovered. We will see applications of this in the implementation of digital modulation systems, in particular multilevel phase shift keying (PSK) and quadrature amplitude modulation (QAM).

Orthogonality can also exist right across a set of signals, with every pair in the set being orthogonal. An example is the set of harmonic sinusoids. Given an interval of duration T_s, then every sinusoidal signal that completes an integer number of cycles in this interval belongs to this set. Thus this is an infinite set of sinusoids having frequencies

$$f_n = nf_s; \quad n = 1, 2, 3, \ldots; \quad f_s = 1/T_s$$

The orthogonality of this set is illustrated in Fig. 9.2 for a pair with $n = 2$ cycles and $m = 3$ cycles in T_s. Note that the total area of the bottom plot is zero since every positive area is cancelled out by a corresponding negative area. Fig. 9.3 similarly illustrates the orthogonality of half-harmonic sinusoids. Here the set consists of all sinusoids that complete an integer number of half cycles in the interval T_s, and therefore have frequencies

$$f_n = n\frac{f_s}{2}; \quad n = 1, 2, 3, \ldots; \quad f_s = 1/T_s$$

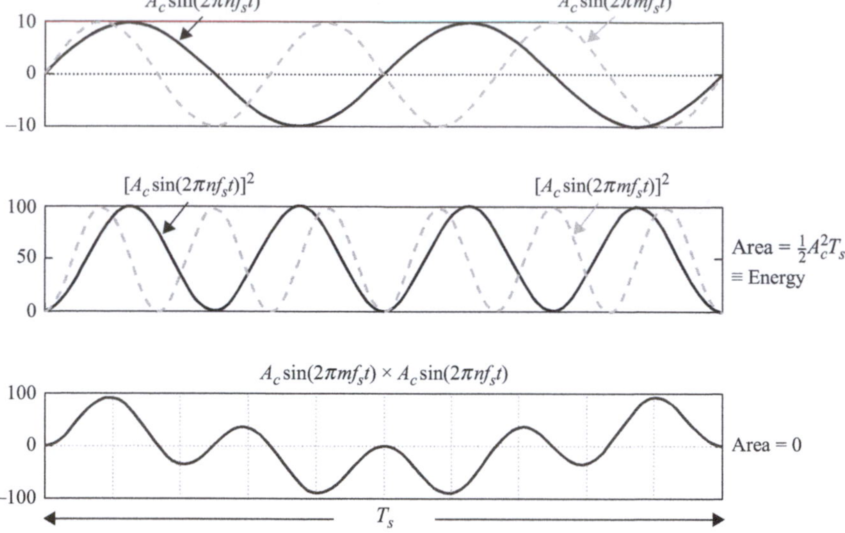

Fig. 9.2 Orthogonality of harmonic sinusoids: $m \neq n$; $f_s = 1/T_s$

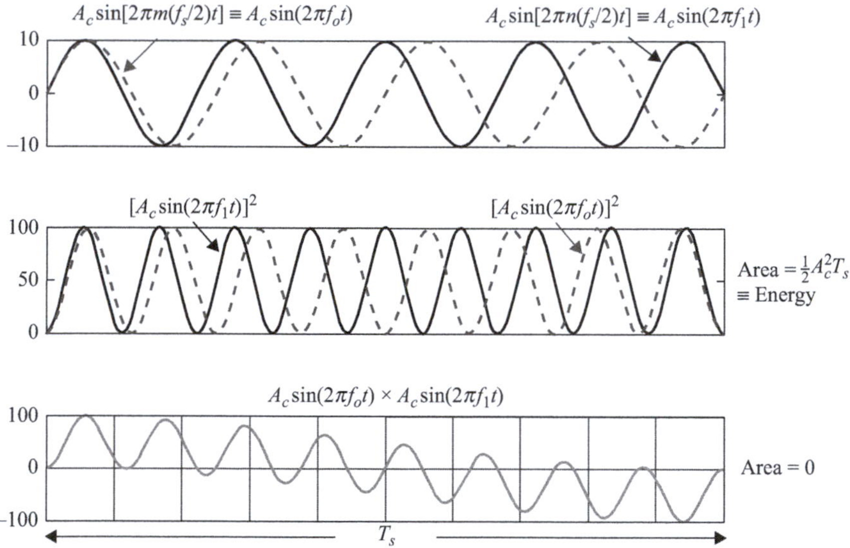

Fig. 9.3 Orthogonality of half-harmonic sinusoids: $f_s = 1/T_s$; $f_o = mf_s/2$; $f_1 = nf_s/2$

It is the orthogonal set of harmonic sinusoids that leads to the Fourier series summarised in Appendix C. We will see later that the orthogonality of half-harmonic sinusoids is the basis of some frequency shift keying (FSK) systems.

Finally, Fig. 9.4 illustrates the orthogonality of shifted sinc functions. Here the set comprises the following signals:

$$g_n(t) = A_c \operatorname{sinc}\left(\frac{t - nT_s}{T_s}\right); \quad n = \ldots, -3, -2, -1, 0, 1, 2, 3, \ldots \tag{9.1}$$

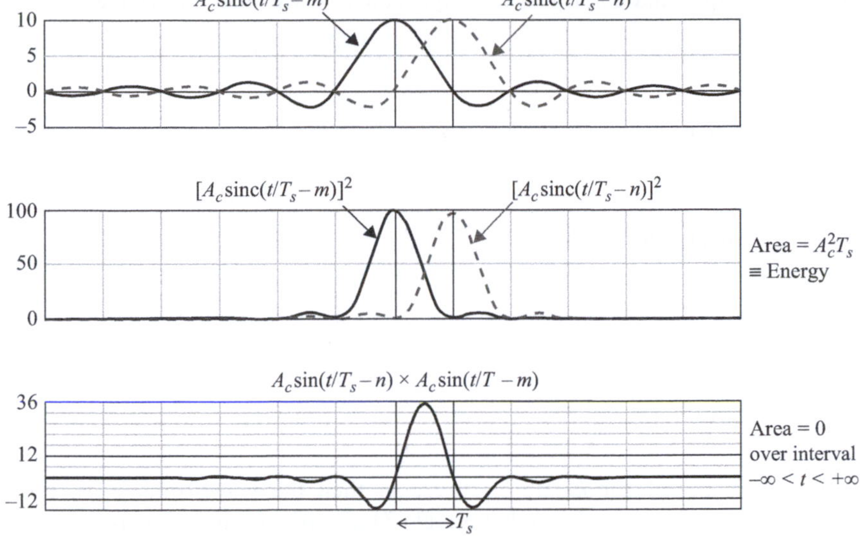

Fig. 9.4 Orthogonality of shifted sinc functions: $m \neq n$

which are sinc pulses of main lobe duration $2T_s$ shifted in time relative to each other by integer multiples of T_s. As shown graphically in Fig. 9.4 and later proved mathematically in Worked Example 9.2, these signals are orthogonal over the interval $-\infty < t < +\infty$. It turns out (see Section 4.6) that this orthogonal set of sinc pulses provides the interpolation functions for reconstructing a base band (i.e. low pass) analogue signal $g(t)$ of maximum frequency component f_m from its samples $g(nT_s)$ taken at regular intervals T_s. That is

$$g(t) = \sum_{n=-\infty}^{\infty} g(nT_s)\,\text{sinc}\left(\frac{t - nT_s}{T_s}\right); \quad \text{provided } T_s \leq \frac{1}{2f_m} \tag{9.2}$$

This equation happens to be a statement of the sampling theorem which was discussed in detail in Chapter 4. Other examples of a complete orthogonal set include Legendre polynomials and Bessel functions.

Let us now return to the main business of this section with a more precise discourse involving some mathematics.

Energy signals $g_1(t)$, $g_2(t)$, $g_3(t), \ldots, g_N(t)$, each of duration T_s, are said to be orthogonal with respect to each other if

$$\int_0^{T_s} g_k(t) g_m(t)\,dt = \begin{cases} 0, & k \neq m \\ E_k, & k = m \end{cases} \tag{9.3}$$

where E_k, the energy of $g_k(t)$, is non-zero and positive. When two energy signals $g_1(t)$ and $g_2(t)$ are orthogonal, then their energies add independently. That is, the energy E of the sum signal $g(t) = g_1(t) + g_2(t)$ is given by the sum of the energies

E_1 and E_2 of $g_1(t)$ and $g_2(t)$, respectively. You can see that this is the case by observing that

$$\begin{aligned}E &= \int_0^{T_s} g^2(t)dt \\ &= \int_0^{T_s} [g_1(t) + g_2(t)]^2 dt \\ &= \int_0^{T_s} g_1^2(t)dt + \int_0^{T_s} g_2^2(t)dt + 2\int_0^{T_s} g_1(t)g_2(t)dt \\ &= E_1 + E_2 + 0\end{aligned}$$

If in Eq. (9.3) $E_k = 1$ for $k = 1, 2, 3, \ldots, N$, then the waveforms $g_1(t)$, $g_2(t)$, $g_3(t), \ldots, g_N(t)$ are said to be orthonormal. Thus orthonormal signals are unit-energy orthogonal signals.

Orthogonality can also be defined for periodic power signals. Power signals $g_1(t)$, $g_2(t)$, $g_3(t), \ldots,$ of period T are said to be orthogonal with respect to each other if

$$\frac{1}{T}\int_{-T/2}^{T/2} g_k(t)g_m(t)dt = \begin{cases}0, & k \neq m \\ P_k, & k = m\end{cases} \tag{9.4}$$

If the power signal is nonperiodic, then Eq. (9.4) is applied in the limit $T \to \infty$.

The following examples of orthogonal signal sets are of particular interest in digital communications. It is easy to show the orthogonality of each of the sets given below simply by evaluating Eq. (9.3) for any two functions $g_k(t)$ and $g_m(t)$ in the set to show that the integration yields zero in all cases except when $k = m$. A demonstration of this verification is given for the final set in Eq. (9.9):

- Harmonically related sinusoidal signals

$$\begin{aligned}g_k(t) &= A_k \cos(2\pi k f_o t + \phi_k) \\ f_o &= 1/T; \quad k = 1, 2, 3, \ldots\end{aligned} \tag{9.5}$$

  form an orthogonal set.
- Non-overlapping pulses form an orthogonal set. An example is the set of rectangular pulses of duration τ shown in Fig. 9.5 (for the case $\gamma = 0$) and given by

$$g_k(t) = A_k \operatorname{rect}\left(\frac{t - (k + \gamma)\tau}{\tau}\right) \tag{9.6}$$

  where $k = \ldots, -2, -1, 0, 1, 2, 3, \ldots,$ and γ is a fraction, $0 \leq \gamma < 1$. To convert $g_k(t)$ into an orthonormal set, we assign each pulse an amplitude A_k that gives it unit energy:

$$E_k = A_k^2\tau = 1; \quad \text{whence } A_k = \sqrt{1/\tau} \tag{9.7}$$

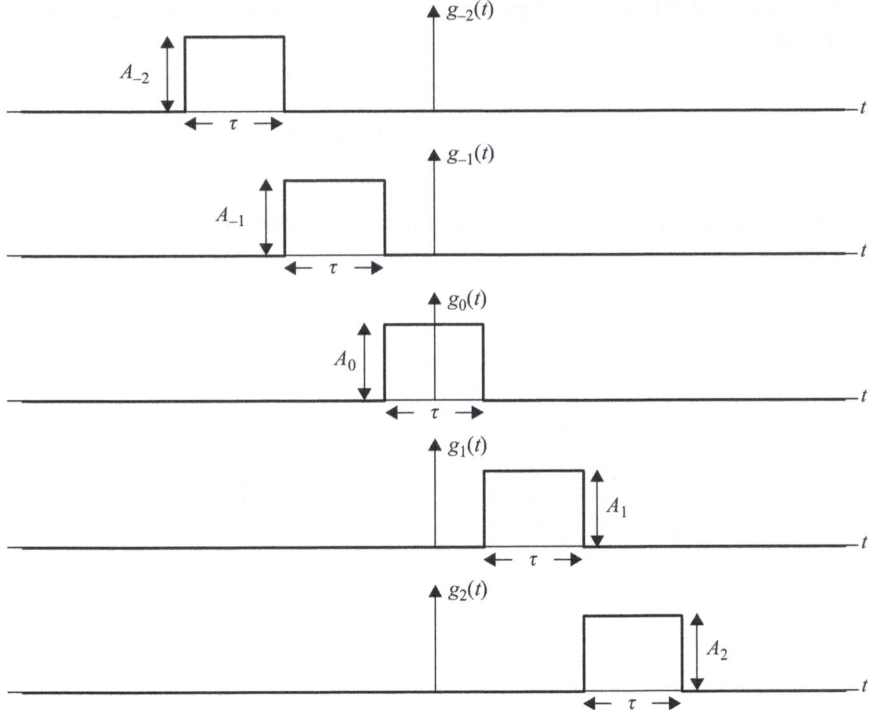

Fig. 9.5 Orthogonal set of pulses: $g_k(t) = A_k\,\mathrm{rect}([t - k\tau]/\tau)$, $k = \cdots -2, -1, 0, 1, 2, \ldots$

- The cosine and sine pulses $\alpha_0(t)$ and $\alpha_1(t)$ given below, having an integer number n of cycles within their duration T_s, are orthonormal:

$$\alpha_0(t) = \sqrt{2/T_s}\cos(2\pi f_c t)\mathrm{rect}\left(\frac{t - \frac{1}{2}T_s}{T_s}\right)$$

$$\alpha_1(t) = -\sqrt{2/T_s}\sin(2\pi f_c t)\mathrm{rect}\left(\frac{t - \frac{1}{2}T_s}{T_s}\right) \tag{9.8}$$

$$f_c = n/T_s$$

You should be able to show that

$$\int_0^{T_s} \alpha_0^2(t)dt = \int_0^{T_s} \alpha_1^2(t)dt = 1; \quad \int_0^{T_s} \alpha_0(t)\alpha_1(t)dt = 0$$

- Finally, the sinusoidal pulses given below, of amplitude $\sqrt{2/T_s}$, which complete an integer number nk (where n is an integer ≥ 1) of half cycles within their duration T_s, form an orthonormal set

$$\alpha_k(t) = \sqrt{2/T_s}\cos[2\pi k(f_c/2)t]\mathrm{rect}\left(\frac{t - \frac{1}{2}T_s}{T_s}\right) \tag{9.9}$$

$$k = 1, 2, 3, \ldots; \quad f_c = n/T_s$$

To verify that this is indeed an orthogonal set, consider any two members of the set

$$\begin{aligned} \alpha_k(t) &= \sqrt{2R_s}\cos[2\pi k(nR_s/2)t] \\ \alpha_m(t) &= \sqrt{2R_s}\cos[2\pi m(nR_s/2)t] \end{aligned} \qquad 0 \le t \le \frac{1}{R_s}$$

Evaluating the integral of their product over the symbol interval $0 \to T_s = 1/R_s$ yields

$$\begin{aligned} \int_0^{1/R_s} \alpha_k(t)\alpha_m(t)dt &= 2R_s \int_0^{1/R_s} \cos\left(2\pi k\frac{nR_s}{2}t\right)\cos\left(2\pi m\frac{nR_s}{2}t\right)dt \\ &= R_s\left[\int_0^{1/R_s} \cos n\pi R_s(k+m)tdt + \int_0^{1/R_s} \cos n\pi R_s(k-m)tdt\right] \\ &= R_s\left[\frac{\sin n\pi(k+m)}{n\pi R_s(k+m)} + \frac{\sin n\pi(k-m)}{n\pi R_s(k-m)}\right] \\ &= \operatorname{sinc}[n(k+m)] + \operatorname{sinc}[n(k-m)] \\ &= \begin{cases} 1, & k=m \\ 0, & \text{Otherwise} \end{cases} \end{aligned} \tag{9.10}$$

where we made use of trigonometric identity Eq. (B.6) (in Appendix B) in the second line. Since n, k and m are integers, so are the products $n(k+m)$ and $n(k-m)$. But when N is an integer, then $\operatorname{sinc}(N) = 0$ everywhere except at $N = 0$ when it equals 1, hence the above result which verifies the orthogonality of the entire set according to Eq. (9.3).

9.2.2 Signal space

Each of the M symbols of an M-ary transmission system can be expressed as a linear combination of N orthonormal basis functions $\alpha_0(t)$, $\alpha_1(t), \ldots, \alpha_{N-1}(t)$, where $N \le M$ and $0 \le t \le T_s$:

$$g_k(t) = s_{k0}\alpha_0(t) + s_{k1}\alpha_1(t) + \cdots + s_{k,N-1}\alpha_{k,N-1}(t) = \sum_{n=0}^{N-1} s_{kn}\alpha_n(t) \quad \text{for } k = 0, 1, \ldots, M-1 \tag{9.11}$$

Since

$$\int_0^{T_s} \alpha_k(t)\alpha_m(t)dt = \begin{cases} 0, & k \ne m \\ 1, & k = m \end{cases} \tag{9.12}$$

it follows that the energy of $g_k(t)$ is given by the expression

$$\begin{aligned}
E_k &= \int_0^{T_s} g_k^2(t)dt \\
&= \int_0^{T_s} \left[s_{k0}\alpha_0(t) + s_{k1}\alpha_1(t) + \cdots + s_{k,N-1}\alpha_{N-1}(t)\right]^2 dt \\
&= s_{k0}^2 \int_0^{T_s} \alpha_0^2(t)dt + s_{k1}^2 \int_0^{T_s} \alpha_1^2(t)dt + \cdots + s_{k,N-1}^2 \int_0^{T_s} \alpha_{N-1}^2(t)dt \\
&\quad + 2\sum_{n=1}^{N-1} s_{k,n} \sum_{m=0}^{n-1} s_{k,m} \int_0^{T_s} \alpha_m(t)\alpha_n(t)dt \\
&= s_{k0}^2 + s_{k1}^2 + \cdots + s_{k,N-1}^2 = \sum_{n=0}^{N-1} s_{k,n}^2
\end{aligned} \tag{9.13}$$

Therefore, by analogy with Pythagoras' theorem, we may represent the kth symbol as a point (called transmitted state S_k) in N-dimensional Euclidean space, which consists of N mutually perpendicular axes $\alpha_0, \alpha_1, \ldots, \alpha_{N-1}$, and is called the signal space or constellation diagram. Signal spaces with $N > 3$ cannot be visualised or sketched in real-life space which is limited to three dimensions, but they remain an important mathematical concept. Distances in this space represent the square root of energy. In particular, the square of the distance of a point from the origin gives the energy of the transmitted symbol corresponding to that point.

To illustrate, we show signal space diagrams in Fig. 9.6, with $N = 1$, $M = 2$ in (a); $N = 2$, $M = 8$ in (b); and $N = 3$, $M = 4$ in (c). Signal space diagrams in general convey the following information about the transmission systems that they represent.

(a)

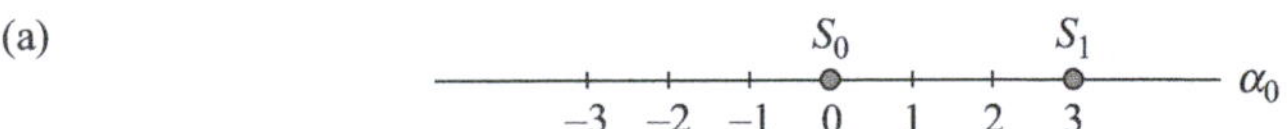

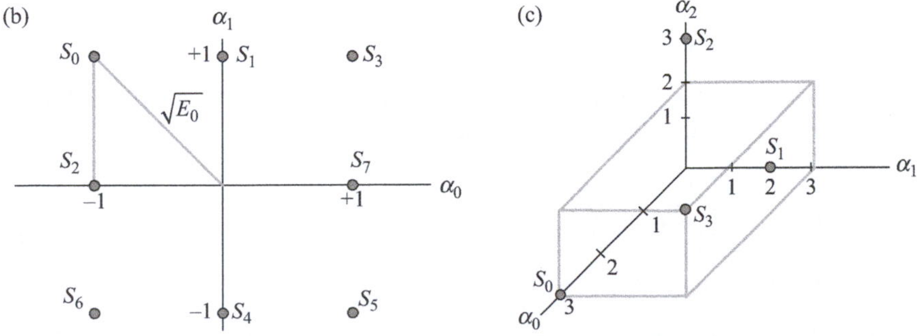

Fig. 9.6 Signal space diagrams: (a) N = 1, M = 2; (b) N = 2, M = 8; (c) N = 3, M = 4

- *Number of transmitted states*: Information (in the form of a string of 0s and 1s, i.e. bit stream) is conveyed using 2, 8 and 4 distinct symbols or transmitted signal states in (a), (b) and (c), respectively. The assignment of states may be as follows. In Fig. 9.6a, state S_0 represents bit 0, and S_1 represents bit 1. In Fig. 9.6b, each state represents a block of three bits; for example, S_0 represents 000, S_1 represents 001,..., and S_7 represents 111. Similarly, each state in Fig. 9.6c represents a block of two bits, with S_0 representing 00, S_1 representing 01, and so on. Generally, in a signal space of M states, each state represents $\log_2(M)$ bits.
- *Energy E_k of each transmitted state S_k*: The energy of each state equals the square of its distance from the origin. Using arbitrary energy units for the moment, we have in Fig. 9.6a

$$\begin{aligned} E_0 &= 0 \\ E_1 &= 3^2 = 9 \end{aligned}$$

In Fig. 9.6b,

$$\begin{aligned} E_0 &= 1^2 + 1^2 = 2 \\ E_1 &= 1^2 = 1, \quad \text{and so on.} \end{aligned}$$

And in Fig. 9.6c

$$\begin{aligned} E_1 &= 2^2 = 4 \\ E_3 &= 3^2 + 3^2 + 2^2 = 22, \quad \text{and so on.} \end{aligned}$$

- *Transmitted symbols*: The symbol $g_k(t)$ transmitted for each state S_k follows from Eq. (9.11), with the coefficients equal to the distance moved along (or parallel to) each axis in order to get to point S_k starting from the origin. Thus, in Fig. 9.6a

$$g_0(t) = 0; \quad g_1(t) = 3\alpha_0(t)$$

In Fig. 9.6b

$$\begin{aligned} g_0(t) &= -\alpha_0(t) + \alpha_1(t) \\ g_1(t) &= \alpha_1(t); \quad \text{etc.} \end{aligned}$$

And in Fig. 9.6c

$$\begin{aligned} g_1(t) &= 2\phi_1(t) \\ g_3(t) &= 3\phi_0(t) + 3\phi_1(t) + 2\phi_2(t); \quad \text{etc.} \end{aligned}$$

- *Average energy per bit E_b*: The average energy per bit can be determined from the signal space diagram using the relations

$$\begin{aligned} E_s &= \frac{1}{M}\sum_{k=0}^{M-1} E_k = \frac{E_{space}}{M} \\ E_b &= E_s/\log_2(M) \end{aligned} \tag{9.14}$$

where E_s is the average energy per symbol, E_{space} is the total energy of all the M states in the signal space, and it is assumed that each state is devoted to a unique group of $\log_2(M)$ bits. For example, in Fig. 9.6a, $M=2$ and $E_{space}=9$, giving $E_b=4.5$. And in Fig. 9.6b, $M=8$ and $E_{space}=12$, giving $E_b=0.5$. The parameter E_b is important in the design and analysis of digital communication systems, since along with noise power per unit bandwidth N_o it determines the bit error rate of a given modulation scheme through the ratio E_b/N_o.

- *Signal power*: If we know the transmission symbol rate R_s (in symbols per second or baud) or bit rate R_b ($\equiv$ bits per second) of the system, then the signal power P ($\equiv$ energy per second) can be determined using the average energy per symbol E_s or average energy per bit E_b calculated above:

$$P = E_s R_s = E_b R_b \tag{9.15}$$

We have expressed transmitted symbols, of duration T_s, in terms of orthonormal basis functions $\{\alpha_k(t)\}$. Signal space diagrams do not usually explicitly show the functional form of their basis functions, which determines whether the system is base band or modulated. In general the basis functions of base band systems are rectangular pulses. The basis functions of modulated systems, namely amplitude shift keying (ASK), phase shift keying (PSK), frequency shift keying (FSK), and quadrature amplitude modulation (QAM) are sinusoidal pulses. Both systems will however always shape these pulses in a suitable filter prior to transmission as discussed in detail in Chapter 8.

Worked Example 9.1: Signal Vector

Each symbol $g_k(t)$, $k = 0, 1, 2, 3, \ldots, M-1$, of an M-ary transmission system corresponds to a point S_k in the signal space diagram. This point is identified by a vector $\boldsymbol{s}_k$ called the signal vector having N elements $s_{k0}, s_{k1}, \ldots, s_{kN-1}$, which are the components of S_k along the N mutually perpendicular axes.

(a) Derive an expression giving the components of S_k in terms of $g_k(t)$ and the basis functions $\alpha_0(t), \alpha_1(t), \ldots, \alpha_{N-1}(t)$.

(b) Hence suggest an arrangement which can be employed for symbol detection at the receiver.

(a) Multiplying both sides of Eq. (9.11) by one of the basis functions $\alpha_n(t)$, where $n = 0, 1, 2, 3, \ldots, N-1$, and integrating over the interval $(0, T_s)$ yields

$$\int_0^{T_s} g_k(t)\alpha_n(t)dt$$

$$= \int_0^{T_s} \left[s_{k0}\alpha_0(t) + s_{k1}\alpha_1(t) + \cdots + s_{kn}\alpha_n(t) + \cdots s_{k,N-1}\alpha_{N-1}(t)\right]\alpha_n(t)dt$$

$$= \int_0^{T_s} [s_{kn}\alpha_n(t)]\alpha_n(t)dt$$

$$= s_{kn}\int_0^{T_s} \alpha_n^2(t)dt = s_{kn}$$

by virtue of the orthogonality of the basis functions and their unit energies as expressed in Eq. (9.12). Thus the desired expression for the elements of signal vector s_k, or equivalently the components of transmitted state S_k in the N-dimensional signal space of the transmission system, is given by

$$s_{kn} = \int_0^{T_s} g_k(t)\alpha_n(t)dt, \quad n = 0, 1, 2, \ldots, N-1 \tag{9.16}$$

It will be useful to note here that in general given a physically realisable signal $g(t)$ defined over the interval (a, b) and N functions $\beta_0(t)$, $\beta_1(t), \ldots, \beta_{N-1}(t)$ that form a complete orthogonal set in the interval, then we may represent $g(t)$ as an orthogonal series

$$g(t) = \sum_{n=0}^{N-1} c_n\beta_n(t) \tag{9.17}$$

with the coefficients c_n given by

$$c_n = \frac{1}{E_n}\int_a^b g(t)\beta_n^*(t)dt; \quad n = 0, 1, 2, \ldots, N-1 \tag{9.18}$$

where E_n is the energy of $\beta_n(t)$, and the asterisk denotes complex conjugation. Note that if the set consists only of real signals, then $\beta_n^*(t) = \beta_n(t)$.

(b) Since extracting the signal vector s_k uniquely identifies the corresponding transmitted state S_k and symbol $g_k(t)$, we can employ for symbol detection an arrangement suggested by the previous solution — Eq. (9.16), and shown in Fig. 9.7. To determine the signal vector s_k, we feed an incoming symbol $g_k(t)$ as a common input to a bank of N correlators each of which is supplied with its own basis function. The N outputs of this arrangement are the N elements of the vector s_k.

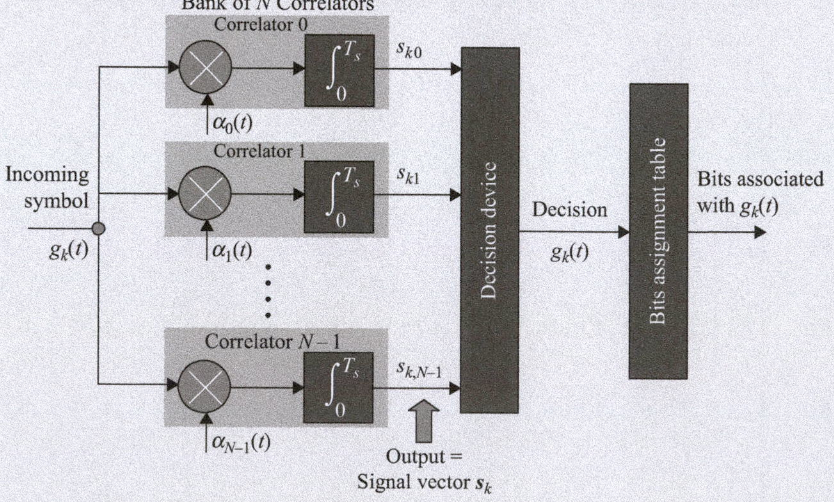

Fig. 9.7 Worked Example 9.1b: Arrangement for symbol detection

Worked Example 9.2: Orthogonality of Sinc Functions

We wish to show that the shifted sinc functions given in Eq. (9.1) form an orthogonal set in the interval $(-\infty, \infty)$.

The easiest way to prove this orthogonality is by making use of Parseval's theorem which states that

$$\int_{-\infty}^{\infty} g_n(t)g_m^*(t)dt = \int_{-\infty}^{\infty} G_n(f)G_m^*(f)df \tag{9.19}$$

Letting $g_n(t) = A_c \operatorname{sinc}[(t - nT_s)/T_s]$ in Eq. (9.19), it follows from the table of Fourier transforms (FTs) in Section C.2 in Appendix C that the FT $g_n(f)$ of $g_n(t)$ is

$$g_n(f) = A_c T_s \operatorname{rect}(fT_s)e^{-j2\pi fnT_s} \tag{9.20}$$

And since in this case $g_n(t)$ is a real function, it follows that $g_m^*(t) = g_m(t)$ so that Eq. (9.19) becomes

$$\begin{aligned}\int_{-\infty}^{\infty} g_n(t)g_m(t)dt &= \int_{-\infty}^{\infty} [A_c T_s \operatorname{rect}(fT_s)e^{-j2\pi fnT_s}][A_c T_s \operatorname{rect}(fT_s)e^{j2\pi fmT_s}]df \\ &= A_c^2 T_s^2 \int_{-1/2T_s}^{1/2T_s} e^{-j2\pi fT_s(n-m)}df \\ &= \begin{cases} A_c^2 T_s, & n = m \\ 0, & n \neq m \end{cases}\end{aligned} \tag{9.21}$$

Thus the sinc functions in Eq. (9.1) form an orthogonal set, with each function having energy $E_n = A_c^2 T_s$.

Worked Example 9.3: Signal Space Diagram

A digital modulated system transmits the following equally likely symbols

$$g_k(t) = A_k \cos\left[10^{10}\pi t + \frac{\pi}{4}(1 - n + 2m)\right]\operatorname{rect}(10^6 t)$$

where $A_k = 100\left[2n + \sqrt{2}(1 - n)\right]$;

$$m = 0, 1, 2, 3; \quad n = 0, 1; \quad k = m + 4n$$

(a) Sketch the signal space diagram of the transmission system.
(b) Determine the transmitted energy per bit and carrier power.

(a) Note that there are eight possible values of k from 0 to 7, which is the result of four values of m combining with two values of n. Thus the system transmits symbols

$$g_k(t) = A_k \cos\left[10^{10}\pi t + \frac{\pi}{4}(1 - n + 2m)\right]\text{rect}(10^6 t)$$
$$\equiv A_k \cos[2\pi f_c t + \phi_k]\text{rect}(t/T_s); \quad k = 0, 1, 2, \ldots, 7$$

having the following parameters:

Amplitude: $A_k = 100\left[2n + \sqrt{2}(1 - n)\right]$
Phase: $\phi_k = \pi(1 - n + 2m)/4$
Duration: $T_s = 10^{-6}$ s
Energy: $E_k = A_k^2 T_s/2$
Distance from signal space origin: $D_k = \sqrt{E_k}$

The values of these parameters are tabulated in Table 9.1. Note that the transmission is at a carrier frequency $f_c = 5$ GHz, which means that the spectrum of each symbol is centred at this frequency. Based on each transmitted state's distance D_k from the origin and its angle ϕ_k with the horizontal direction, the signal space diagram is sketched as shown in Fig. 9.8. We will learn in the next section that this is a star

Table 9.1 Worked Example 9.3

>m	n	$k=m+4n$	A_k	ϕ_k (deg)	E_k	D_k
0	0	0	$100\sqrt{2}$	45°	1/100	1/10
1	0	1	$100\sqrt{2}$	135°	1/100	1/10
2	0	2	$100\sqrt{2}$	225°	1/100	1/10
3	0	3	$100\sqrt{2}$	315°	1/100	1/10
0	1	4	200	0°	2/100	$(\sqrt{2})/10$
1	1	5	200	90°	2/100	$(\sqrt{2})/10$
2	1	6	200	180°	2/100	$(\sqrt{2})/10$
3	1	7	200	270°	2/100	$(\sqrt{2})/10$

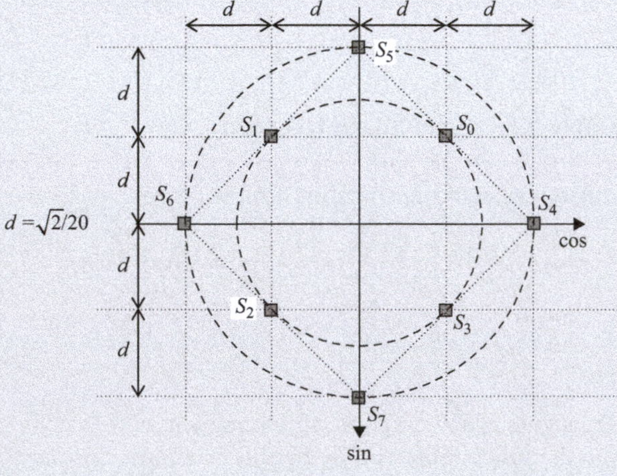

Fig. 9.8 Worked Example 9.3: Signal space diagram

8-QAM constellation in which states of equal energy are separated by at least 90°.

(b) From Eq. (9.14), the average energy per symbol E_s is the average of the E_k column in Table 9.1. Thus $E_s = 3/200$ J. Dividing this by the number of bits per symbol $\log_2 M$, where $M = 8$ in this case, yields the average energy per bit $E_b = 1/200 = 5$ mJ.

Using Eq. (9.15), the average power is

$$P = E_s R_s = E_s/T_s = \frac{3}{200} \times 10^6 = 15 \text{ kW}$$

9.3 Symbols in digital transmission systems

9.3.1 Base band systems

Basis functions in base band systems consist of fractional-width rectangular pulses, one for each time interval between transitions in the voltage level of transmitted symbols. There is an important exception however: If the system employs two opposite polarity (or antipodal) symbols $g_0(t)$ and $g_1(t)$, then only one basis function $\alpha_0(t)$ is required irrespective of the number of level transitions in one symbol. The basis function in this case is the symbol $g_0(t)$ appropriately scaled to unit energy. The two symbols are then given by $g_0(t) = d\alpha_0(t)$ and $g_1(t) = -d\alpha_0(t)$, where d is the signal space distance of each state from the origin and d^2 is the energy of each symbol.

Fig. 9.9 is an example based on the coded mark inversion (CMI) line code discussed in Section 7.5, which conveys information using the symbols $g_0(t)$, $g_1(t)$ and $g_2(t)$ shown. Bit 0 is conveyed using $g_0(t)$ and bit 1 is conveyed alternately by $g_1(t)$ and $g_2(t)$. We see that within the symbol interval T_s, a voltage level transition occurs only at $t = T_s/2$ in one of the symbols $g_0(t)$. Thus we require two basis functions $\alpha_0(t)$ and $\alpha_1(t)$ as shown in Fig. 9.9b, one for the interval $(0, T_s/2)$ and the other for the interval $(T_s/2, T_s)$. Note that these basis functions come from the set in Eq. (9.6) with $\tau = T_s/2$ and $\gamma = 0.5$, and their amplitude is given by Eq. (9.7) as $A_\alpha = \sqrt{2/T_s}$. It is clear from Fig. 9.9a,b that we can write the transmitted symbols in terms of $\alpha_0(t)$ and $\alpha_1(t)$:

$$\begin{aligned} g_0(t) &= -\frac{A}{A_\alpha}\alpha_0(t) + \frac{A}{A_\alpha}\alpha_1(t) \\ &= A\sqrt{T_s/2}[-\alpha_0(t) + \alpha_1(t)] \\ g_1(t) &= A\sqrt{T_s/2}[\alpha_0(t) + \alpha_1(t)] \\ g_2(t) &= -A\sqrt{T_s/2}[\alpha_0(t) + \alpha_1(t)] \end{aligned} \tag{9.22}$$

The signal space diagram is therefore as shown in Fig. 9.9c. Note that the transmitted symbols all have the same energy

$$E_0 = E_1 = E_2 = A^2 T_s$$

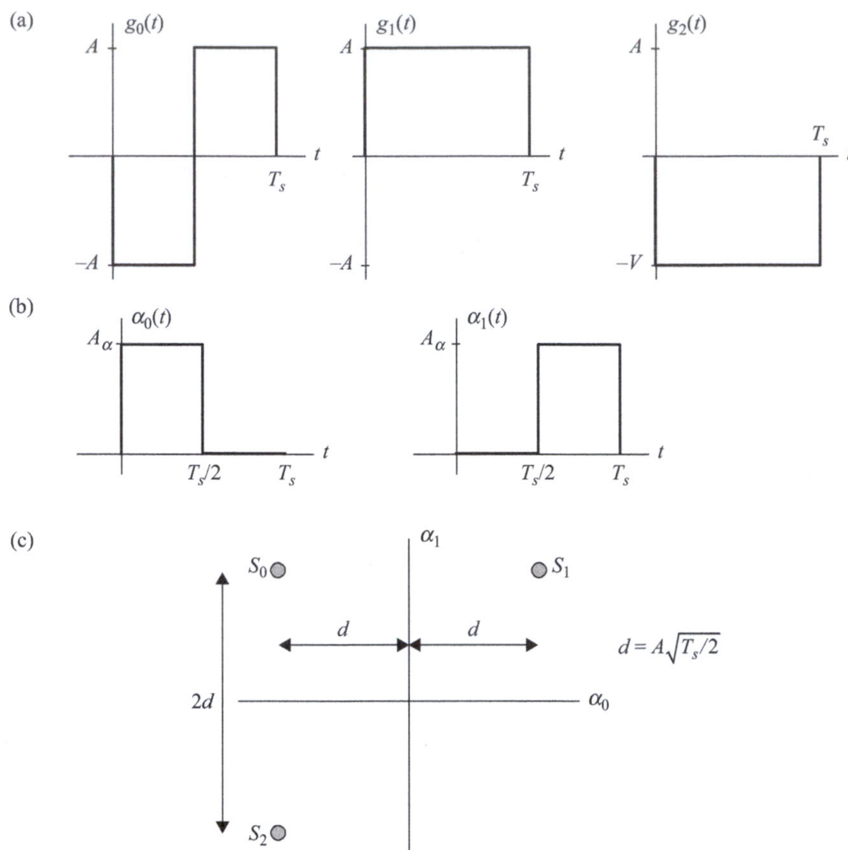

Fig. 9.9 CMI base band system: (a) transmitted symbols; (b) basis functions; (c) signal space diagram

9.3.2 Modulated systems

9.3.2.1 M-ary ASK symbols

M-ary ASK systems have a one-dimensional signal space involving just one basis function $\alpha_0(t)$, given by Eq. (9.8). This means that the transmitted states S_k are all arranged along the α_0 axis with equal spacing d as shown in Fig. 9.10a. It follows that the kth transmitted symbol $g_k(t)$ of M-ary ASK is

$$g_k(t) = dk\alpha_0(t) = dk\sqrt{2/T_s}\cos(2\pi f_c t)\text{rect}\left(\frac{t - \frac{1}{2}T_s}{T_s}\right) \qquad (9.23)$$
$$k = 0, 1, 2, 3, \ldots, M - 1$$

Symbol $g_k(t)$ given above results from shifting (i.e. modulating) the amplitude of a carrier of frequency f_c in direct proportion to the number 0, 1, 2,..., $M-1$, corresponding to the decimal value of the information bit stream taken $\log_2 M$ bits at a time in successive intervals T_s, but the bits actually assigned to $g_k(t)$ is the gray code of decimal number k. This arrangement (see, e.g. Fig. 8.3) is to ensure that adjacent states represent a group of bits that differ in only one bit position in order

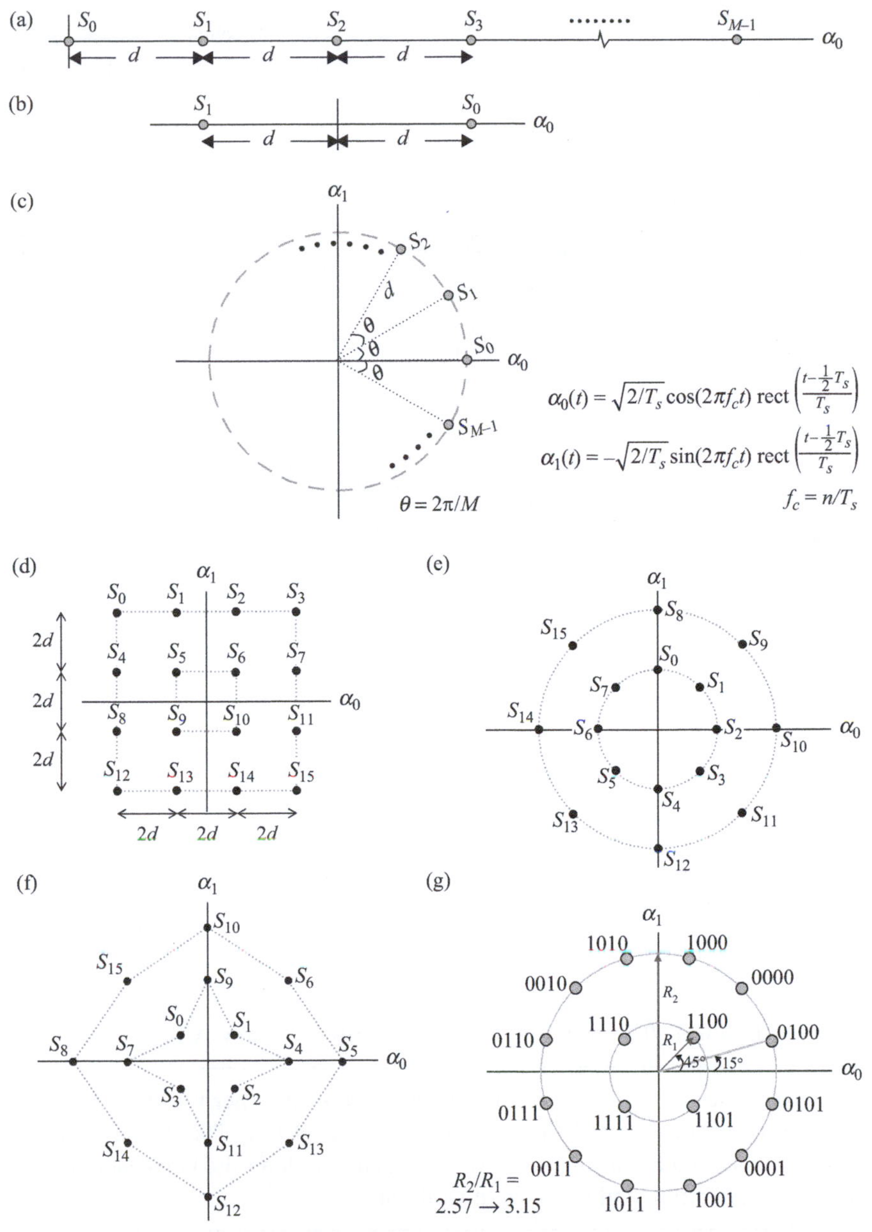

Fig. 9.10 Signal space diagrams of various modulated transmissions: (a) M-ary ASK; (b) BPSK; (c) M-ary PSK, $M \geq 4$ (d) Square 16-QAM; (e) circular 16-QAM; (f) star 16-QAM; (g) 16-APSK (Continued)

to minimise the number of bit errors incurred when there is a symbol error, which usually involves mistaking between adjacent states.

The average energy per bit of *M*-ary ASK can be shown (see Worked Example 9.4) to be

$$E_b = \frac{(M-1)(2M-1)d^2}{6\log_2 M} \tag{9.24}$$

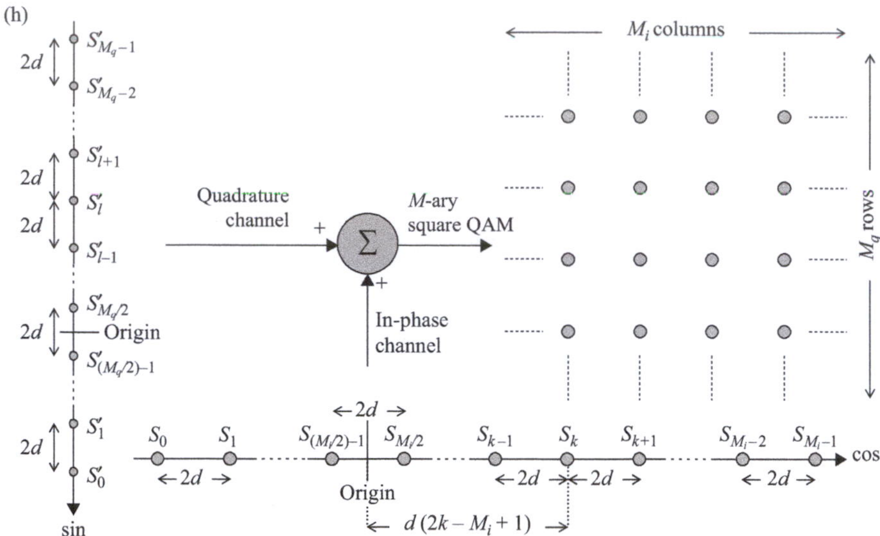

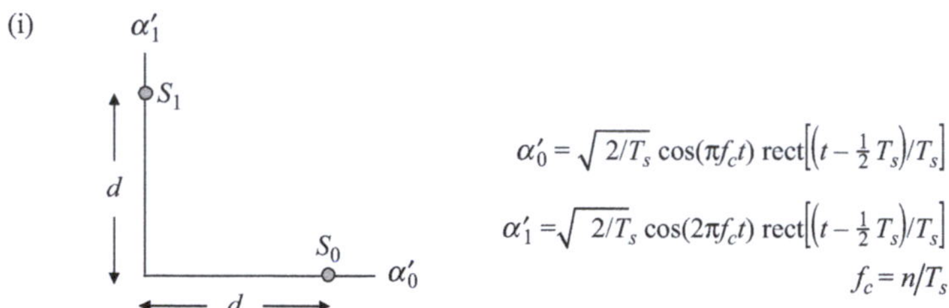

Fig. 9.10 (Continues) Signal space diagrams of various modulated transmissions: (h) M-ary square QAM as a sum of two orthogonal channels (i) BFSK

Worked Example 9.4: Average Energy Per Bit of *M*-ary ASK

We wish to derive an expression for the average energy per bit of an M-ary ASK system with constellation as shown in Fig. 9.10a. The M states are equally likely and are located at uniform spacing d along the horizontal axis starting at the origin and going rightward only.

Note that in solving this problem we will need to make use of the following standard expression for the sum of squares:

$$\sum_{k=0}^{n} k^2 = 1^2 + 2^2 + 3^2 + \cdots + n^2 = \frac{1}{6}n(n+1)(2n+1) \tag{9.25}$$

Counting the states in Fig. 9.10a from 0 to $M - 1$, the distance D_k of the kth state from the signal space origin is given below along with the energy E_k of this state:

$$D_k = kd; \quad E_k = D_k^2 = k^2 d^2$$

The average energy per bit of this system then follows from Eq. (9.14):

$$
\begin{aligned}
E_b &= \frac{1}{M \log_2 M} \sum_{k=0}^{M-1} E_k = \frac{1}{M \log_2 M} \sum_{k=0}^{M-1} k^2 d^2 \\
&= \frac{d^2}{M \log_2 M} \sum_{k=0}^{M-1} k^2 = \frac{d^2}{M \log_2 M} \left\{ 1^2 + 2^2 + \cdots + (M-1)^2 \right\} \\
&= \frac{d^2}{M \log_2 M} \left\{ \frac{1}{6}(M-1)[(M-1)+1][2(M-1)+1] \right\} \\
&= \frac{d^2}{6M \log_2 M} \{(M-1)M(2M-1)\} \\
&= \frac{(M-1)(2M-1)d^2}{6 \log_2 M}
\end{aligned}
$$

9.3.2.2 *M*-ary PSK symbols

In M-ary PSK systems the transmitted states are all of the same energy and are arranged in a circular constellation of radius d centred at the origin. The signal space is two-dimensional and involves the two basis functions $\alpha_0(t)$ and $\alpha_1(t)$ of Eq. (9.8). See Fig. 9.10c. A uniform angular spacing of states $\theta = 2\pi/M$ is employed, which in the special binary case ($M = 2$, Fig. 9.10b) reduces to a one-dimensional signal space employing antipodal symbols $g_1(t) = d\alpha_0(t)$ and $g_0(t) = -d\alpha_0(t)$. The M-ary PSK scheme shifts the phase of a carrier of frequency f_c in direct proportion to the value of $\log_2 M$ bits in successive intervals T_s of the message. The actual bit assignment is again gray-coded. It follows from the signal space diagram that the transmitted symbols are given by

$$
\begin{aligned}
g_k(t) &= d\cos(\theta_k)\alpha_0(t) + d\sin(\theta_k)\alpha_1(t) \equiv A_{kI}\alpha_0(t) + A_{kQ}\alpha_1(t) \\
&= d\sqrt{2/T_s}\cos(2\pi f_c t + \theta_k)\mathrm{rect}\left[\left(t - \tfrac{1}{2}T_s\right)/T_s\right]
\end{aligned}
$$

where (9.26)

$$
\begin{aligned}
A_{kI} &= d\cos(\theta_k); \quad A_{kQ} = d\sin(\theta_k); \quad \theta_k = \theta_o + 2\pi k/M \\
k &= 0,\ 1,\ 2,\ 3, \ldots,\ M-1
\end{aligned}
$$

θ_0 is the angular offset of the constellation, equal to zero in Fig. 9.10b,c. The average energy per bit of M-ary PSK is determined from the constellation radius d as

$$
E_b = \frac{d^2}{\log_2 M} \tag{9.27}
$$

9.3.2.3 *M*-ary QAM Symbols

M-ary QAM (also called APSK for amplitude and phase shift keying) systems have a two-dimensional signal space involving the same basis functions $\alpha_0(t)$ and $\alpha_1(t)$ as above. However, the transmitted states are more freely distributed and not constrained to a circular arrangement. A completely general expression for the transmitted symbols in M-ary QAM is given by

$$
\begin{aligned}
g_k(t) &= D_k\sqrt{2/T_s}\cos(2\pi f_c t + \phi_k)\text{rect}\left[\left(t - \tfrac{1}{2}T_s\right)/T_s\right] \\
&\equiv [D_{k1}\cos(2\pi f_c t) + D_{k2}\sin(2\pi f_c t)]\sqrt{2/T_s}\,\text{rect}\left[\frac{t - \frac{1}{2}T_s}{T_s}\right] \qquad (9.28)
\end{aligned}
$$

$$
D_k = \sqrt{D_{k1}^2 + D_{k2}^2}; \quad \phi_k = -\tan^{-1}(D_{k2}/D_{k1}); \quad k = 0, 1, \ldots, M-1
$$

where D_k is the distance of state S_k from the signal space origin, ϕ_k is the angle of S_k with respect to the horizontal axis α_0, and the energy of state S_k is $E_k = D_k^2$. Note that this scheme is the result of modulating the carrier both in amplitude and phase. Fig. 9.6b is an example of 8-QAM. For the same E_b value, M-ary QAM has a wider spacing of states in its signal space than M-ary PSK and this reduces the likelihood that the effects of additive noise will cause the receiver to misidentify the symbols. There is an infinite variety of M-ary QAM constellations depending on how the M states are arranged in the two-dimensional signal space. Fig. 9.10d–g show four examples for $M=16$, named square-, circular-, star-QAM and 16-APSK, respectively. The star-QAM has a minimum phase difference of 90° between states of the same energy and is therefore thought to perform better than square QAM in transmission media with a predominance of phase distortion. The 16-APSK constellation in Fig. 9.10g has been adopted for satellite communications because its states are more spread out than 16-PSK and, having just two amplitudes levels, it copes better with nonlinearity than the square constellation of Fig. 9.10d that uses three amplitude levels. Square QAM constellations in general comprise states arranged in a regular grid of uniform spacing $2d$ along each direction. This arrangement lends itself very well to the analysis of noise effects and derivation of an expression for bit error rate that is applicable for all M. Note that square-QAM and PSK are identical when $M=4$.

An important view of M-ary QAM is suggested by the equivalent expression in the second line of Eq. (9.28): At any given interval T_s we take a block of $\log_2(M)$ bits and divide it into two sub-blocks $\log_2(M_i)$ and $\log_2(M_q)$, where $M_iM_q=M$. We then use one sub-block to modulate a cosine carrier, which yields the in-phase channel. The modulation is a type of amplitude modulation that includes 180° phase reversal of the carrier. This means that D_{k1} in Eq. (9.28) can take on a range of both positive and negative values. We also use the second sub-block to similarly modulate a sine carrier, which constitutes the quadrature channel. The two modulated carriers are added together and simultaneously transmitted. Their sum corresponds to line one of Eq. (9.28) which represents a point in the M-ary QAM constellation and conveys $\log_2(M_i)+\log_2(M_q)=\log_2(M)$ bits. As explained in Section 9.2, the two channels are orthogonal, and this allows each sub-block of bits to be separately recovered at the detector to yield $\log_2(M)$ received bits per symbol interval T_s.

Fig. 9.10h depicts M-ary square QAM as the sum of two orthogonal channels, with M_i states in the in-phase channel and M_q states in the quadrature channel, where

$$
M = M_iM_q \qquad (9.29)
$$

For the best system performance in the presence of noise, an equitable loading of both channels must be approached as closely as possible. So the values of M_i and M_q are set as follows:

$$\begin{aligned} M_i &= M_q = \sqrt{M}, && \text{For } M \text{ an even power of 2} \\ M_i &= \sqrt{2M}; \quad M_q = \sqrt{M/2}, && \text{For } M \text{ an odd power of 2} \end{aligned} \tag{9.30}$$

Note that when M is an odd power of 2 (e.g. $M=8$, 32, 128, ...) it is not significant which channel carries more symbols. Eq. (9.30) assigns double the number of symbols to the in-phase channel, but this could have been assigned to the quadrature channel with absolutely no effect on system performance.

The in-phase channel has a one-dimensional signal space involving just one basis function $\alpha_0(t)$, given by Eq. (9.8). All M_i transmitted states are arranged as shown along the in-phase axis with equal spacing $2d$, so that the kth in-phase symbol $g_k(t)$ is given by

$$\begin{aligned} g_k(t) &= d(2k - M_i + 1)\alpha_0(t); \quad k = 0, 1, 2, 3, \ldots, M_i - 1 \\ &= d(2k - M_i + 1)\sqrt{2/T_s}\cos(2\pi f_c t)\text{rect}\left(\frac{t - \frac{1}{2}T_s}{T_s}\right) \end{aligned} \tag{9.31}$$

and has energy

$$E_k = d^2(2k - M_i + 1)^2 \tag{9.32}$$

It can be shown that the average energy per bit of the in-phase channel is given by

$$E_b = \frac{(M_i^2 - 1)d^2}{3\log_2(M_i)} \tag{9.33}$$

Similarly the quadrature channel has just one basis function $\alpha_1(t)$, given by Eq. (9.8), and all its M_q states are arranged as shown in Fig. 9.10g along the quadrature axis with equal spacing $2d$, so that the lth quadrature symbol $g_l(t)$ is given by

$$\begin{aligned} g_l(t) &= -d(2l - M_q + 1)\sqrt{2/T_s}\sin(2\pi f_c t)\text{rect}\left(\frac{t - \frac{1}{2}T_s}{T_s}\right) \\ l &= 0, 1, 2, 3, \ldots, M_q - 1 \end{aligned} \tag{9.34}$$

The energy of this symbol and the average energy per bit of the quadrature channel are given by expressions similar to Eqs. (9.32) and (9.33), respectively.

It can be shown (see Worked Example 9.5) that the average energy per bit E_b of the entire square M-ary QAM constellation is given by the simple expression

$$\begin{aligned} E_b &= \frac{2d^2(aM - 1)}{3\log_2 M} \\ \text{where, } a &= \begin{cases} 1, & M \text{ an even power of 2} \\ 1.25, & M \text{ an odd power of 2} \end{cases} \end{aligned} \tag{9.35}$$

We should emphasise that Eqs. (9.31) to (9.35) apply only to the arrangement of in-phase and quadrature states shown in Fig. 9.10h which leads to the very popular square-QAM constellation having a uniform two-dimensional grid of states shown in

Fig. 9.10d for $M = 16$. However, all types of M-ary QAM work on the same principle that during each symbol interval, the in-phase channel puts out a symbol $g_k(t)$ which represents a displacement D_k along the in-phase axis, and the quadrature channel puts out a symbol $g_l(t)$ corresponding to a displacement D_l along the quadrature axis. The sum of these two symbols corresponds to a point (D_k, D_l) in two-dimensional signal space. Since there are M_i possible values of D_k and M_q possible values of D_l, there will be $M = M_iM_q$ possible points, which defines the particular M-ary QAM constellation.

It is worth mentioning at this point that when dealing with a two-dimensional signal space employing a cosine pulse basis function for the $+x$-axis and a negative sine pulse for the $+y$-axis (as in PSK and QAM), a frequently used approach is to adopt complex numbers to represent the states. In this case, the x-axis is the real or in-phase axis and the y-axis is the imaginary (j) or quadrature axis. Thus the kth state S_k in M-ary PSK may be specified as

$$S_k = A_{kI} + jA_{kQ} \tag{9.36}$$

with A_{kI} and A_{kQ} given by Eq. (9.26).

Worked Example 9.5: Average Energy Per Bit of Square M-ary QAM

We wish to derive an expression for the average energy per bit of a square M-ary QAM system with constellation as shown in Fig. 9.10h, assuming all states are equally likely.

In solving this problem we will, in addition to Eq. (9.25), make use of the following relation for the sum of all positive integers up to n:

$$\sum_{k=0}^{n} k = 1 + 2 + 3 + \cdots + n = \frac{1}{2}n(n+1) \tag{9.37}$$

As shown in Fig. 9.11, the constellation is a matrix of M_q by M_i states with a uniform spacing $2d$ and with the signal space origin at the centre. Let us

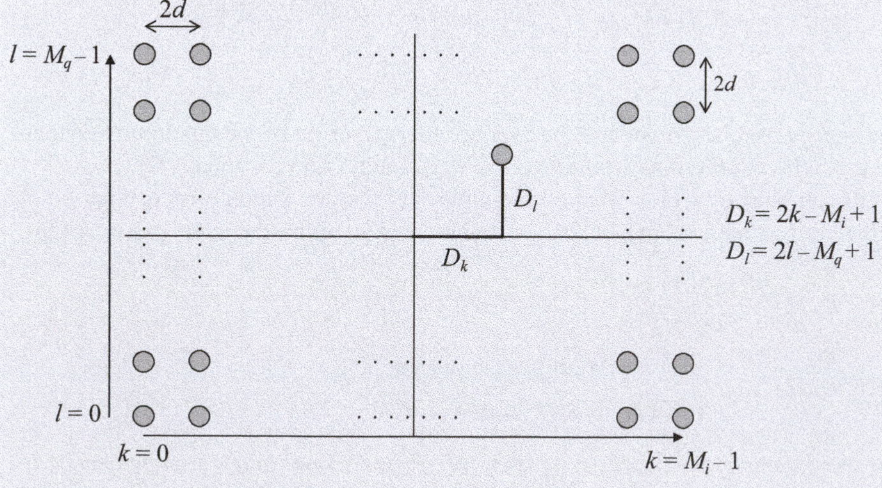

Fig. 9.11 Worked Example 9.5: Numbering scheme used

number the states from the bottom left corner, going rightward from $k=0$ to $k=M_i-1$, and upward from $l=0$ to M_q-1 to cover the entire constellation. The coordinates (D_k, D_l) of a state at location (k, l) and the energy E_{kl} of this state are given by

$$\begin{aligned}
D_k &= d(2k - M_i + 1) = -d(M_i - 1 - 2k) \\
D_l &= -d(M_q - 1 - 2l) \\
E_{kl} &= D_k^2 + D_l^2 = d^2[(M_i - 1 - 2k)^2 + (M_q - 1 - 2l)^2]
\end{aligned}$$

Averaging E_{kl} over all M_iM_q states yields the average energy per symbol E_s:

$$\begin{aligned}
E_s &= \frac{d^2}{M_iM_q}\sum_{k=0}^{M_i-1}\sum_{l=0}^{M_q-1}\left[(M_i - 1 - 2k)^2 + (M_q - 1 - 2l)^2\right] \\
&= \frac{d^2}{M_iM_q}\sum_{k=0}^{M_i-1}\left[M_q(M_i - 1 - 2k)^2 + \sum_{l=0}^{M_q-1}(M_q - 1 - 2l)^2\right] \\
&= \frac{d^2}{M_iM_q}\left[M_q\sum_{k=0}^{M_i-1}(M_i - 1 - 2k)^2 + M_i\sum_{l=0}^{M_q-1}(M_q - 1 - 2l)^2\right]
\end{aligned}$$

Making the substitution $n \equiv M_q - 1$, and $m \equiv M_i - 1$ yields

$$\begin{aligned}
E_s &= \frac{d^2}{M_iM_q}\left[M_q\sum_{k=0}^{m}(m - 2k)^2 + M_i\sum_{l=0}^{n}(n - 2l)^2\right] \\
&= \frac{d^2M_q}{M_iM_q}\left[(m+1)m^2 - 4m\sum_{k=0}^{m}k + 4\sum_{k=0}^{m}k^2\right] \\
&\quad + \frac{d^2M_i}{M_iM_q}\left[(n+1)n^2 - 4n\sum_{l=0}^{n}l + 4\sum_{l=0}^{n}l^2\right] \\
&= \frac{d^2M_q}{M_iM_q}\left[(m+1)m^2 - 4m\frac{1}{2}m(m+1) + \frac{4}{6}m(m+1)(2m+1)\right] \\
&\quad + \frac{d^2M_i}{M_iM_q}\left[(n+1)n^2 - 4n\frac{1}{2}n(n+1) + \frac{4}{6}n(n+1)(2n+1)\right]
\end{aligned}$$

Since $M_q(m+1) = M_qM_i$, and $M_i(n+1) = M_iM_q$, the common factor M_iM_q cancels out in the last step above to yield

$$\begin{aligned}
E_s &= d^2\left[m^2 - 2m^2 + \frac{2}{3}m(2m+1) + n^2 - 2n^2 + \frac{2}{3}n(2n+1)\right] \\
&= \frac{d^2}{3}[m(m+2) + n(n+2)] \\
&= \frac{d^2}{3}[(M_i - 1)(M_i + 1) + (M_q - 1)(M_q + 1)] \\
&= \frac{d^2}{3}\left[M_i^2 + M_q^2 - 2\right]
\end{aligned}$$

Using Eq. (9.30) to express M_i and M_q in terms of M gives

$$\begin{aligned} E_s &= \frac{d^2}{3}(M + M - 2), \quad M \text{ even integer power of 2} \\ &= \frac{2d^2}{3}(M - 1) \end{aligned}$$

and

$$\begin{aligned} E_s &= \frac{d^2}{3}\left(2M + \frac{M}{2} - 2\right), \quad M \text{ odd integer power of 2} \\ &= \frac{2d^2}{3}(1.25M - 1) \end{aligned}$$

Finally, dividing by $\log_2 M$ yields Eq. (9.35), the desired expression for average energy per bit in a square M-ary QAM constellation.

9.3.2.4 *M*-ary FSK symbols

M-ary FSK systems have an M-dimensional signal space involving the M basis functions given by Eq. (9.9) for $k = 1, 2, 3, \ldots, M$. The kth transmitted symbol of M-ary FSK is therefore given by

$$g_k(t) = d\sqrt{2/T_s}\cos[2\pi k(f_c/2)t]\text{rect}\left(\frac{t - \frac{1}{2}T_s}{T_s}\right) \tag{9.38}$$

$$k = 1, 2, 3, \ldots, M; \quad f_c = n/T_s$$

Fig. 9.10i shows the signal space diagram of binary FSK. The constellation of M-ary FSK is an M-dimensional Euclidean sphere of radius d centred at the origin, having M mutually orthogonal axes and one state on each axis at a distance d from the origin. Thus the average energy per bit of M-ary FSK is also given by Eq. (9.27).

9.4 Complex representation of signals

Most transmitted signals in modern communication systems are band pass signals, containing significant frequency components centred around a frequency $f_c > B$, where B is the bandwidth of the signal and f_c is referred to as the carrier frequency. To facilitate the simulation of these systems it is more convenient to represent the transmitted band pass signal $g(t)$ using an equivalent base band signal $\tilde{g}(t)$ which contains only frequencies centred around 0 Hz and which can therefore be accurately represented within a digital computer using samples taken at a rate $f_s \geq 2B$ (in order to satisfy the sampling theorem).

The transmitted band pass signal is usually the result of varying (i.e. modulating) the amplitude and/or angle of a sinusoidal signal (called the carrier signal) of frequency f_c by a base band or message signal $m(t)$. The most convenient form for $\tilde{g}(t)$ will therefore be complex, having imaginary and real parts which elegantly encapsulate both amplitude and phase variations. We will first introduce the Hilbert transform of signals, and then apply this to obtain the canonical and envelope forms of $\tilde{g}(t)$ for both transmitted signals and noise.

9.4.1 The Hilbert transform

Delaying (i.e. reducing the phase of) every frequency component of a signal $g(t)$ by 90° yields what is known as the Hilbert transform of the signal, denoted by $\hat{g}(t)$. It is worth emphasising at the outset that both $g(t)$ and $\hat{g}(t)$ are time-domain signals, and the Hilbert transform $\hat{g}(t)$ should not be confused with the FT $G(f)$, which gives the frequency domain representation of $g(t)$.

Writing $g(t)$ in terms of its Fourier series (see Appendix C)

$$g(t) = \sum_{n=0}^{\infty} A_n \cos(2\pi n f_o t + \phi_n), \quad \text{where } \phi_0 = 0°$$

it follows that the Fourier series of the Hilbert transform of $g(t)$ is given by

$$\hat{g}(t) = \sum_{n=0}^{\infty} A_n \cos(2\pi n f_o t + \phi_n - 90°) = \sum_{n=0}^{\infty} A_n \sin(2\pi n f_o t + \phi_n) \tag{9.39}$$

This equation represents the effect of the Hilbert transform operation on the single-sided spectrum: The phase of every positive frequency component is reduced by 90°, which changes every cosine component into sine. But the frequency domain is double-sided, comprising pairs of positive and negative frequency components, so we need to see the effect of this operation on the double-sided spectrum of $g(t)$. Recall that since $\cos(\theta) = \cos(-\theta)$, then

$$\cos(2\pi f_c t - 90°) = \frac{1}{2}\cos[2\pi(f_c)t - 90°] + \frac{1}{2}\cos[2\pi(-f_c)t + 90°]$$

Clearly then obtaining the Hilbert transform of $g(t)$ involves reducing the phase of every positive frequency component of $g(t)$ by 90° and increasing the phase of every negative frequency component by 90°. We already know that advancing the phase of a signal by 180° is equivalent to multiplying the signal by -1. What then is the effect of advancing the phase by 90°? Let us say that it is equivalent to multiplying the signal by a number denoted by j. Since increasing the phase twice by 90° equals a 180° increase, it follows that

$$j \times j = -1; \quad \text{or} \quad j = \sqrt{-1} \tag{9.40}$$

which means that reducing the phase of a signal by 90° is equivalent to multiplying the signal by $-j$, and increasing a signal's phase by 90° multiplies the signal by $+j$, where j denotes the imaginary number of unit amplitude. Note the following useful identities involving j:

$$\begin{aligned}
j^2 &= -1 \\
\frac{1}{j} &= \frac{1}{j} \times \frac{j}{j} = \frac{j}{j^2} = -j \\
j^3 &= j^2 \times j = -j \\
j^4 &= j^2 \times j^2 = -1 \times -1 = 1 \\
j^5 &= j^4 \times j = j
\end{aligned} \tag{9.41}$$

Returning to the problem at hand, we see that if $G(f)$ is the FT of $g(t)$, then the FT of $\hat{g}(t)$ — denoted by $\hat{G}(f)$ — is obtained by multiplying the positive-frequency region of $G(f)$ by $-j$ and the negative-frequency region by $+j$. That is

$$\hat{G}(f) = G(f) \times -j\,\mathrm{sgn}(f) \tag{9.42}$$

The RHS of the above equation makes use of the signum function introduced in Fig. 2.2b and multiplies two FTs, namely $G(f)$ which is the FT of $g(t)$, and $-j\,\mathrm{sgn}(f)$ which from Section C.2 (in Appendix C) is the FT of $1/\pi t$. Since multiplication in the frequency domain is equivalent to convolution in the time domain, it follows by taking the inverse FT of both sides of Eq. (9.42) that $\hat{g}(t)$ is the result of convolving $g(t)$ with $1/\pi t$, which is written mathematically as

$$\hat{g}(t) = \frac{1}{\pi}\int_{-\infty}^{\infty} \frac{g(\tau)}{t-\tau}\,d\tau \tag{9.43}$$

This equation is the formal definition of the Hilbert transform. Thus there are two ways to compute the Hilbert transform of $g(t)$. One method is indirect and follows the FT route suggested by Eq. (9.42): We determine the FT of $g(t)$, multiply this by $-j\,\mathrm{sgn}(f)$, and then obtain $\hat{g}(t)$ as the inverse FT of this product. The second is a more direct method where we simply evaluate the integral of Eq. (9.43). This is a convolution operation discussed in Chapter 2 and evaluated using a graphical approach in Worked Example 2.1.

Given the Hilbert transform $\hat{g}(t)$, the original signal $g(t)$ may be recovered from $\hat{g}(t)$ simply by removing the phase changes introduced, which amounts to obtaining the inverse Hilbert transform. This means multiplying the negative-frequency region of $\hat{g}(f)$ by $-j$ and the positive-frequency region by $+j$. In other words

$$G(f) = \hat{G}(f) \times j\,\mathrm{sgn}(f) \tag{9.44}$$

By a reasoning similar to that after Eq. (9.42), it follows that $g(t)$ is the result of convolving $\hat{g}(t)$ with $-1/\pi t$. Hence the inverse Hilbert transform of $\hat{g}(t)$ is given by the integral

$$g(t) = -\frac{1}{\pi}\int_{-\infty}^{\infty} \frac{\hat{g}(\tau)}{t-\tau}\,d\tau \tag{9.45}$$

The Hilbert transform (HT) has the following self-evident properties:

- A signal and its HT have the same amplitude spectrum.
- The HT of $\hat{g}(t)$ is $-g(t)$.
- $g(t)$ and $\hat{g}(t)$ are orthogonal over the entire time interval $(-\infty, \infty)$.

Fig. 9.12 shows the HT of various signals, namely the cosine waveform, triangular waveform, sinc pulse and the rectangular pulse. In each case the signal $g(t)$, its HT $\hat{g}(t)$ and the product $g(t)\hat{g}(t)$ are plotted. It can be seen that in all cases the total area of this product waveform is zero, which is an indication of the orthogonality between $g(t)$ and $\hat{g}(t)$ referred to above.

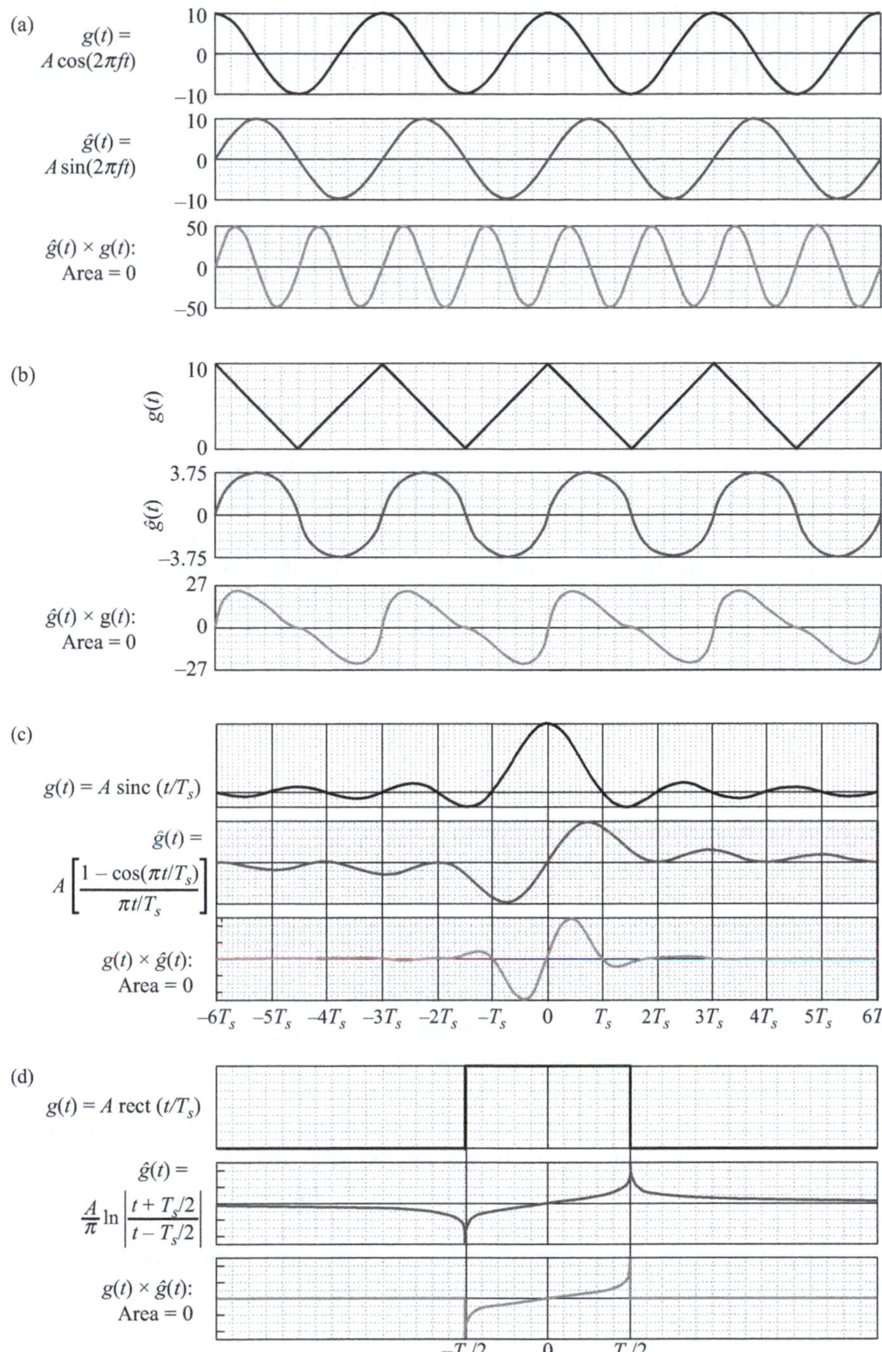

Fig. 9.12 Hilbert transform examples: (a) sinusoidal and (b) triangular signals (c) Sinc pulse and (d) rectangular pulse, and their HTs

9.4.2 Canonical and envelope representations

The pre-envelope or analytic signal of a real band pass signal $g(t)$ is a complex signal $g_+(t)$ whose real part is the signal $g(t)$ and whose imaginary part is the Hilbert transform of $g(t)$. That is

$$g_+(t) = g(t) + j\hat{g}(t) \tag{9.46}$$

Taking the FT of both sides and making use of Eq. (9.42) yields

$$\begin{aligned} G_+(f) &= G(f) + j\hat{G}(f) = G(f) + \text{sgn}(f)G(f) \\ &= \begin{cases} 2G(f), & f > 0 \\ G(0), & f = 0 \\ 0, & f < 0 \end{cases} \end{aligned} \tag{9.47}$$

where $G_+(f)$ is the FT of $g_+(t)$. This equation shows that the pre-envelope of a signal $g(t)$ results from suppressing the entire negative-frequency region of the signal's spectrum, doubling the amplitude of the positive-frequency region, but leaving the DC component $G(0)$ unchanged. Note that there will be no DC component if $g(t)$ is a band pass signal. Eq. (9.46) is the formal definition of the pre-envelope and provides a straightforward way to obtain $g_+(t)$ if the Hilbert transform $\hat{g}(t)$ is known. However, if $G(f)$ is known, then the frequency domain method suggested by Eq. (9.47) can be used, with $g_+(t)$ obtained by taking the inverse FT (FT^{-1}) of $G_+(f)$

$$\begin{aligned} g_+(t) &= \text{FT}^{-1}[G_+(f)] = \int_{-\infty}^{\infty} G_+(f)e^{j2\pi ft}df \\ &= 2\int_{0}^{\infty} G(f)e^{j2\pi ft}df \end{aligned} \tag{9.48}$$

These two methods are illustrated in Fig. 9.13. Note from Fig. 9.13b that $G_+(f)$ and $G(f)$ are identical in bandwidth B and shape — the triangular shape used here is merely illustrative, serving to show that the spectrum of the analytic signal $g_+(t)$ is a scaled and translated but undistorted version of the spectrum of the original signal $g(t)$. Thus the amplitude and phase variations, and hence information, contained in $g(t)$ are faithfully represented in $g_+(t)$. We can therefore obtain an equivalent base band representation of the band pass signal $g(t)$ by translating $G_+(f)$ to the left through f_c from its band pass location at $f = f_c$ to a base band location at $f = 0$. Denoting this equivalent base band signal as $\tilde{g}(t)$, we note from Appendix C that this frequency translation is accomplished in the time domain by multiplying $g_+(t)$ by $\exp(-j2\pi f_c t)$. That is

$$\begin{aligned} \tilde{g}(t) &= g_+(t)e^{-j2\pi f_c t} \\ &\equiv g_I(t) + jg_Q(t) \quad \text{Cartesian form} \\ &\equiv a(t)e^{j\phi(t)} \quad \text{Polar form} \end{aligned} \tag{9.49}$$

The spectral relationships among $g(t)$, $g_+(t)$ and $\tilde{g}(t)$ are illustrated in Fig. 9.14. The base band equivalent $\tilde{g}(t)$ is also known as the complex envelope of $g(t)$. Being a complex signal, $\tilde{g}(t)$ can be represented as shown in Eq. (9.49) in Cartesian form

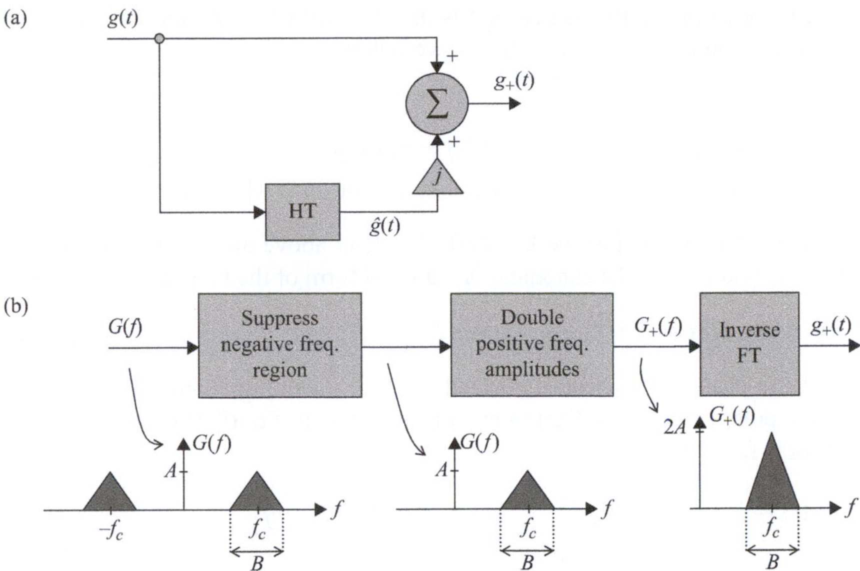

Fig. 9.13 Pre-envelope of g(t): (a) time-domain method; (b) frequency-domain method

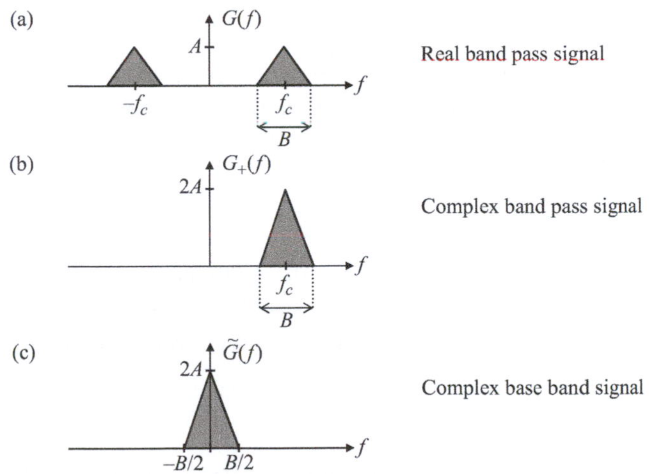

Fig. 9.14 Spectra of (a) band pass signal g(t); (b) pre-envelope of g(t); (c) complex envelope of g(t)

having a real part $g_I(t)$ and an imaginary part $g_Q(t)$, or in polar form with magnitude $a(t)$ and phase $\phi(t)$, where

$$\begin{aligned} a(t) &= \sqrt{g_I^2(t) + g_Q^2(t)} \\ \phi(t) &= \tan^{-1}\left[g_Q(t)/g_I(t)\right] \end{aligned} \tag{9.50}$$

The magnitude $a(t)$ is known as the natural envelope or simply envelope of $g(t)$, whereas $\phi(t)$ specifies the phase of $g(t)$.

Recalling from Eq. (9.46) that $g(t)$ is the real part of $g_+(t)$, and making $g_+(t)$ the subject of the first line of Eq. (9.49), we obtain

$$\begin{aligned} g(t) &= \mathrm{Re}\{g_+(t)\} \\ &= \mathrm{Re}\{\tilde{g}(t)e^{j2\pi f_c t}\} = \mathrm{Re}\{[g_I(t) + jg_Q(t)]e^{j2\pi f_c t}\} \\ &= \mathrm{Re}\{[g_I(t) + jg_Q(t)][\cos(2\pi f_c t) + j\sin(2\pi f_c t)]\} \end{aligned} \tag{9.51}$$

Expanding the terms within the last curly brackets above and taking its real part yields what is known as the canonical or standard form of the band pass signal $g(t)$:

$$g(t) = g_I(t)\cos(2\pi f_c t) - g_Q(t)\sin(2\pi f_c t) \tag{9.52}$$

$g_I(t)$ is the in-phase component of $g(t)$ and $g_Q(t)$ the quadrature component. Making use of the polar (rather than Cartesian) form for $\tilde{g}(t)$ in Eq. (9.51) yields an alternative form for $g(t)$:

$$\begin{aligned} g(t) &= \mathrm{Re}\{[a(t)e^{j\phi(t)}]e^{j2\pi f_c t}\} = \mathrm{Re}\{a(t)e^{j(2\pi f_c t + \phi(t))}\} \\ &= a(t)\cos[2\pi f_c t + \phi(t)] \end{aligned} \tag{9.53}$$

Eq. (9.53) is the envelope representation of the band pass signal $g(t)$. In this form it is easy to see that $g(t)$ results from a hybrid modulation of the amplitude and angle of a sinusoidal carrier $A_c\cos(2\pi f_c t + \phi_c)$ by a base band or message signal $m(t)$. It includes amplitude modulation, phase modulation and frequency modulation as special cases:

- *Amplitude modulation* (AM): The phase $\phi(t)$ of $g(t)$ is a constant, usually set to zero as a phase reference. Information is conveyed in the envelope $a(t)$ of $g(t)$, which varies in direct proportion to the message signal $m(t)$. That is

$$\begin{aligned} \phi(t) &= \phi_c \equiv 0 \\ a(t) &= A_c + k_a m(t) \end{aligned} \tag{9.54}$$

  where k_a is the amplitude sensitivity of the AM modulator in volt/volt.
- *Phase modulation* (PM): The envelope $a(t)$ is a constant A_c. Information is conveyed in the phase $\phi(t)$ which varies in direct proportion to $m(t)$. Thus

$$\begin{aligned} a(t) &= A_c \\ \phi(t) &= k_p m(t) \end{aligned} \tag{9.55}$$

  where k_p is the phase sensitivity of the phase modulator in rad/volt.
- *Frequency modulation* (FM): The envelope $a(t)$ is again a constant A_c. Information is conveyed in the variations of the instantaneous frequency f_i of the carrier, which varies from f_c in direct proportion to $m(t)$. A little thought will show that this causes the phase $\phi(t)$ to vary in direct proportion to the integral of $m(t)$. That is

$$\begin{aligned} a(t) &= A_c \\ f_i &= f_c + k_f m(t) \\ \phi(t) &= 2\pi k_f \int_0^t m(t)dt \end{aligned} \tag{9.56}$$

  where k_f is the frequency sensitivity of the frequency modulator in Hz/volt.

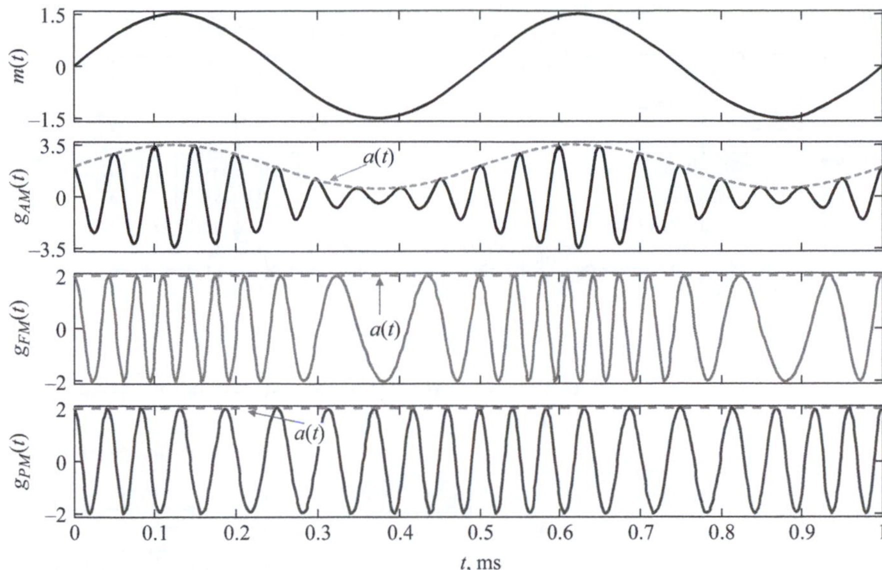

Fig. 9.15 Special cases of band pass signals $g_{AM}(t)$, $g_{FM}(t)$ and $g_{PM}(t)$ resulting respectively from AM, FM and PM modulation of a 20 kHz carrier by a 2 kHz sinusoidal message signal m(t)

Fig. 9.15 shows the above special cases of $g(t)$ for a simple sinusoidal message signal $m(t)$ of frequency 2 kHz and amplitude 1.5 V; carrier of frequency 20 kHz and amplitude $A_c = 2$ V; and modulation parameters $k_a = 1$ volt/volt, $k_p = \pi/2$ rad/volt, and $k_f = 8$ kHz/volt. The envelope $a(t)$ of the band pass signal is shown on each plot.

To conclude, note that we have discussed a number of important signals associated with a band pass signal $g(t)$, namely

- The signal $g(t)$, a *real band pass signal* of bandwidth B centred at f_c as shown in Fig. 9.14a.
- The *Hilbert transform* of $g(t)$, denoted by $\hat{g}(t)$. This is also a *real band pass signal* of the same bandwidth B and amplitude spectrum as $g(t)$.
- The *pre-envelope* or *analytic signal* of $g(t)$, denoted by $g_+(t)$. This is a *complex band pass signal* of bandwidth B, and spectrum as represented in Fig. 9.14b. Its real part is $g(t)$ and its imaginary part is $\hat{g}(t)$.
- The *complex envelope* or *base band equivalent* of $g(t)$, denoted by $\tilde{g}(t)$. This is a *complex base band signal* of bandwidth $B/2$ as shown in Fig. 9.14c. It is vital to note for system analysis and computer simulation purposes that the occupied bandwidth B_{bb} of a base band equivalent signal is always *half* the occupied bandwidth B_{bp} of the band pass signal that it represents:

$$B_{bb} = B_{bp}/2 \tag{9.57}$$

- The *in-phase component* of $g(t)$, denoted by $g_I(t)$. This is a *real base band signal* of bandwidth $B/2$. It is the *real part* of $\tilde{g}(t)$.
- The *quadrature component* of $g(t)$, denoted by $g_Q(t)$. This is a *real base band signal* of bandwidth $B/2$. It is the *imaginary part* of $\tilde{g}(t)$.

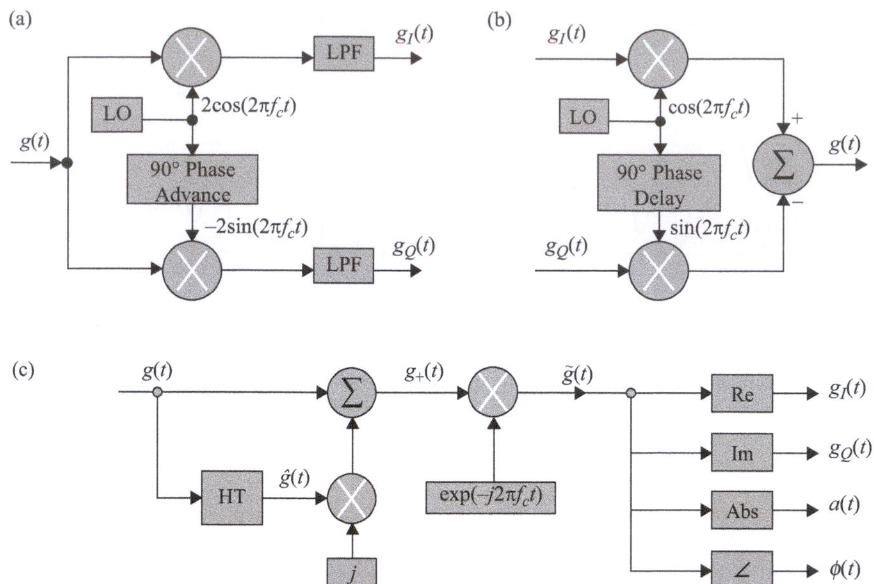

Fig. 9.16 (a) Obtaining the in-phase and quadrature components of a band pass signal g(t); (b) reconstructing a band pass signal g(t) from its in-phase and quadrature components; (c) computation steps

- The *envelope* of $g(t)$, denoted by $a(t)$. It is a *real base band signal* and is the *magnitude* of $\tilde{g}(t)$.
- The *phase* of $g(t)$, denoted by $\phi(t)$. This is also a *real base band signal* and is the *angle* of $\tilde{g}(t)$.

Fig. 9.16 summarises the relationships among these signals. Given a band pass signal $g(t)$, we can derive its in-phase and quadrature components using the routine shown in Fig. 9.16a. Conversely, given $g_I(t)$ and $g_Q(t)$ we can reconstruct $g(t)$ by the scheme shown in Fig. 9.16b. Once $g_I(t)$ and $g_Q(t)$ are known, the envelope $a(t)$ and phase $\phi(t)$ of $g(t)$ follow from Eq. (9.50), and the base band equivalent or complex envelope $\tilde{g}(t)$ of $g(t)$ is obtained using either the Cartesian or polar form of Eq. (9.49). Fig. 9.16c summarises suitable computation steps that can be employed to derive from $g(t)$ all of the signal components discussed above.

9.4.3 Application to noise

It is useful at this stage to express noise in a similar form to the results of the previous section. This will be useful when we later analyse the effect of noise on the symbol detection task of a receiver. A digital communication receiver usually consists of several processing stages, but the simple model shown in Fig. 9.17 will serve our purpose here of representing the noise signal at the detection point of a receiver. The band pass filter (BPF) is designed to pass the wanted signal which is usually a band pass signal of bandwidth B centred at frequency f_c, such as the signal $g(t)$ discussed above, and to prevent unwanted signals at other frequencies from reaching the detector. The receiver will however inevitably admit a portion of the additive white Gaussian noise (AWGN) $w(t)$ which is present at its input, so that at

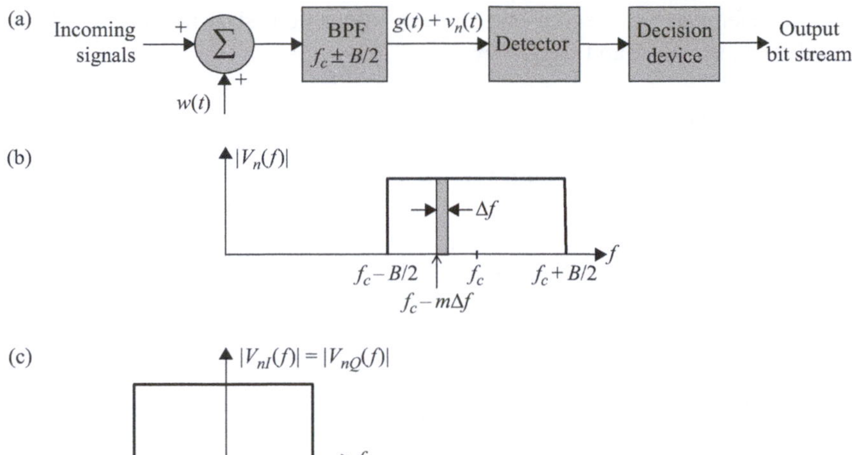

Fig. 9.17 (a) Simple digital communication receiver model showing white noise w(t) at receiver input, and wanted signal g(t) plus narrow band noise $v_n(t)$ at detection point; (b) amplitude spectrum of $v_n(t)$; (c) amplitude spectrum of in-phase and quadrature components of $v_n(t)$

the detection point there is not only the wanted signal $g(t)$ but also narrow band noise $v_n(t)$ of normalised power

$$P_n = N_o B \tag{9.58}$$

where N_o is noise power per unit bandwidth.

Clearly $v_n(t)$ is also a band pass signal of bandwidth B centred at frequency f_c, and therefore (from the previous Section) it has a base band equivalent

$$\tilde{v}_n(t) = v_{nI}(t) + jv_{nQ}(t) \tag{9.59}$$

as well as a canonical representation

$$v_n(t) = v_{nI}(t)\cos(2\pi f_c t) - v_{nQ}(t)\sin(2\pi f_c t) \tag{9.60}$$

where $v_{nI}(t)$ and $v_{nQ}(t)$ are the in-phase and quadrature components, derivable from the narrow band noise $v_n(t)$ using the scheme of Fig. 9.16. Note that since the processes involved in transforming the zero-mean Gaussian noise $w(t)$ into $v_n(t)$ in Fig. 9.17, and $v_n(t)$ into $v_{nI}(t)$ and $v_{nQ}(t)$ in Fig. 9.16 are all linear, it follows that the signals $v_n(t)$, $v_{nI}(t)$ and $v_{nQ}(t)$ are also Gaussian random signals with zero-mean. Since the mean-square-value (≡ normalised power) and variance of a signal are identical if the signal has zero mean, it follows that P_n in Eq. (9.58) is also the variance σ^2 of $v_n(t)$. So what is the variance, and hence normalised powers P_{nI} and P_{nQ}, of $v_{nI}(t)$ and $v_{nQ}(t)$? To answer this question let us take a closer look at the spectrum of $v_n(t)$.

Fig. 9.17b shows the (continuous) amplitude spectrum $|V_n(f)|$ of the band limited noise $v_n(t)$. Let us approximate $v_n(t)$ using a discrete Fourier series by dividing $|V_n(f)|$ into infinitesimally small frequency intervals Δf, and placing in each interval a sinusoid of power equal to the noise power $N_o\Delta f$ of that interval. Thus each sinusoid has amplitude

$$A = \sqrt{2N_o\Delta f} \tag{9.61}$$

Using f_c as reference, the spectral line of the mth sinusoid is shown in Fig. 9.17b and has random phase $\phi_m(t)$ and frequency

$$f_m = f_c - m\Delta f \tag{9.62}$$

So we may express $v_n(t)$ as a Fourier series

$$v_n(t) = \sum_{m=-M/2}^{M/2} A\cos[2\pi f_m t + \phi_m(t)] \tag{9.63}$$

The summation involves M sinusoids (for $m = -M/2, -M/2+1, \ldots, -2, -1, 1, 2, \ldots, M/2-1, M/2$) in the bandwidth B, where

$$\Delta f = \frac{B}{M} \tag{9.64}$$

Note in particular that $m \neq 0$. The approximation of Eq. (9.63) becomes exact as $M \to \infty$ and $\Delta f \to 0$. Substituting Eq. (9.62) into Eq. (9.63) and expanding yields

$$\begin{aligned} v_n(t) &= \sum_m A\cos[2\pi(f_c - m\Delta f)t + \phi_m(t)] \\ &= \left[\sum_{m=-M/2}^{M/2} A\cos(-2\pi m\Delta f t + \phi_m)\right]\cos(2\pi f_c t) \\ &\quad - \left[\sum_{m=-M/2}^{M/2} A\sin(-2\pi m\Delta f t + \phi_m)\right]\sin(2\pi f_c t) \end{aligned} \tag{9.65}$$

We see that this equation has the canonical form of Eq. (9.60) with the in-phase and quadrature components of $v_n(t)$ given by the expressions

$$\begin{aligned} v_{nI}(t) &= \sum_{m=-M/2}^{M/2} A\cos(-2\pi m\Delta f t + \phi_m) \\ v_{nQ}(t) &= \sum_{m=-M/2}^{M/2} A\sin(-2\pi m\Delta f t + \phi_m) \end{aligned} \tag{9.66}$$

The properties of $v_{nI}(t)$ and $v_{nQ}(t)$ are now more evident. We examine some of these below:

- Eq. (9.66) is a Fourier series with no DC component (since $m \neq 0$). Thus both $v_{nI}(t)$ and $v_{nQ}(t)$ have zero mean as earlier stated.
- From their Fourier series approximations given by Eq. (9.66), we see that $v_{nI}(t)$ and $v_{nQ}(t)$ have the same amplitude spectrum, shown in Fig. 9.17c, which is simply the amplitude spectrum of the band pass noise $v_n(t)$ translated from f_c to base band. However, because one is a cosine series whereas the other is a sine series, the phase spectra of $v_{nI}(t)$ and $v_{nQ}(t)$ differ by 90°.

- $v_{nI}(t)$ and $v_{nQ}(t)$ have exactly the same normalised power P_n ($=$variance σ^2) as the band pass noise $v_n(t)$. You can see this by noting that their Fourier series representations contain M sinusoids each of amplitude A. Thus

$$P_{nI} = P_{nQ} = \frac{A^2}{2}M = \frac{2N_o\Delta f}{2}M = N_oM\Delta f = N_oB$$

where we have made use of Eqs. (9.61) and (9.64). To emphasise, we note that

$$P_{nI} = P_{nQ} = P_n \equiv \sigma^2 = N_oB \tag{9.67}$$

Similarly to Eq. (9.53), $v_n(t)$ may also be expressed in envelope form as

$$v_n(t) = r(t)\cos[2\pi f_ct + \psi(t)] \tag{9.68}$$

where

$$\begin{aligned} r(t) &= \sqrt{v_{nI}^2(t) + v_{nQ}^2(t)} \\ \psi(t) &= \tan^{-1}\left[v_{nQ}(t)/v_{nI}(t)\right] \end{aligned} \tag{9.69}$$

The envelope $r(t)$ and phase $\psi(t)$ of $v_n(t)$ given above are random functions which result from non-linear operations, specified in Eq. (9.69), on two independent zero-mean Gaussian random signals of variance $\sigma^2 = N_oB$. The result of each processing is a new random signal that follows a non-Gaussian distribution. The probability density function (PDF) of $r(t)$ is given by

$$p(r) = \begin{cases} \dfrac{r}{\sigma^2}\exp\left(-\dfrac{r^2}{2\sigma^2}\right), & r \geq 0 \\ 0, & \text{elsewhere} \end{cases} \tag{9.70}$$

This equation defines what is known as a Rayleigh PDF. Thus the envelope of $v_n(t)$ is Rayleigh-distributed.

The phase $\psi(t)$ of v_n(t) is uniformly distributed between $-\pi$ and π *radians*. That is, the phase of narrow band noise $v_n(t)$ has a uniform PDF given by

$$p(\psi) = \begin{cases} \dfrac{1}{2\pi}, & -\pi \leq \psi \leq \pi \\ 0, & \text{elsewhere} \end{cases} \tag{9.71}$$

9.5 Signal correlation

The correlation coefficient ρ of two signals is a dimensionless number in the range from -1 to $+1$ which gives a measure of the similarity between the two signals. For two energy signals $g_0(t)$ and $g_1(t)$ of duration T_s and respective energies E_0 and E_1 we define ρ as

$$\rho = \frac{\int_0^{T_s} g_0(t)g_1(t)dt}{\text{Average energy}} = \frac{2\int_0^{T_s} g_0(t)g_1(t)dt}{E_0 + E_1} \tag{9.72}$$

If the two signals are periodic power signals of period T and respective powers P_0 and P_1 then their correlation coefficient is given by

$$\rho = \frac{\frac{1}{T}\int_{-T/2}^{T/2} g_0(t)g_1(t)dt}{\text{Average power}}$$

$$= \frac{2}{(P_0 + P_1)T}\int_{-T/2}^{T/2} g_0(t)g_1(t)dt \tag{9.73}$$

If the two power signals are nonperiodic then the above definition is applied in the limit $T \to \infty$. It is a straightforward matter to show that the correlation coefficient of two energy signals $g_k(t)$ and $g_m(t)$ represented as states in N-dimensional signal space may be expressed in terms of the components of the two signals along the N mutually perpendicular axes $\alpha_0, \alpha_1, \ldots, \alpha_{N-1}$ as follows:

$$\rho = \frac{2\sum_{n=0}^{N-1} s_{kn}s_{mn}}{\sum_{n=0}^{N-1} s_{kn}^2 + \sum_{n=0}^{N-1} s_{mn}^2} \tag{9.74}$$

where s_{kn} is the component of $g_k(t)$ — i.e. state S_k — along axis α_n.

It is worth pointing out that Eq. (9.74) is a very general expression for the correlation coefficient of two states S_k and S_m in N-dimensional signal space. However in the vast majority of cases the signal space is at most two-dimensional, and this equation will reduce to an expression for ρ which is more straightforward to apply. The signal space diagrams of all ASK, PSK and QAM systems as well as binary FSK systems fall in this category. Denoting the two signal space axes as x- and y-axes and components of state S_i along these axes as x_i and y_i respectively, the correlation coefficient of two states S_k and S_m is given by the expression

$$\rho = \frac{2\left(\overrightarrow{S_k} \cdot \overrightarrow{S_m}\right)}{\overrightarrow{S_k} \cdot \overrightarrow{S_k} + \overrightarrow{S_m} \cdot \overrightarrow{S_m}} = \frac{2(x_k x_m + y_k y_m)}{\left(x_k^2 + y_k^2\right) + \left(x_m^2 + y_m^2\right)} \tag{9.75}$$

where $\overrightarrow{S_i}$ is a position vector drawn in signal space from origin to the location of state S_i; and the operator $(\cdot)$ denotes scalar product.

To illustrate the use of these equations, we will consider the correlation coefficients among symbols in the digital modulation schemes shown in Fig. 9.10.

9.5.1 Correlation of M-ary ASK symbols

There is only one axis α_0. The components of the kth state S_k and m^{th} state S_m are $s_{k0} = kd$ and $s_{m0} = md$, respectively. Eq. (9.74) yields the correlation coefficient of the two states:

$$\rho = \frac{2km}{k^2 + m^2}, \quad k, m = 0, 1, 2, \ldots, M - 1 \tag{9.76}$$

Note that we can also obtain this result by employing the simpler Eq. (9.75) with $x_k = kd$, $x_m = md$, $y_k = 0$, and $y_m = 0$. It should be emphasised that Eq. (9.76) applies to all combinations of values of k and m, except $k = m = 0$ which involves comparing state S_0 with itself. In this case a direct use of Eq. (9.72) yields $\rho = 1$. All other self-comparisons also yield $\rho = 1$. When $m = k + 1$ then we are comparing adjacent states and these turn out to be the most positively correlated, except for S_0 and S_1 which have zero correlation. In fact S_0 has zero correlation with every other state.

9.5.2 Correlation of M-ary PSK symbols

In binary PSK there is only one axis α_0 with two states S_0 having component $s_{00} = d$, and S_1 having component $s_{10} = -d$. Or to use the simpler notation of Eq. (9.75), the state S_0 has components $x_0 = d$, $y_0 = 0$; and state S_1 has components $x_1 = -d$, $y_1 = 0$. The correlation coefficient of the two antipodal states is therefore

$$\rho = \frac{2(x_0 x_1 + y_0 y_1)}{(x_0^2 + y_0^2) + (x_1^2 + y_1^2)} = \frac{2(d)(-d)}{(d)^2 + (-d)^2} = \frac{-2d^2}{2d^2} = -1$$

Note that two antipodal signals always have a correlation coefficient $\rho = -1$. In the general M-ary PSK, there are two axes α_0 and α_1, and M states given by Eq. (9.26). In this case, any state S_i has the components

$$x_i = d\cos(\phi_i); \quad y_i = d\sin(\phi_i)$$

where $\phi_i = \theta_o + 2\pi i/M$ is the angle of the ith state. We can now apply Eq. (9.75) to obtain the correlation coefficient of states S_k and S_m:

$$\begin{aligned}\rho &= \frac{2[d^2\cos(\phi_k)\cos(\phi_m) + d^2\sin(\phi_k)\sin(\phi_m)]}{2d^2} \\ &= \cos(\phi_k)\cos(\phi_m) + \sin(\phi_k)\sin(\phi_m)\end{aligned}$$

Using trigonometric identity Eq. (B.4) in Appendix B and substituting $\phi_k = \theta_o + 2\pi k/M$ and $\phi_m = \theta_o + 2\pi m/M$ yields the final result

$$\rho = \cos(\phi_k - \phi_m) = \cos[2\pi(k - m)/M] \tag{9.77}$$

This states that the correlation coefficient of any two states in M-ary PSK equals the cosine of the angular separation of the states and does not depend on the angular offset θ_o of the constellation. It applies to all values of M including the binary case where it yields $\rho = -1$ as expected from the previous result. Since the cosine of smaller angles is larger, we see that adjacent states (for which $|k - m| = 1$) will be the most positively correlated, and the extent of this correlation will increase as M increases. For example, it follows from Eq. (9.77) that the correlation coefficient of adjacent states is $\rho = -1$ in BPSK; $\rho = 0$ in QPSK; $\rho = 0.7071$ in 8-PSK; and $\rho = 0.9239$ in 16-PSK.

9.5.3 Correlation of M-ary FSK symbols

In binary FSK there are two axes which we identify as α_0 and α_1 for notational consistency. And there are two states

$$\begin{aligned}S_0 &\equiv s_{00}\alpha_0 + s_{01}\alpha_1 = d\alpha_0 + 0\alpha_1 \\ S_1 &\equiv s_{10}\alpha_0 + s_{11}\alpha_1 = 0\alpha_0 + d\alpha_1\end{aligned}$$

Eq. (9.74) yields the correlation coefficient of the two orthogonal states as

$$\rho = \frac{2[s_{00}s_{10} + s_{01}s_{11}]}{s_{00}^2 + s_{01}^2 + s_{10}^2 + s_{11}^2} = \frac{2[d \times 0 + 0 \times d]}{d^2 + 0^2 + 0^2 + d^2} = 0$$

A similar procedure will show that the correlation coefficient of any two states in the M-ary FSK scheme of Eq. (9.38) is zero. This was earlier verified in Eq. (9.10). In general, the correlation coefficient of two orthogonal energy signals is zero. Thus

$$\rho = 0, \quad \text{for } M\text{-ary FSK} \tag{9.78}$$

The M symbols of M-ary FSK are sinusoids (called carriers) at unique frequencies located at a regular frequency spacing of Δf. The result of Eq. (9.78) is for the situation where the carriers are orthogonal. To determine what frequency spacing Δf guarantees this orthogonality and gain some insight into the correlation coefficient that will be in force at other non-orthogonal frequency spacings, let us consider any two of the carriers $g_1(t)$ and $g_2(t)$ (normalised in amplitude for unit energy) and separated in frequency by k spacings, $k = 1, 2, 3, \ldots$. The two carriers are

$$\left.\begin{aligned} g_1(t) &= \sqrt{2R_s}\cos(2\pi f_1 t) \\ g_2(t) &= \sqrt{2R_s}\cos(2\pi f_2 t) \end{aligned}\right\}; \quad 0 \le t \le \frac{1}{R_s}, \quad f_2 = f_1 + k\Delta f$$

with correlation coefficient

$$\begin{aligned} \rho &= 2R_s \int_0^{1/R_s} \cos(2\pi f_1 t)\cos(2\pi f_2 t)dt \\ &= R_s\left[\int_0^{1/R_s} \cos 2\pi(f_1 + f_2)t dt + \int_0^{1/R_s} \cos 2\pi(f_2 - f_1)t dt\right] \\ &= \frac{\sin[2\pi(2f_1 + k\Delta f)/R_s]}{2\pi(2f_1 + k\Delta f)/R_s} + \frac{\sin(2\pi k\Delta f/R_s)}{2\pi k\Delta f/R_s} \\ &= \text{sinc}(2k\Delta f/R_s) + \text{sinc}[2(2f_1 + k\Delta f)/R_s] \end{aligned}$$

Notice that now ρ is no longer guaranteed to be zero. Neither is it guaranteed to be the same for all pairs of carriers, rather it depends on f_1 (the reference carrier frequency), Δf the carrier spacing, R_s the symbol rate, and k the separation count (where $k = 1$ indicates nearest neighbour, $k = 2$ indicates next nearest neighbour etc.). To focus on the effect of k and Δf, let us express both f_1 and Δf in units of symbol rate so that $f_1 \equiv nR_s$, and $\Delta f = \beta R_s$. Thus

$$\left.\begin{aligned} \rho &= \text{sinc}(2k\beta) + \text{sinc}(4n + 2k\beta) \\ \beta &= \Delta f/R_s, \quad n = f_1/R_s \end{aligned}\right\} \quad (M\text{-ary FSK}) \tag{9.79}$$

This result is plotted in Fig. 9.18 for four neighbours $k = 1, 2, 4, 8$; and for $f_1 = 3R_s$ (i.e. $n = 3$). Note that n must be at least 0.5 since carrier frequency usually exceeds bandwidth B (and in any case must be at least $B/2$), and we know from our discussion in Chapter 8 that band pass channel bandwidth $B \ge R_s$. We see from Fig. 9.18 that the smallest frequency spacing that guarantees zero correlation or orthogonality among all pairs of M-ary FSK symbols is $\beta = 0.5$ or $\Delta f = R_s/2$. To see

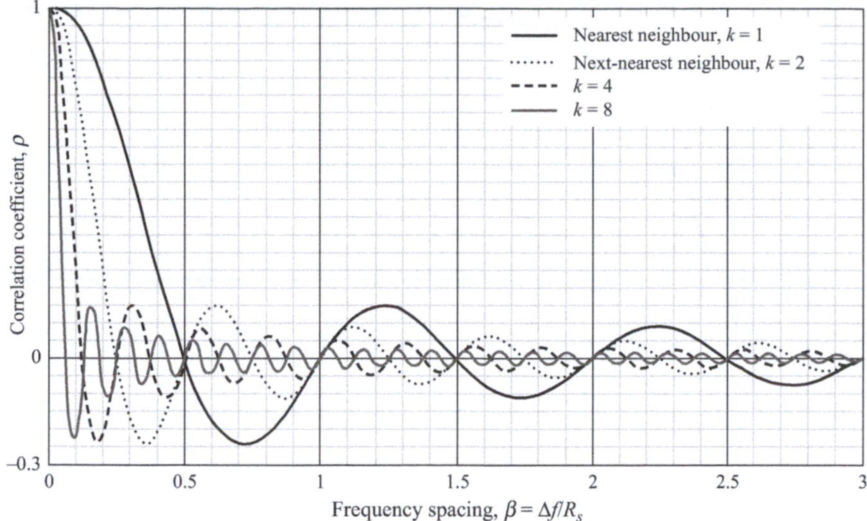

Fig. 9.18 Correlation coefficient between an M-ary FSK symbol at carrier frequency $f_1 = 3R_s$ and four neighbours (at frequencies $f_1 + k\Delta f$, for $k = 1, 2, 4, 8$) plotted against carrier frequency spacing Δf expressed in multiples of R_s

this, follow the curve for $k = 1$ and note that it first crosses zero at $R_s/2$ and then subsequently at intervals of $R_s/2$. So if the first symbol frequency is f_1, the other $M - 1$ symbols would have frequencies $f_2 = f_1 + \Delta f$, $f_3 = f_1 + 2\Delta f$, $f_4 = f_1 + 3\Delta f$, and so on, which would all be zero-correlated if $\Delta f = R_s/2$. Notice also that below $\Delta f = R_s/2$, the symbol at frequency f_1 becomes increasingly positively correlated with the symbol at frequency $f_2 = f_1 + \Delta f$.

An FSK scheme that employs this minimum frequency spacing of $\Delta f = R_s/2$ as well as having continuous phase is given the special name minimum shift keying (MSK). You may also notice from Fig. 9.18 that certain values of Δf give negative correlation (which is even better than zero correlation in terms of noise performance). For example, at $\Delta f = 0.718R_s$, $\rho = -0.24$. However, be careful to note that this correlation coefficient value is only applicable between f_1 and its nearest neighbour f_2, so this spacing would suit binary FSK that only has two symbols at frequencies f_1 and f_2. When there are more than two symbols (i.e. $M \geq 4$) then other pairs of symbols will have a different correlation coefficient value than –0.24 at this frequency spacing ($\Delta f = 0.718R_s$); e.g. correlation with the next-nearest neighbour ($f_3 = f_1 + 2\Delta f$) is $\rho = 0.052$. In general if the M-ary FSK system is non-orthogonal (i.e. $\Delta f \neq R_s/2$ or its integer multiples), the correlation coefficient between any pair of symbols will tend to zero in an oscillatory manner as the frequency separation between the two symbols increases. In what follows we assume an orthogonal M-ary FSK implementation in which all pairs of symbols have zero correlation.

9.5.4 Impact of symbol correlation

The susceptibility of a digital transmission to noise (assumed AWGN) is controlled by two factors, namely the spacing of adjacent states in the transmission's signal space and their correlation coefficient. The first factor relates to average energy E_s per received symbol. For a given E_s, symbol error (i.e. misidentification) is more likely as ρ increases from -1 towards $+1$. As an interesting thought, if you

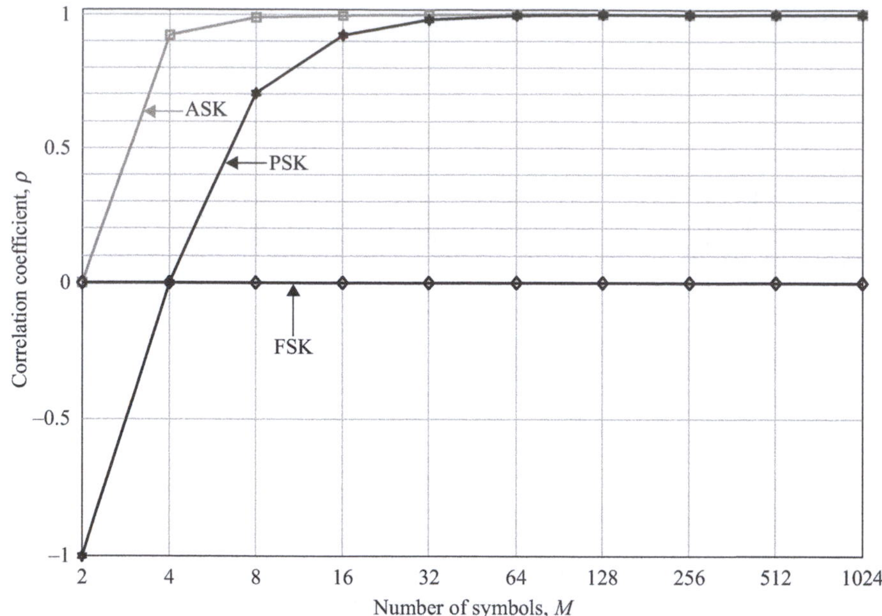

Fig. 9.19 Correlation coefficient of adjacent symbols in M-ary ASK, M-ary PSK and orthogonal M-ary FSK

designed a system that employs M identical symbols to represent each of the $\log_2 M$ groups of bits, ρ would be $+1$, and the probability of bit error, called bit error rate (BER) would be ½ irrespective of E_s or the level of transmitted power. The receiver would do just as well to switch off the detection circuitry and base its decisions purely on random guesses.

Fig. 9.19 shows the correlation coefficient of adjacent symbols in M-ary ASK, PSK and FSK transmissions for values of M from 2 to 1024. Note that this is simply a plot of Eqs. (9.76)–(9.78). The M-ary ASK correlation shown is for the outermost two states corresponding to the two symbols transmitted with the largest and second largest amplitudes respectively, i.e. $k = M - 1$ and $m = M - 2$. These are the most positively correlated states in the linear M-ary ASK constellation. Subject to this clarification, we see that the correlation of adjacent ASK symbols noticeably exceeds that of adjacent PSK symbols for all $M \leq 32$, and only negligibly for larger M. It is for this reason that, for a given average energy per transmitted symbol, M-ary PSK has a lower BER than the corresponding M-ary ASK. Furthermore, it can be seen that the correlation of (adjacent) ASK and PSK symbols increases rapidly towards $+1$ as M increases, with $\rho > 0.99$ for $M \geq 64$ in both systems, whereas $\rho = 0$ for all M in FSK transmission. An important practical consequence of this feature will be brought out in the next section.

9.6 Coherent detection

It was noted in Section 1.3.7 (see Fig. 1.21 and the discussion surrounding it) that a coherent demodulator, matched filter and correlation receiver (or simply correlator) are in fact equivalent. We then presented a mathematical proof of the equivalence between the matched filter and the correlator in Section 8.4.3. In the coherent demodulation of analogue signals the aim is to obtain a faithful reproduction of the original analogue waveform, whereas in digital communications the aim is to obtain a reliable indication of whether or not a matched pulse is present at each decision instant. For this reason it is more appropriate to use the term coherent demodulation when referring to analogue receivers but coherent detection in digital communications. Coherent detection provides the best immunity to noise when compared to non-coherent methods, but it requires accurate knowledge of the phase and frequency of the incoming pulse. Any phase error leads to a reduction in correlator output which reduces the noise margin from the value derived in the discussion that follows. Frequency and phase information is usually extracted from the incoming signal using a clock extraction circuit and this makes coherent detectors more complex than their non-coherent counterparts which do not require phase synchronisation.

Our focus below will be on coherent detection using a bank of correlators. A highly readable treatment of coherent and non-coherent demodulation, phase synchronisation circuitry, and the impact of phase error on coherent receivers can be found in [1], whereas a more detailed discussion of digital modulation and detection techniques than we have space for in this book is given in [2].

9.6.1 Correlation receiver

Signal correlation finds extensive applications in digital communications. One incoming symbol $g_{in}(t)$ is received during each interval T_s, and the correlation receiver — a multiplier followed by an integrator (see Fig. 9.20a) — simply computes

$$\int_0^{T_s} g_{in}(t) g_k(t) dt \equiv y_o(T_s) \tag{9.80}$$

(a)

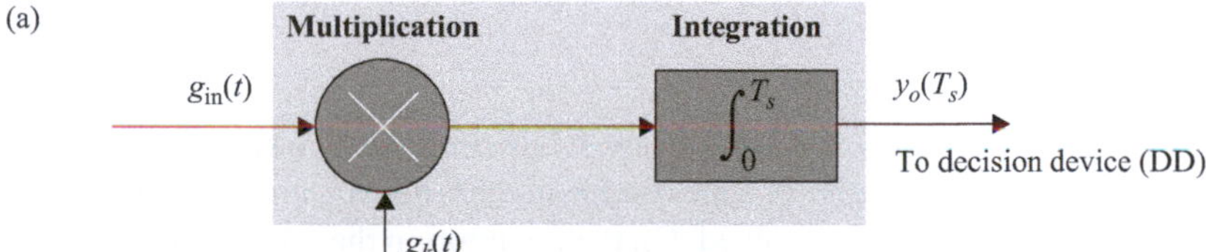

(b)

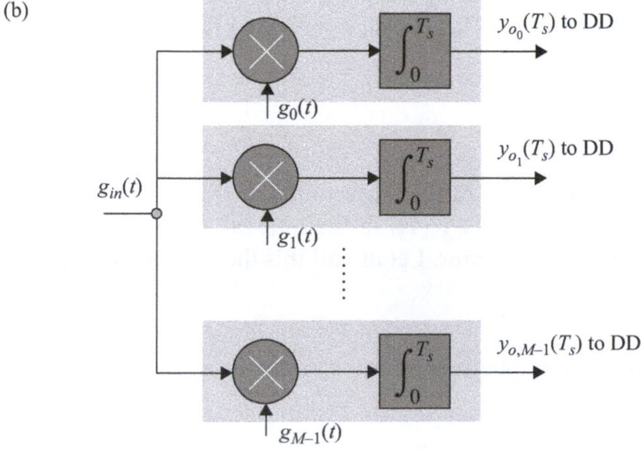

Fig. 9.20 Correlation receiver: (a) binary system; (b) M-ary system

If the incoming symbol is $g_k(t)$, then $y_o(T_s)$ equals E_k, the energy of symbol $g_k(t)$. If however $g_{in}(t)$ is some other symbol say $g_m(t)$ of energy E_m, where $k \neq m$, then the output is proportional to the correlation coefficient of $g_k(t)$ and $g_m(t)$. Specifically

$$y_o(T_s) = \begin{cases} E_k, & \text{if } g_{in}(t) = g_k(t) \\ \rho E_s = \rho(E_k + E_m)/2, & \text{if } g_{in}(t) = g_m(t) \end{cases} \tag{9.81}$$

where E_s is the average energy of symbols $g_k(t)$ and $g_m(t)$, ρ is their correlation coefficient, and we have made use of the definition of ρ in Eq. (9.72). This computation result may therefore be fed into a decision device with a decision threshold E_{th}. If the result exceeds E_{th} the receiver decides that $g_k(t)$ (and hence the bits it represents) was transmitted. If the result is below E_{th} the receiver decides that the incoming symbol is *not* $g_k(t)$. In a binary system this negative decision would of course mean a positive decision for the only other symbol that could have been transmitted; but in a multilevel ($M > 2$) system, there are other possibilities and a bank of M correlators shown in Fig. 9.20b is required, each branch checking for one of the M symbols or states $S_0, S_1, \ldots, S_{M-1}$. In this case the 'other' symbol $g_m(t)$ used in Eq. (9.81) to set the decision threshold for the kth branch is the one that lies adjacent to S_k in signal space. In orthogonal M-ary FSK all states are mutually adjacent in the sense that they are equally correlated (with $\rho = 0$) so the decision is very much simplified: The incoming symbol is identified as belonging to whichever branch of the bank of M correlators has the largest output. That is

if

$$y_{oj}(T_s) > y_{ok}(T_s) \quad \text{for all } k = 0, 1, \ldots, M-1; \quad j \neq k \qquad (M\text{-ary FSK}) \tag{9.82}$$

then

$$g_{in}(t) = g_j(t)$$

But for M-ary ASK, M-ary PSK and M-ary APSK in general where the correlation coefficient varies among pairs of states and is highest between neighbouring states, the decision is slightly more involved. On the assumption that all the M symbols are equally likely to be transmitted, the maximum likelihood rule dictates that the decision threshold lies halfway between E_k and $\rho(E_m + E_k)/2$, the respective outputs when it is $g_k(t)$ and when it is $g_m(t)$. This decision threshold is shown in Fig. 9.21. Thus

$$E_{th} = \frac{\rho + 2}{4} E_k + \frac{\rho}{4} E_m \tag{9.83}$$

The gap $E_k - E_{th}$ gives an indication of the protection that the system has against noise. Noise contribution to the output $y_o(T_s)$ in the interval T_s has to exceed this amount before it can lead to symbol error. Let us call this the noise margin E_{NM}:

$$E_{NM} = E_k/2 - \rho(E_m + E_k)/4 \tag{9.84}$$

In M-ary PSK and FSK, all symbols have the same energy E_s and the above expressions simplify to

$$E_{th} = (1 + \rho)\frac{E_s}{2}; \quad E_{NM} = (1 - \rho)\frac{E_s}{2} \tag{9.85}$$

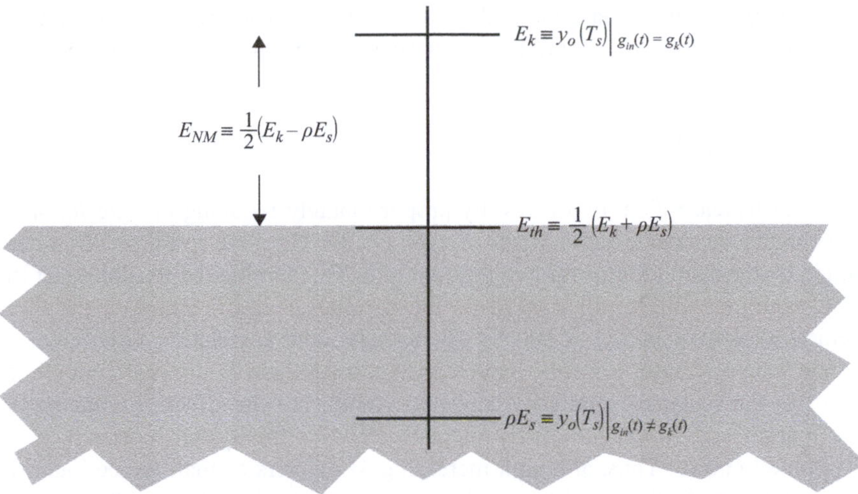

Fig. 9.21 Decision threshold E_{th} and noise margin E_{NM} applicable to the detection (by a correlation receiver) of state S_k corresponding to symbol $g_k(t)$ of energy E_k. $E_s = (E_k + E_m)/2$, where E_m is the energy of an adjacent state S_m, and ρ is the correlation coefficient of S_k and S_m

For example, in binary FSK (Fig. 9.10i) $\rho = 0$ and therefore $E_{th} = E_s/2$ and $E_{NM} = E_s/2$. But in binary PSK (Fig. 9.10b), $\rho = -1$ so that $E_{th} = 0$ and $E_{NM} = E_s$, which is the biggest margin achievable and is double that of binary FSK. Eq. (9.85) indicates that for a fixed average energy per symbol E_s the noise margin decreases towards zero as ρ approaches $+1$, and the effect of this increasingly positive correlation is to increase the likelihood of the detection threshold being crossed due to contributions from noise.

9.6.2 Noise rejection in M-ary FSK

An important strength of the correlation-aided detection process discussed above is that the effect of AWGN — which inevitably accompanies the incoming symbol $g_{in}(t)$ — is averaged out by the integrator. You will recall from the discussion in Section 8.4.3 that the correlation receiver is an implementation of the matched filter that optimises signal detection in the presence of noise. In fact since the mean value of AWGN tends to zero as the averaging time increases, we can reduce the contribution of AWGN to the output $y_o(T_s)$ of the correlation receiver by increasing T_s. In the limit $T_s \to \infty$ it is clear that $y_o(T_s)$ and hence the detection process is unaffected by noise.

It is instructive to also take a frequency domain view of how the effect of noise is reduced through increasing T_s. Recall that T_s is the duration of each transmitted symbol. By virtue of the inverse relationship between time and frequency parameters, it follows that as T_s is increased the bandwidth B of the transmitted symbol reduces by the same factor. The frequency opening or receiver bandwidth B needed to pass each symbol thereby reduces, and hence the noise power P_n admitted by the receiver in the process of passing each symbol through to detection point reduces proportionately since $P_n = N_oB$. In summary, we see that when $T_s \to \infty$, then $B \to 0$ and $P_n \to 0$, which minimises the effect of noise on the symbol detection process.

But how does an increase in T_s affect other aspects of our transmission system? To answer this question we note that T_s is related to bit duration T_b and hence bit rate $R_b = 1/T_b$ by

$$T_s = T_b \log_2 M = \frac{\log_2 M}{R_b} \tag{9.86}$$

So we can increase T_s in two ways: by proportionately reducing bit rate R_b or by exponentially increasing the number of symbols M. The first option is often constrained by the need to transmit at a high bit rate. To examine the feasibility of the second option recall the effect of M on the correlation coefficient ρ of the three schemes, as shown in Fig. 9.19. As M increases, the correlation coefficient of adjacent states increases towards +1 in both M-ary PSK and M-ary ASK. Eq. (9.85) shows that this increased similarity between symbols has the effect of reducing the noise margin and will significantly increase BER as the system becomes more susceptible to noise. Thus, although increasing M delivers a small (since logarithmic) increase in T_s, the benefit gained is small compared to the penalty incurred from the resulting higher symbol correlations.

Orthogonal M-ary FSK is however very different since $\rho = 0$ for all M. So we can increase M indefinitely without any penalty of increased symbol correlation. And in this way the BER of M-ary FSK actually decreases as M increases. But this increased robustness to noise comes at a price. With M carriers at a minimum spacing of $R_s/2$, the minimum transmission bandwidth of M-ary FSK is

$$B_{FSK} = \frac{M+1}{\log_2 M}\frac{R_b}{2} \tag{9.87}$$

which increases with M. For example, the bandwidth of 1024-FSK is 41 times the bandwidth of 4-FSK. We conclude that M-ary FSK allows us to trade bandwidth for an improved noise performance in a way that is not possible with M-ary PSK and M-ary ASK. This unique feature of M-ary FSK makes it a very useful technique in deep space communication where bandwidth is plentiful (making bandwidth efficiency an unimportant issue) and the design priority is the reliable detection of extremely weak signals in the presence of noise.

9.7 Digital modulators

Block diagrams that generate the M-ary ASK, PSK, QAM and FSK signals discussed in Section 9.3.2 are introduced in this section, including the QPSK and binary modulators as special cases.

9.7.1 M-ary ASK modulator

Fig. 9.22a shows a block diagram of an M-ary ASK modulator. The serial to parallel converter takes $k = \log_2 M$ bits from the input bit stream and presents them at a time (i.e. simultaneously and in parallel) to the M-ary unipolar coder. The k bits cover k bit intervals over a duration $T_s = kT_b$, where T_b = bit interval and T_s is the duration of each pulse generated by the coder. The unipolar coder generates rectangular pulses, ideally impulses, having M different weights corresponding to normalised amplitudes

$$A_i = \{0, \quad 1/(M-1), \quad 2/(M-1), \quad 3/(M-1), \ldots, \quad 1\}$$

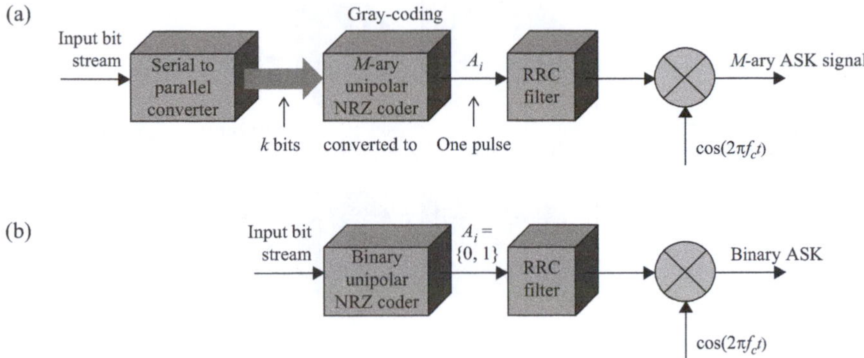

Fig. 9.22 M-ary ASK modulator: (a) general case M; (b) special case, M = 2

in response to each k-bit group of bits taken from the input stream. The amplitudes are arranged in gray code order as shown in Fig. 8.3 to ensure that adjacent levels represent k bits that differ in only one bit position. For example, if $M = 16$, then $A_i = 0$ for input 0000, 1/15 for 0001, 2/15 for 0011, 3/15 for 0010 and so on. The generated pulses, each of duration T_s, are filtered by a square root raised cosine (RRC) filter as discussed in Chapter 8 before multiplying a sinusoidal carrier of frequency f_c. Thus, apart from the filtering operation, this circuit generates the M-ary pulses exactly as specified in Eq. (9.23). The special binary case is shown in Fig. 9.22b, where the coder is the binary unipolar NRZ coder discussed in Chapter 7. There are only two levels, $A_i = 0$ which turns the carrier off and $A_i = 1$ which turns the carrier on, giving the so-called on-off keying (OOK) signal.

9.7.2 M-ary PSK modulator

Fig. 9.23a shows a suitable block diagram for M-ary PSK generation based on quadrature modulation involving two branches that are combined to yield the M-ary PSK signal. In the upper branch the bipolar coder pulses are filtered before being multiplied by a cosine carrier of frequency f_c. This branch is referred to as the in-phase channel. In the lower branch, called the quadrature channel, the filtered pulses are multiplied by a negative sine carrier of the same frequency as the carrier in the in-phase branch, but having a phase lead of 90°. This is consistent with the structure of an M-ary PSK signal which has a two-dimensional signal space based upon the cosine and sine basis functions given in Eq. (9.8). Each state of the signal consists of in-phase and quadrature components of amplitudes A_I and A_Q specified in Eq. (9.26). The serial to parallel converter works as previously described, feeding $k = \log_2 M$ bits in each pulse interval to the M-ary bipolar coder in both branches.

Assuming constellation angular offset $\theta_o = 0$ (its value does not affect transmission system performance) in Eq. (9.26), the bipolar coders generate pulses of normalised amplitudes A_I and A_Q whose values depend on the k-bit input during the symbol interval and are selected to achieve a gray code arrangement of states in the circular constellation. The values of A_I and A_Q for $M = 8$ are tabulated in Fig. 9.23b along with the resulting constellation. Note that a characteristic of M-ary PSK is

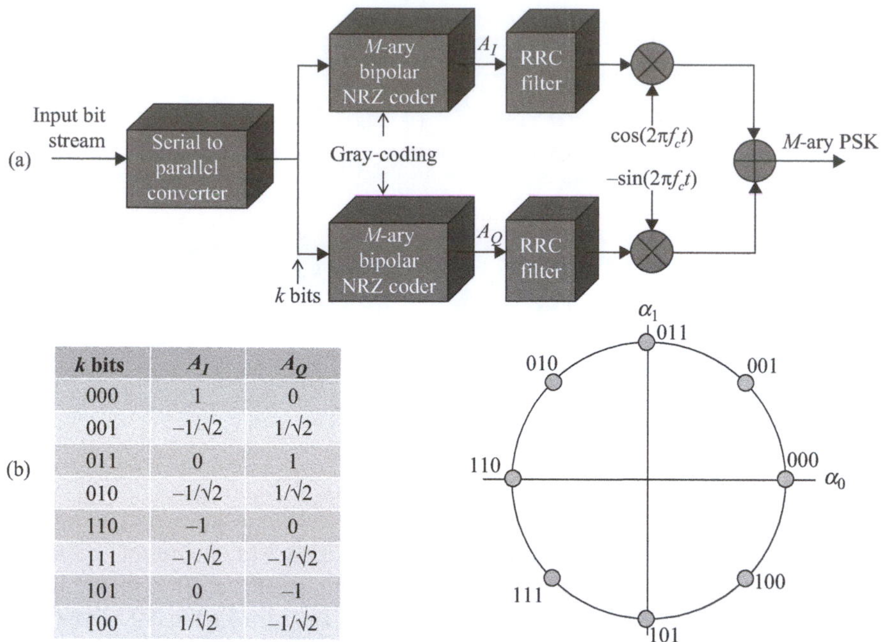

k bits	A_I	A_Q
000	1	0
001	$-1/\sqrt{2}$	$1/\sqrt{2}$
011	0	1
010	$-1/\sqrt{2}$	$1/\sqrt{2}$
110	−1	0
111	$-1/\sqrt{2}$	$-1/\sqrt{2}$
101	0	−1
100	$1/\sqrt{2}$	$-1/\sqrt{2}$

Fig. 9.23 (a) M-ary PSK modulator; (b) example for M = 8: normalised amplitudes of A_I and A_Q and resulting constellation

that the sum of the squares of A_I and A_Q has unit value (normalised) for all k-bit inputs. That is

$$\sqrt{A_I^2 + A_Q^2} = \text{Constant} \tag{9.88}$$

This is essential to ensure that states are all located on a circle centred at the origin in signal space.

When $M = 4$, the block diagram simplifies to that shown in Fig. 9.24. Now the coder is a single binary bipolar coder, which ideally generates a positive impulse $\delta(t)$ for binary 1 and a negative impulse $-\delta(t)$ for binary 0. The serial to parallel converter takes two bits (i.e. bipolar pulses) at a time and connects the first to the in-phase branch and the second to the quadrature branch. The waveforms generated at each point along the block diagram are shown. For simplicity the effect of filtering is not reflected in the sketched waveforms. Notice how the in-phase channel produces two states at 0° and 180°, the quadrature channel produces two states at −90° and 90°, and the sum of the two channels results in four states at 45°, 135°, −135° and −45°, which correspond to the QPSK constellation also shown in Fig. 9.24.

9.7.3 *M-ary QAM modulator*

The M-ary QAM modulator works on exactly the same principle as M-ary PSK using two branches with orthogonal (cosine and negative sine) carriers at the same frequency. The only difference is Eq. (9.88) is now relaxed so that the amplitude levels A_I and A_Q produced by the two M-ary bipolar NRZ coders in response to each k-bit input can take on the pair of values necessary to generate the desired constellation. Fig. 9.25 shows the M-ary QAM modulator along with tabulated

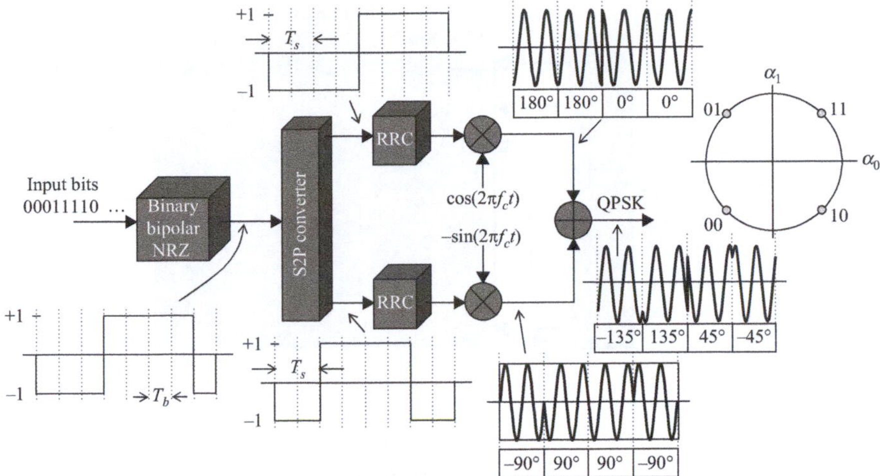

Fig. 9.24 QPSK modulator, waveforms at each processing stage (excluding filter effect) and signal constellation

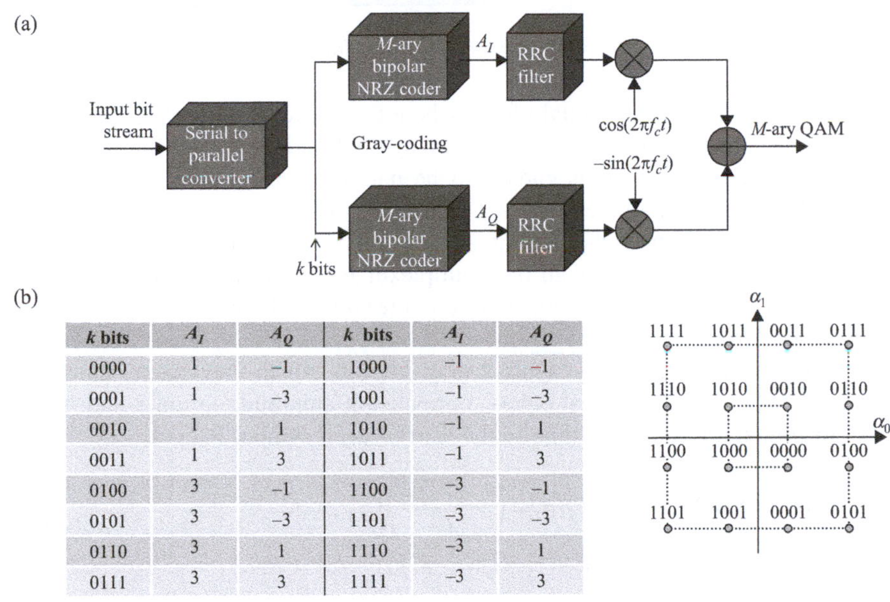

k bits	A_I	A_Q	k bits	A_I	A_Q
0000	1	–1	1000	–1	–1
0001	1	–3	1001	–1	–3
0010	1	1	1010	–1	1
0011	1	3	1011	–1	3
0100	3	–1	1100	–3	–1
0101	3	–3	1101	–3	–3
0110	3	1	1110	–3	1
0111	3	3	1111	–3	3

Fig. 9.25 (a) M-ary QAM modulator; (b) example for $M = 16$: normalised amplitudes of A_I and A_Q and resulting constellation

values of normalised A_I and A_Q amplitudes for $M = 16$ necessary to generate the square 16-QAM constellation shown. Notice that the levels have been selected to achieve a gray code arrangement of states.

9.7.4 M-ary FSK modulator

Fig. 9.26 shows an arrangement for generating M-ary FSK using a frequency synthesizer. The synthesizer generates M continuously running and available frequencies

$$f_o, \quad f_o + \Delta f, \quad f_o + 2\Delta f, \quad f_o + 3\Delta f, \ldots, \quad f_o + (M-1)\Delta f$$

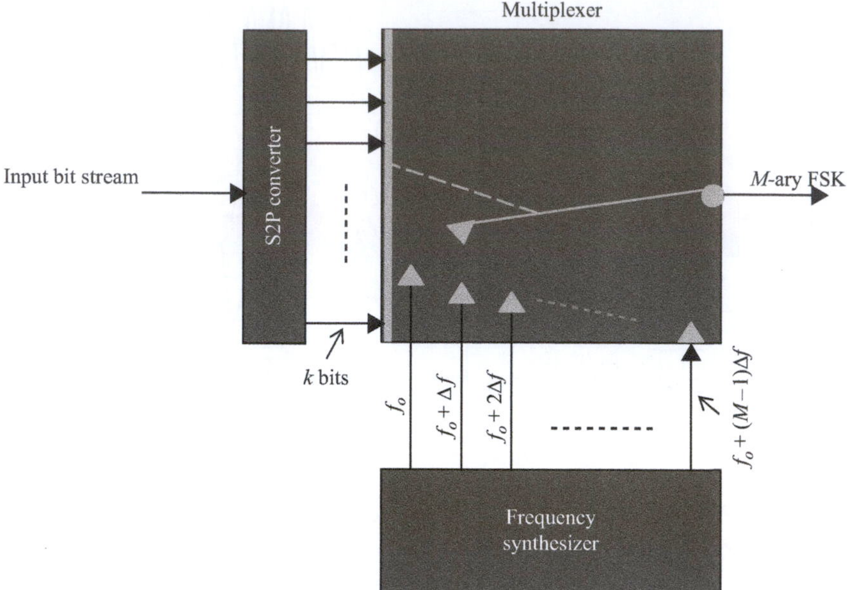

Fig. 9.26 M-ary FSK modulator

all of which have the same initial phase, where the carrier frequency f_o completes an integer number of half cycles within one symbol period T_s, and the frequency spacing between carriers Δf is chosen to be half the symbol rate ($\Delta f = R_s/2$) to ensure orthogonality of all the carriers. During each symbol interval T_s, the serial to parallel (S2P) converter fetches a k-bit group of bits from the input bit stream, and this is used as a control signal in the multiplexer to select one of the carrier frequencies and connect it to the modulator output terminal. There are M possible k-bit combinations from all-zeros to all-ones and these respectively select frequencies from f_o to $f_o + (M-1)\Delta f$ in the symbol interval T_s in which they occur. Thus, the generated M-ary signal is given below, normalising amplitude to unity and initial phase to zero and referencing time to the start of each symbol interval

$$g_k(t) = \cos[2\pi(f_o + k\Delta f)t]\text{rect}\left(\frac{t - T_s/2}{T_s}\right)$$
$$k = 0, 1, 2, 3, \ldots, M-1; \quad f_o = nR_s/2; \quad \Delta f = R_s/2 \tag{9.89}$$

Note that a gray code arrangement is irrelevant in this case since all symbols are orthogonal.

9.8 Summary

We have studied the important concepts of signal orthogonality (leading to signal space diagrams) and correlation which are central to signal detection in digital communications. The correlation coefficient of two signals is a number ρ in the range $(-1, +1)$ which gives a measure of the similarity of the two signals, with $\rho = +1$ indicating that the two signals are perfectly identical, $\rho = 0$ if they are orthogonal, and $\rho = -1$ if they are perfect opposites (i.e. antipodal). The main function of a digital communication receiver is to decide during each symbol interval which one of a set of M symbols was sent, and hence to generate the group of bits associated with that symbol. The detection process can be based for example on the use of a bank of M detectors (e.g. matched filter or correlation receiver) each of which produces significant output for only one of the pulses as input. In the presence of noise this ability of the receiver to distinguish between the M symbols depends not only on the energy of the symbols but also crucially on their correlation. For a given transmitted signal power, the greatest immunity to noise (called noise margin), indicated by the gap between the output produced in each detector by the symbol to which it is matched and the closest output produced by all the other symbols, is provided by pulses having $\rho = -1$. The noise margin degrades towards zero as ρ increases towards $+1$. However the choice of $\rho = -1$ is only possible in binary transmission involving only two symbols, leading to binary phase shift keying (BPSK) and binary bipolar base band (BBB). In all other cases ρ exceeds -1.

M-ary ASK, PSK and QAM systems in particular have correlation coefficients that increase as M increases resulting in a significantly increased probability of detection error, or the need for a considerable increase in transmitted signal power if detection accuracy is to remain at the same level as in binary transmission. The unique ability of M-ary FSK to maintain a fixed correlation coefficient among its transmitted symbols as M increases, but with the symbols spread out over a transmission bandwidth that increases with M, allows us to trade bandwidth for improved detection of extremely weak signals, where for example noise power might exceed signal power.

We have also studied in detail the mathematical expressions for the symbols generated for transmission in digital communication systems, and the geometric and complex representations of these signals and noise. You now fully understand the base band equivalent representation of band pass signals and noise, and various signal components such as signal envelope, and the in-phase and quadrature components. You are therefore well equipped to carry out a complete analysis of the effects of noise in digital communication systems — our task in Chapter 10. Having studied the characterisation of linear systems (Chapter 2) and nonlinear systems (Chapter 3) and having worked through the material in Chapter 8 on the constraints posed by noise and channel bandwidth, the next chapter will complete the instruction necessary to enable you to competently engage in the simulation and analysis of digital communication systems and to exploit in your system design the trade-offs involved among the key design parameters, namely transmission bandwidth, signal power, noise power, transmission bit rate, and bit error rate.

9.9 References

[1] Otung, I. (2001). *Communication engineering principles*. Basingstoke: Palgrave Macmillan

[2] Xiong, F. (2000). *Digital modulation techniques*. Norwood, MA: Artech House

Chapter 10

Noise impact in digital transmission

In Chapter Ten

- ✓ Brief discussion of random signals and their characterisation using probability density function (PDF), distribution function, moments, etc. The concepts of stationarity and ergodicity are also introduced for application to noise in communication systems, and the statistical characterisation of additive white Gaussian noise (AWGN) is reviewed.
- ✓ Quantification of noise in communication systems: Brief but lucid discussion of system noise temperature, noise factor and noise power, and the calculation of carrier-to-noise ratio and related parameters.
- ✓ Noise effects in coherent base band and modulated digital communication systems: Detailed and easy-to-follow analysis of the bit error rate (BER) of coherent binary and *M*-ary systems based on a signal space diagram approach. Analytic expressions are derived for the BER of the different systems and graphs are provided from which values can be read for system design. Systems analysed include coherent amplitude shift keying (ASK), phase shift keying (PSK), quadrature amplitude modulation (QAM) and frequency shift keying (FSK).

10.1 Introduction

Noise is inherent and in fact inescapable in communication systems. Any piece of conducting material at a physical temperature T in kelvin will be subject to thermal agitation of its electrons which produces available noise power of amount kT watt per unit hertz of bandwidth over which the conductor is used, where k is Boltzmann's constant. To completely eliminate thermal noise requires reducing the temperature of all conductors in the communication system to absolute zero ($T = 0$ K), which is a thermodynamic state that cannot be reached in a finite number of steps. The use of liquid helium can greatly limit thermal noise by reducing T from $\sim$290 K to 4.2 K, but this is prohibitively expensive and impractical, and would produce a communication system that requires extremely delicate physical handling. Thermal noise aside, every radio antenna will pick up the wanted signal as well as electromagnetic energy radiated by absorbers in the atmosphere and by other sources in the vicinity of the antenna. These also constitute unavoidable noise. And transmissions in optical fibre, although not based on electron flow, are nevertheless subject to fluctuations in photon flow which also constitute noise.

The communication system designer must therefore be equipped not only to characterise and quantify noise but also to understand and evaluate its impact on system performance in order to be able to make judicious use of scarce system resources (in particular signal power and transmission bandwidth) to deliver the desired quality of service. The aim of this chapter is to give you the necessary tools in this regard with a focus on only one measure of quality of service, namely the bit error rate (BER). Following a bare bones review of the theory of random signals, we characterise the prevalent type of communication system noise described as additive white Gaussian noise (AWGN) and identify its impulsive autocorrelation function and hence flat power spectral density. We then learn how the noisiness of device components are specified and apply this understanding to characterise whole communication receivers and to determine carrier-to-noise ratio and other relevant parameters at the receiver's reference point. The rest of the chapter is then devoted to a detailed study of the effect of noise on the symbol detection process, so that given the carrier-to-noise ratio at the reference point of a binary or M-ary digital transmission system that employs coherent detection you will be able to obtain a reliable estimate of BER. Conversely, you will be able to specify the required level of carrier-to-noise ratio to achieve a desired BER.

By way of a brief introduction to the task we will face in evaluating the impact of noise on the symbol detection process, consider that an incoming symbol $g_{in}(t)$ arriving at the detection point of an M-ary digital communication receiver (see Fig. 9.17) will be accompanied by noise $v_n(t)$, assumed to be band limited AWGN. Thus

$$g_{in}(t) = g_i(t) + v_n(t); \quad i = 0, 1, 2, \ldots, M-1 \tag{10.1}$$

and the output of the correlation receiver corresponding to the kth symbol $g_k(t)$ is no longer as given by Eq. (9.81) but is

$$y_o(T_s) = \int_0^{T_s} [g_i(t) + v_n(t)]g_k(t)dt = \begin{cases} E_k + \varepsilon_n, & i = k \\ \rho E_s + \varepsilon_n, & i \neq k \end{cases}$$
$$\text{where} \quad \varepsilon_n = \int_0^{T_s} v_n(t)g_k(t)dt \tag{10.2}$$

We see that noise introduces a random shift ε_n in this output. In view of the noise margin E_{NM} illustrated in Fig. 9.21, there will be a symbol error if

1. Symbol $g_k(t)$ was transmitted but $\varepsilon_n \le -E_{NM}$, or
2. An adjacent symbol $g_m(t)$ was transmitted but $\varepsilon_n \ge E_{NM}$.

The first condition causes the receiver to decide in favour of the adjacent state S_m, although S_k was transmitted. The probability of this event is denoted by $\Pr[S_m|S_k]$. The second condition causes the receiver to decide in favour of S_k when in fact it was the adjacent state S_m that was transmitted. Its probability is denoted by $\Pr[S_k|S_m]$.

As an example, Fig. 10.1 shows the result of a computer simulation of a square 16-QAM transmission system that illustrates the shift between transmitted and received states due to noise. There were 5000 repeated transmissions of state S_o through a noisy channel. The plot shows that the received states are shifted by some distance from S_o. We see that in eight instances this shift is large enough for the received state to be actually nearer to some other state than it is to S_o, which causes the detector to decide in favour of that other state. That is, of the 5000 symbols received in this transmission 8 are in error.

Because of the random nature of noise, we can only talk of the probability P_e that a symbol error will occur, but cannot predict with certainty the interval when

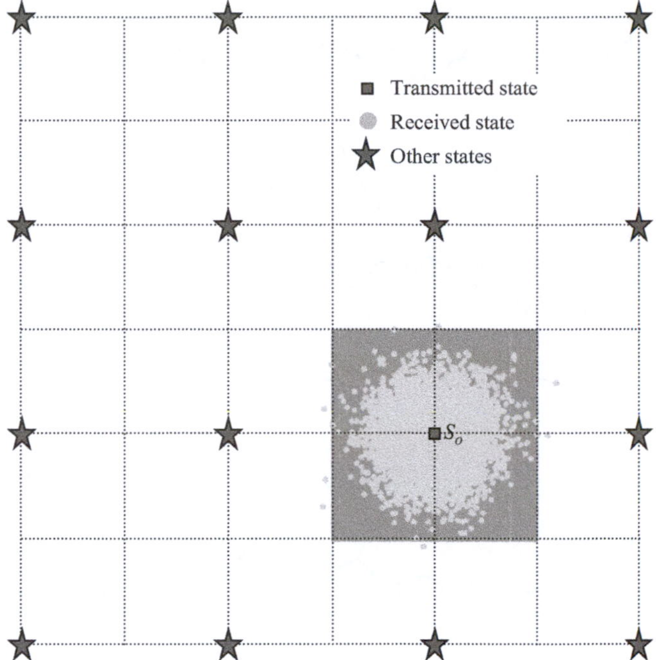

Fig. 10.1 *Illustration of noise effect on symbol detection. State S_o of* 16-*QAM was transmitted* 5000 *times through an AWGN channel. The round dots show each of the* 5000 *received states*

this error will occur. For example, a *probability of symbol error* $P_e = 0.001$ means that on average one symbol in a thousand will be incorrectly received. Note that this does not imply that there will always be one and only one error in every 1000 transmitted symbols. In fact, there may well be some periods of time during which 10,000 or more symbols are received without a single error, and others in which there are two or more errors in 1000 symbols. What this statement means is that if we observe the transmission over *a sufficiently long time*, then we will find that the ratio of the number of symbols in error to the total number of symbols transmitted is $P_e = 0.001$. This probability is therefore also referred to as *symbol error rate* (SER). To illustrate, a rough estimate for the SER of the system of Fig. 10.10 is SER $\approx 8/5000 = 0.0016$. A more reliable estimate of SER for this system requires observation of the error rate over a much larger number of received symbols up to several millions.

10.2 Random signals

Random signals play a central role in digital communications. For example, the information signal is always treated as a random bit stream since the receiver cannot know ahead of time what bit will be in a given interval. Furthermore, noise from a variety of sources is inherent in all communication systems and is by nature random. We wish to provide in this section the basic tools and concepts necessary for characterising random signals in communication systems.

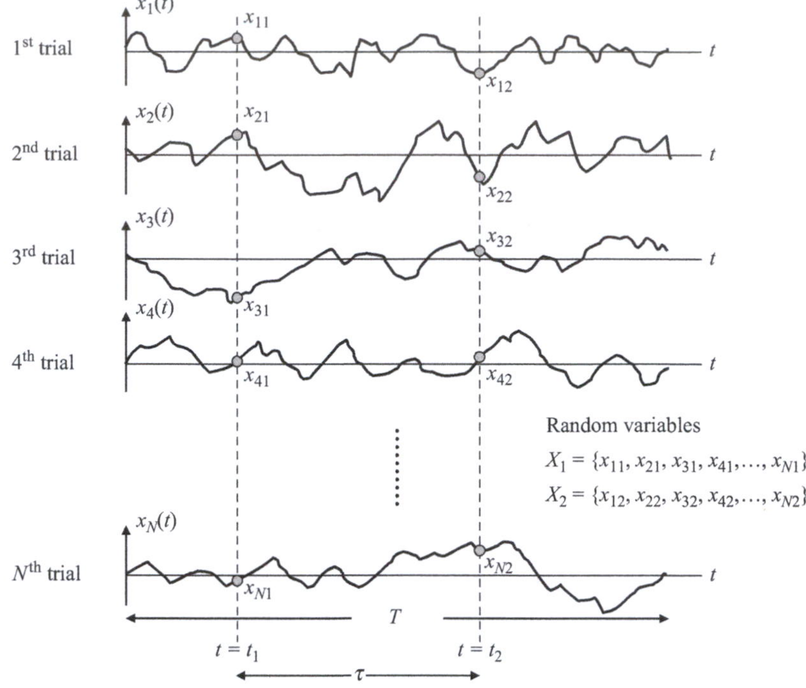

Fig. 10.2 Random process

10.2.1 Random process

A random process is a collection or *ensemble* of *sample functions* as illustrated in Fig. 10.2. Each sample function $x_k(t)$, $k = 1, 2, 3, \ldots, N$, is a function of time and is the result of a particular trial of the random experiment (i.e. observation of the random process) over a time interval T. In general no two sample functions have an identical sequence of values or waveform. Taking the value of each of the sample functions at a fixed time $t = t_1$ within the observation interval results in a set of numbers $\{x_{11}, x_{21}, x_{31}, x_{41}, \ldots, x_{N1}\}$ that constitutes a *random variable*, say X_1. Repeating the exercise at another time $t = t_2$ produces another random variable $X_2 = \{x_{12}, x_{22}, x_{32}, x_{42}, \ldots, x_{N2}\}$.

10.2.2 Random signal parameters

We cannot know with certainty in advance what value a random variable will take, but we can put forward a *positive number* on a scale from 0 to 1 that expresses the likelihood (i.e. *probability*) of the value being in a specified range. A probability of 0 or 1 expresses certainty. Zero means the value of the random variable will not be in the given range, whereas one means it will. More generally there is some uncertainty and this is indicated in the extent of the departure of the probability from the extreme ends of the scale. In particular, the probability that a random variable X will have a value lying in the infinitesimal range $(x, x + dx)$ is denoted by $\Pr(x \leq X \leq x + dx)$ and given by

$$\Pr(x \leq X \leq x + dx) = p_X(x)dx \tag{10.3}$$

The parameter $p_X(x)$ is known as the *probability density function* (PDF) of the random variable X. It is in general not constant but varies with x in a way that reflects the relative likelihood of each x value. Eq. (10.3) suggests that if we plot a graph of $p_X(x)$ versus x, then the probability that X lies in a sizeable range (x_1, x_2) will be the area under the PDF curve in that range. In other words

$$\Pr(x_1 \leq X \leq x_2) = \int_{x_1}^{x_2} p_X(x)dx$$

The probability that X takes on some value less than or equal to x gives the *distribution function* $F_X(x)$ of the random variable. Thus

$$F_X(x) = \Pr(-\infty < X \leq x) = \int_{-\infty}^{x} p_X(x)dx \tag{10.4}$$

The distribution function of a random variable is therefore the integral of its PDF, or equivalently, $p_X(x)$ is the derivative of $F_X(x)$. Since by definition probability is a positive number between 0 and 1, and we are *certain* that X has a value in the range $(-\infty, \infty)$, we can make the following characterising statements about $p_X(x)$ and $F_X(x)$:

- $p_X(x) \geq 0$
- $\int_{-\infty}^{\infty} p_X(x)dx = 1$. The total area under a PDF curve is unity.
- $0 \leq F_X(x) \leq 1$
- $F_X(\infty) = 1$
- $F_X(-\infty) = 0$
- $F_X(x_2) \geq F_X(x_1)$ for $x_2 \geq x_1$. Thus $F_X(x)$ is *monotonically non-decreasing*.

We can characterise a random variable X using various aggregate quantities, called *moments* of the random variable. The nth moment of X is the expected value of X^n, denoted by $\mathbf{E}\{X^n\}$, and is obtained by adding all the possible values of X raised to power n, each addition being weighted by the relative likelihood of occurrence of the value. Thus

$$\mathbf{E}\{X^n\} = \int_{-\infty}^{\infty} x^n p_X(x)dx \tag{10.5}$$

The first two moments are the most useful and give the mean A_o, and mean-square value (i.e. *power*) P of the random variable respectively:

$$\begin{aligned} A_o = \mathbf{E}\{X\} &= \int_{-\infty}^{\infty} x p_X(x)dx \\ P = \mathbf{E}\{X^2\} &= \int_{-\infty}^{\infty} x^2 p_X(x)dx \end{aligned} \tag{10.6}$$

Another important characterising parameter is the expected value of $(X - A_o)^2$, called the *variance* and denoted by σ^2. Since the expected value operator is linear, we note that

$$\begin{aligned}\sigma^2 &= \mathbf{E}\left\{(X - A_o)^2\right\} = \mathbf{E}\{X^2 - 2XA_o + A_o^2\} \\ &= \mathbf{E}\{X^2\} - 2A_o\mathbf{E}\{X\} + A_o^2 = \mathbf{E}\{X^2\} - 2A_o^2 + A_o^2 \\ &= \mathbf{E}\{X^2\} - A_o^2 \equiv \text{Total power} - \text{DC power}\end{aligned} \tag{10.7}$$

Thus the second moment is the total power of the random variable, the variance specifies the AC power, whereas the square of the mean gives the DC power. The square root of the variance is called the *standard deviation* σ. To further emphasise the physical significance of the variance, note that if σ^2 is zero it means that X has a constant value A_o, and its PDF is a unit impulse function located at A_o. A non-zero value of σ^2 is an indication that X does take on values other than A_o. And the larger the value of σ^2, the more the spread of the values of X around its mean, and hence the broader is the PDF curve as illustrated in Fig. 10.3 using the well-known Gaussian PDF:

$$p_X(x) = \frac{1}{\sqrt{2\pi}\sigma}\exp\left[\frac{(x-\mu)^2}{2\sigma^2}\right] \tag{10.8}$$

where σ is the standard deviation and μ the mean of the Gaussian random variable X.

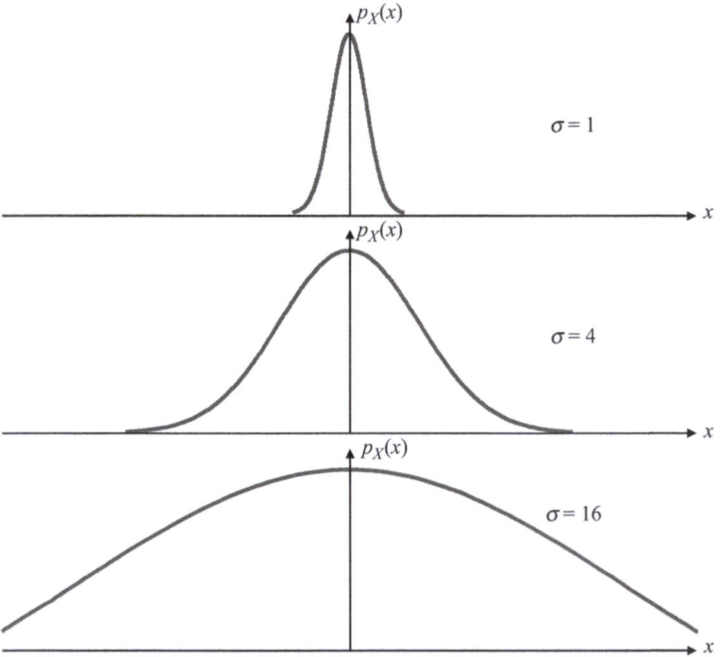

Fig. 10.3 Significance of variance σ^2. All three plots are Gaussian PDF and use the same scale on the x-axis but different scales on the y-axis. Note that as PDF spreads out with increasing σ, its values decrease since the total area under a PDF curve is always unity

It is sometimes necessary in system design to make provision for the peak value of a random variable. For example, what is the maximum height of a human being to be allowed for in the design of doorways? Many random variables such as the Gaussian random variable do not have a finite *peak value* or *amplitude* since any value up to infinity is theoretically possible. We may however define the peak value A_p of a random variable X as the point at which the distribution function of X is less than say 0.99. That is $F_X(A_p) \leq 0.99$, which means that X takes on values less than this 'peak' with a probability of 0.99. Thus the 'peak value' so defined will be exceeded 1% of the time on average.

10.2.3 Stationarity and ergodicity

The task of characterising a random process is greatly simplified if the process is *stationary*. A random process is said to be *strict-sense stationary* if its statistics do not change with time. However, if only two of its statistics are time-independent, namely its *mean* and *autocorrelation function* — defined as the expected value of the product of two random variables of the process — then the process is said to be *wide-sense stationary* (WSS). For example, if the random process of Fig. 10.2 is WSS, then

$$\begin{aligned} \mathbf{E}\{X_1\} &= \mathbf{E}\{X_2\} = \text{Mean of process} \\ \mathbf{E}\{X_1X_2\} &= R_X(\tau) \end{aligned} \tag{10.9}$$

and the autocorrelation function $R_X(\tau)$ depends only on the shift τ, but not on time t_1.

A strict-sense stationary process may have a further attribute, although its sample functions cannot be identical in waveform, it may turn out that they have identical statistical characterisations, and we can obtain the statistical properties of the random process by observing any one of its sample functions. Such a random process is said to be *ergodic*. In this case we can replace the more difficult task of *ensemble averaging* (to obtain say the mean of the process) with time-averaging over any one of the observed sample functions. For example, if the random process of Fig. 10.2 is ergodic then the mean A_o and autocorrelation function $R_X(\tau)$ — discussed in Chapter 7 — can be obtained as follows:

$$\begin{aligned} A_o &= \lim_{T\to\infty} \frac{1}{T} \int_{-T/2}^{T/2} x_k(t)dt \\ R_x(\tau) &= \lim_{T\to\infty} \frac{1}{T} \int_{-T/2}^{T/2} x_k(t)x_k(t+\tau)dt \\ k &= 1, 2, 3, \ldots, \text{or } N \end{aligned} \tag{10.10}$$

Notice that the random process can now be treated simply as a *nonperiodic power signal*, which allows us to apply relevant analysis tools developed in earlier chapters. This property of ergodicity applies in particular to AWGN, the type of random noise signal usually assumed in communication systems.

10.2.4 Additive white Gaussian noise

Noise in communication systems comes from various sources including thermal noise, quantisation noise, radio or sky noise, shot noise, partition noise, quantum noise and man-made noise. Thermal noise for example results from the thermal agitation of electrons in conductors, which produces random voltage fluctuations across the conductor. Each of the large number of electrons moves randomly under thermal agitation about an equilibrium position. The resulting noise voltage therefore has *zero mean* and is equally likely to be positive or negative. Furthermore, this thermal agitation is an ever-present background event that goes on in addition to any general drift of the electrons in one direction caused by the presence of an externally applied voltage signal. Thus the noise is described as *additive*.

The *central limit theorem* indicates that under circumstances where noise is the result of a large number of statistically independent and random contributions, the noise voltage v_n has a *Gaussian* distribution with PDF

$$p(v_n) = \frac{1}{\sigma\sqrt{2\pi}}\exp\left(-\frac{v_n^2}{2\sigma^2}\right) \tag{10.11}$$

where σ^2 is the *variance* of the noise voltage v_n. Because the mean is zero in this case, σ^2 is also the mean square noise voltage or (*normalised*) *noise power* P_n. Note that the above Gaussian PDF (see Fig. 10.3) has even symmetry about $v_n = 0$, indicating that the area under the curve in the interval $(-\infty, 0)$ equals the area in the interval $(0, \infty)$. Each of the two areas must be 0.5 since the total area is unity, confirming our earlier statement that the noise voltage takes on positive and negative values with equal probability. The probability that v_n exceeds a certain value z is given by

$$\begin{aligned}
\Pr(v_n \geq z) &= \int_z^{\infty} p(v_n)dv_n \\
&= \int_z^{\infty} \frac{1}{\sigma\sqrt{2\pi}}\exp\left(-\frac{v_n^2}{2\sigma^2}\right)dv_n = \frac{1}{\sqrt{\pi}}\int_{z/\sigma\sqrt{2}}^{\infty} \exp(-y^2)dy \\
&= \frac{1}{2}\operatorname{erfc}\left(\frac{z}{\sigma\sqrt{2}}\right)
\end{aligned} \tag{10.12}$$

where erfc is the *complementary error function* defined by

$$\operatorname{erfc}(x) = \frac{2}{\sqrt{\pi}}\int_x^{\infty} \exp(-y^2)dy \tag{10.13}$$

Assuming ergodicity, we can determine the autocorrelation function $R_n(\tau)$ of this noise from the noise voltage $v_n(t)$ measured over a long observation interval $T \to \infty$. If this noise comes from a large number of statistically independent contributions and we invoke the *idealisation* of perfect randomness, meaning that the noise generating mechanism has zero recovery time so that it can change in negligible time from one value to any other value within its range, then for $\tau \neq 0$, the sign of the product $v_n(t)v_n(t+\tau)$ involved in the computation of $R_n(\tau)$ is equally likely to be positive or negative and will therefore have zero mean when aggregated over a long interval T. The autocorrelation function integral — Eq. (10.10) — for this idealised noise therefore evaluates to zero for all values of τ except $\tau = 0$ where it evaluates to the noise power P_n by definition. That is, $R_n(\tau)$ is an impulse function

$$R_n(\tau) = \begin{cases} P_n, & \tau = 0 \\ 0, & \text{Elsewhere} \end{cases}$$

and therefore has a constant Fourier transform, which from Eq. (7.8) gives the PSD $S_n(f)$ of white noise. Furthermore, from Section C.1.12 in Appendix C, the total area under $R_n(\tau)$ — i.e. the weight of the impulse — equals $S_n(0)$, which is also the value of this flat PSD at all frequencies. Denoting this constant value as $N_o/2$, we can say that the autocorrelation function of white noise is an impulse function of weight $N_o/2$ and (necessarily infinite) height $R_n(0) = P_n$, and we write

$$R_n(\tau) = \frac{N_o}{2}\delta(\tau) \rightleftharpoons S_n(f) = \frac{N_o}{2} \tag{10.14}$$

The above functions are shown in Fig. 10.4. The constant N_o is therefore the *noise power per unit bandwidth*, whereas $N_o/2$ is the noise power per unit frequency or noise power spectral density, and is given by

$$N_o = kT_{sys} \tag{10.15}$$

where $k = 1.38 \times 10^{-23}$ J/K is Boltzmann's constant, and T_{sys} (in Kelvin) is the *physical temperature* of the conductor — in the case of thermal noise, or more generally the *equivalent noise temperature* of the system — for all classes of noise usually likened to thermal noise for convenience. We see from Fig. 10.4b that this noise contains all frequencies from $f = 0$ to ∞ at the same power level. For this reason it is therefore called *white noise*, in analogy with white light, which contains equal amounts of all wavelengths in the visible spectrum. Furthermore, the value of $N_o/2$ for the uniform PSD applies to a double-sided spectrum. The height of a single-sided PSD of white noise is therefore N_o, and the total noise power within a bandwidth B is given by

$$P_{nB} = N_o B = kT_{sys} B \tag{10.16}$$

It is worth emphasising that white noise (with perfect randomness and infinite power P_n) is not realisable in practice. Nevertheless, the concept of white noise is a

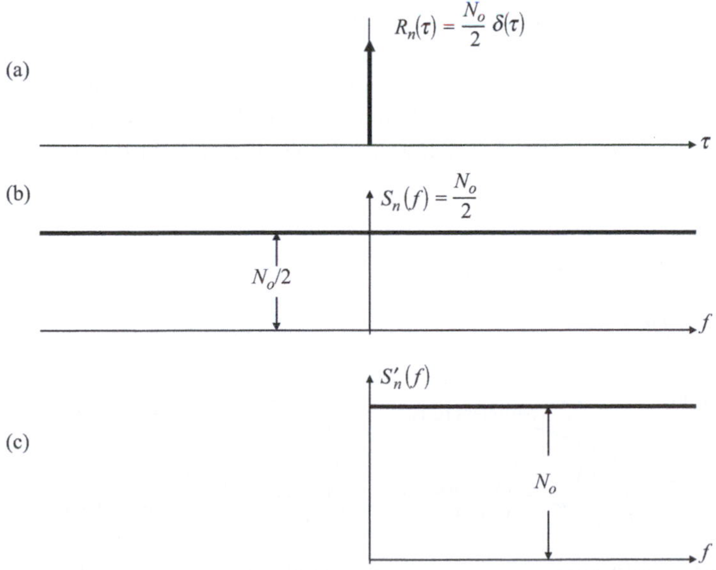

Fig. 10.4 Characterisation of white noise: (a) autocorrelation function, (b) double-sided power spectral density (PSD), and (c) single-sided PSD

simplifying idealisation that satisfactorily models real noise at the input of band limited real systems. Within the limited bandwidth of interest, the spectrum of such real noise looks flat and hence white. And we are not concerned with the inevitable tailing off of the spectrum of real noise as $f \to \infty$, since the pass band of our real system does not extend to those frequencies. For these reasons we will assume in our analysis in this chapter that input noise is additive, white and Gaussian, hence the term additive white Gaussian noise (AWGN).

10.3 System noise quantification

We present here a summary of system noise power specification and calculations. If you require a more in-depth discussion, including a review of the physical sources of noise in communication systems and derivation of some of the results quoted below, Chapter 9 of [1], which also features numerous worked examples, is a good starting point. A worked example that demonstrates the application of some of the link power budget concepts discussed here in the design of a mobile radio link will be presented in Chapter 12.

10.3.1 Noise temperature and noise factor

Thermal agitation of electrons in conductors produces random voltage fluctuations having mean square value measured across the terminals of the conductor given by

$$\overline{v_n^2} = 4kTBR \tag{10.17}$$

where k is Boltzmann's constant ($= 1.38 \times 10^{-23}$ J/K); R is the resistance of the conductor in ohm; T is the absolute temperature of the conductor in kelvin (K); and B is the bandwidth of the measurement in Hz. This noisy conductor of resistance R may be modelled as a noiseless resistor R in series with a noise voltage source of rms voltage $\sqrt{\overline{v_n^2}}$ as shown in Fig. 10.5a. The noisiness of this device is actually determined by the amount of noise power P_n that it delivers to a matched load connected across its terminals as shown in Fig. 10.5b. Matching ensures that there is no reflection of power at load back towards source and hence that there is a maximum power transfer to the load, and is attained when load resistance equals source resistance as shown. Analysing this simple circuit and using Eq. (10.17) for $\overline{v_n^2}$ we obtain

$$\begin{aligned} P_n = i_n^2 R = \frac{\overline{v_n^2}}{4R^2} R = \frac{4kTRB}{4R^2} R \\ = kTB \end{aligned} \tag{10.18}$$

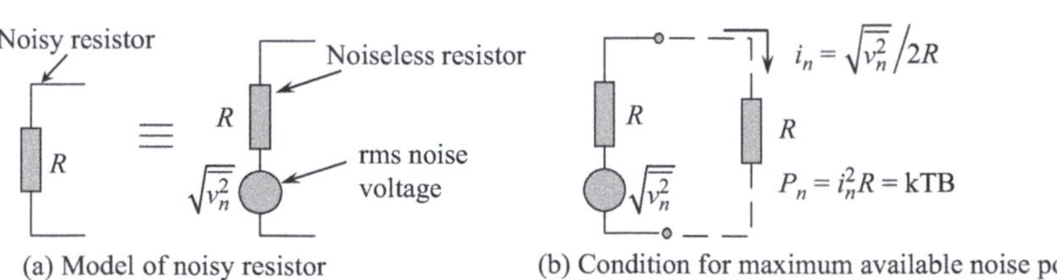

Fig. 10.5 Maximum available thermal noise power

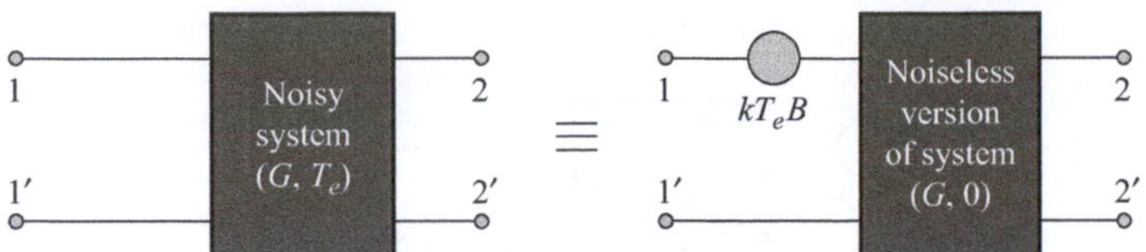

Fig. 10.6 Model of a noisy system

Notice that thermal noise power depends only on the physical temperature of the resistor and operating bandwidth, but not on the resistance R.

Not all noise in communication systems are of a thermal origin. However, in analogy with thermal noise the noisiness of every device (whatever the origin of its internally generated noise) is characterised by specifying an equivalent noise temperature T_e (which is not to be confused with the physical temperature of the device) defined as follows:

The equivalent noise temperature T_e of a device or system is the absolute temperature at which a noisy resistor has to be maintained so that when it is connected to the input of a noiseless version of the device, the available noise power observed at the output of the device is exactly the same as that produced by all the noise sources in the actual device.

Therefore a noisy system can be modelled as shown in Fig. 10.6: If the system has an equivalent noise temperature T_e, it means that the system internally generates a total noise power kT_eB at its input. The noise power at the system's output will be GkT_eB. Conversely, if a system of gain G delivers noise power $P_{n,output}$ over a bandwidth B to a matched load connected at its output terminals, then the equivalent noise temperature of the system is

$$T_e = \frac{P_{n,output}}{GkB} \tag{10.19}$$

Another parameter that is often used to quantify the noisiness of devices, especially amplifiers, is the *noise factor F* defined as follows:

The noise factor F of a system is the ratio of actual noise power output when the input is a standard noise source (of equivalent noise temperature $T_e = 290$ K, denoted by T_o) to the noise power output that exists under the same conditions if the system is noiseless.

This leads to the following equivalent definition:

The noise factor F of a system is the signal to noise ratio SNR_i at the system input divided by the signal to noise ratio SNR_o at the system output when the input is a standard noise source $T_o = 290$ K.

The system configuration stipulated in the above definition is shown in Fig. 10.7. The standard noise source v_s having internal resistance R_s and producing noise power kT_oB is connected to an impedance-matched (input resistance $\equiv R_s$) noisy system of equivalent noise temperature T_e. This system has input noise power kT_eB (due to its own internally generated noise) and output resistance R_o. Together with the standard noise source input the entire configuration delivers noise power

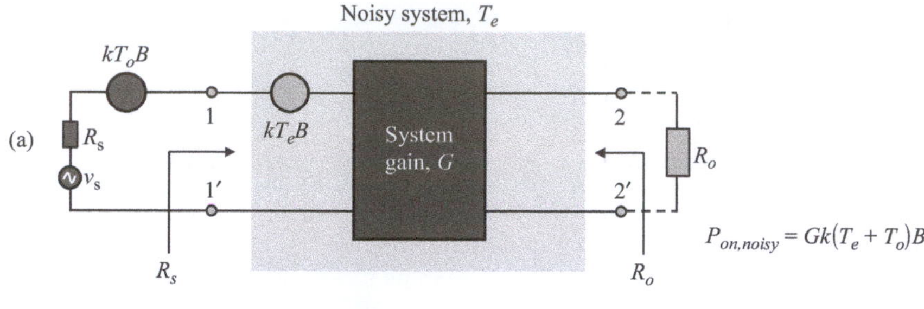

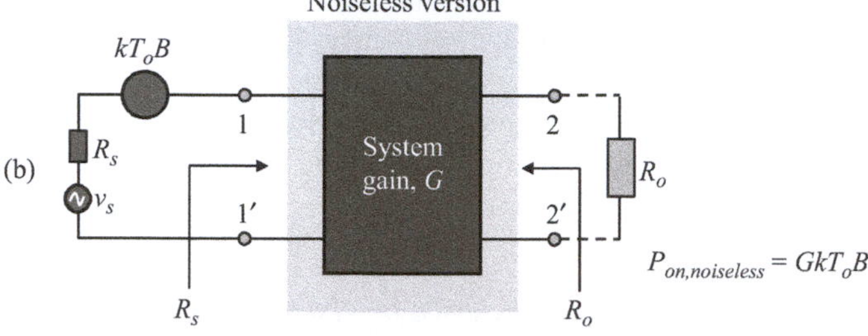

Fig. 10.7 Configurations for noise factor definition: (a) noisy system with standard noise input; (b) noiseless version of system with same standard noise input

$P_{on,noisy}$ to an impedance-matched load. Replacing the system with its noiseless version, means that noise power component kT_eB is absent at the input as shown, so that the noise power delivered by the configuration to a matched load is now $P_{on,noiseless}$. It follows from the definition of noise factor given above that

$$
\begin{aligned}
F &= \frac{P_{on,noisy}}{P_{on,noiseless}} = \frac{Gk(T_e + T_o)B}{GkT_oB} \\
&= 1 + \frac{T_e}{T_o} \\
&\Rightarrow T_e = (F - 1)T_o \qquad (10.20)
\end{aligned}
$$

The noisiness of a device can therefore be quantified either by specifying its equivalent noise temperature or its noise factor, with Eq. (10.20) providing a conversion between the two parameters. When expressed in dB, noise factor is usually referred to as noise figure. Also, equivalent noise temperature is usually referred to simply as noise temperature.

A communication system receiver usually consists of a cascade connection of multiple stages as illustrated in Fig. 10.8 for three stages with respective noise temperatures T_1, T_2 and T_3, gains G_1, G_2 and G_3, and noise factors F_1, F_2 and F_3. To quantify the noisiness of such a receiver, we must determine the noise temperature T or noise factor F of the overall cascade connection. Note that a cascade connection is a series connection with impedance matching between stages which ensures that output power from one stage is delivered in its entirety to the next stage without any reflections. Applying the definition of noise factor, following the hint

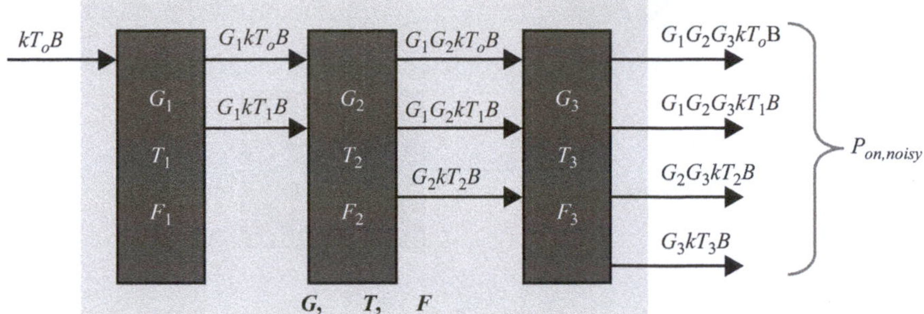

Fig. 10.8 Noise factor and noise temperature of cascade connection of n stages (n = 3)

in Fig. 10.8, we obtain expressions for the overall noise temperature and noise factor of a cascade connection of n stages as

$$\begin{aligned} T &= T_1 + \frac{T_2}{G_1} + \frac{T_3}{G_1G_2} + \frac{T_4}{G_1G_2G_3} + \cdots + \frac{T_n}{G_1G_2G_3\cdots G_{n-1}} \\ F &= F_1 + \frac{F_2 - 1}{G_1} + \frac{F_3 - 1}{G_1G_2} + \frac{F_4 - 1}{G_1G_2G_3} + \cdots + \frac{F_n - 1}{G_1G_2G_3\cdots G_{n-1}} \end{aligned} \tag{10.21}$$

It can be seen that the overall noise factor is dominated by the first stage. If gain G_1 of the first stage is high, then $F \approx F_1$ regardless of how noisy subsequent stages might be. For this reason many radio receiver systems will locate a high gain, low noise amplifier (LNA) in a sealed box right under the outdoor antenna. This arrangement reduces the noise impact of subsequent stages in the radio reception system, which usually includes a lossy waveguide or coaxial cable feed and a noisy mixer.

10.3.2 Overall system noise temperature

When determining the overall noise temperature of a communication receiver, care must be taken not only to include all contributions but also to correctly combine them to yield the system noise temperature at the reference point. This section briefly summarises the noise components in a radio reception system and provides a formula for the overall system noise temperature based on Eq. (10.21).

Fig. 10.9 shows a radio reception system comprising an outdoor antenna of noise temperature T_a connected by a feed having loss L_f to an indoor receiver unit of noise temperature T_e. The overall system noise temperature T_{sys} is calculated at a specified reference point, usually chosen to be the LNA input of the receiver.

The antenna noise temperature T_a may contain contributions from the following sources:

- Extra-terrestrial noise, T_{xn}: This comprises galactic noise which ranges from around 1000 K at 100 MHz to a negligible value at 2 GHz and above; solar noise which ranges from around 10^6 K at 100 MHz to around 10,000 K above 10 GHz under the quiet Sun; and cosmic background noise which is 2.7 K at all frequencies.
- Atmospheric noise, T_{an}: This is due to absorption of electromagnetic energy by hydrometeors (i.e. rain, ice, snow and cloud water droplets), aerosols and

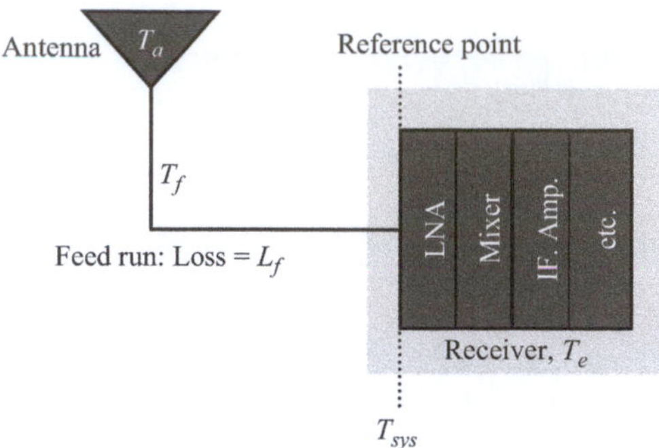

Fig. 10.9 Overall noise temperature T_{sys} of a radio reception system

gases in the atmosphere, which under thermal equilibrium will incoherently re-radiate all absorbed energy as noise. T_{an} is given in terms of total absorptive attenuation L_{atm} (as a loss factor, denoted by A when expressed in dB) in the atmosphere, and the mean radiating temperature T_m of the atmosphere, which is approximately 275 K but may be determined at any location on earth using an empirical relation with local surface temperature T_{surf}:

$$\begin{aligned} T_{an} &= T_m(1 - 1/L_{atm}) \\ L_{atm} &= 10^{A/10} \\ T_m &= 1.12T_{surf} - 50 \end{aligned} \tag{10.22}$$

Note that the first line is based on the fact that the noise factor of a passive attenuator equals its loss. Therefore, treating the atmosphere as a passive attenuator with loss L_{atm}, its noise factor $F = L_{atm}$, and, by Eq. (10.20), its noise temperature is $(L_{atm} - 1)T_o$. This noise is referred to the input into the atmosphere and will be reduced by the loss factor L_{atm} by the time it gets to the output of the atmosphere (which is the antenna input port where T_a is specified) to $(L_{atm} - 1)T_o/L_{atm} = (1 - 1/L_{atm})T_o$. We set $T_o \sim T_m$ to obtain Eq. (10.22).

- Earth surface noise, T_{en}: Noise radiated by the Earth is also picked up by the antenna and is given by

$$T_{en} = \varepsilon_\phi T_{surf}/L_{sl} \tag{10.23}$$

where ε_ϕ is the emissivity of the Earth, T_{surf} is the surface temperature (~300 K), and L_{sl} is the sidelobe reduction factor of the receive antenna. Typically, the Earth is assumed to be a blackbody radiator for which $\varepsilon_\phi = 1$. If the antenna is omnidirectional, as in cellular mobile phones, then $L_{sl} = 1$. For directional antennas pointed at a transmitter located at some height above the Earth's surface (e.g. satellite and radio relay links), the Earth's noise contribution enters through a sidelobe whose gain is much lower than the gain of the boresight through which the wanted signal enters. In that case, L_{sl} could be a factor of 100 or more.

Combining all the above three components gives antenna noise temperature approximately as

$$T_a \approx 10^{-A/10}T_{xn} + 275(1 - 10^{-A/10}) + 300/L_{sl} \tag{10.24}$$

It is important to note that the Sun makes a very large noise contribution $> \sim 10{,}000$ K to T_{xn}. Therefore pointing the receive antenna at the Sun will cause immediate link outage. Note also that T_{xn} passes through the atmosphere to get to the antenna and therefore is reduced by the dB amount of atmospheric absorption. If operating frequency permits galactic noise to be assumed negligible then T_{xn} can be set to 2.7 K in the above equation.

Returning to Fig. 10.9, the cable or waveguide feed has loss L_f and hence (see comments made above on atmospheric noise temperature) has noise temperature T_f referred to its input and $T_{f,ref}$ referred to its output which coincides with the reference point, given by

$$\begin{aligned} T_f &= 290(L_f - 1) \\ T_{f,ref} &= 290(1 - 1/L_f) \end{aligned} \tag{10.25}$$

The receiver noise temperature T_e is obtained by combining the noise temperatures of the receiver stages (LNA, mixer, IF amplifier, etc.) according to Eq. (10.21). These component noise temperatures (or noise factor) can be read from manufacturer's data sheets for the components. T_e is likely to be dominated by the LNA noise temperature as discussed earlier.

Finally, the overall system noise temperature T_{sys} at the reference point (LNA input) is obtained by passing the antenna noise temperature T_a (given by Eq. (10.24)) through loss factor L_f in order to bring it to the reference point, and adding this to $T_{f,ref}$ and T_e (both of which are already specified at the reference point). Thus

$$T_{sys} = T_e + T_a/L_f + 290(1 - 1/L_f) \tag{10.26}$$

If the feed has negligible loss ($L_f = 1$) or the LNA is co-located with the antenna inside a low noise block (LNB) then

$$T_{sys} = T_e + T_a \tag{10.27}$$

10.3.3 Carrier-to-noise ratio

The BER performance of a digital communication receiver depends primarily on the ratio between received signal power P_r and the sum of noise power P_n and interference signal power P_i at the receiver. Contributions to interference signal power P_i may come from a variety of sources such as (i) crosstalk in metallic lines due to inductive- and capacitive-coupling, or in radio systems (called *co-channel interference*) when two transmitters share the same radio frequency and polarisation but are insufficiently separated in distance; (ii) lightning and ignition spark induced voltages; (iii) inadequate filtering of individual channels in multiplexed systems; (iv) scattering induced coupling when two radio links intercept a common scattering volume aloft; (v) depolarisation in dual-polarised radio systems in which two signals are independently transmitted at the same frequency but on orthogonal polarisations. If the propagation path includes an anisotropic medium such as atmospheric rain or ice, a small fraction of the energy in one polarisation will be converted into the orthogonal polarisation thereby contributing to interference.

Evaluation of interference to obtain a reliable estimate of P_i can be a time-consuming exercise which is in any case beyond the scope of this book. We will therefore assume that through careful system design, interference has been reduced to a negligible level so that $P_i \sim 0$, allowing us to focus on received signal power P_r and noise power P_n.

The ratio between received signal power P_r and noise power P_n is called the carrier-to-noise ratio (denoted by C/N or CNR, and sometimes called signal-to-noise ratio (denoted by S/N or SNR), although this latter term ought to be reserved for the post-demodulation stage when the signal has been removed from the carrier). It is a crucial parameter in the design and analysis of digital communication systems, and through Eqs. (8.52) and (10.16) may be related to E_b/N_o another equally important parameter, which is the ratio between received energy per bit E_b and noise power per unit bandwidth N_o as follows:

$$\begin{aligned} C/N &= \frac{P_r}{P_n} \\ &= \frac{E_b}{N_o} \cdot \frac{R_b}{B} \end{aligned} \tag{10.28}$$

where R_b is the transmission bit rate and B is the noise equivalent bandwidth (discussed in Section 2.6 and derived in Eq. (2.37) for the raised cosine filter). Since the raised cosine filter is universally employed in the design of digital communication systems (see Section 8.2), we make use of this relationship in Eq. (2.37) between B and symbol rate R_s as well as the relationship given in Eq. (8.3) between bit rate R_b and symbol rate R_s in an M-ary transmission system to write the relationship between C/N and E_b/N_o in the form

$$\frac{E_b}{N_o} = \frac{C}{N} \cdot \frac{(1 - \alpha/4)}{\log_2 M} \tag{10.29}$$

where α is the roll-off factor of the filter. Thus given C/N we may obtain E_b/N_o and vice versa for a given M-ary transmission. The roll-off factor of the filter should be known, but if not a value of $\alpha = 0$ can be assumed. This has the effect of slightly exaggerating system noise power which is safe for system design.

In a radio system the noise power is given by Eq. (10.16) which is usually expressed in logarithmic units as

$$\begin{aligned} P_n &= 10 \log_{10}(kT_{sys}B) \\ &= -228.6 + 10 \log_{10}(T_{sys}) + 10 \log_{10}(B) \\ &\equiv -228.6 + T_{sys} + B \end{aligned} \tag{10.30}$$

There is a tacit understanding in the last line, by virtue of the summation rather than multiplication involved, that T_{sys} is in dBK and B is in dBHz. The overall system noise temperature T_{sys} is evaluated at the reference point as described in Eq. (10.26). To determine C/N the received signal power P_r must be calculated at precisely the same reference point. In fact C/N will be the same at any point in the digital receiver system up to the detector input provided the ratio is based on values of P_r and P_n calculated at the same point. The determination of P_r can be an involved process depending on the transmission system. We discuss only two systems below, first a satellite communication link and then a mobile radio system.

Satellite link: On a satellite link and other line-of-sight radio systems where free space propagation can be assumed we obtain P_r by extending Eq. (2.54) to include feed loss at the receiver L_{fr}, additional losses L_a due to atmospheric attenuation, antenna mispointing, polarisation mismatch and any other miscellaneous losses. Thus, in logarithmic units

$$P_r = \text{EIRP} - L_s - L_a + G_r - L_{fr} \tag{10.31}$$

We subtract Eq. (10.30) from Eq. (10.31) to obtain C/N and manipulate to obtain expressions for E_b/N_o and C/N_o as follows:

$$\begin{aligned} C/N &= P_r - P_n \\ &= \text{EIRP} - L_s - L_a + G_r - L_{fr} + 228.6 - T_{sys} - B \\ &= \text{EIRP} + 228.6 + G/T_{sys} - L_s - L_a - B \\ C/N_o &= \frac{C}{N} + B = \text{EIRP} + 228.6 + G/T_{sys} - L_s - L_a \\ E_b/N_o &= \frac{C}{N_o} - R_b = \text{EIRP} + 228.6 + G/T_{sys} - L_s - L_a - R_b \end{aligned} \tag{10.32}$$

where $\text{EIRP} = P_t + G_t - L_{ft}$ is the effective isotropically radiated power in dBW; P_t is the transmitted power in dBW; G_t is the transmit antenna gain in dB; L_{ft} is the transmit feed loss in dB; L_s is the free space path loss in dB given by Eq. (2.51); $G = G_r - L_{fr}$ is the effective gain of the receive antenna; G/T_{sys} in dB/K is called a *figure of merit* of the receiver; R_b is the bit rate in dBHz; and C/N_o is also in dBHz.

Mobile radio system: The mean received signal power P_r can be determined by using Eq. (2.58) to estimate path loss L_p. Alternatively (as illustrated in Fig. 10.10), path loss can be estimated by assuming free space propagation (with path loss of 20 dB per decade) up to a break point at a distance d_1 (km) from the transmitter, and path loss of $10n$ dB per decade beyond d_1 up to the receiver location at a

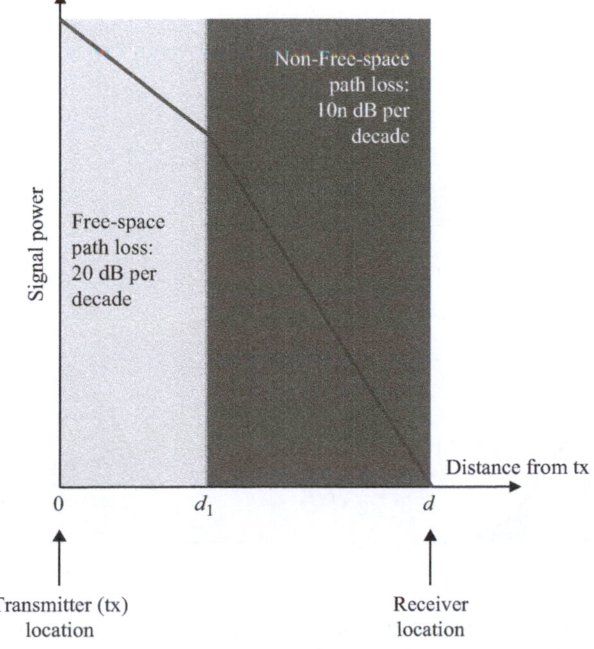

Fig. 10.10 Practical calculation of mean path loss on mobile radio link

distance d (km) from the transmitter, where n is typically between 3.5 and 4.5. The mean path loss at a distance $d > d_1$ is therefore

$$\begin{aligned} L_p &= 92.44 + 20\log_{10}f_{\text{GHz}} + 20\log_{10}d_1 + 10n\log_{10}(d/d_1) \\ &= 92.44 + 20\log_{10}f_{\text{GHz}} + 10n\log_{10}d - 10(n-2)\log_{10}d_1 \end{aligned} \tag{10.33}$$

The values of d_1 and n are usually chosen to give best fit to available measurement data recorded in a comparable terrain and radio frequency. The carrier-to-noise ratio of the mobile radio link may then be obtained by replacing L_s in Eq. (10.32) with the value of L_p computed here, and using an estimate of additional losses L_a that is applicable to the mobile radio link. In this regard, it is noted that rain attenuation, the predominant additional loss on high-availability satellite links at frequencies above 10 GHz, will be negligible as an additional loss on a mobile radio link operating at a frequency below 2 GHz.

10.4 Basic analysis of noise impact

We wish to derive important expressions for the probability of symbol error in binary transmission systems that employ coherent detection — which the correlation receiver discussed in Sections 8.4 and 9.6 implements in full. Consider a binary transmission system which transmits at a symbol rate $R_s = 1/T_s$, where T_s equals symbol duration. The system makes use of two states S_0 and S_1 that are separated in signal space by $\sqrt{E}$ as shown in Fig. 10.11a. The decision boundary of the detector is the perpendicular bisector of the line joining S_0 and S_1. We will employ base band equivalent parameters in the ensuing analysis, which means that results obtained will be applicable to both base band and band pass (i.e. modulated) transmission systems.

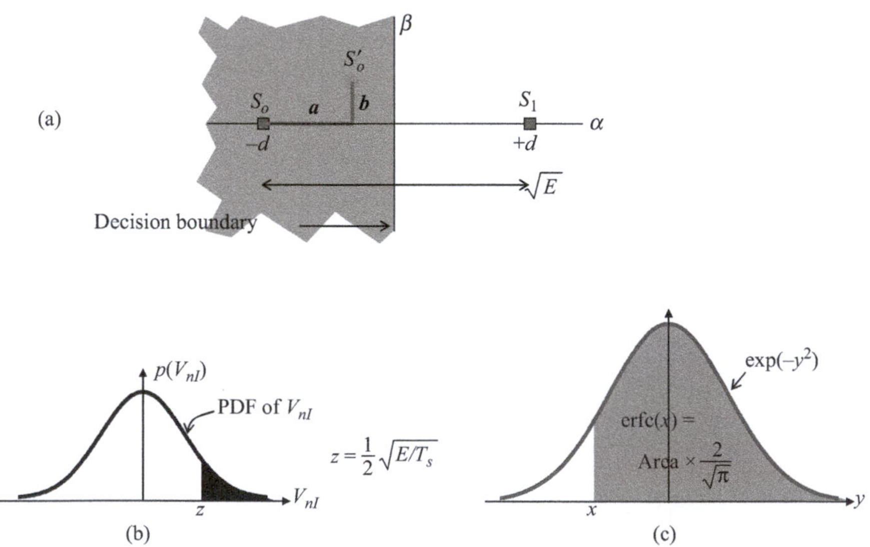

Fig. 10.11 (a) Basic binary signal space; (b) probability of symbol error equals shaded area; (c) the complementary error function erfc(x) is the area under the curve of exp($-y^2$) from $y=x$ to $y=\infty$, multiplied by $2/\sqrt{\pi}$

Assuming raised cosine filtering with zero roll-off factor, the base band equivalent bandwidth of this system is

$$B = R_s/2 = \frac{1}{2T_s}$$

which means that the variance σ^2 of the Gaussian noise $v_n(t)$ at the detection point is

$$\sigma^2 = P_n = N_o B = N_o/2T_s \tag{10.34}$$

A sample of this noise has in-phase and quadrature components v_{nI} and v_{nQ} with respective energies E_{nI} and E_{nQ} over one symbol duration given by

$$E_{nI} = v_{nI}^2 T_s; \quad E_{nQ} = v_{nQ}^2 T_s$$

Recalling that distance in signal space is the square root of energy, it is easy to see as shown in Fig. 10.11 that a transmitted state S_0 will be received at point S_0' having been displaced by distance a along the in-phase axis α and distance b along the quadrature axis β, where

$$\begin{aligned} a &= \sqrt{E_{nI}} = v_{nI}\sqrt{T_s} \\ b &= \sqrt{E_{nQ}} = v_{nQ}\sqrt{T_s} \end{aligned} \tag{10.35}$$

Thus when S_0 is sent, the transmitted symbol is

$$g_0(t) = -d\alpha(t), \quad \text{where } d = \sqrt{E}/2$$

but the received symbol is given by

$$g_0'(t) = [-d + a]\alpha(t) + b\beta(t)$$

There is symbol error if S_0' lies to the right of the decision boundary shown in Fig. 10.11a, which will be the case if $a > d$, or

$$v_{nI}\sqrt{T_s} > \sqrt{E}/2; \Rightarrow v_{nI} > \frac{1}{2}\sqrt{\frac{E}{T_s}} \equiv z$$

Since v_{nI} is a Gaussian random variable of zero-mean and variance σ^2 given by Eq. (10.34), it follows that the probability P_{e0} of symbol error given that S_0 is sent is the shaded area of Fig. 10.11b, which from Eq. (10.12) is given by

$$\begin{aligned} P_{e0} &= \Pr[S_1|S_0] = \Pr[v_{nI} > z] \\ &= \frac{1}{2}\operatorname{erfc}\left(\frac{z}{\sigma\sqrt{2}}\right) \end{aligned} \tag{10.36}$$

where erfc is the *complementary error function* whose values are provided in an extensive table in Appendix D. Note from Fig. 10.11c that erfc(x) is the area under the curve of $\exp(-y^2)$ in the region from $y = x$ to $y = \infty$, multiplied by the factor $2/\sqrt{\pi}$. For large x, say $x > 4$, the following approximation applies:

$$\operatorname{erfc}(x) \simeq \frac{\exp(-x^2)}{\sqrt{\pi}\,x} \tag{10.37}$$

From the symmetry of the problem in Fig. 10.11a, it is clear that the probability P_{e1} of an error occurring when S_1 is sent is the same as P_{e0}. A transmission channel that satisfies this condition, $P_{e1} = P_{e0}$, is referred to as a *binary symmetric channel*. Therefore the probability P_e of an error occurring in the detection of any symbol is given by Eq. (10.36), which when we substitute the expressions for z and σ given earlier yields

$$P_e = \frac{1}{2}\operatorname{erfc}\left(\frac{1}{2}\sqrt{\frac{E}{N_o}}\right). \tag{10.38}$$

To reiterate, Eq. (10.38) gives the SER in a binary transmission system where (i) band limited white Gaussian noise of power per unit bandwidth N_o is the only source of degradation, and (ii) the two transmitted states are separated by a distance $\sqrt{E}$ in signal space. In this basic analysis the transmitted states were located along a single axis in signal space. In the next Section we extend the analysis to obtain the BER of all binary transmission systems directly in terms of the received average energy per bit E_b.

It is useful to note that error probability is often expressed in some literature in terms of the *Gaussian integral*, called the *Q-function* $Q(x)$, which is the probability that a zero-mean, unit variance Gaussian random variable has a value that exceeds x. Thus

$$Q(x) = \frac{1}{\sqrt{2\pi}}\int_x^\infty \exp\left(-z^2/2\right)dz \tag{10.39}$$

By comparing this with the definition of erfc(x) given in Eq. (10.13) or Fig. 10.11c, we see that $Q(x)$ and erfc(x) are related by

$$\begin{aligned} Q(x) &= \frac{1}{2}\operatorname{erfc}\left(x/\sqrt{2}\right) \\ \operatorname{erfc}(x) &= 2Q\left(\sqrt{2}\,x\right) \end{aligned} \tag{10.40}$$

All our results will be expressed exclusively in terms of erfc, the values of which can be obtained from Appendix D. You may use Eq. (10.40) to convert these into equivalent Q-function expressions if desired.

10.5 BER of binary transmission systems

Consider Fig. 10.12 where two states S_0 and S_1 are arbitrarily located in a two-dimensional signal space having axes denoted by α and β. This signal space

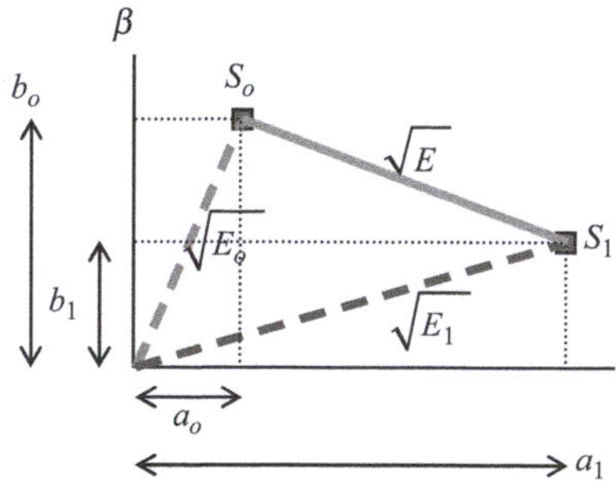

Fig. 10.12 Signal space diagram of an arbitrary binary transmission system

diagram is applicable to all binary systems with an appropriate choice of coefficient values (a_0, b_0, a_1, b_1) and basis functions $\alpha(t)$ and $\beta(t)$. We also know that since there is one bit per symbol the BER is given by Eq. (10.38) with $\sqrt{E}$ the distance between points S_0 and S_1. This system employs two symbols

$$\begin{aligned} g_0(t) &= a_0\alpha(t) + b_0\beta(t), \quad \text{for binary 0} \\ g_1(t) &= a_1\alpha(t) + b_1\beta(t), \quad \text{for binary 1} \end{aligned}$$

The respective energies of the symbols are

$$E_0 = a_0^2 + b_0^2; \quad E_1 = a_1^2 + b_1^2 \tag{10.41}$$

The average energy per bit E_b is given by

$$E_b = (E_0 + E_1)/2 \tag{10.42}$$

And the correlation coefficient of the two symbols follows from Eq. (9.75):

$$\begin{aligned} \rho &= \frac{2(a_0a_1 + b_0b_1)}{(a_0^2 + b_0^2) + (a_1^2 + b_1^2)} = \frac{2(a_0a_1 + b_0b_1)}{E_0 + E_1} \\ &= (a_0a_1 + b_0b_1)/E_b \end{aligned} \tag{10.43}$$

Finally, applying Pythagoras' rule in Fig. 10.12 allows us to express the energy E in terms of E_b and ρ as follows:

$$\begin{aligned} \left(\sqrt{E}\right)^2 &= (a_1 - a_0)^2 + (b_1 - b_0)^2 \\ &= \left(a_0^2 + b_0^2\right) + \left(a_1^2 + b_1^2\right) - 2(a_0a_1 + b_0b_1) \end{aligned}$$

Replacing each term on the RHS with its equivalent from Eqs. (10.41) to (10.43) yields the following important relation:

$$\begin{aligned} E &= E_0 + E_1 - 2\rho E_b = 2E_b - 2\rho E_b \\ &= 2E_b(1 - \rho) \end{aligned} \tag{10.44}$$

Substituting this relation into Eq. (10.38) gives the BER of *any* binary transmission system (assumed to employ a coherent detector such as a correlation receiver and to have binary symmetry):

$$\text{BER} = \frac{1}{2}\text{erfc}\left(\sqrt{\frac{E_b(1-\rho)}{2N_o}}\right) \tag{10.45}$$

We emphasise that Eq. (10.45) applies to *all* coherent binary symmetric systems, whether modulated or base band. A few special cases will help demonstrate the wide applicability of this important equation.

10.5.1 PSK and bipolar base band

In binary phase shift keying (BPSK) and binary bipolar base band (BBB) systems, two antipodal symbols $g_0(t)$ and $-g_0(t)$ are employed so that $\rho = -1$. Eq. (10.45) yields

$$\text{BER} = \frac{1}{2}\text{erfc}\left(\sqrt{E_b/N_o}\right) \tag{10.46}$$

10.5.2 ASK and unipolar base band

In binary amplitude shift keying (BASK) and binary unipolar base band (BUB) systems two symbols are used that differ only in their amplitudes $A_0 < A_1$, which are of course *unequal* positive numbers. It follows from Eq. (9.76) that in this case $\rho \geq 0$ and is given by

$$\rho = \frac{2A_0A_1}{A_0^2 + A_1^2} \tag{10.47}$$

We see from Eq. (10.45) that for a given E_b, the lowest BER is obtained when $A_0 = 0$, giving $\rho = 0$. For all other values of A_0, the correlation coefficient ρ has a positive value between 0 and unity. This reduces the argument of the complementary error function and leads to a larger BER. Setting $A_0 = 0$ gives what is known as on-off keying (OOK). It is therefore clear that OOK gives BASK and BUB their best (i.e. lowest) possible BER. Assigning non-zero values to both A_0 and A_1 always results in a poorer BER compared to an OOK of the same average energy per bit. Thus assuming this OOK design and setting $\rho = 0$ in Eq. (10.45) yields the following expression for the BER of BASK and BUB systems:

$$\text{BER} = \frac{1}{2}\text{erfc}\left(\sqrt{\frac{E_b}{2N_o}}\right) \quad \text{(OOK)} \tag{10.48}$$

10.5.3 FSK

In a binary frequency shift keying (BFSK) system that employs two orthogonal symbols, $\rho = 0$ as in the previous case. Thus the BER of an orthogonal BFSK system is also

$$\text{BER} = \frac{1}{2}\text{erfc}\left(\sqrt{\frac{E_b}{2N_o}}\right) \quad \text{(Orthogonal BFSK)} \tag{10.49}$$

It is worth noting that non-orthogonal binary FSK implementations are common especially in old modems in which the frequencies of the two sinusoidal pulses are chosen to give $\rho < 0$. In such cases the correct value of ρ must be used in Eq. (10.45) to determine the BER. The value of ρ can be read from Fig. 9.18 (bold solid curve) for a given frequency spacing Δf. For example, if the two frequencies are spaced apart by $\Delta f = 0.718R_s$, then $\rho = -0.24$ which gives BER:

$$\text{BER} = \frac{1}{2}\text{erfc}\left(\sqrt{0.62\frac{E_b}{N_o}}\right) \quad (\text{BFSK with } \Delta f = 0.718R_s) \tag{10.50}$$

10.5.4 Identical symbols

If $g_0(t) = g_1(t)$, then $\rho = 1$, and Eq. (10.45) gives

$$\text{BER} = \frac{1}{2}\text{erfc}\left(\sqrt{\frac{E_b(1-1)}{2N_o}}\right) = \frac{1}{2}\text{erfc}(0) = \frac{1}{2}$$

Think how ridiculous it would be to use the same symbol to convey both binary 1 and 0. The resulting BER of 0.5 is the same as would be obtained by basing each decision entirely on the result of flipping a fair coin. The receiver does not gain any information from detecting the incoming symbols, and should not even bother.

10.6 BER of *M*-ary transmission systems

We can readily extend the above analysis of noise effects, specifically Eq. (10.45), to *M*-ary digital transmission systems involving $M > 2$ symbols, each representing $\log_2(M)$ bits. We will assume that errors involve the mistaking of one symbol for its nearest neighbours only, which means that noise contributions are not excessive but fall within a range that will cause the received state to cross *no more than one* decision boundary in any direction. We will further assume that all *M* symbols are equally likely to be sent. In addition to these assumptions, we must make important clarifications regarding applying Eq. (10.45) to *M*-ary systems and how bits are assigned to states in these systems.

- Eq. (10.45) was derived for binary systems where each symbol represents one bit, so it is also the symbol error rate (SER).
- The parameter E_b in Eq. (10.45) is the average energy per bit only in the context of binary systems. In applying this equation to *M*-ary systems we must note that this parameter is actually the average energy E_{km} of the two adjacent states S_k and S_m with correlation coefficient ρ. Thus we must rephrase Eq. (10.45) as

$$P_{e_{km}} = \frac{1}{2}\operatorname{erfc}\left(\sqrt{\frac{E_{km}(1-\rho)}{2N_o}}\right) \tag{10.51}$$

where $P_{e_{km}}$ is the probability of mistaking S_k for S_m. The average energy per bit that is applicable to *M*-ary systems is given by Eq. (9.14).

10.6.1 Gray coding and BER

Care is usually taken to ensure that adjacent states in the signal space of ASK, PSK and QAM systems represent bits that differ in only one bit position. In the event of a symbol error — usually involving mistaking adjacent states, this arrangement minimises the number of bit errors. In an *M*-ary system, one received symbol delivers a block of $\log_2(M)$ bits, and one symbol error causes δ bit errors; δ being the number of bit positions in which the sent block differs from the mistakenly received block. This is called the *code distance*. For example, the code distance between 0101 and 1001 is $\delta = 2$, and the code distance between 0101 and 1101 is $\delta = 1$. Thus in *M*-ary systems, BER is given by

$$\begin{aligned} \text{BER} &= \frac{\text{SER} \times \overline{\delta}}{\log_2 M}; \\ \overline{\delta} &= 1 \text{ with a Gray code scheme} \end{aligned} \tag{10.52}$$

where $\overline{\delta}$ is the mean code distance between nearest-neighbour states in the constellation. A *gray code* scheme is one in which bits are assigned to states in such a way that neighbouring states represent a block of bits or codeword that differ in only one bit position. Fig. 8.3 shows the sequence of codewords in a gray code scheme for $M = 2$, 4, 8, 16 and 32. Note that the sequence is 0110... for the rightmost bit or LSB; 00111100... for the next-rightmost, 0000111111110000... for the next, and so on, where one pattern is derived from a previous pattern by duplicating every bit. This scheme is *always employed* in all two-dimensional constellations (to the fullest possible extent) because it gives the smallest possible mean code distance $\overline{\delta} = 1$, which minimises BER. For all other arrangements $\overline{\delta}$ in Eq. (10.52) exceeds unity and is computed as a weighted mean that takes into account the fact that errors between adjacent states in the particular constellation may not be equally probable.

Fig. 10.13 shows the gray coding of square 16-QAM. This is achieved by following a gray code sequence in the horizontal direction for the leftmost two bits (conveyed by the in-phase channel), and a gray code sequence in the vertical direction for the rightmost two bits which are conveyed by the quadrature channel. By starting the sequence at different points of the gray code cycle various realisations of a gray code arrangement can be achieved which places a specific group of four bits at any desired location. This procedure can be generalised as follows for square QAM constellations of all values of M: Note that each state represents $\log_2 M$ bits of which $\log_2 M_i$ are conveyed by the in-phase channel and $\log_2 M_q$ by the quadrature channel — see Eqs. (9.29) and (9.30), and the constellation is a matrix of M_q rows and M_i columns. So you simply follow a gray code sequence in the horizontal direction for the leftmost $\log_2 M_i$ bits, and a gray code sequence in the vertical direction for the rightmost $\log_2 M_q$ bits.

10.6.2 M-ary ASK

M-ary ASK has a linear constellation as shown before in Fig. 9.10a. An error will occur in the detection of state S_k in Fig. 10.14 if this state is mistaken for either of

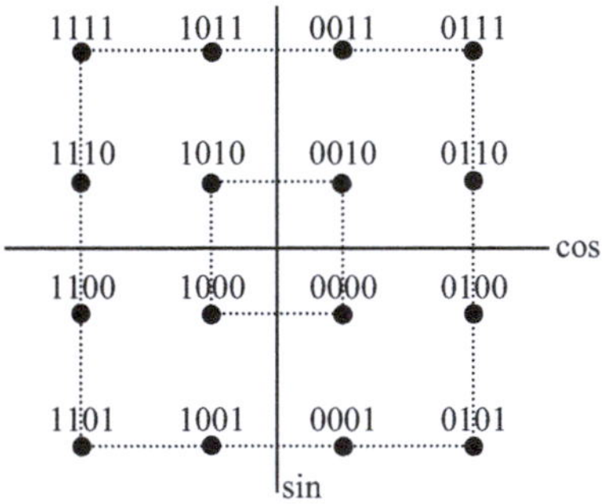

Fig. 10.13 Gray coding of square 16-QAM

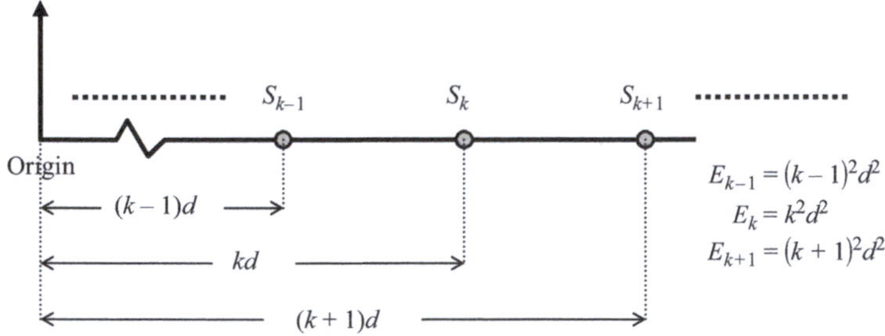

Fig. 10.14 Adjacent states in M-ary ASK

its nearest neighbours S_{k-1} or S_{k+1}. Denoting the probability of error in state S_k as P_{e_k}, the probability that S_k is mistaken for S_{k-1} as $P_{e_{k-}}$, and the probability that it is mistaken for S_{k+1} as $P_{e_{k+}}$, we may write

$$P_{e_k} = P_{e_{k-}} + P_{e_{k+}} \tag{10.53}$$

$P_{e_{k+}}$ follows from Eq. (10.51) with E_{km} given by — see Fig. 10.14:

$$E_{km} = \frac{E_k + E_{k+1}}{2} = \frac{k^2 d^2 + (k+1)^2 d^2}{2} = \frac{d^2}{2}[2k(k+1)+1]$$

and ρ given by Eq. (9.76) with $m = k + 1$, so that

$$1 - \rho = 1 - \frac{2k(k+1)}{k^2 + (k+1)^2} = \frac{1}{2k(k+1)+1}$$

Thus

$$\begin{aligned} P_{e_{k+}} &= \frac{1}{2}\mathrm{erfc}\left(\sqrt{\frac{d^2[2k(k+1)+1]}{2} \times \frac{1}{[2k(k+1)+1]} \times \frac{1}{2N_o}}\right) \\ &= \frac{1}{2}\mathrm{erfc}\left(\sqrt{\frac{d^2}{4N_o}}\right) \equiv P_{ea} \end{aligned} \tag{10.54}$$

This result is independent of k, and is therefore the probability (denoted by P_{ea}) of mistaking *any* symbol for an adjacent symbol. It follows from Eq. (10.53) that the probability of error in S_k is

$$P_{e_k} = 2P_{ea}$$

and this applies for all $k = 1, 2, 3, \ldots, M-2$. Symbols S_0 and S_{M-1} however have only one immediate neighbour, and hence

$$P_{e_1} = P_{e_{M-1}} = P_{ea}$$

Since all symbols are equally likely to be transmitted, the desired probability of symbol error or SER in the M-ary ASK detector is obtained by simply averaging these errors over all symbols. Thus

$$\begin{aligned} \mathrm{SER} &= \frac{1}{M}\left[P_{ea} + \sum_{k=1}^{M-2} 2P_{ea} + P_{ea}\right] = \frac{2(M-1)P_{ea}}{M} \\ &= \frac{(M-1)}{M}\mathrm{erfc}\left(\sqrt{\frac{d^2}{4N_o}}\right) \end{aligned} \tag{10.55}$$

It is more useful to express this equation in terms of the average energy per bit E_b in the M-ary ASK system. Eq. (9.24) gives an expression for E_b in terms of d^2 from which we obtain

$$d^2 = \frac{E_b \times 6\log_2 M}{(M-1)(2M-1)}$$

so that

$$\text{SER} = \frac{(M-1)}{M}\text{erfc}\left(\sqrt{\frac{3E_b\log_2 M}{2N_o(M-1)(2M-1)}}\right)$$

Finally, Eq. (10.52) yields the desired BER of M-ary ASK systems as

$$\text{BER} = \frac{(M-1)}{M\log_2 M}\text{erfc}\left(\sqrt{\frac{3\log_2 M}{2(M-1)(2M-1)}\frac{E_b}{N_o}}\right) \tag{10.56}$$

This is a remarkable equation. It gives the BER of M-ary ASK explicitly in terms of M and our now familiar E_b/N_o. Note that when $M=2$ this equation reduces nicely to Eq. (10.48) for binary ASK, as expected. Fig. 10.15 shows a plot of BER against E_b/N_o for various values of M. We see that BER increases rapidly with M. For example, at $E_b/N_o = 14$ dB, the BER is 2.7×10^{-7} for $M=2$, but increases dramatically to 2.8×10^{-3} for $M=4$. To put it in another way, we need to increase transmitted signal power very significantly in order to obtain the same BER in multilevel ASK as in binary ASK. For example, to obtain the same BER of 1×10^{-7} in both binary ASK and 32-ASK the transmitted power in the 32-ASK system must be increased by 27.7 dB (which includes a 20.7 dB increase in E_b/N_o and a factor of 5 ($\equiv$ 7 dB) increase in bit rate). This finding is, however, not at all surprising. 32-ASK uses the same bandwidth as binary ASK but increases transmission bit rate by a factor of 5 (since it delivers five bits per received symbol). The information capacity law (discussed in Chapter 8) stipulates that there is a price to pay in terms of signal power for this improved bandwidth efficiency.

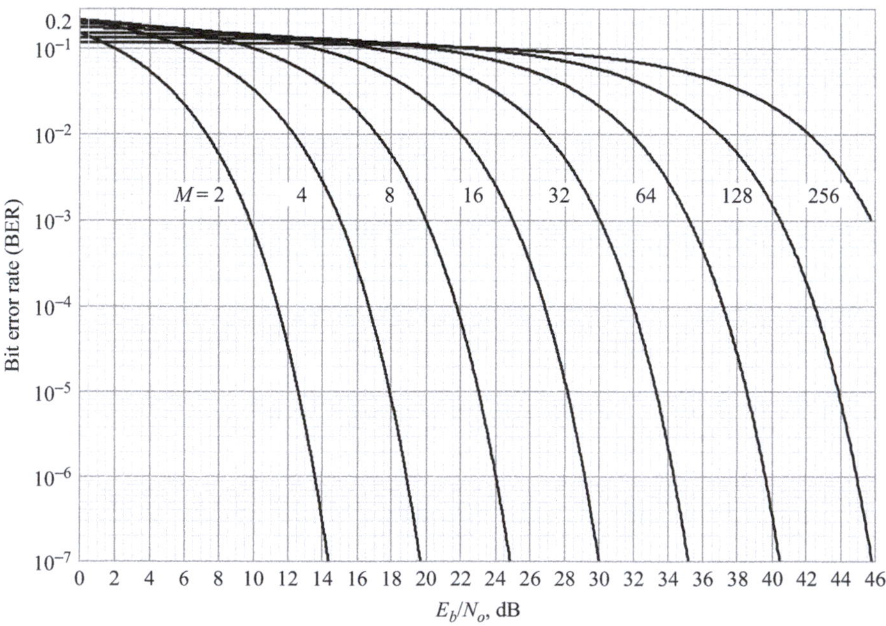

Fig. 10.15 BER of M-ary ASK for $M=2$, 4, 8, 16, 32, 64, 128, 256

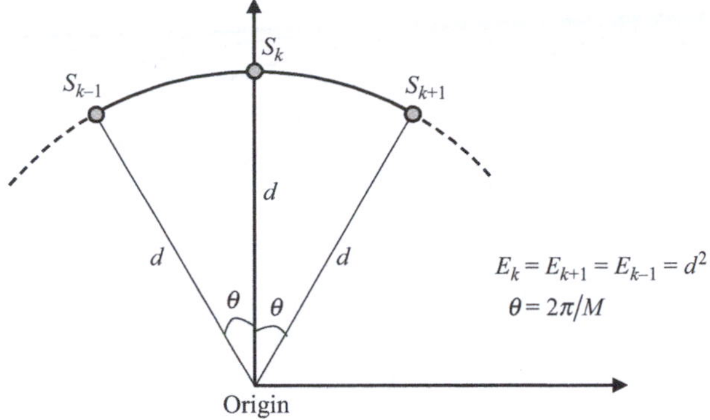

Fig. 10.16 Adjacent states in M-ary PSK

10.6.3 M-ary PSK

The circular constellation of M-ary PSK was shown before in Fig. 9.10c. Fig. 10.16 shows state S_k and its nearest neighbours S_{k-1} and S_{k+1}. An error occurs in the detection of S_k when this state is mistaken for either S_{k-1} or S_{k+1}. Using a similar notation and reasoning as in the previous Section, the probability of error P_{ek} in state S_k is as given by Eq. (10.53).

Now the probability $P_{e_{k+}}$ of mistaking S_k for S_{k+1} follows from Eq. (10.51) with $E_{km} = d^2$ (see Fig. 10.16), and $\rho = \cos(2\pi/M)$ from Eq. (9.77) so that

$$\begin{aligned} P_{e_{k+}} &= \frac{1}{2}\operatorname{erfc}\left(\sqrt{\frac{d^2[1 - \cos(2\pi/M)]}{2N_o}}\right) \\ &= \frac{1}{2}\operatorname{erfc}\left(\sqrt{\frac{E_b[1 - \cos(2\pi/M)]\log_2 M}{2N_o}}\right) \equiv P_{ea} \end{aligned} \tag{10.57}$$

where we have made use of Eq. (9.27) to eliminate d^2. Notice that this result is independent of k and is therefore the probability (denoted by P_{ea}) of mistaking *any* symbol for an adjacent symbol. It follows from Eq. (10.53) that the probability of error in S_k is

$$P_{e_k} = \begin{cases} 2P_{ea}, & M \geq 4 \text{ (Two nearest neighbours)} \\ P_{ea}, & M = 2 \text{ (Only one nearest neighbour)} \end{cases} \tag{10.58}$$

and this applies for all $k = 0, 1, 2, \ldots, M-1$, and is therefore the probability of error in any M-ary PSK symbol. That is, symbol error rate SER $= P_{e_k}$ as given above, so that using Eq. (10.52) yields the desired BER:

$$\text{BER} = \begin{cases} \dfrac{1}{\log_2 M}\operatorname{erfc}\left(\sqrt{\dfrac{\log_2 M}{2}\left[1 - \cos\left(\dfrac{2\pi}{M}\right)\right]\dfrac{E_b}{N_o}}\right), & M \geq 4 \\ \dfrac{1}{2}\operatorname{erfc}\left(\sqrt{E_b/N_o}\right), & M = 2 \end{cases} \tag{10.59}$$

Note that $M = 2$ corresponds to the BPSK case previously solved in Eq. (10.46). Fig. 10.17 provides plots of Eq. (10.59) for various values of M. It can be seen that

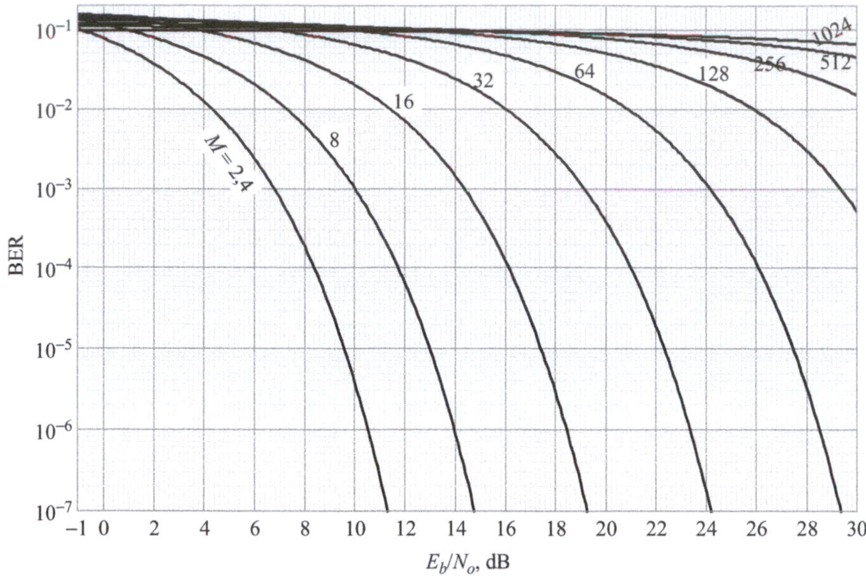

Fig. 10.17 BER of M-ary PSK for $M = 2, 4, 8, 16, 32, 64, 128, 256, 512, 1024$

the BER curves for $M = 2$ and $M = 4$ are coincident. This means that 4-PSK — more commonly referred to as *quaternary PSK* (QPSK) — has exactly the same BER as BPSK for a given E_b/N_o, but it transmits at *twice* the bit rate of BPSK for a given bandwidth since it conveys two bits per symbol. Put another way, QPSK uses half the bandwidth required by BPSK to deliver the same bit rate R_b. However, there is a price for this doubling of bandwidth efficiency. If you look at Eq. (9.15) you will see that transmitted power $P = E_bR_b$, which means that for the same E_b/N_o and hence BER, the power in QPSK is (as bit rate R_b) double that of BPSK. From this and the rest of Fig. 10.17 we see that for a given bit rate, the transmitted power in an M-ary PSK system must be increased in order to maintain the error rate obtainable with BPSK. However, transmission bandwidth is reduced by a factor of $\log_2 M$. So this is a case of trading power for bandwidth.

By comparing Figs. 10.16 and 10.17 it is clear that the noise performance of M-ary PSK is significantly superior to that of M-ary ASK. In fact 8-PSK has about the same BER as a BASK system that transmits the same average energy per bit. You would, for example, have to increase the transmitted power in a 4-ASK system by a factor of 7 (or 8.5 dB) to achieve the same BER as in a QPSK system. Since M-ary ASK and M-ary PSK require the same bandwidth, there is really no reason to choose multilevel ASK, except for the simplicity of its modulation and detection circuits.

10.6.4 M-ary QAM

You may wish to quickly review the treatment of M-ary QAM in Section 9.3.2.3 before studying this Section. We know from that discussion that the $\log_2 M$ bits delivered per symbol interval in M-ary QAM are conveyed using two orthogonal and hence independent channels, with the in-phase channel carrying $r_i = \log_2 M_i$

bits and the quadrature channel carrying $r_q = \log_2 M_q$ bits, where $r_i + r_q = \log_2 M$. If BER_i and BER_q are the BERs of the in-phase and quadrature channels respectively, it means that in one symbol interval we will receive r_i bits with $r_i \times \text{BER}_i$ bit errors on average through the in-phase channel, and r_q bits with $r_q \times \text{BER}_q$ bit errors on average through the quadrature channel, so that the BER of the combined transmission (i.e. *M*-ary QAM system) is given by

$$\begin{aligned}\text{BER} &= \frac{r_i \times \text{BER}_i + r_q \times \text{BER}_q}{r_i + r_q} = \frac{\log_2(M_i) \times \text{BER}_i + \log_2(M_q) \times \text{BER}_q}{\log_2(M)} \\ &= \frac{\text{SER}_i + \text{SER}_q}{\log_2(M)} \end{aligned} \tag{10.60}$$

where we have made use of Eq. (10.52) in the last line. Determining the BER of an *M*-ary QAM system therefore reduces nicely to the less complicated task of calculating the symbol error rates SER_i and SER_q of M_i-ary and M_q-ary transmissions both of which have a one-dimensional signal space. It is not possible to obtain a single expression for BER that is applicable to all types of *M*-ary QAM constellations, so we will tailor our analysis in this section specifically to the popular *M*-ary square QAM which was introduced in Section 9.3.2.3 and broken down into its orthogonal channels structure in Fig. 9.10h. Each constellation has states arranged along a single axis with spacing $2d$, and with half the states located on either side of the signal space origin.

Eq. (10.60) shows the importance of equal loading of the two channels, otherwise the value of BER will be dominated by one of the two terms in the numerator of the RHS. However, equal loading is only possible when M is an even power of 2, e.g. $M = 4, 16, 64, 256, \ldots$. In that case, $M_i = M_q = \sqrt{M}$, and the two channels have identical BERs, so that Eq. (10.60) simplifies to

$$\text{BER} = \frac{\frac{1}{2}\log_2(M) \times \text{BER}_i + \frac{1}{2}\log_2(M) \times \text{BER}_i}{\log_2(M)} = \text{BER}_i \tag{10.61}$$

Since $M_i = M_q = \sqrt{M}$

That is, the BER of the *M*-ary QAM system is given by the BER of either channel. But when M is an odd power of 2, e.g. $M = 8, 32, 128, \ldots$, then the closest we can get to equitable loading is by setting

$$M_i = \sqrt{2M}; \quad M_q = \sqrt{M/2} \tag{10.62}$$

and using these values in the evaluation of Eq. (10.60).

Referring to the linear constellation of the in-phase channel in Fig. 9.10h, let us determine the probability that noise will cause the detector to mistake state S_k for S_{k+1}. This is given by Eq. (10.51), which requires knowledge of the average energy E_{km} and correlation coefficient ρ of states S_k and S_{k+1}. It is clear from Fig. 9.10h that the first parameter is

$$\begin{aligned} E_{km} &= \frac{1}{2}\left\{d^2[2k - M_i + 1]^2 + d^2[2(k+1) - M_i + 1]^2\right\} \\ &= d^2\left[p^2 + (p+2)^2\right], \quad \text{where } p \equiv 2k - M_i + 1 \end{aligned} \tag{10.63}$$

Next, Eq. (9.75) gives the correlation coefficient of the two states as

$$\rho = \frac{2d[2k - M_i + 1]d[2(k+1) - M_i + 1]}{d^2[2k - M_i + 1]^2 + d^2[2(k+1) - M_i + 1]^2}$$

$$= \frac{p(p+2)}{p^2 + 2p + 2}, \quad \text{where } p \equiv 2k - M_i + 1 \tag{10.64}$$

so that the term $E_{km}(1-\rho)$ in Eq. (10.51) is given by

$$E_{km}(1-\rho) = d^2\left[p^2 + (p+2)^2\right]\left[1 - \frac{p(p+2)}{p^2+2p+2}\right]$$

$$= 2d^2$$

Substituting this into Eq. (10.51) yields the probability of mistaking S_k for S_{k+1}:

$$P_{e_{k+}} = \frac{1}{2}\,\text{erfc}\left(\sqrt{d^2/N_o}\right) \equiv P_{ea} \tag{10.65}$$

This result is independent of k and is therefore the probability (denoted by P_{ea}) of mistaking *any* state for an adjacent state. Note that each of the $M_i - 2$ inner states of the linear constellation has two adjacent states and hence a probability of error equal to $2P_{ea}$, whereas the end states, namely S_0 and S_{M_i-1} have only one adjacent state and therefore a probability of error equal to P_{ea}. The probability of symbol error SER_i in the in-phase channel is the average of these probabilities:

$$\text{SER}_i = \frac{1}{M_i}[P_{ea} + (M_i - 2) \times 2P_{ea} + P_{ea}]$$

$$= \frac{2(M_i - 1)}{M_i}P_{ea} = \frac{(M_i - 1)}{M_i}\,\text{erfc}\left(\sqrt{d^2/N_o}\right) \tag{10.66}$$

Recalling the expression for the average energy per bit of a square M-ary QAM constellation given in Eq. (9.35) and making d^2 the subject of this expression, we see that in Eq. (10.66)

$$d^2 = \frac{3E_b\log_2 M}{2(aM - 1)}$$

and hence the SER of the in-phase channel is given by

$$\text{SER}_i = \frac{(M_i - 1)}{M_i}\,\text{erfc}\left(\sqrt{\frac{3E_b\log_2 M}{2(aM-1)N_o}}\right) \tag{10.67}$$

Similarly, the SER of the quadrature channel is given by

$$\text{SER}_q = \frac{(M_q - 1)}{M_q}\,\text{erfc}\left(\sqrt{\frac{3E_b\log_2 M}{2(aM-1)N_o}}\right) \tag{10.68}$$

The BER of the in-phase channel now follows from Eq. (10.67):

$$\text{BER}_i = \frac{\text{SER}_i}{\log_2 M_i} = \frac{(M_i - 1)}{M_i\log_2 M_i}\,\text{erfc}\left(\sqrt{\frac{3E_b\log_2 M}{2(aM-1)N_o}}\right) \tag{10.69}$$

Recalling Eqs. (10.61) and (9.35), it is clear that the BER of M-ary QAM system *when M is an even power of* 2 is given by Eq. (10.69) with $a=1$ and $M_i = \sqrt{M}$. Thus

$$\text{BER} = \frac{2}{\log_2 M}\left(\frac{\sqrt{M}-1}{\sqrt{M}}\right)\text{erfc}\left(\sqrt{\frac{3\log_2 M}{2(M-1)}\frac{E_b}{N_o}}\right) \tag{10.70}$$

Notice that when $M=4$ this equation reduces to the same expression as the M-ary PSK equation (10.59) for $M=4$. This is expected since QPSK and square 4-QAM are identical.

For M an odd power of 2, the unequal loading stipulated in Eq. (10.62) applies, and BER is given by Eq. (10.60) with SER_i as in Eq. (10.67), SER_q as in (10.68), and $M_i = \sqrt{2M}$, $M_q = \sqrt{M/2}$, $a = 1.25$. Thus

$$\begin{aligned}\text{BER} &= \frac{1}{\log_2 M}\left[\frac{\sqrt{2M}-1}{\sqrt{2M}} + \frac{\sqrt{\frac{M}{2}}-1}{\sqrt{\frac{M}{2}}}\right]\text{erfc}\left(\sqrt{\frac{3\log_2 M}{2(1.25M-1)}\frac{E_b}{N_o}}\right)\\ &= \frac{2}{\log_2 M}\left[\frac{\sqrt{M}-3\sqrt{2}/4}{\sqrt{M}}\right]\text{erfc}\left(\sqrt{\frac{3\log_2 M}{2(1.25M-1)}\frac{E_b}{N_o}}\right)\end{aligned} \tag{10.71}$$

In summary, the BER of any M-ary square QAM for $M \geq 4$ is given by

$$\text{BER} = \frac{2}{\log_2 M}\left[\frac{\sqrt{M}-u}{\sqrt{M}}\right]\text{erfc}\left(\sqrt{\frac{3\log_2 M}{2(aM-1)}\frac{E_b}{N_o}}\right)$$

where (10.72)

$a = u = 1$, for M an even integer power of 2

$a = 1.25$; $u = \frac{3\sqrt{2}}{4}$, for M an odd integer power of 2

The above result is shown graphically in Fig. 10.18 from which values of E_b/N_o required to achieve a desired BER level in various square QAM systems can be read. Notice the closeness between the BER curves for M an odd power of 2 and the next higher M (an even power of 2). For example, the horizontal shift between the BER curves for 8-ary and 16-ary QAM is only ~0.9 dB, compared to a shift of ~3 dB between the BER curves for 8-ary and 4-ary QAM. This is because errors in a 2^{2n-1}-ary system, $n=2, 3, 4, \ldots$, are dominated by contributions from its more heavily loaded channel which carries the same number of bits per symbol as each channel of the next higher 2^{2n}-ary system. A word of caution is in order here. This closeness should not be interpreted to mean for example that 8-QAM and 16-QAM systems require near-equal transmitted powers in order to achieve the same BER, just as this could not be said about BPSK and QPSK in spite of the perfect coincidence of their BER curves (see Fig. 10.17). In fact, in view of Eq. (9.15), the power required to achieve a specified BER in an M_2-ary QAM transmission will exceed that required to achieve the same BER in an M_1-ary transmission of the same bandwidth by

$$\Delta P = 10\log_{10}\left[\frac{\log_2 M_2}{\log_2 M_1}\right] + \Delta_{E_b/N_o}\ \text{dB} \tag{10.73}$$

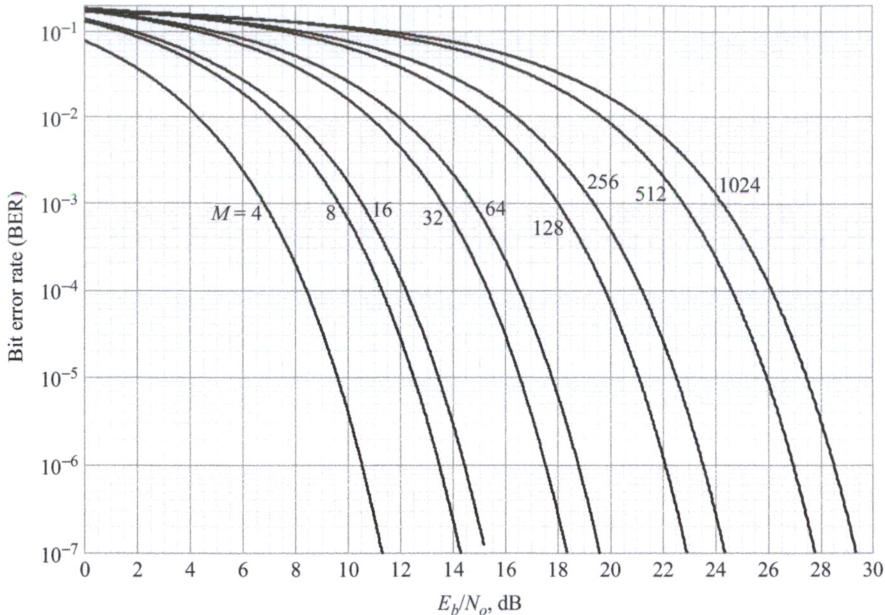

Fig. 10.18 BER of square M-ary QAM for $M = 4$, 8, 16, 32, 64, 128, 256, 512, 1024

where $M_2 > M_1$, and Δ_{E_b/N_o} is the horizontal shift between their BER curves. For example, as already quoted above, $\Delta_{E_b/N_o} \approx 0.9$ dB for $M_2 = 16$ and $M_1 = 8$; and $\Delta_{E_b/N_o} \approx 3$ dB for $M_2 = 8$ and $M_1 = 4$.

By comparing Fig. 10.18 with Fig. 10.17 it is clear that, excluding the case $M = 4$ for which the two systems are identical, then M-ary square QAM has a better AWGN noise performance than its corresponding M-ary PSK system. For example, to achieve a BER of 1×10^{-7}, a 16-PSK system requires 4 dB more transmitted power than 16-QAM, whereas 64-PSK requires 10 dB more power than 64-QAM. Note however that we have assumed band limited white noise as the only source of impairment in the channel. If there are other channel impairments, in particular amplitude distortions, then M-ary QAM will not perform as well as depicted in Fig. 10.18.

10.6.5 M-ary FSK

Orthogonal M-ary FSK transmits M mutually orthogonal symbols each of which has the same symbol energy E_s and conveys $\log_2 M$ bits. We obtain the probability of symbol error P_{es} by substituting $\rho = 0$ and $E_{km} \equiv E_s = E_b \log_2 M$ into Eq. (10.51):

$$P_{es} = \frac{1}{2} \operatorname{erfc}\left(\sqrt{\frac{E_b \log_2 M}{2N_o}}\right) \tag{10.74}$$

Symbols are located at the same distance from the origin along M mutually orthogonal axes in an M-dimensional signal space. Thus the symbols are equidistant and each symbol has $M-1$ adjacent states for which it can be mistaken, which gives the SER:

$$\mathrm{SER} = (M-1)P_{es} = \left(\frac{M-1}{2}\right)\mathrm{erfc}\left(\sqrt{\frac{E_b \log_2 M}{2N_o}}\right) \tag{10.75}$$

This is actually a worst case estimate which gives an upper bound for SER because we have tacitly assumed in summing the $M-1$ probabilities that when an error occurs the received state is nearer to only one other state than the transmitted state, which is true only in one-dimensional and two-dimensional signal spaces that describe ASK, PSK and QAM. When the dimension exceeds 2 as in M-ary FSK with $M \geq 4$, then there will be some situations in which a received state is nearer to two or more other states than the transmitted state, particularly so at small values of E_s/N_o where noise energy is comparable to symbol energy. In that case, there will be regions of intersection which are counted more than once by the summing done in Eq. (10.75). At $M=2$ however there is only one adjacent state and Eq. (10.75) is an exact result. It should therefore be noted that Eq. (10.77) derived below for the BER of M-ary FSK does exaggerate BER slightly for small values of E_b/N_o at modest $M \geq 4$, but quite significantly at large M as multiple counting becomes excessive. The use of this equation above $M = 1024$ is therefore to be discouraged.

What remains is for us to convert *SER* into *BER* using Eq. (10.52). This requires $\bar{\delta}$, the average code distance between the bits of a given state and the bits of adjacent states. Because states are equidistant and represent all k-bit combinations, where $k = \log_2 M$, this parameter will be the same for all states. We therefore focus on calculating $\bar{\delta}$ for the all-zero-bits state which reduces the problem nicely to that of calculating the average number of 1s in all possible combinations of k bits, excluding the all-zero combination. Thus, noting that there are $\binom{k}{m}$ arrangements of k bits that contain m number of 1s, or that there are a total of $k \times 2^k$ bits in the signal space half of which are 1s, whichever thinking you prefer leads to

$$\bar{\delta} = \frac{1}{2^k - 1}\sum_{m=1}^{k} m\binom{k}{m} = \frac{1}{M-1}\sum_{m=1}^{k} m\binom{k}{m} = \frac{2^{k-1}k}{M-1} = \frac{M \log_2 M}{2(M-1)}$$

$$\text{where} \quad \binom{k}{m} = \frac{k!}{m!(k-m)!}; \quad k = \log_2 M \tag{10.76}$$

Using Eq. (10.76) for $\bar{\delta}$ and Eq. (10.75) for *SER* in Eq. (10.52) yields

$$BER \approx \frac{M}{4}\mathrm{erfc}\left(\sqrt{\frac{E_b \log_2 M}{2N_o}}\right) \quad \begin{matrix}(M\text{-ary FSK})\\ M \leq 1024\end{matrix} \tag{10.77}$$

When $M=2$ this equation reduces to Eq. (10.49) for binary FSK as expected. Results from Eq. (10.77) are plotted in Fig. 10.19 for values of M from 2 to 1024 against E_b/N_o in dB. It can be seen that at any given E_b/N_o (above $\sim$ 1 dB), BER decreases as M increases. This is a manifestation of noise rejection by M-ary FSK as discussed in Section 9.6.2. Alternatively, as M increases, the same BER is achieved at a lower E_b/N_o, which is the result of a trade-off of bandwidth for improved noise performance. Comparing the BER curves for M-ary FSK with those

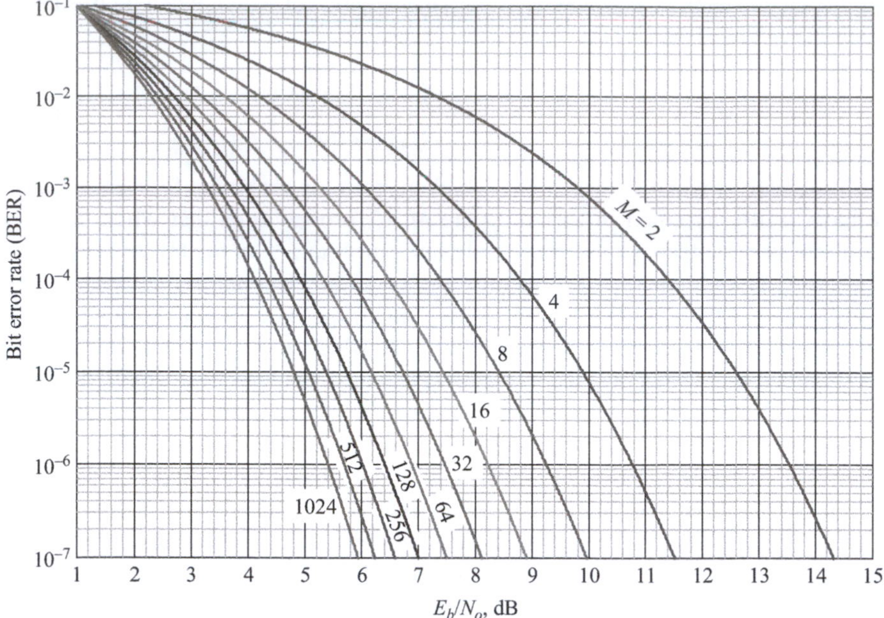

Fig. 10.19 BER of orthogonal M-ary FSK for $M = 2$, 4, 8, 16, 32, 64, 128, 256, 512, 1024

of M-ary ASK, PSK and QAM it can be seen that these latter exhibit an opposite trend in that as M increases a higher value of E_b/N_o is needed to maintain the same BER, confirming their trading of signal power for improved bandwidth efficiency.

You may recall that the bandwidth of M-ary FSK increases with M as given by Eq. (9.87) so that bandwidth $B \to \infty$ as $M \to \infty$. Thus the trend observed in Fig. 10.19 is an indication that the technique of M-ary FSK provides a mechanism that allows the system designer to use more and more bandwidth in order to continue operating reliably with less and less signal power. We learnt however in Chapter 8 that this exchange cannot carry on indefinitely. According to the Shannon-Hartley information capacity law, signal power (relative to noise power) cannot be less than the Shannon limit $E_b/N_o = -1.6$ dB if reliable operation is to continue, even if infinite bandwidth is provided.

10.7 Summary

Noise is inherent in all communications systems and the system designer does not need to incur huge system costs in pursuit of a goal of system noise reduction. In fact a *zero-noise* goal is not only unattainable; it is absolutely unnecessary. Given the right knowledge and skills, we can design very reliable transmission systems through noisy channels.

In this chapter we have learnt the salient points about the statistical characterisation of random signals in general and AWGN in particular. We have also learnt how to quantify the noisiness of devices and systems and have applied this knowledge to communication system receivers, particularly radio reception systems, and have been able to calculate noise power per unit bandwidth at the reference point of such systems. We also learnt how to calculate signal power at the same point and hence to determine the ratio between carrier signal power and noise power C/N as well as the ratio between signal energy per bit and the noise power per unit bandwidth E_b/N_o. We also performed extensive analysis of the impact of noise on various digital transmission systems and derived expressions for the BER of such systems when coherent symbol detection is employed. We found that in each system BER depends entirely on E_b/N_o and it monotonically decreases as E_b/N_o increases.

In other words, it is the system designer rather than noise that calls the shots. Given a modulation or line code scheme the desired BER that is acceptable to the particular communication service can be achieved simply by providing a sufficient level of E_b/N_o. And if that is inconvenient, expensive or even impractical then the designer may simply switch modulation schemes in order to trade bandwidth for a reliable operation at an affordable lower E_b/N_o. What if that still leaves us short of the target BER? Then we have one more weapon in our design arsenal, namely error control coding, which is the subject of the next chapter.

10.8 Reference

[1] Otung, I. (2001). *Communication engineering principles*. Basingstoke: Palgrave Macmillan, Chapter 9

Chapter 11
Error control coding

In Chapter Eleven

- ✓ Purpose and classes of error control coding.
- ✓ Automatic repeat request (ARQ): We discuss four types of ARQ and present a detailed analysis of data throughput and link utilisation in systems employing this technique.
- ✓ Efficacy of error detection: We quantify the reduction in probability of bit error achievable through error detection in general, and present an example based on single parity check.
- ✓ Concept of forward error correction: A non-mathematical and graphical introduction to the 'magic' of actually knowing which bit is in error in a received unknown message. Three techniques are discussed, namely product code, Hamming code and repetition code. We also highlight the potential danger of mis-correction.
- ✓ Interleaving and concatenation: We discuss the use of interleaving sandwiched between an inner and an outer code to improve coding gain.
- ✓ Linear block codes: We discuss the mathematical computations involved in linear block codes and the elegant circuit implementation of the cyclic variety using shift registers. The operation of polynomial multiplication and division circuits are also discussed in detail.
- ✓ Reed-Solomon (RS) code: We introduce the theory of finite fields and discuss the principles involved in RS coding and decoding supported by worked examples.

11.1 Introduction

The Shannon-Hartley information capacity law (discussed in Chapter 8) lays down the rule for the exchange between transmitted signal power and required transmission bandwidth in the design of reliable communication through a noisy channel. It also provides a benchmark against which practical systems can be compared, which states the minimum E_b/N_o (ratio between signal energy per bit and noise power per unit bandwidth) required for reliable communication at a given bandwidth efficiency (bit rate per hertz of bandwidth). Although Shannon's channel coding theorem (on which this law is based) interprets 'reliable communication' to mean practically *error-free* transmission, the interpretation and aim in design is always *error-rare* transmission in which the bit error rate (BER) is finite but falls below a threshold BER_{th} set according to the particular service. For example, the threshold might be set at $BER_{th} = 10^{-4}$ for voice communication and at 10^{-7} for most data services.

One way of exchanging between bandwidth and signal power is through a choice of modulation scheme, the subject of the previous chapter. For example, if $BER_{th} = 10^{-4}$ and the service requires transmission at a constant bit rate of 64 kb/s, then Fig. 10.17 shows that (assuming AWGN channel and ideal Nyquist filtering in which band pass transmission bandwidth B = symbol rate R_s) we can choose the one-bit-per-symbol BPSK modulation scheme to deliver this service at the cost of $E_b/N_o = 8.4$ dB and $B = 64$ kHz. We do, however, have a multiplicity of options that allow us to use less bandwidth but more signal power for this service. For example, this service can be delivered at $E_b/N_o = 16.2$ dB, $B = 16$ kHz using the four bits per symbol 16-PSK modulation scheme; and at $E_b/N_o = 15.3$ dB, $B = 12.8$ kHz using the five bits per symbol 32-QAM scheme (see Fig. 10.18). We could even choose to go in the opposite direction and use less signal power but more bandwidth. For example, the 1024-FSK modulation scheme allows us to deliver this particular service at $E_b/N_o = 4.1$ dB, $B = 3280$ kHz. To see this, you will need to consult Fig. 10.19 for the E_b/N_o value and Eq. (9.87) for the bandwidth value.

The choice of modulation scheme does truly give us a theoretically limitless (albeit practically constrained) set of options in exchanging between bandwidth and signal power. See also the Sun transmitter comments in Section 8.3.3. However, the performance of all these modulation schemes falls short (and sometimes significantly so) of the Shannon-Hartley benchmark. To compare just two of the examples given above, the 1024-FSK scheme transmits 64 kb/s using 3280 kHz of bandwidth, which corresponds to bandwidth efficiency $\eta = 64/3280 \approx 0.02$ b/s/Hz. Referring to Fig. 8.15, we see that the minimum required E_b/N_o is –1.56 dB, but 1024-FSK uses 5.7 dB more than this minimum. And our BPSK, operating at $\eta = 1$ b/s/Hz uses $E_b/N_o = 8.4$ dB, which (referring to Fig. 8.15) is 8.4 dB higher than the minimum possible. To design more efficient digital communication systems that are closer to the Shannon-Hartley benchmark we must find more efficient ways of exchanging between bandwidth and signal power, and that is the reason for this chapter.

Error control coding was introduced in Section 1.3.3 as the main task of channel coding. Prior to modulation or line coding, the output bits from a source encoder are further processed in a channel encoder which introduces redundant bits that can be used at the receiver to detect and/or correct channel-induced bit errors. This arrangement is shown in Fig. 11.1, where it should be noted that the blocks labelled modulator or line coder at the transmitter and detector at the receiver include the filtering operations discussed in Chapter 8. For every k bits from the source encoder the channel encoder adds $n - k$ redundant bits to produce n coded bits which are then converted into transmitted symbols by the modulator or line coder. The channel encoder is said to have code rate $r = k/n$, and its code is described as an (n, k) code.

To see how error control coding supports the exchange between signal power and transmission bandwidth, consider again the 64 kb/s transmission discussed above and imagine that we are able to design a channel encoder of code rate ½ that is able to correct all errors in a received coded bit stream provided they do not exceed seven bit errors in 100 bits on average. Combining this coding scheme with BPSK modulation, we would need to transmit at 128 kb/s (to cover 64 kb/s of information plus 64 kb/s of redundant bits) and hence require 128 kHz of bandwidth. Using 128 kHz in order to send 64 kb/s of information means that our bandwidth efficiency is now $\eta = 0.5$ b/s/Hz. Furthermore, we now only need to deliver a BER of 0.07 to the channel decoder input in order to have reliable

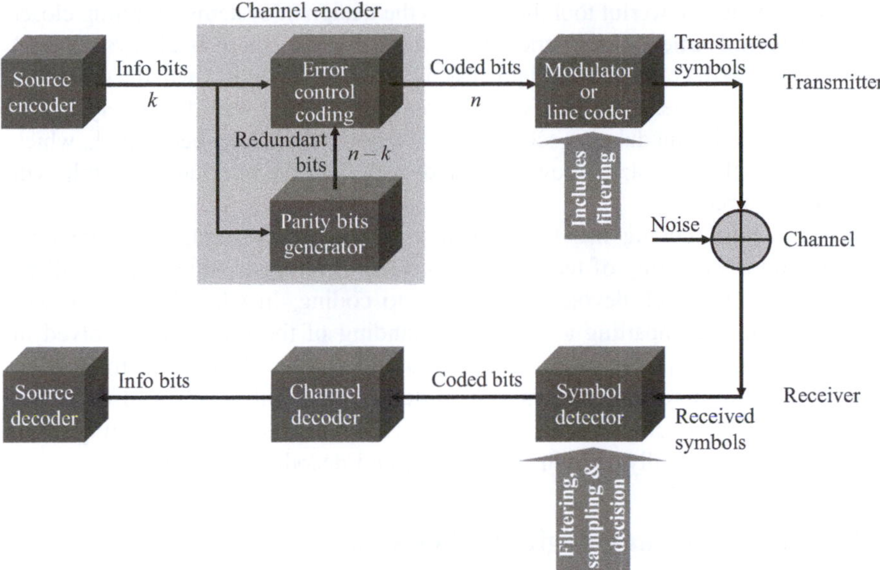

Fig. 11.1 Error control coding

communication, the decoder being able to correct all errors when its input BER ≤ 0.07, and this (from the BPSK curve of Fig. 10.17) means a required E_b/N_o of 0.37 dB. Therein lies the signal power versus bandwidth exchange facilitated entirely by error control coding without recourse to modulation scheme switching: The uncoded BPSK in this example has operating point $E_b/N_o = 8.4$ dB, $B = 64$ kHz; whereas the coded BPSK operates at $E_b/N_o = 0.37$ dB, $B = 128$ kHz.

So without changing modulation scheme, we have made some savings in signal power at the expense of bandwidth by using error control coding. But how much power have we really saved? It would be incorrect to subtract the two values of E_b/N_o to arrive at a saving of 8 dB in power since bit rates are different in the two cases — we are sending more bits per second in the coded scheme and that will eat into our power saving. To obtain a correct measure of power saving, we need the parameter C/N_o for both systems, which, working in logarithmic units, is

$$\frac{C}{N_o} = \frac{E_b}{N_o} + R_b = \begin{cases} 8.4 + 10\log_{10}\left(64 \times 10^3\right) = 56.46 \text{ dBHz} & \text{(Uncoded BPSK)} \\ 0.37 + 10\log_{10}\left(128 \times 10^3\right) = 51.44 \text{ dBHz} & \text{(Coded BPSK)} \end{cases}$$

Thus, the coded BPSK scheme has enabled us to trade a doubling in bandwidth for a 5 dB saving in transmitted signal power. What about our other interest in operating closer to the benchmark specification of Shannon-Hartley? We earlier saw that our uncoded BPSK system operates at 8.4 dB worse than the specification. The coded BPSK system on the other hand operates at bandwidth efficiency $\eta = 0.5$, which according to Shannon-Hartley (see Fig. 8.15) needs a minimum E_b/N_o of -0.82 dB. Thus the coded system, operating at $E_b/N_o = 0.37$ dB, is within 1.2 dB of the benchmark, which is much closer than uncoded BPSK can manage while maintaining reliable communication in a noisy channel.

The above discussion and examples are meant to highlight the fact that error control coding facilitates an efficient trade-off between signal power and transmission bandwidth and is a powerful tool that enables the design of systems operating closer to the Shannon-Hartley benchmark. The small price (yes there is always a price) is increased system complexity and processing delay. Used on its own, the swapping of modulation schemes will also facilitate this trade-off but will do so very inefficiently. An uncoded modulation scheme will always be far off the benchmark, which means that much more signal power or bandwidth than the minimum possible will always be required.

Error control coding has been an area of intense research for more than 50 years and a vast array of techniques has been developed, which cannot all be covered even in a book devoted exclusively to coding. In what follows we will primarily focus on imparting a clear understanding of the principles involved in error control coding in general, and on the parameters and design constraints that govern their use in communication system design. We will however also discuss linear binary block coding and the non-binary Reed-Solomon code, including a few worked examples to fully explain their coding and decoding algorithms.

11.2 General terminologies and classes

The smaller the number of redundant bits required to control a given number of errors the more *efficient* the coding scheme is said to be. To see the importance of code efficiency, consider again our 64 kb/s transmission and two coding schemes both of which can correct all errors in a 32-QAM transmission provided error rate does not exceed one bit error in 10 received bits. One coder (coder *a*) works at a code rate $r_1 = ½$, whereas coder *b* has code rate $r_2 = 1/3$. Thus we can aim to deliver a BER of 0.1 to each decoder, which (from the 32-QAM curve in Fig. 10.18) requires $E_b/N_o = 4.32$ dB. Coder *a* involves bit rate $R_b = 128$ kb/s, bandwidth $B = 25.6$ kHz, and bandwidth efficiency $\eta = 64/25.6 = 2.5$ b/s/Hz; whereas coder *b* involves $R_b = 192$ kb/s, $B = 38.4$ kHz and $\eta = 64/38.4 = 1.67$ b/s/Hz. Recall that 32-QAM conveys five bits per symbol, so symbol rate and hence bandwidth in this assumed Nyquist filtered channel is obtained by dividing each bit rate by 5. According to the Shannon-Hartley law, (Fig. 8.15) the minimum E_b/N_o is 2.7 dB for coder *a*, and 1.16 dB for coder *b*. Thus the more efficient coder is closer to the benchmark, being only 1.62 dB off the mark, whereas the less efficient coder is 3.16 dB off. Furthermore, computing the C/N_o of the two systems as earlier done, we can establish that the signal power required by the less efficient coder exceeds that required by the more efficient coder by an amount equal to the ratio of their code rates expressed in dB, which in this case is $10 \log_{10}(r_1/r_2) =$ 1.76 dB. It should therefore be clear that the insertion of redundant bits by the channel encoder is an overhead that places an extra demand on both transmission bandwidth and transmitted signal power. Efficient codes are therefore always preferred.

The use of error control coding allows us to reduce the E_b/N_o required to achieve a specified BER. The amount of this reduction is known as *coding gain* (expressed in dB). The concept of coding gain is shown in Fig. 11.2, which features notional BER versus E_b/N_o curves for coded and uncoded transmissions both using the same modulation scheme. To achieve a specified bit error rate $\text{BER} = P_{eb}$, the uncoded scheme requires $E_b/N_o = E_2$, whereas the coded scheme requires $E_b/N_o = E_1$, which gives a coding gain

$$G_c = \left(\frac{E_b}{N_o}\right)_{\text{uncoded}} - \left(\frac{E_b}{N_o}\right)_{\text{coded}} = E_2 - E_1 \quad \text{(dB)} \tag{11.1}$$

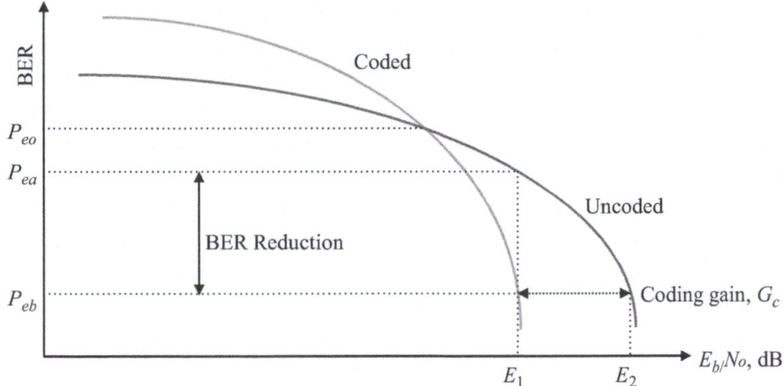

Fig. 11.2 Coding gain

Coding gain varies with BER, being lower at higher BER until it reaches zero at some BER (denoted by P_{eo} in Fig. 11.2) above which the decoder makes more wrong corrections than right ones and ends up inserting more bit errors than what was originally induced by the channel so that the BER of the coded transmission is actually higher than the uncoded. Error control coding is therefore not a panacea that can remedy badly designed links. As noted in Chapter 8 for example, there is no coder in the world that can deliver reliable communication through a binary symmetric channel with bit error probability 0.5. Every coding algorithm is limited in the number of errors that it can detect or rightly correct. A code which can correct all patterns of a given number of errors is referred to as a *perfect code*. What this description means is that if the code is able to correct up to a maximum of t errors in a block of n bits, then it will do so no matter where the errors appear or how they are distributed within the block. If the number of errors in the block exceeds t, then this code though perfect will be unable to correct the errors. We will assume perfect codes throughout this chapter as this is not only valid for most codes but also enables us to derive analytic expressions describing link performance.

The use of error control coding also enables us to reduce the BER that is achievable with a specified E_b/N_o. This can be seen in Fig. 11.2, where $E_b/N_o = E_1$ achieves a lower BER of P_{eb} on the coded link but a higher BER of P_{ea} on the uncoded link. The BER reduction is given by the ratio P_{ea}/P_{eb}, which can be several orders of magnitude.

Error control coding may be implemented as *automatic repeat request* (ARQ) in which the receiver performs error detection only and requests from the transmitter a retransmission of any errored block, or it may be used for *forward error correction* (FEC) where the receiver attempts to correct detected errors. FEC decisions are less reliable than ARQ decisions since it is possible to wrongly correct an error, whereas it is not possible to wrongly detect an error. An (n, k) code, where $n > k$, has 2^k possible transmitted codewords and 2^n possible received sequences. The transmitted codeword is assumed to be the one nearest to (in the sense of differing in the fewest bit positions from) the received coded sequence. However, if the number of errors exceeds the correcting capability of the code, then the received sequence will have been changed in so many bit positions that it ends up being nearer to another codeword than the one that was actually transmitted, and as a result the decoder will wrongly correct the bit errors by picking the wrong codeword.

Fig. 11.1 shows a situation in which the channel decoder input is a bit stream. In this case the symbol detector makes a firm or hard decision on each received symbol sampled at regular intervals T_s, and potentially useful information on the closeness of each sample to a decision threshold is not passed to the decoder, which is then said to perform *hard-decision decoding*, sometimes described as *algebraic decoding*. If on the other hand the detector does not make a decision but passes a quantised value of each of its output samples to the decoder input, the decoder can significantly improve its error correction performance by using the relative-location information received from the detector. The decoder is then said to perform *soft-decision decoding*, also described *as probabilistic decoding*. Soft-decision decoding is more complex to implement than hard decision. With signal level v_r (denoted by $y(T_s)$ in Chapter 8) at the detector output, the value passed to a soft-decision decoder is the log-likelihood ratio

$$\gamma = \log\left(\frac{p(1|v_r)}{p(0|v_r)}\right)$$

Since

$$p(a|b)p(b) = p(b|a)p(a)$$

It follows that

$$\begin{aligned} p(1|v_r)p(v_r) &= p(v_r|1)p(1) \quad \text{(i)} \\ p(0|v_r)p(v_r) &= p(v_r|0)p(0) \quad \text{(ii)} \end{aligned}$$

and dividing (i) by (ii) and setting $p(1)=p(0)$ yields

$$\frac{p(1|v_r)}{p(0|v_r)} = \frac{p(v_r|1)}{p(v_r|0)}$$

and hence

$$\gamma = \log\left(\frac{p(v_r|1)}{p(v_r|0)}\right)$$

For binary transmission using antipodal pulses of energy E_b in a Gaussian noise channel of noise power per unit bandwidth N_o, we may carry out an analysis similar to what was done in Section 10.4 (Fig. 10.11) to evaluate the above log ratio which finally yields

$$\gamma = \frac{4v_r}{\sqrt{E_b}}\frac{E_b}{N_o} \tag{11.2}$$

The value of γ that is passed to the decoder is typically expressed using eight-level quantisation (i.e. three quantisation bits) which gives a good compromise between complexity and performance.

Most codes, and in fact all the ones used for error detection in ARQ systems, are block codes in which the coder generates a unique n-bit codeword for each unique group of k information bits from the source encoder. The coder is said to be an (n, k) block code with code rate $r=k/n$. A convolutional coder, whose decoding algorithm makes it suitable only for FEC application, works on an entirely different basis: The information bits from the source encoder are demultiplexed and passed serially through one or more branches of a coder. The output of each branch is obtained by convolution of the input stream with the impulse response of the branch, hence the name of the encoder. These branch outputs are then interleaved to give the coded bit sequence, the length of which exceeds input sequence length, leading to a code rate $r < 1$.

11.3 ARQ techniques

ARQ techniques group data into blocks of n bits, which may also be referred to as a frame or codeword, and transmit frame by frame until the end of the data is reached. There are different types of ARQ depending on how the transmitter steps through sending the sequence of data frames and how it responds to a notification of error from the receiver. We discuss each type below and derive expressions for the data throughput and link utilisation of each technique.

11.3.1 Stop and wait ARQ

In stop and wait ARQ the transmitter waits for a response from the receiver after sending each data frame. If the response is a positive confirmation of correct reception (denoted by ACK), then the transmitter moves on to send the next frame of n bits. But if the response is an indication of errored reception, called negative acknowledgement and denoted by NAK, the transmitter re-sends the same frame. This process is illustrated in Fig. 11.3 and involves idle time 2τ while the transmitter waits for a response from the receiver, where τ is the propagation delay — the time it takes for a transmitted bit to reach the detection point of the receiver. This idle time will have a significant impact on the *data throughput R* of the transmission system defined as *the number of bits successfully delivered to a destination divided by the total connection time*. Notice in Fig. 11.3 that the sequence of transmitted frames is numbered modulo-2, which requires only one bit to be dedicated to sequence numbering and therefore minimises protocol overhead. The receiver usually communicates an ACK or NAK simply by setting a sequence number field in the frame that it sends back to the transmitter to the number of the next frame expected. Upon receipt of this frame, the transmitter transmits the requested frame, whether it is the next frame in the sequence (meaning the receiver had sent an ACK) or it is the previously sent frame (meaning the receiver had sent a NAK).

Other parameters that affect data throughput include link bit error rate (BER) which determines the probability of detecting an error in the frame and hence requiring a retransmission, and the frame length n. A large frame length on the one

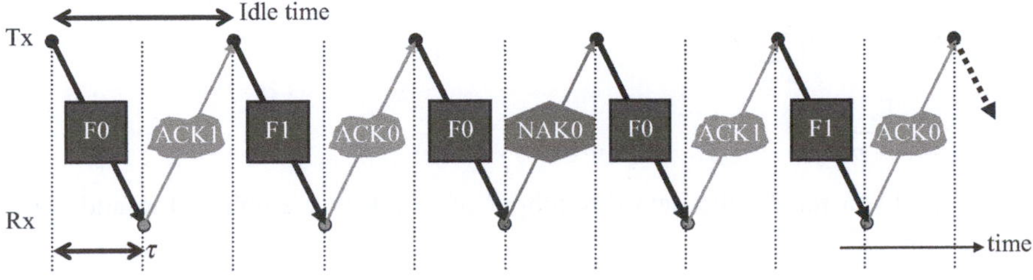

Fig. 11.3 Stop and wait ARQ on a dedicated path where retransmission is due to channel-induced errors only

hand reduces the number of waits (since the data is in that case broken into a smaller number of frames) but on the other hand increases the likelihood of a bit error occurring which would necessitate retransmission. There will therefore be an optimum frame size at which throughput is maximised. To obtain an expression for throughput R_{sw} in a stop and wait ARQ system that encapsulates these parameters, we consider a link that has the following parameters: one-way path delay τ(s), bit error rate BER, transmission bit rate R_b (bits/s), and frame size n (bits). The system employs an (n, k) block code that can detect a maximum of t bit errors in the n-bit frame. To determine the probability p of error being detected in the frame, we note that the probability of a bit being in error is BER and the probability of a bit not being in error is 1 – BER. The event that one and only one bit in the frame is in error occurs when one bit is in error AND the other $n - 1$ bits are not in error. This event can occur in $\binom{n}{1}$ mutually exclusive ways, which is the number of ways of picking one error position out of n, each with probability$(\text{BER})(1 - \text{BER})^{n-1}$. Similarly, two bit errors can occur in $\binom{n}{2}$ mutually exclusive ways, each with probability $(\text{BER})^2(1 - \text{BER})^{n-2}$, and so on, up to a maximum of t detectable errors. Summing these probabilities yields the probability p of error being detected in the frame as

$$p = \sum_{i=1}^{t} \binom{n}{i} (\text{BER})^i (1 - \text{BER})^{n-i}; \quad \binom{n}{i} = \frac{n}{i} \times \frac{n-1}{i-1} \times \cdots \times \frac{n-i+1}{1} \tag{11.3}$$

Note that if every error can be detected in the frame of n bits, then $t = n$ and Eq. (11.3) simplifies to

$$p = 1 - (1 - \text{BER})^n \tag{11.4}$$

The time it takes to send one frame is

$$T_o = 2\tau + \frac{n}{R_b} \tag{11.5}$$

which includes the time to complete serially putting all n bits on a link that has bit rate R_b plus the idle time 2τ. But how long will it actually take to successfully send a frame? It depends on how many times we have to send it before it is successfully received. It could be once, which is with probability $(1 - p)$, this being the probability of no frame error. Or it could be two times (which happens if there was error the first time AND no error the second time), with probability $p(1 - p)$. Or it could be three times, with probability $p^2(1 - p)$, and so on. The average time T that it takes to successfully send one frame is obtained by aggregating these times weighted by their probability of occurrence. Thus,

$$\begin{aligned} T &= T_o(1 - p) + 2T_o p(1 - p) + 3T_o p^2(1 - p) + \cdots \\ &= T_o(1 - p)\sum_{k=1}^{\infty} kp^{k-1} \\ &= \frac{T_o}{1 - p} \end{aligned}$$

Using Eq. (11.5) for T_o, the data throughput of a link that employ stop and wait ARQ is thus

$$R_{sw} = \frac{n}{T} = \frac{n(1 - p)R_b}{n + 2\tau R_b} \quad \text{(bits/s)} \tag{11.6}$$

Let us recast this important expression in a more useful format in terms of the following parameters which we define below:

> Link utilisation U *is the data throughput of a link expressed as a fraction of the link bit rate. Thus* $0 \le U \le 1$ *and is given by*

$$U = \frac{\text{Data throughput}}{R_b} \tag{11.7}$$

> *The* round trip time *(RTT) of a link is the time it takes for a bit to travel from transmitter to receiver and back again. Thus*

$$\text{RTT} = 2\tau \tag{11.8}$$

> *Bandwidth delay product (BDP), expressed in units of bits, is the product of RTT and the bit rate R_b of the bottleneck link in the transmission path. Thus*

$$\text{BDP} = 2\tau R_b \tag{11.9}$$

Using these definitions in Eq. (11.6), we obtain the link utilisation of a stop and wait ARQ system as

$$U_{sw} = \frac{n(1-p)}{n + \text{BDP}} \tag{11.10}$$

We see that link utilisation decreases as BDP increases. On satellite links where $2\tau \approx 500$ ms, BDP will be in excess of 10^6 bits for links with bit rate ≥ 2 Mb/s, and U_{sw} will be extremely small. U_{sw} also depends on link BER through probability of frame error p, and approaches zero as BER increases causing p to approach unity. Furthermore, U_{sw} depends on frame size n, but does so in a more complicated manner than is immediately obvious from Eq. (11.10). Increasing frame size n towards infinity would have increased U_{sw} towards unity, but for the fact that as n increases so also does p increase towards unity if BER is non-zero, according to Eq. (11.3), and this reduces throughput. There is therefore an optimum value of n that maximises U_{sw}, and depends on both link BER and BDP. Fig. 11.4 shows link utilisation plotted against frame size n for link BER from 10^{-2} to 10^{-5} and BDP from 100 to 2×10^6 bits. It is assumed that all frame errors can be detected, allowing the use of Eq. (11.4) for frame error probability p. It can be seen that as link quality improves (i.e. as BER decreases), larger frame sizes can be used and link utilisation improves. The critical factor however is the link BDP. Even at low BER the utilisation of the link with a BDP of 2×10^6 bits remains below 2%, whereas the low-BDP link has utilisation in excess of 90% for a range of frame sizes between 1000 and 10,000. It can therefore be concluded that stop and wait ARQ is unsuitable for high-BER or high BDP links such as satellite links.

It is important to point out that the link utilisation analysis presented above is the best case scenario based on the depiction in Fig. 11.3, where retransmission is caused by channel errors only. Such a scenario is however only applicable when there is a dedicated link between transmitter and receiver. If transmission is over a communication network, then network problems such as congestion will introduce

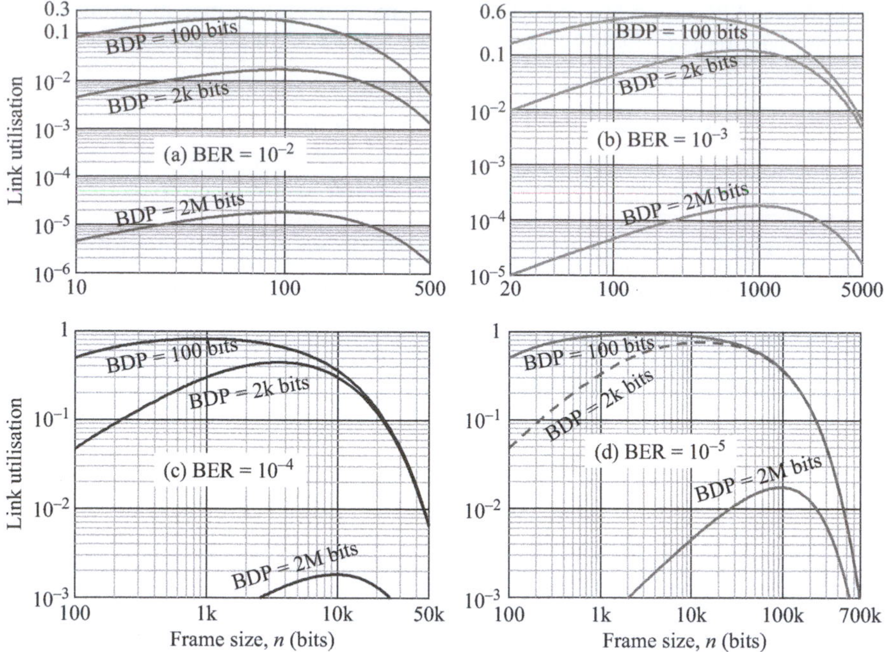

Fig. 11.4 Link utilisation of stop and wait ARQ versus frame size at various BDP and BER

further degradations where for example a data frame from transmitter or ACK from receiver might be lost in transit. If that happens, then there will be a congestion-induced reduction in link utilisation, which is not included in Eq. (11.10). The transmitter handles this more general situation by incorporating a timer into its operation. It starts the timer upon sending each frame, and if there is timeout before an ACK (or NAK) is received then the transmitter simply assumes a NAK and re-sends the previous frame. Fig. 11.5 depicts this situation and shows two timeout incidents, one caused by the loss of a frame and the other caused by the loss of an ACK.

11.3.2 Go-back-N ARQ

In this technique the transmitter transmits the sequence of n-bit frames continuously, moving from one frame to the next without pause until it receives a NAK from the receiver requesting the retransmission of an identified frame, say N. At that point the transmitter backtracks to frame N and retransmits every frame from N onwards. This technique is also called *continuous ARQ with pullback*. The disadvantage of go-back-N ARQ is that the transmitter cannot discard a sent frame, but must keep it in a buffer (just in case it needs to re-send) until an ACK has been received for the frame or for a subsequent (higher numbered) frame. However, go-back-N has an improved data throughput compared to the stop-and-wait protocol, as shown in the analysis below.

Fig. 11.6 illustrates the operation of go-back-N, first for the case where the receiver sends ACK at a slower rate than the transmitter sends frames, and second for the case where the receiver sends ACK at the same rate as the transmitter sends frames. Notice that the frames are numbered modulo-8, which allows the transmitter to send multiple uniquely identified frames without pausing while limiting

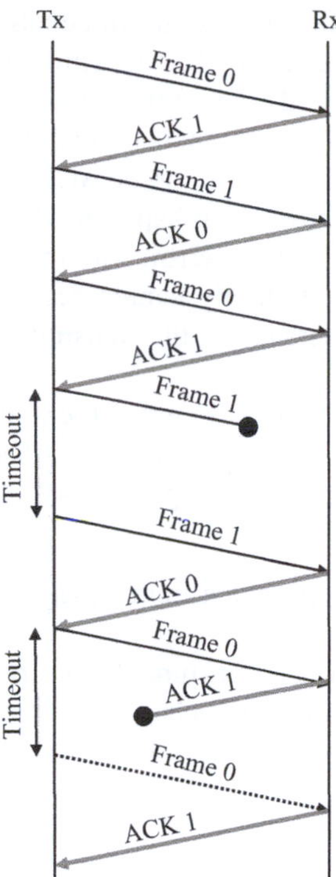

Fig. 11.5 Stop and wait ARQ over a communication network where congestion and other network problems might lead to frame loss in transit

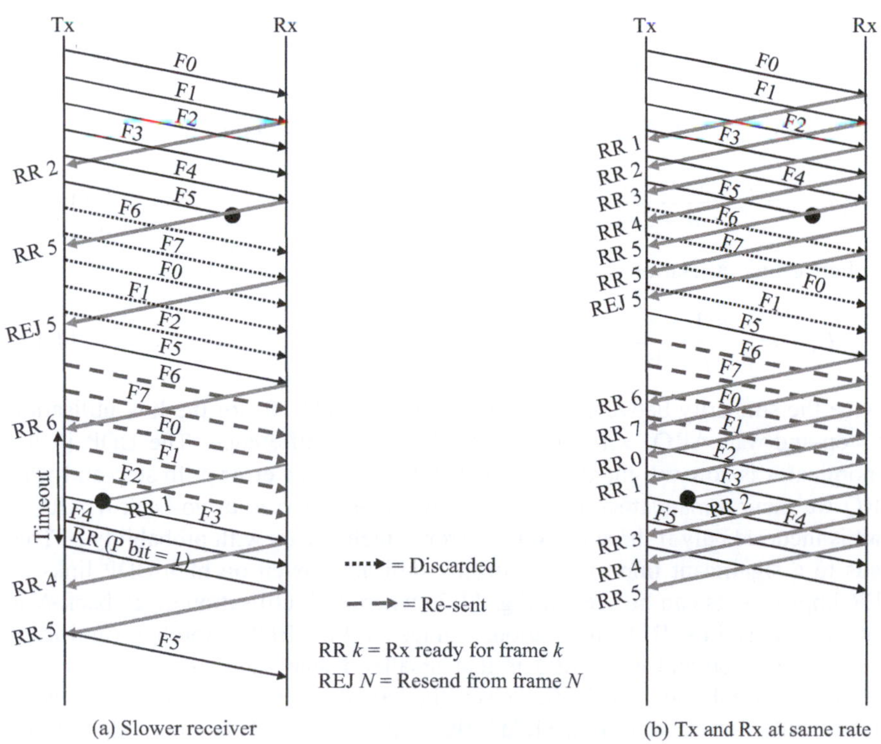

Fig. 11.6 Operation of go-back-N ARQ over a communication network

the overhead spent on frame numbering to just three bits. Part (a) of the figure shows how a timeout situation is handled: The transmitter polls the receiver (using P bit = 1) and the receiver responds stating the frame that it expects next, which the transmitter then sends. Note that the analysis that follows specifically excludes timeouts, which usually arises due to frame or ACK loss caused by network congestion. The illustration also shows what happens when a frame (e.g. F5) does not arrive at the receiver or is received in error: The receiver discards all subsequent frames (F6, F7, F0, ...) since it strictly cannot receive frames out of order, and sends a REJ N (e.g. REJ 5) message to the transmitter telling it to retransmit all frames starting at N.

By going back to N when there is an error, the transmitter effectively behaves as though it had not moved on at all from frame N. Therefore a little thought will show that, ignoring timeout incidents, the go-back-N ARQ scheme basically works as follows:

- If there is no transmission error in a frame, the frame is sent (without waiting) in a time $T_o = n/R_b$.
- When there is error, a time of 2τ expires before re-sending.
- Each frame is sent repeatedly until successfully received before the transmitter moves on to sending the next frame.

Therefore (recalling the explanation under stop-and-wait ARQ), the average time T taken to send each frame is

$$\begin{aligned} T &= T_o(1-p) + [T_o + 2\tau + T_o]p(1-p) + [T_o + 2(2\tau + T_o)]p^2(1-p) + \cdots \\ &= T_o(1-p)\sum_{k=1}^{\infty} kp^{k-1} + 2\tau(1-p)\sum_{k=1}^{\infty}(k-1)p^{k-1} \\ &= \frac{T_o + 2\tau p}{1-p} \end{aligned} \tag{11.11}$$

which, with $T_o = n/R_b$ as stated above, leads to data throughput

$$R_{gbn} = \frac{n}{T} = \frac{n(1-p)R_b}{n + 2\tau R_b p} \quad \text{bits/s} \tag{11.12}$$

and hence link utilisation,

$$U_{gbn} = \frac{n(1-p)}{n + p \times \text{BDP}} \tag{11.13}$$

Notice the similarity between this equation and Eq. (11.10) for the link utilisation of stop-and-wait ARQ, with one absolutely crucial difference: The BDP in the denominator is now weighted by the probability of frame error p. In stop-and-wait ARQ we incur a mandatory wait after each frame, whereas in go-back-N ARQ a wait is incurred only if there is a frame error, which occurs with probability p. This leads to a significant improvement in link utilisation even on high-BDP links as BER improves, as can be seen in Fig. 11.7c where link utilisation of go-back-N is plotted against link BER for various values of link BDP. Another significant change from stop-and-wait ARQ is that smaller frame sizes deliver better link utilisation since the smaller the frame size the less likely there is to be a frame error which would cause a wait. Fig. 11.7a,b shows this trend and also highlight the fact

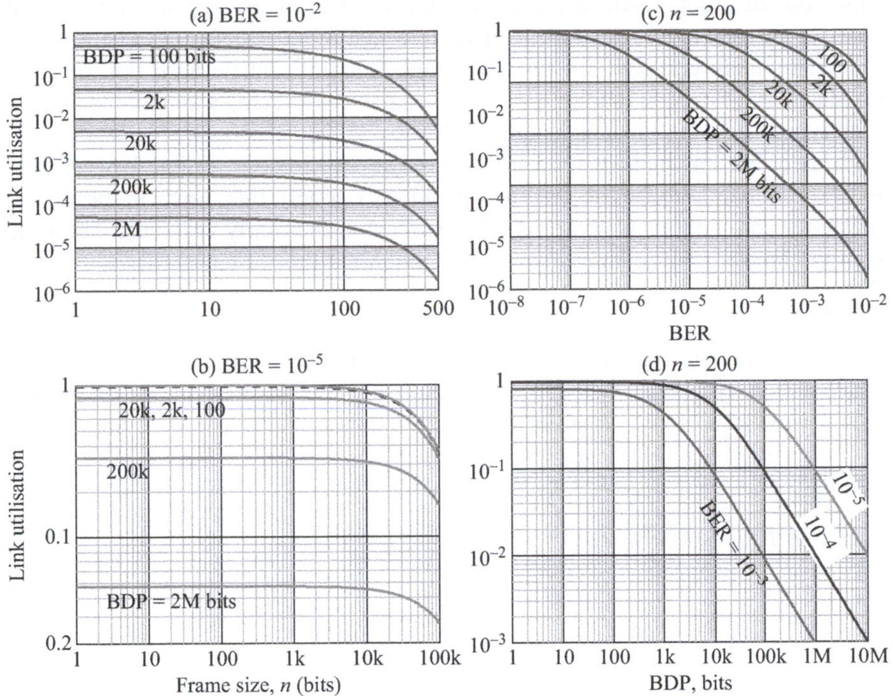

Fig. 11.7 Link utilisation of go-back-N ARQ versus frame size, BER and BDP

that a high BER still causes poor link utilisation especially on high BDP links. Note however that very small frame sizes are to be avoided in order to maintain a high frame efficiency and ensure that frame overhead (such as bits for ARQ management, etc.) constitutes only a small fraction of overall frame size.

Finally, although go-back-*N* ARQ delivers a significant improvement in link utilisation compared to stop-and-wait ARQ, Fig. 11.7d shows that it is still a poor choice for error control on high BDP links. As earlier discussed, one of the aims of error control is to transform an incoming high-BER bit stream into a low-BER one for delivery to the destination application. Go-back-*N* achieves that aim by ensuring that errored frames are re-sent until correctly received, but it does so at a cost: Fig. 11.7d shows that if the incoming bit stream was at a BER of 10^{-3} and the link was a satellite link (with BDP usually in excess of 1M bits), then error control would be achieved with a link utilisation below 0.1% if go-back-*N* ARQ were used. Stop-and-wait ARQ would fare much worse with utilisation ~0.02% on such links at best.

11.3.3 Continuous ARQ with selective repeat

This technique, also referred to simply as *selective ARQ*, works like go-back-*N* when there is no frame error. The transmitter transmits the sequence of *n*-bit frames continuously, moving from one frame to the next without pause. But when it receives a NAK from the receiver requesting the retransmission of an identified frame, say frame *N*, then, unlike go-back-*N*, only the corrupted frame is retransmitted. This means that not only does the transmitter need to maintain a buffer as in go-back-*N*, but now the receiver must also keep a buffer and must be able to receive frames out of order, retain the valid frames and selectively replace any invalid frames at the right order once they have been re-sent by the transmitter. The protocol is therefore more complex, but the advantage is significant because link delay is eliminated as a factor in link utilisation, making selective ARQ a suitable technique on all types of duplex links regardless of delay.

Link utilisation analysis for selective ARQ is much more straightforward than the previous two because the scheme operation is simply this: Each frame is sent repeatedly and without waiting until it has been successfully received. The time taken to complete the sending of an n-bit frame on a link of bit rate R_b is

$$T_o = n/R_b$$

Averaging over the event of sending the frame k times ($k = 1, 2, 3, \ldots, \infty$) before it is successfully received, weighted by the probability of each event, we obtain the average time taken to successfully send each frame as

$$\begin{aligned} T &= T_o[1-p] + 2T_o[p(1-p)] + 3T_o\left[p^2(1-p)\right] + \cdots \\ &= T_o(1-p)\sum_{k=1}^{\infty} kp^{k-1} \\ &= \frac{T_o}{1-p} \end{aligned} \tag{11.14}$$

Using the expression for T_o in the above yields the following data throughput R_{sr} and link utilisation U_{sr} on a link employing selective repeat:

$$\begin{aligned} R_{sr} &= \frac{n}{T} = R_b(1-p) \\ U_{sr} &= 1-p \end{aligned} \tag{11.15}$$

We see that link utilisation now depends only on p, which in turn depends on frame size n and link BER through Eqs. (11.3) or (11.4). In particular, link BDP is no longer a factor. Assuming that the error detection algorithm can detect all frame errors, Eq. (11.4) applies and the above result simplifies further to

$$U_{sr} = (1 - \text{BER})^n \tag{11.16}$$

Fig. 11.8 shows a plot of this link utilisation result versus frame size n at various values of BER from which it may be concluded that selective repeat ARQ can be used for error control on high-BER links regardless of their BDP without imposing an intolerable penalty on link utilisation provided frame size is kept small. However, the requirement of a small frame size by the protocol must be balanced against the conflicting need for high frame efficiency which can only be satisfied by a large frame size.

11.3.4 Sliding window flow control

The ARQ techniques discussed above are actually two extremes of a protocol that can fulfil the dual functions of error and flow controls. Stop and wait ARQ can only send one unacknowledged frame. This is one extreme of the protocol. And the two continuous ARQ techniques (go-back-N and selective repeat) can send an indefinite number of unacknowledged frames. This is the other extreme. The first needlessly and severely limits link utilisation, whereas the second leaves the receiver with no protection against being overwhelmed by a fast transmitter.

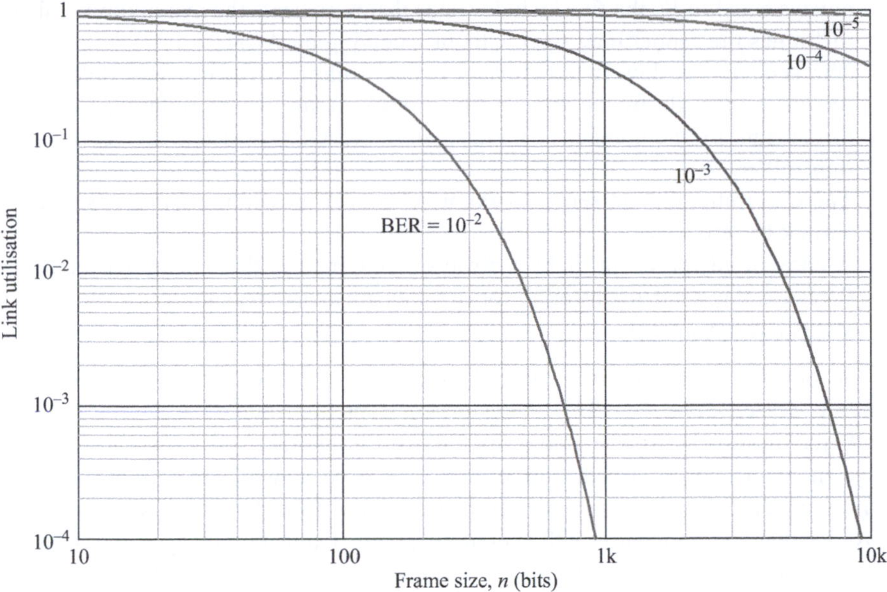

Fig. 11.8 Link utilisation of selective repeat ARQ versus frame size at various BER

We may improve ARQ performance while still giving the receiver control over the rate at which the transmitter sends frames by introducing a window size parameter W that specifies the maximum number of frames that the sender may send without receiving an ACK. The transmitter maintains a sliding *sending window* while the receiver also maintains a sliding *receiving window* as illustrated in Fig. 11.9, hence the scheme is called *sliding window flow control*. The transmitter's operation is controlled as follows:

- The transmitter can only send the frames within its sending window in numerical order.
- The sum of the number of frames in flight (i.e. transmitted frames whose ACKs have not yet been received) and the size of the sending window is

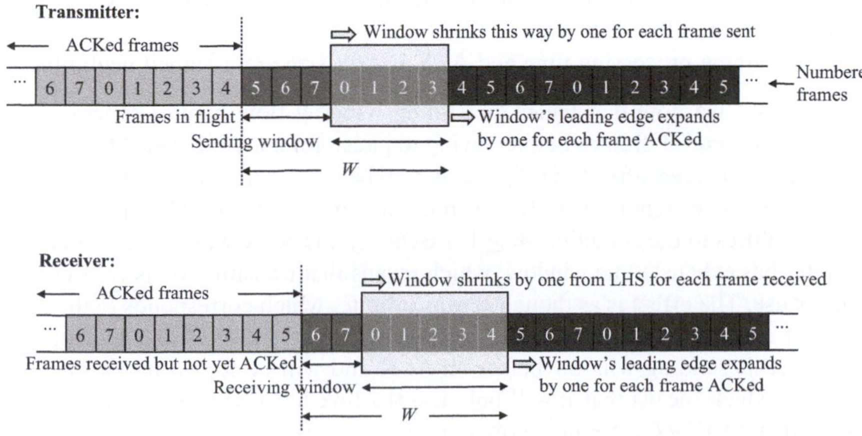

Fig. 11.9 Sliding window flow control

always equal to W, the fixed window size that represents the maximum number of frames the sender may send without receiving an ACK.

- The sending window is therefore a sort of sliding window whose left edge moves to the right by the number of frames sent and whose right edge moves to the right by the number of frames whose ACKs have been received by the transmitter.
- The transmitter cannot send any frame whenever its sending window size has shrunk to zero.
- A copy of all frames in flight is always maintained in the transmitter's buffer (because the need might arise for a retransmission, depending on the ACK status later received).

At the other end, the receiver can only receive the frames within its receiving window. For example, Fig. 11.9 shows the receiver in a state of expecting frames numbered from 0 to 4. If a frame arrived with a sequence number outside this range, it would be discarded. Furthermore, the pace at which ACKs are sent is entirely under the receiver's control. If the receiver withholds or delays the sending of ACKs then the transmitter will eventually reach W unacknowledged frames and therefore be forced to stop sending any more frames. In Fig. 11.9 the receiver has sent an ACK for frame 5, which the transmitter is yet to receive. When the receiver detects a frame error it reacts as earlier discussed and sends either a go-back-N request to the transmitter or a selective repeat request, whichever scheme is followed. The sliding window flow control scheme is therefore just as effective in error control as the other three ARQ schemes previously discussed. What needs to be examined is its impact on link utilisation.

First of all, note that stop and wait ARQ is simply a special case of sliding window flow control with $W=1$, whereas go-back-N and selective repeat ARQs are sliding window flow control at the other end of the scale with $W \to \infty$. So we may derive results for link utilisation in sliding window flow control by following a similar argument as previously done, and making the following assumptions:

- All errors are due to frames arriving with one or more bit errors, and every such error is detected. Timeout incidents due to a frame failing to arrive are specifically excluded.
- Every ACK packet that is sent is received without loss and without error. This is a reasonable assumption to make since the size of an ACK packet is very small compared to a data frame.
- Finally, frame processing time and ACK packet size are assumed negligible.

On these bases we note that under the sliding window flow control scheme, the transmitter can send W frames before having to pause for an ACK. The ACK of the first frame is received after time $T_1 = 2\tau + t_f$, where $t_f = n/R_b$ is the transmission time for one frame, ignoring ACK transmission time as stated. Thus if $Wt_f > T_1$ (which simplifies to the condition $W \geq 1 + \text{BDP}/n$), this ACK will arrive before the transmitter has exhausted its window, which means that transmission is continuous with no pause. The effect is as though W was infinite, which corresponds to the two continuous ARQ schemes earlier analysed. However if $Wt_f < T_1$, the transmitter will exhaust its sending window before an ACK and will be forced into a mandatory pause, which means that it will only use the link for a fraction of the time, a fraction equal to Wt_f/T_1. We may now introduce the effect of frame errors as was done earlier and derive link utilisations for the two methods of remedying the errors, namely go-back-*N* and selective repeat. The result for go-back-*N* is

$$U_{gbn,W} = \begin{cases} \dfrac{n(1-p)}{n+p\times \mathrm{BDP}}, & W \geq 1+\mathrm{BDP}/n \\ \dfrac{nW(1-p)}{n+(1+pW)\times \mathrm{BDP}}, & W < 1+\mathrm{BDP}/n \end{cases} \tag{11.17}$$

And for selective repeat it is

$$U_{sr,W} = \begin{cases} 1-p, & W \geq 1+\mathrm{BDP}/n \\ \dfrac{W(1-p)}{1+\mathrm{BDP}/n}, & W < 1+\mathrm{BDP}/n \end{cases} \tag{11.18}$$

We see that in order to obtain the selective repeat ARQ performance discussed in the previous section in which link utilisation is independent of link delay, the transmitter and receiver must agree on a window size *W* that is no less than $1+\mathrm{BDP}/n$. For example, on a satellite link with $\mathrm{BDP}=5\times 10^6$ bits, if the frame size used is $n=16{,}200$ bits, then the window size must be at least $W=309$ frames. However, in view of Fig. 11.8, a frame size of 16,200 would lead to an extremely low link utilisation if BER is high. Assuming $\mathrm{BER}\sim 10^{-3}$, then to obtain a link utilisation of around 90% we must choose $n=100$, which therefore requires a window size *W* that is at least 50,001. Finally, *W* needs to be one less than an integer power of 2 (i.e. $W=2^k-1$) in order to make full use of a *k*-bit frame sequence number field in the frame header. For example, *W* can be 1, 7, 127, ... In this case, we would need to choose $W=65{,}535$. Fig. 11.10 shows the link utilisation of a satellite link with

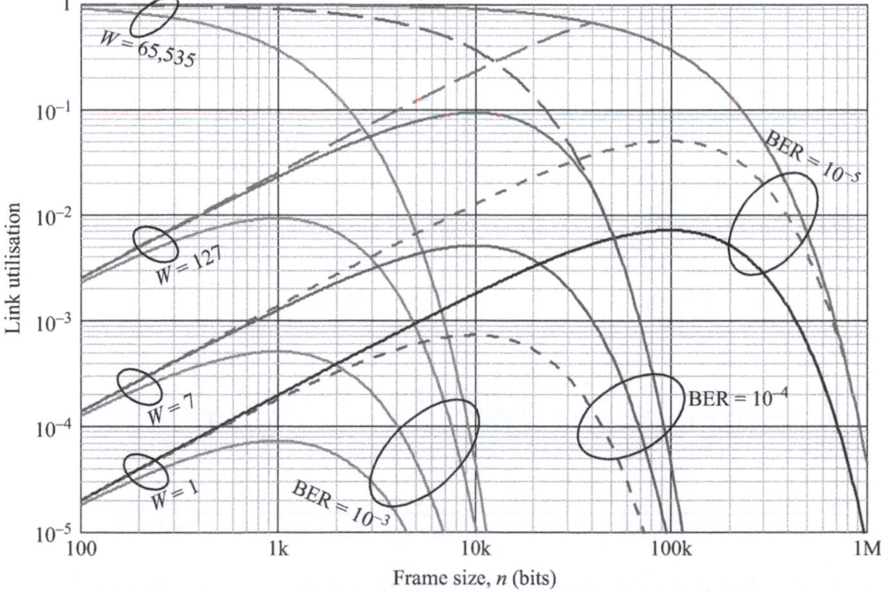

Fig. 11.10 Link utilisation of sliding window flow control with selective repeat error correction for a satellite link with $\mathrm{BDP}=5\times 10^6$ *bits versus frame size n, window size W and link bit error rate BER*

BDP $= 5 \times 10^6$ bits that employs sliding window flow control with selective repeat plotted against frame size at various values of link BER and for window sizes $W = 1$, 7, 127 and 65,535. It should be noted that for a sliding window flow control with $W = 1$ the option of selective repeat is identical to stop and wait ARQ.

11.4 Effectiveness of error control codes

The ARQ techniques discussed above rely on some means of detecting the presence of error in the n-bit frame or codeword, but how effective are these methods and how do they work? Before discussing specific classes of error control codes it is useful to examine the principle behind error detection and error correction and to quantify the error rate reduction capability of selected simple codes.

11.4.1 Error detection

The ability to decide with certainty that there is some error in a message which is not known in advance is neither complicated nor psychic. If we simply arrange to send an unknown message using only a subset of codewords, meaning that the remainder of the set is forbidden, then a receiver can know that there has been some error if it receives a forbidden codeword at any time.

11.4.1.1 Single parity check code

Consider the single parity check (SPC) error control method shown in Fig. 11.11, where the coder takes k message bits $m_0 m_1, \ldots, m_{k-1}$ and appends a

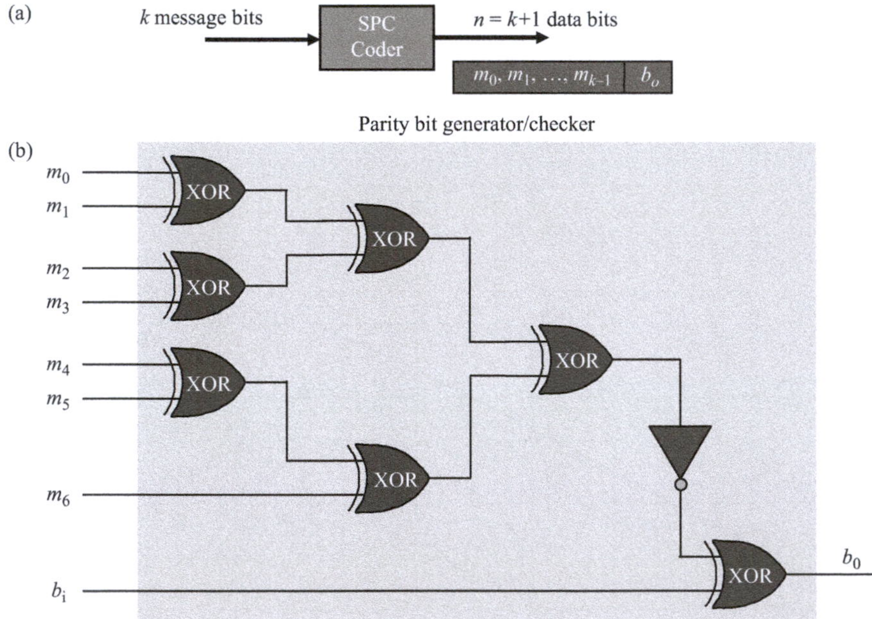

Fig. 11.11 Single parity check: (a) concept; (b) circuit for generating and checking parity: b_i is set to 0(1) for even(odd) parity bit generation. Used as a checker, $m_0m_1m_2m_3m_4m_5m_6b_i$ is the received eight-bit codeword and $b_o = 0(1)$ indicates invalid codeword in even(odd) parity check

parity bit b_o to produce a $(k + 1)$-bit codeword $m_0m_1, \ldots, m_{k-1}b_o$, where the value of the parity bit is set to ensure that every codeword contains an even number of 1s (for the even parity check system) or an odd number of 1s (for the odd parity check version). This is an (n, k) block code with $n = k + 1$ and code rate $r = k/(k + 1)$. Valid codewords differ in at least two bit locations, so one bit flip in a codeword will always produce an invalid codeword, and it takes at least two bit flips to change a valid codeword into another valid codeword. The number of bit locations at which two codewords differ is called the *distance* between the two codewords, sometimes also referred to as *Hamming distance*. And the minimum distance of a code is the smallest distance between any pair of codewords in the entire set of valid codewords used by the code. Thus SPC code has a minimum distance of 2.

Part (b) of the figure shows a circuit implementation used in asynchronous transmission of ASCII characters in which $k = 7$. The circuit consists of 7 exclusive-or (XOR) gates and one inverter (NOT) gate, and doubles up as both a parity bit generator at the transmitter and a parity bit checker at the receiver. To use it to generate an even parity bit, set input b_i to a permanent 1 (high) and connect the 7 message bits m_0 to m_6 to the 7 input pins indicated. The output b_o gives the required parity bit, which will be 1 if m_0 to m_6 have an odd number of 1s and will be 0 otherwise. To use the circuit to generate an odd parity bit, set b_i permanently to 0 (ground), and the output b_o will be 1 if the k-bit input message block contains an even number of 1s, and will be 0 otherwise. For parity bit checking at the receiver, the input $m_0m_1m_2m_3m_4m_5m_6b_i$ is the received codeword, and the output b_o is a flag that indicates an error when it is 0 (1) for even (odd) parity system.

Returning to the concept, notice that there are 2^{k+1} codewords to pick from, one for each of the 2^k possible message blocks. Therefore a sub-set equal to half of the possible $(k + 1)$-bit combinations are unused or forbidden. In the case of even parity check, the forbidden half are the $(k + 1)$-bit combinations that contain an odd number of 1s, and vice versa for odd parity check. By this simple arrangement, if any single bit of a transmitted codeword is flipped due to channel-induced error, this changes the received codeword into a forbidden one. Therefore the receiver will know that there has definitely been some error since the transmitter could not under any circumstance have transmitted a forbidden codeword. However, the receiver does not know which one or more of the eight bits were flipped. So by this arrangement error detection is possible but not error correction.

But there is a problem, and this is common to all coding methods: Error declaration (i.e. rejecting a codeword or quarantining it for corrective treatment) is the only certainty, whereas acceptance is merely an assumption in hope. In this single parity check scheme, an even number of bit errors in a codeword is undetectable because the corrupted codeword will arrive as one of the valid codewords. To quantify the effectiveness of the single parity check (SPC) scheme for error control, let us compare the number of errored characters that are delivered to the destination application with and without SPC error control. We call the ratio between the probability of errored character without SPC and the probability of errored character with SPC the *character error rate reduction factor* (*CERRF*).

Let the probability of bit error on the link be very small, i.e. BER $\ll 1$ (which is usually the case in a well-designed link). Without SPC coding, we send $(n - 1)$-bit characters, and the probability p_u that any one or more of the bits will be in error and hence the character will be received as an errored character is

$$\begin{aligned} p_u &= 1 - (1 - \text{BER})^{n-1} \\ &\approx 1 - [1 - (n-1)\text{BER}] \\ &\approx (n-1)\text{BER} \end{aligned} \tag{11.19}$$

It is easy to see the above result by noting that p_u is the complement of the probability of no error in all $n-1$ bits, which is the event: No error in the 1st AND no error in the 2nd AND no error in the 3rd, and so on, having probability $(1-\text{BER})^{n-1}$. When SPC is used, the characters are sent as n-bit codewords and will be received in error only if there was an even number of bit errors. Thus, the probability p_c of a character error when coding is used in this way is

$$\begin{aligned} p_c &= \sum_{j=2,4,\cdots}^{n} \binom{n}{j} (\text{BER})^j (1-\text{BER})^{n-j} \\ &\approx \binom{n}{2} (\text{BER})^2 \\ &\approx \frac{1}{2} n(n-1)(\text{BER})^2 \end{aligned} \tag{11.20}$$

CERRF is given by the ratio between p_u and p_c:

$$\begin{aligned} \text{CERRF} &= \frac{1-(1-\text{BER})^{n-1}}{\sum_{j=2,4,\cdots}^{n} \binom{n}{j} (\text{BER})^j (1-\text{BER})^{n-j}} \\ &\approx \frac{2}{n(\text{BER})} \end{aligned} \tag{11.21}$$

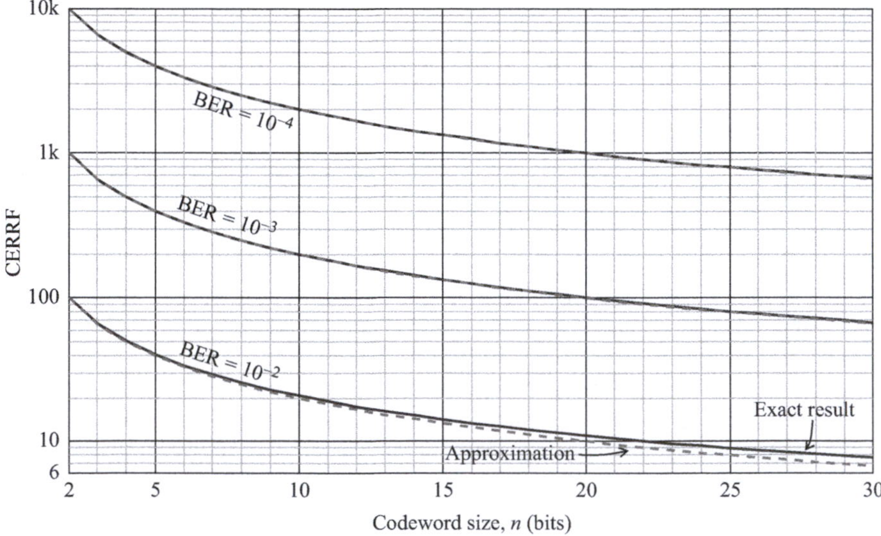

Fig. 11.12 Character error rate reduction factor (CERRF) due to single parity check versus codeword size at various values of link bit error rate

We see that SPC effectiveness is inversely proportional to codeword size n. The smallest codeword size $n = 2$, gives the best performance but at the price of the lowest code rate $r = ½$; whereas the largest codeword sizes $n \to \infty$ give the highest code rates $r \to 1$ but at the price of the lowest effectiveness CERRF $\to$ 1 at high BER. This result is shown in Fig. 11.12 for codeword sizes 2 to 30 and BER 10^{-2}, 10^{-3} and 10^{-4}, using both the exact expression (1st line) and approximate expression (2nd line) of Eq. (11.21). It can be seen that the approximate expression gives excellent results that are indistinguishable from the exact expression except at large n and high BER. The following worked example demonstrates the practical impact of SPC.

Worked Example 11.1: Practical Impact of Single Parity Check

An 800-page textbook having on average 1250 characters per page is to be transmitted using seven-bit ASCII code on a communication link operating at a bit rate of 2048 kb/s. The bit error rate on the link is 10^{-4}. Determine

---

(a) The expected number of character errors in the received textbook and the time taken to complete transmission when there is no error control.
(b) The expected number of character errors in the received textbook if a single parity check is used for error control.
(c) The average time taken to complete the transmission in (b) using each of the 3 ARQ schemes of stop-and-wait, go-back-*N* and selective repeat if the link is a terrestrial link of one-way path delay 1 ms.

Total number of characters in the book $N_c = 800 \times 1250 = 1 \times 10^6$.

(a) With no SPC coding, the probability of character error p_u is given by Eq. (11.19) with $n - 1 = 7$, and the expected number of character errors N_{eu} is given by the product of p_u and N_c. That is,

$$p_u = (n-1)\text{BER} = 7 \times 10^{-4}$$
$$N_{eu} = N_c p_u = 1 \times 10^6 \times 7 \times 10^{-4} = 700$$

Time T_u taken to complete transmission is

$$T_u = \frac{\text{No. of bits}}{\text{Link bit rate}} = \frac{7N_c}{2{,}048{,}000} = \frac{7 \times 10^6}{2{,}048{,}000}$$
$$= 3417.97 \text{ ms}$$

(b) With SPC coding, the probability of character error p_c is given by Eq. (11.20), and the expected number of character errors N_{ec} is given by the product of p_c and N_c:

$$p_c = \frac{1}{2} n(n-1)(BER)^2 = 28 \times 10^{-8}$$

$$N_{ec} = N_c p_c = 1 \times 10^6 \times 28 \times 10^{-8} = 0.28$$

You may observe that solely by using the simple SPC coding, we have improved from 700 channel-induced typos per download of the book on average to only 0.28 per download on average, which means that on average we would download the book 3.6 times before the first channel-induced typo appeared. That is a dramatic improvement, but at what price? We are still operating at the same link bit rate, so the bandwidth used has not changed, but we will be using the link for longer as the next part confirms.

(c) First we need the probability p that error is detected in the eight-bit codeword, and then we need the link utilisation of each ARQ scheme. An odd number of errors is detected, so p is given by Eq. (11.3) applied to odd numbers of errors as follow:

$$p = \sum_{j=1,3,5,7} \binom{8}{j} (10^{-4})^j (1-10^{-4})^{8-j}$$
$$= 7.9944 \times 10^{-4} + 5.5972 \times 10^{-11} + 5.5983 \times 10^{-19} + 7.9992 \times 10^{-28}$$
$$= 7.9944 \times 10^{-4}$$

This (for BER $\ll 1$) is practically equal to the probability of one bit error occurring in the codeword which is roughly $p \approx n \times \text{BER}$. With link BDP given by

$$\text{BDP} = 2\tau R_b = 2 \times 1 \times 10^{-3} \times 2048 \times 10^3$$
$$= 4096 \text{ bits}$$

we obtain the link utilisation of the three schemes using Eqs. (11.10), (11.13) and (11.15) as follows:

$$U_{sw} = \frac{n(1-p)}{n+BDP} = \frac{8(1-7.9944\times 10^{-4})}{8+4096} = 1.9478 \times 10^{-3}$$
$$U_{gbn} = \frac{n(1-p)}{n+p\times BDP} = 0.709$$
$$U_{sr} = 1 - p = 0.9992$$

Transmission time is the coded book size in bits divided by the effective transmission bit rate R_{eff}, which is the product of link utilisation and link bit rate. Thus average book transmission time T_{av} under each protocol is

$$T_{av} = \frac{8N_c}{R_{eff}} = \begin{cases} \dfrac{8\times 10^6}{R_b U_{sw}} = \dfrac{8\times 10^6}{2048\times 10^3 \times 1.9478\times 10^{-3}} = 2005.51 \text{ s} \\ \dfrac{8\times 10^6}{R_b U_{gbn}} = \dfrac{8\times 10^6}{2048\times 10^3 \times 0.709} = 5.5095 \text{ s} \\ \dfrac{8\times 10^6}{R_b U_{sr}} = \dfrac{8\times 10^6}{2048\times 10^3 \times 0.9992} = 3.9094 \text{ s} \end{cases}$$

This underlines the importance of employing the right protocol when implementing ARQ error control. Under stop-and-wait the book takes over 30 minutes to download compared to only 3.91 seconds with selective repeat. Nevertheless, given the right protocol (i.e. selective repeat) we still need

14.4% more transmission time with error control than without. Note that the bulk of this increase, 14.3% to be precise, is due to the coded book being longer, and the remaining 0.1% is for a few codewords being sent more than once due to error. Which would you prefer, to download a book with around 700 typos in 3.42s or to wait slightly longer and perhaps pay slightly more to download the same book with hardly any typos in 3.91s?

11.4.1.2 Perfect (*n*, *k*) block code

The above discussion was applied to SPC coding, but in general if a perfect (n, k) code can detect up to t bit errors in a codeword then the effectiveness of the code can be expressed in terms of a CERRF given by

$$\text{CERRF} = \frac{1 - (1 - \text{BER})^k}{1 - \sum_{j=0}^{t} \binom{n}{j} (\text{BER})^j (1 - \text{BER})^{n-j}} \tag{11.22}$$

where character is defined as the k-bit grouping of bits extracted from the n-bit received codeword, the numerator is the probability of any error in the character when transmission is without error control coding, and the denominator is the probability of an undetected error in the received n-bit codeword. The above result is a somewhat conservative estimate because there may be occasions in which all the $t+1$ or more undetected errors are confined to the parity section of the codeword so that the k message bits are unaffected. Eq. (11.22) should therefore be seen as giving the minimum improvement factor in a perfect code if the number of parity bits that it uses is at least $t+1$.

To illustrate, the (7, 4) Hamming code can detect up to two bit errors in its seven-bit codeword. If used for ARQ error control on a link of $\text{BER} = 10^{-4}$, Eq. (11.22), with $n=7$, $k=4$ and $t=2$, indicates that it would deliver an improvement factor of

$$\begin{aligned}
\text{CERRF} &= \frac{1 - (1 - \text{BER})^4}{1 - \sum_{j=0}^{2} \binom{7}{j} (\text{BER})^j (1 - \text{BER})^{7-j}} \\
&= \frac{3.9994 \times 10^{-4}}{1 - \left[(1 - 10^{-4})^7 + 7(10^{-4})(1 - 10^{-4})^6 + 21(10^{-4})^2(1 - 10^{-4})^5\right]} \\
&= 1.143 \times 10^7
\end{aligned}$$

For comparison, Fig. 11.12 shows that an SPC control based on the same character size ($k=4$, and hence $n=5$) on the same link ($\text{BER} = 10^{-4}$) would achieve an improvement factor of 4000. Thus the (7, 4) Hamming code is more effective than SPC, but this comes at price: Its code rate $r = 4/7 = 0.57$, compared to SPC code rate $r = 4/5 = 0.8$.

11.4.2 Error correction

Error correction is also made possible by the systematic use of redundancy. When an invalid codeword arrives at the receiver, the decoder knows with certainty that some error has occurred. If error detection were all that the decoder does then it would simply issue an error flag, a decision which would be totally reliable. To make it possible to detect all possible incidents of one bit error it must be the case that flipping one bit of a valid codeword does not turn it into any other valid codeword. That is, the code must have a minimum distance of 2. If we wish to go a step further and design a code with the capability to correct one bit error, then it must be the case that flipping one bit of a valid codeword does not make it more like some other valid codeword than the original codeword. That is, an invalid codeword produced by flipping any one bit of a valid codeword should still be nearer in distance to the original valid codeword than to any other valid codeword. So the minimum distance of the code should be at least 3. If this condition is satisfied then the decoder can make decisions on each received codeword based entirely on choosing the most likely event to have led to the arrival of that codeword. This is in keeping with the *maximum likelihood rule* which guides all decisions in digital communications whenever there is uncertainty.

To illustrate, consider a seven-bit code that has a minimum distance 3 and is used on a link with BER 10^{-4}. We list in column 2 of Table 11.1 the probability of each number of bit errors, from 0 (meaning no bit error in the received codeword) to 7 (meaning all bits in the codeword are flipped). Notice that the event 'no bit error' is significantly more likely than the event 'one bit error' which is significantly more likely than 'two bit errors', and so on. This is always the order of probabilities in a well-designed link, and is the order assumed by all decoders. Therefore if the received codeword is a valid codeword, the decoder chooses the most likely of the eight events (from no error to seven bit errors) to have led to the arrival of this valid codeword. It could not have been one or two bit errors (because the code's minimum distance is 3), so it had to be either no bit error or three or more bit errors. The decoder decides in favour of no bit error simply because that is the more likely cause. It therefore assumes that the received valid codeword was indeed the transmitted codeword. Although we cannot say that the decoder cannot be wrong in this decision, on a link with BER 10^{-4} where the probability of three bit flips in 7 is $\sim 3.5 \times 10^{-11}$ the decoder will be extremely rarely wrong.

The decoder follows a similar reasoning when dealing with the arrival of an invalid codeword. It knows for sure that there was some error and it assumes that the cause was one bit error (because this is more likely than two bit errors, which is

Table 11.1 Probability of occurrence of each number of bit errors in a seven-bit codeword

Number of bit errors	**Probability of occurrence**		
	When link BER = 10^{-4}	**When link BER = 0.1**	**When link BER = 0.5**
0	0.9993	0.4782969	0.0078125
1	6.9958×10^{-4}	0.3720087	0.0546875
2	2.09895×10^{-7}	0.1240029	0.1640625
3	3.4986×10^{-11}	0.0229635	0.2734375
4	3.49895×10^{-15}	0.0025515	0.2734375
5	2.09958×10^{-19}	0.0001701	0.1640625
6	6.9993×10^{-24}	0.0000063	0.0546875
7	1×10^{-28}	0.0000001	0.0078125

more likely than three bit errors, and so on). The decoder therefore first looks for a valid codeword that differs in only one bit location from the received invalid codeword. If a code has the capability of correcting one bit error, then no more than one valid codeword can match an invalid codeword in all but one bit location. If such a codeword is found the receiver assumes that it was the transmitted codeword. Of course it could be the case that it was four bit flips rather than one bit flip that led to the invalid codeword, but the former is less likely, so the decoder decides (as always) in favour of the more likely cause. But if no such valid codeword is found and the decoder does not have the capability of correcting more than one bit error, it means that there was more than one bit flip, and the error though detected cannot be corrected. If however the decoder can correct two bit errors then only one valid codeword can be at a distance of 2 or less from an invalid codeword. The decoder would in that case next proceed to look for a valid codeword that can be arrived at through flipping two bit locations of the invalid codeword and assume (if it finds one) that that was the transmitted codeword.

Column 3 of Table 11.1 also lists the probability of bit errors in a very high BER link with BER = 0.1. Notice that although the order of probability is still as discussed above for column 2, the event of three bit errors has a probability of 0.0229635 or a 1 in 44 chance of occurring. This means that the decoder has a 1 in 44 chance of being wrong each time it decides in favour of 'no bit error' given the arrival of a valid codeword. Finally, column 4 shows an error-crippled link with BER = 0.5. Here the event of three bit errors is much more likely than 'no bit error', so most of the decoder's decisions in favour of 'no bit error' will be wrong even when a valid codeword is received. And when an invalid codeword is received, most of the decisions in favour of 'one bit error' will also be wrong seeing that four bit flips is much more likely than one bit flip. A link with BER = 0.5 cannot be remedied by any coding scheme in the world. See Section 8.3 for further thoughts on this. Let us now see how these principles are applied in the realisation of simple error correcting codes.

11.4.2.1 Repetition block code

The simplest, albeit most inefficient, error correcting code is the $(n, 1)$ repetition code in which the encoder takes the message bits one at a time and simply repeats it n times to form the codeword. Thus there are only two codewords, the all-zeros 000 . . . 0 and the all-ones 111 . . . 1. The codeword length n must be an odd integer, i.e. $n = 2m + 1$, where $m \geq 1$ is some positive integer. The code rate of the $(n, 1)$ repetition code is $r = 1/n$, the minimum distance is n, the maximum number of errors per codeword guaranteed to be detected is $n - 1$, and the maximum number of errors per codeword guaranteed to be corrected is $t = \lfloor ½(n - 1) \rfloor = m$. The decoder is guided in its decisions by the maximum likelihood principle discussed above: It assumes the most likely cause (i.e. number of bit flips, from zero to n in order of decreasing probability) to have led to the arrival of the received codeword. In a repetition code this reduces to a *majority vote rule*, which is stated as follows:

> *If the received codeword contains more* 1*s than* 0*s the decision is in favour of binary* 1 *as the message bit, otherwise the decision is in favour of binary* 0*. The codeword length must be odd to prevent a tie.*

The above rule translates into choosing whichever of the two valid codewords has the smallest distance from the received codeword. This process is illustrated in Fig. 11.13 for the (5, 1) repetition code, where all possible outcomes in the transmission of codeword 11111 are examined. Note that what matters is the bit error count and not their locations within the codeword, so Fig. 11.13 shows bit errors in the leftmost bit locations. It is clear that, whether due to wrong correction or non-detection, there is an error whenever the number of bit errors in the $(2m+1)$-bit codeword exceeds m.

The performance of the repetition code can be assessed under two criteria: When used for error detection (for example to aid ARQ implementation) on a link with bit error rate given by BER, an undetected error occurs only if all n bits are flipped which occurs with probability $(\text{BER})^n$. Without coding, bit error would occur with probability BER. Therefore the CERRF, which in this case of one-bit characters corresponds to a reduction in bit error rate due to the use of repetition coding for error detection is given by

$$\text{CERRF} = (\text{BER})^{1-n} \qquad (n,\, 1) \text{ code error detection} \tag{11.23}$$

Thus by making n large, we can make the repetition code an extremely effective error detector whatever the BER < 1. For example, if BER = 0.1, then CERRF = 10^4 at $n = 5$, increasing to 10^9 at $n = 10$, and 10^{19} at $n = 20$, ad infinitum. There is a heavy price, however, as code rate decreases directly as n increases.

The other assessment criterion is when the repetition code is used for error correction. As earlier stated an error will occur whenever there are $(n+1)/2$ or more bit errors in the n-bit codeword, so that the probability of error is

$$\begin{aligned} p_c &= \sum_{j=(n+1)/2}^{n} \binom{n}{j} (\text{BER})^j (1-\text{BER})^{n-j} \\ &= 1 - \sum_{j=0}^{(n-1)/2} \binom{n}{j} (\text{BER})^j (1-\text{BER})^{n-j} \end{aligned} \tag{11.24}$$

(a) Received sequence 11111 (no bit error)

Codeword	Distance	Decision	Output/Comment
00000	5	Loser	Output = Bit 1:
11111	0	Winner	No error

(b) Received sequence 01111 (one bit error)

Codeword	Distance	Decision	Output/Comment
00000	4	Loser	Output = Bit 1:
11111	1	Winner	One bit error corrected

(c) Received sequence 00111 (two bit errors)

Codeword	Distance	Decision	Output/Comment
00000	3	Loser	Output = Bit 1:
11111	2	Winner	Two bit errors corrected

(d) Received sequence 00011 (three bit errors)

Codeword	Distance	Decision	Output/Comment
00000	2	Winner	Output = Bit 0: Three bit errors
11111	3	Loser	detected but wrongly corrected

(e) Received sequence 00001 (four bit errors)

Codeword	Distance	Decision	Output/Comment
00000	1	Winner	Output = Bit 0: Four bit errors
11111	4	Loser	detected but wrongly corrected

(f) Received sequence 00000 (five bit errors)

Codeword	Distance	Decision	Output/Comment
00000	0	Winner	Output = Bit 0:
11111	5	Loser	Five bit errors undetected

Fig. 11.13 Decoder decisions in (5, 1) repetition code when codeword 11111 *is transmitted*

This equation indicates that the probability of error can be made as small as one wishes by increasing n, provided BER < 0.5. For example, reliable communication at a probability of error $\sim 2 \times 10^{-8}$ can be achieved on a noisy link with BER = 0.01 using (9, 1) repetition code; and if one wants to achieve an even lower error probability say 10^{-15} using the same link, then a (19, 1) repetition code may be used; or if the link degrades further to BER = 0.1, reliable communication at error probability $\sim 2 \times 10^{-8}$ can still be achieved by increasing the codeword length of the repetition code to $n = 29$. However, if the link has BER = 0.5, then

$$\begin{aligned} p_c &= \sum_{j=(n+1)/2}^{n} \binom{n}{j} \left(\frac{1}{2}\right)^j \left(\frac{1}{2}\right)^{n-j} = \left(\frac{1}{2}\right)^n \sum_{j=(n+1)/2}^{n} \binom{n}{j} = \left(\frac{1}{2}\right)^n 2^{n-1} \\ &= \frac{1}{2} \end{aligned}$$

Note therefore that it is not possible to obtain reliable communication over a link with BER = 0.5 even if a repetition code is used with code length $n \to \infty$. And if a repetition code cannot redeem a link as $n \to \infty$, then a code does not exist that can. The bandwidth requirement of a repetition code increases with code length n, so this code represents a mechanism for trading bandwidth for reduced signal power as discussed in the introduction. You can see that this is the case by noting that for a fixed modulation scheme, E_b/N_o uniquely determines link BER which increases monotonically as E_b/N_o decreases. Thus, in the above example, the increase in BER from 0.01 to 0.1 is due to a reduction in E_b/N_o, and this can be compensated for by increasing the repetition code length from 9 to 29 (i.e. reducing code rate from 1/9 to 1/29) which corresponds to increasing bandwidth usage in order to continue maintaining reliable communications. We need error correction codes that facilitate this exchange between bandwidth and signal power more efficiently than the repetition code.

11.4.2.2 Product code

We may extend the idea of a single parity check in such a way as to make error correction possible if we arrange for each message bit to contribute to setting the value of two different parity bits in a block of bits. A bit error in the block of bits will cause two parity checks to fail, and the decoder can then identify the offending message bit by looking for the one bit that is common to both failures. Again this is a decision guided by the maximum likelihood principle discussed earlier, because there are other possible causes involving multiple bit errors but the most likely is a one bit error.

One arrangement that can be used to cause each bit to contribute to setting the value of two different parity bits is shown in Fig. 11.14 in which a block of data containing both a *vertical redundancy check* (VRC) and *longitudinal redundancy check* (LRC) is formed by transmitting one n-bit parity word after every Q codewords. Each codeword is n bits long and of the form

$$m_{k,0}m_{k,1}m_{k,2}, \ldots, m_{k,n-2}b_{k,0}, \quad k = 0, 1, 2, \ldots, Q-1$$

where the last bit $b_{k,0}$ provides a parity check on the preceding $n-1$ message bits. An n-bit parity word, denoted by

$$b_{Q,0}b_{Q,1}b_{Q,2}, \ldots b_{Q,n-2}b_{Q,n-1}$$

is formed on the Q codewords such that its first bit $b_{Q,0}$ provides a parity check on all first bits of the Q codewords; its second bit $b_{Q,1}$ provides a parity check on all

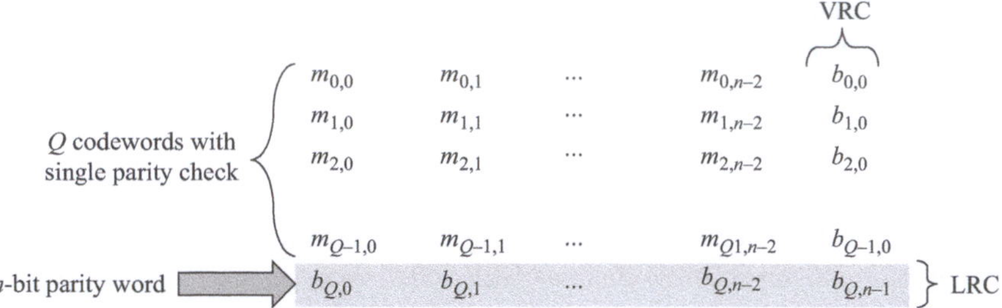

Fig. 11.14 Error correction based on product code

second bits of the Q codewords; and so on. This set of parity bits $b_{Q,0}b_{Q,1}b_{Q,2}, \ldots, b_{Q,n-2}b_{Q,n-1}$ constitutes a longitudinal redundancy check (LRC), whereas the set of parity bits $b_{0,0}b_{1,0}b_{2,0}, \ldots, b_{Q-1,0}$ constitutes a VRC as shown in Fig. 11.14. Notice that by this arrangement every message bit contributes to two parity bits. For example, bit $m_{2,1}$ contributes to the values of parity bits $b_{2,0}$ in the VRC and $b_{Q,1}$ in the LRC. Since there are Q codewords (each containing n bits) plus an n-bit parity word conveying $(n-1)Q$ message bits, this code is an $(n[Q+1], [n-1]Q)$ block code. For example, if $n=8$ and the LRC covers three codewords, then we have a (32, 21) block code with code rate 0.677.

The error correction procedure at the decoder is to write the incoming data into a matrix, with the first Q codewords in the first Q rows and the next n-bit block of data (representing the parity word) in the last row of the matrix. The decoder then performs a parity check of every row and every column of the matrix. If the check fails in the jth row and k^{th} column, then the bit in the $(j, k)^{th}$ cell of the matrix is in error and is simply flipped to correct the error. This scheme will correct every instance of a single bit error in a coded block of $n(Q+1)$ bits. It is therefore only suitable for low-BER simplex links, where errors are rare but must be corrected and ARQ is not an option. Apart from a single bit error, other error patterns can be handled in various ways by the scheme, including

- Any odd number of bit errors in a single row or in a single column can be corrected.
- Any number of multiple bit errors can be corrected provided there is only one error per row or column.
- Some patterns of an even number of bit errors in rows can be detected but not corrected, whereas some will go undetected.
- An odd number of errors in multiple rows or columns can be detected but may be wrongly corrected.

11.4.2.3 (7, 4) Hamming code

Apart from the rectangular format discussed above, a circular arrangement may also be employed to cause each message bit to influence more than one parity bit. Consider three intersecting circles as shown in Fig. 11.15 labelled circle 0, 1 and 2 (going clockwise). We take $k=4$ message bits at a time, denoted by $m_0m_1m_2m_3$ and copy them into the four intersecting regions of the circles as shown. We then determine a parity bit b_j for circle j that ensures that all the bits in each circle have even parity. Thus b_0 parity checks $m_0m_1m_2$; b_1 parity checks $m_0m_1m_3$ and b_2 parity checks $m_1m_2m_3$. The three parity bits so determined are concatenated with the message bits (either as prefix or suffix, both are acceptable) to form the codeword

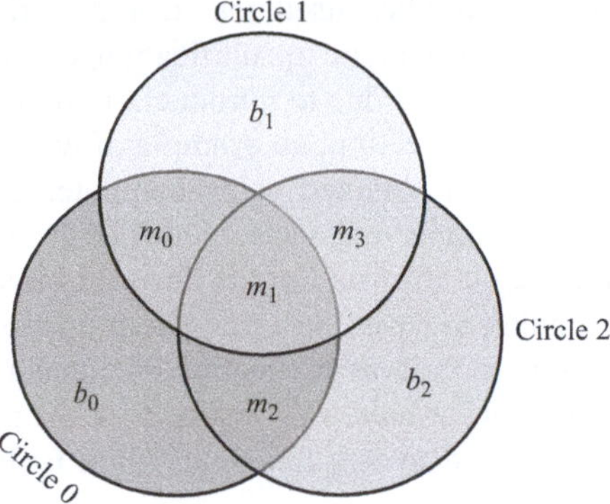

Fig. 11.15 *The (7, 4) Hamming code consisting of four message bits $m_0m_1m_2m_3$ and three parity bits $b_0b_1b_2$. The seven-bit codeword is $b_0b_1b_2m_0m_1m_2m_3$*

of length $n=7$ bits denoted by $b_0b_1b_2m_0m_1m_2m_3$. There are 16 unique message blocks from 0000 to 1111. You may wish to verify that following the procedure just described leads to the complete set of codewords given in the coding table of Table 11.2. By inspecting the set you will notice that this code has a minimum distance of 3 and therefore (as discussed earlier) should be able to correct one bit error while detecting up to two bit errors.

At the receiver, the decoder copies the received codeword into the circles in the same order as above, and performs a parity check of the bits in each circle. A parity check failure in one or more circles indicates error in the bit that is common to the circles in question. More specifically, a parity check failure in circles 1 and 2 implies bit m_3 error, in circles 2 and 0 implies bit m_2 error, in circles 0 and 1 implies bit m_0 error, in all 3 circles implies bit m_1 error, in circle 0 only implies bit b_0 error, and so on. Two bit errors will be detected but wrongly corrected. For example, if bits b_0 and m_1 are both flipped, circle 0 will pass a parity check whereas circles 1 and 2 will fail, which causes the decoder to assume (wrongly) that bit m_3 is in error.

The error-control scheme described here is the popular (7, 4) Hamming code. It has a sound mathematical basis as well as an elegant circuit implementation using shift registers which we introduce in the next section on linear block codes.

Table 11.2 *(7, 4) Hamming code table*

Message	Codeword	Message	Codeword
0000	000 0000	1000	110 1000
0001	101 0001	1001	011 1001
0010	111 0010	1010	001 1010
0011	010 0011	1011	100 1011
0100	011 0100	1100	101 1100
0101	110 0101	1101	000 1101
0110	100 0110	1110	010 1110
0111	001 0111	1111	111 1111

The effectiveness of this code when used for error detection was quantified in Section 11.4.1.2. However, an analytic quantification of error correction effectiveness is a little more complicated due to introduction of errors by the decoder as a result of incidents of wrong correction, an example of which was given above. A reliable quantification has been obtained using computer simulation in which a sufficiently long input bit stream having a specified BER from 0.5 to 10^{-4} was used as the decoder input and the bit error rate of the decoded bit stream was measured. Fig. 11.16 shows the simulation-based bit error rate reduction factor of the Hamming error correction code plotted as a function of input (i.e. link) BER, from which the following empirical formula was obtained for the probability of bit error p_c in (7, 4) Hamming decoded data as a function of link BER:

$$p_c = 6.9638(\mathrm{BER})^{1.9627}, \quad \mathrm{BER} < 0.1 \tag{11.25}$$

11.4.3 Interleaving and concatenation

Error control decoders perform best when channel-induced errors are random and uniformly spread out in the received bit stream. This is the type of error distribution produced by an AWGN channel, which we assume in all our analysis of the effectiveness of error control codes. However, there are many practical situations such as in the mobile radio channel, where errors might occur in bursts due to a deep signal fade or a short-lived spike in noise or interference. A burst of error may also be created by the wrong corrections of an inner coder (involving the selection of the wrong n-bit codeword and hence a burst of bit errors $\leq n$) in a concatenated arrangement of coders discussed later. A burst of error being concentrated in a few codewords is more likely to exceed the error detection or correction capability of a code than if the same number of bit errors was more randomly spread out across a number of codewords. We may ensure that the decoder is more likely to be presented with random errors even when the channel is subject to bursty degradation

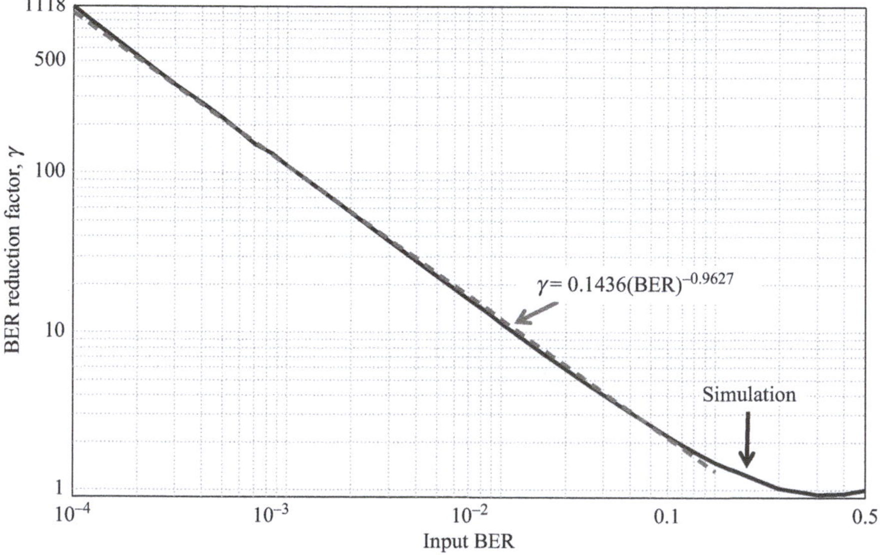

Fig. 11.16 Simulation-based bit error rate reduction factor due to (7, 4) Hamming code error correction as a function of link bit error rate

and hence a bursty pattern of induced errors if we arrange for bits that travelled through the channel in adjacent bit intervals to be presented within different codewords to the decoder. This can be done through a strategy known as *interleaving*, whereby an interleaver is used at the transmitter to disorder the data and a corresponding de-interleaver is used at the receiver for reordering.

At the transmitter, coded data at the output of the error control coder is written column-wise into an $M \times N$ matrix as shown in Fig. 11.17a. Once the matrix is filled the data is immediately read out row-wise and passed on to the line coder or modulator which then produces the filtered pulses that are transmitted through the channel. The writing and reading of the data is done bit by bit if the error control codec is binary and m-bit by m-bit for a non-binary codec (such as Reed-Solomon) that operates on m-bit symbols. The placement of an interleaver at the transmitter and the corresponding de-interleaver at the receiver is therefore as shown in Fig. 11.18. The de-interleaver writes its input data row-wise into a matrix of the same size as shown in Fig. 11.17b. Once the matrix is filled the data is read out column wise. This process at the receiver exactly re-establishes the order of the data as it was at the encoder output, while importantly ensuring that bits that travelled through the channel in adjacent intervals are kept apart by N bit intervals at the decoder input. In this way, if burst error length is $\leq N$ and codeword length $\leq M$, then a long burst of bit errors is distributed singly to separate codewords, which makes it more likely that the error control capability of the decoder will not be exceeded in any received codeword. Furthermore, with this arrangement our assumptions of a memoryless channel and a uniform random distribution of errors in the analysis of coder performance are justifiable even in mobile radio channels.

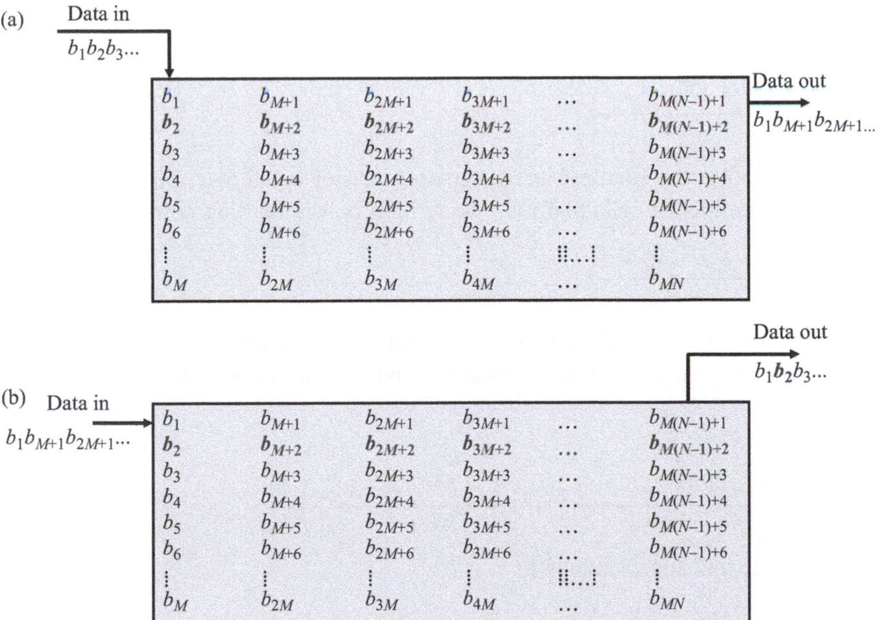

Fig. 11.17 $M \times N$ block interleaver (a) and de-interleaver (b). The interleaver writes data column-wise and reads row-wise, whereas the de-interleaver writes row-wise and reads column-wise

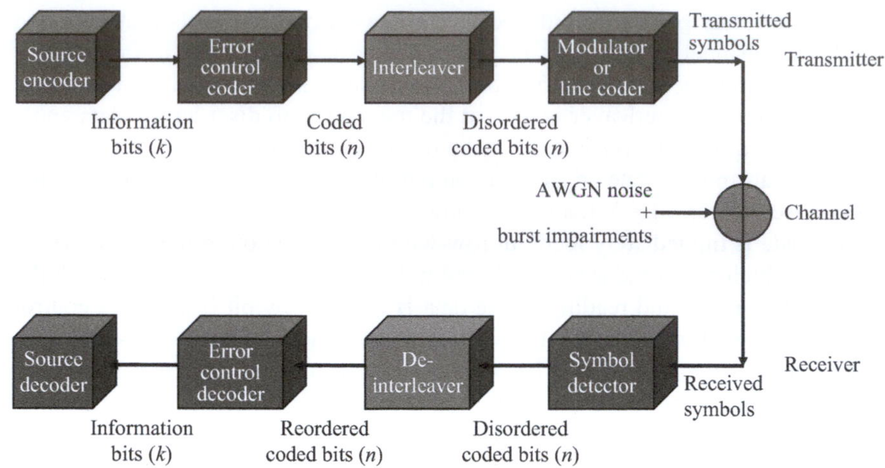

Fig. 11.18 Placement of interleaver and de-interleaver blocks in digital transmission system

This interleaver has depth N, the *depth* of an interleaver being the smallest number of intervals between two symbols at the interleaver output which were adjacent at the input.

Interleaving does not introduce any new redundancy since exactly the same number of bits that are written into the matrix is read out. However, it does introduce delay. If writing and reading are under the control of the same clock and are both carried out at the transmission bit rate R_b, then to avoid starting to read too early which would lead to the reading of a yet-to-be-written cell or waiting too long to start reading, which would cause undue delay, the number of matrix cells that must be written at the transmitter before reading commences is

$$n_t = \left\lceil \frac{1 + M^2(N-1)}{M+1} \right\rceil$$

where $\lceil x \rceil$ denotes the smallest integer greater than or equal to x. Similarly, at the receiver the number of cells that must be written before reading commences is

$$n_r = \left\lceil \frac{1 + N^2(M-1)}{N+1} \right\rceil$$

And with m bits per cell, where m is the number of bits per encoder symbol (=1 for binary encoders), the combined interleaving and de-interleaving delay is

$$\begin{aligned} \tau_i &= (n_r + n_t)\frac{m}{R_b} \\ &= \left(\left\lceil \frac{1 + N^2(M-1)}{N+1} \right\rceil + \left\lceil \frac{1 + M^2(N-1)}{M+1} \right\rceil \right) \frac{m}{R_b} \end{aligned} \tag{11.26}$$

For $M, N \gg 1$, this approximates to

$$\tau_i \approx \frac{m(2MN - M - N)}{R_b} \quad \text{seconds} \tag{11.27}$$

Memory is required to implement interleaving and this may be provided using two buffers of size mMN bits each, deployed in a ping pong manner so that while one buffer is being read from the other is being written into and vice versa. The interleaving strategy which has been discussed here is known as *block interleaving*. There are other strategies that may be used, leading to *random interleaving* (based on a known random permutation of the input bits, sometimes subject to a constraint that guarantees a minimum interleaver depth) and *convolutional interleavers* (where data is written column-wise into a bank of parallel shift registers of successively larger delay, with the first register having zero delay, and data is read column-wise from the output of the registers in synchrony with the writing).

Error correction performance can be significantly improved if the decoded bit stream produced by one decoder is submitted for further decoding in a second decoder of a different design. In this way errors missed or introduced by the first decoder might be detected and corrected by the second decoder, thus further reducing bit error rate and increasing overall coding gain. To allow this type of two-level decoding at the receiver, the transmitter must also employ two level encoding, first by an outer encoder followed in series by another encoder which is called the inner encoder. This arrangement is known as *serial concatenation* and usually includes an interleaver between the outer and inner encoders and a de-interleaver between the inner and outer decoders as shown in Fig. 11.19 for the old DVB-S (digital video broadcasting via satellite) standard featuring a Reed-Solomon outer encoder and a convolutional inner encoder. Interleaving serves to disperse any burst errors introduced by the convolutional inner decoder, which improves the chances of their correction by the outer decoder. Note the order of arrangement of encoders and decoders in a concatenated coding system. The last encoder at the transmitter is the inner encoder and its corresponding decoder must be the first decoder at the receiver. The other encoder being at the outer ends of the coded stream of data between transmitter and receiver is therefore described as the outer encoder.

Code concatenation may also be deployed in parallel in which two encoders are used at the transmitter, one to encode the original bit stream and the other to encode an interleaved version of the same bit stream. The output of the two

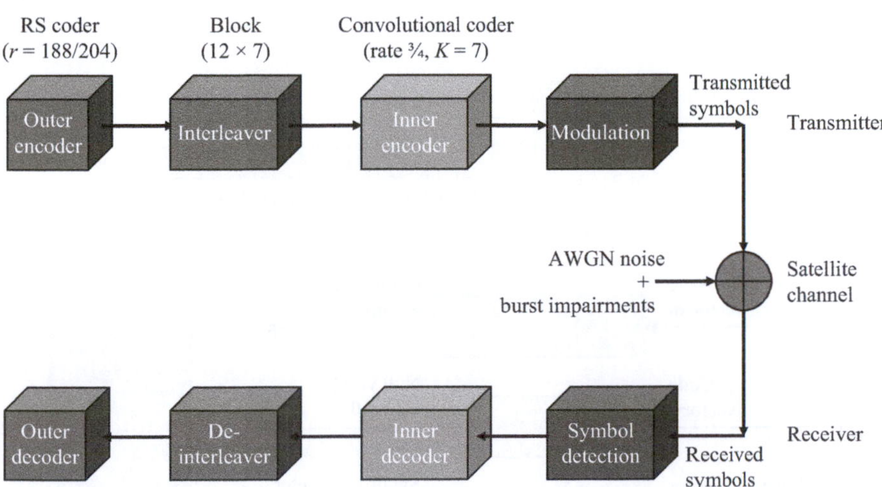

Fig. 11.19 Concatenated coding scheme used in the old DVB-S standard. The new standard (DVB-S2) uses outer BCH and inner LDPC encoders, with the de-/interleaver placed between inner codec and de/modulator

encoders are then multiplexed and transmitted. At the receiver the two coded streams are separated in a demultiplexer before being jointly decoded. The *product code* discussed earlier is in fact a special case of parallel concatenation in which the encoders are two even SPC encoders, and interleaving is dispensed with so that both encoders act on the same bit stream.

11.5 Linear block codes

This section is about linear block codes. We introduce the concept and terminologies of linear codes, adopt matrix notations to mathematically describe various coding and decoding manipulations, and discuss the use of shift registers for implementing the modulo-2 maths (including addition, multiplication and long division) that underpin the operations involved in cyclic linear block codes. Modulo-2 addition and subtraction are identical, and are defined as follows along with modulo-2 multiplication:

$$\begin{array}{ll} 0+0=0 & 0\times 0=0 \\ 0+1=1 & 0\times 1=0 \\ 1+0=1 & 1\times 0=0 \\ 1+1=0 & 1\times 1=1 \end{array}$$

Most of the focus in what follows is on binary codes, but we also discuss the non-binary Reed-Solomon code, using worked examples in coding and decoding to help make the subject clear.

11.5.1 General concepts

An (n, k) block code is generated by taking k information bits at a time, adding $n - k$ parity bits to produce a codeword of length n bits, as illustrated in Fig. 11.20. The code rate is $r = k/n$, and certain rates such as $r = ½$ are preferred for easier clock synchronisation between the coder input and output. Channel data rate is the bit rate R_{bo} at the output of the coder, which is given by the information bit rate R_{bi} at coder input divided by the code rate.

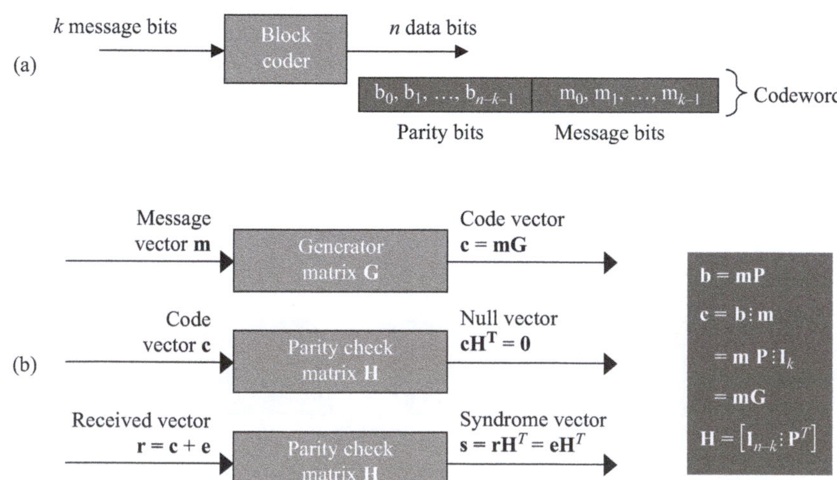

Fig. 11.20 Block code (a) concept; (b) matrix relationships

An (n, k) block code is said to be *linear* if parity bits are given by appropriate modulo-2 sums of message bits. A further necessary condition is that the all-zeros vector must also be a codeword. The latter condition means that the even SPC (single parity check) scheme is a linear block code, whereas odd SPC is not. Linear block codes satisfy the following properties:

- The modulo-2 sum of any two codewords is also a codeword.
- Scaling of a codeword (i.e. multiplying it by a valid scalar) also yields a codeword. Thus the subspace of codewords has *closure*, which means that a non-member cannot be created by summing two or more members or by scaling one member.

A block code is said to be *systematic* if the message bits are transmitted unaltered. That is, the codeword comprises the original message bits followed by parity bits, or parity bits followed by the original message bits. The order makes no difference to code performance. Thus all the codes discussed in Section 11.4 are systematic codes. Furthermore, a code is said to be a *perfect* code if it can correct all possible patterns of t and fewer errors in a codeword, but cannot correct any pattern of $t+1$ or more errors. The (7, 4) Hamming code is a perfect code with $t=1$, and so is the Golay code with $t=3$.

Let us adopt a vector notation to represent the different bit sequences involved in linear block codes. The parity bits are represented as a row vector $\mathbf{b}$ with $n-k$ elements comprising only 1s and 0s,

$$\mathbf{b} = [b_1, \ b_2, \ b_3, \ldots, \ b_{n-k}]$$

Each block of k message bits is a row vector containing k binary elements,

$$\mathbf{m} = [m_1, \ m_2, \ m_3, \ldots, \ m_k]$$

Each codeword is a row vector of n binary elements,

$$\mathbf{c} = [c_1, \ c_2, \ c_3, \ldots, \ c_n]$$

The parity bits are linear sums of the message bits obtained by multiplying the message vector by a matrix $\mathbf{P}$ called the *coefficient matrix*:

$$\begin{aligned} \mathbf{b} &= \mathbf{mP} \\ &= m_1\mathbf{P}_1 + m_2\mathbf{P}_2 + \cdots + m_k\mathbf{P}_k \\ \mathbf{P}_i &= \left[P_{i1}, \ P_{i2}, \ P_{i3}, \ldots, \ P_{i,n-k}\right], \quad i = 1, 2, 3, \ldots, k \end{aligned} \tag{11.28}$$

The relationships between the above vectors and the definition of the *generator matrix* $\mathbf{G}$ and *parity check matrix* $\mathbf{H}$ are summarised in Fig. 11.20b, where it can be seen that the codeword is obtained by multiplying the message vector by a generator matrix $\mathbf{G}$ that completely characterises the coding scheme. Matrix $\mathbf{I}_k$ is the k-by-k *identity matrix*, in which all elements are 0 except along the main diagonal where they are all 1. The received codeword vector $\boldsymbol{r}$ is represented as the sum of the transmitted code vector $\mathbf{c}$ and an error vector $\mathbf{e}$, which is an error pattern that is non-zero at each bit position where an error has occurred in the codeword. The receiver of course does not know $\mathbf{e}$. If it did then it would simply add it to the received code $\boldsymbol{r}$ to obtain the transmitted code $\mathbf{c}$, since (noting that $1+1=0$ under modulo-2 addition)

$$\mathbf{r} + \mathbf{e} = \mathbf{c} + \mathbf{e} + \mathbf{e} = \mathbf{c} + \mathbf{e}(1+1) = \mathbf{c}$$

The decoding process at the receiver involves multiplying the received vector $\boldsymbol{r}$ by a decoding matrix, which is the transpose of the *parity check matrix* **H**. The result of this multiplication is the *error syndrome vector* **s**, which is equal to the **null** vector only if the received vector is a valid code vector **c**. If **s** is not null it indicates an error in the received vector, and if there is error correction capability the value of **s** conveys information about the locations of the error in the received vector. Worked Examples 11.2 and 11.3 illustrate the application of these concepts to error detection (in even SPC code) and error correction (in Hamming code).

Other relevant concepts and terms often encountered in linear block coding include:

- *Weight w*: The Hamming weight, designated $w(\mathbf{c})$, of a codeword **c** is the number of non-zero elements in the codeword.
- *Distance d*: The Hamming distance between two codewords is the number of bit locations in which they differ. The Hamming distance between codewords $\mathbf{c}_x$ and $\mathbf{c}_y$ is denoted by $d(\mathbf{c}_x, \mathbf{c}_y)$.
- *Minimum distance* $d_{\mathbf{min}}$: This is the smallest distance between any pair of codewords in the entire subspace of codewords.
- The maximum number of errors *guaranteed to be detected* in a codeword is $d_{\min}-1$. An (n, k) code will however detect $2^n - 2^k$ number of n-bit error patterns, since there are $2^n - 1$ patterns of which only $2^k - 1$ (identical to valid codewords) will, when added to a codeword, transform it to another codeword and hence pass undetected.
- The maximum number of errors that can be *corrected* in a codeword is $\lfloor ½(d_{\min} - 1)\rfloor$, where $\lfloor x\rfloor$ denotes the integer part of x. For example, $\lfloor 3.95\rfloor = 3$. Of β errors detected in a codeword, α of them can be corrected, provided $d_{\min} \geq \beta + \alpha + 1$.
- The Hamming distance between codewords is the weight of their sum, and hence is the weight of another codeword. Thus

$$d(\mathbf{c}_\mathbf{x}, \mathbf{c}_\mathbf{y}) = w(\mathbf{c}_\mathbf{x} + \mathbf{c}_\mathbf{y}) = w(\mathbf{c}_\mathbf{z}) \tag{11.29}$$

  Furthermore,

$$w(\mathbf{c}) = d(\mathbf{c}, \underline{\mathbf{0}})$$

  where $\underline{\mathbf{0}}$ is the all-zeros codeword. It follows that the minimum distance $d_{\min}$ is given by the minimum Hamming weight found in the subspace of codewords (excluding $\underline{\mathbf{0}}$).

Finally, it can be readily shown that if in an (n, k) code N_w denotes the number of codewords with weight w, and BER is the bit error rate of the uncoded link (assumed to be a binary symmetric channel) then the probability that a codeword is received with an undetectable error is

$$p_c = \sum_{w=1}^{n} N_w(\mathrm{BER})^w(1 - \mathrm{BER})^{n-w} \tag{11.30}$$

For example, Table 11.2 shows that in the (7, 4) Hamming code the number of codewords with respective weights $w = 0$, 3, 4 and 7 are $N_w = 1$, 7, 7 and 1. On a link with BER 10^{-4}, the above equation gives probability of an errored codeword passing undetected as

$$\begin{aligned} p_c &= 7(\text{BER})^3(1-\text{BER})^4 + 7(\text{BER})^4(1-\text{BER})^3 + (\text{BER})^7(1-\text{BER})^0 \\ &= 7[(\text{BER})(1-\text{BER})]^3 + (\text{BER})^7 \\ &= 6.9979 \times 10^{-12} \end{aligned}$$

Worked Example 11.2: Matrix Specification of Even Single Parity Check (SPC) Code

Fig. 11.21 gives a specification of the (8, 7) even SPC in matrix form, including its generator matrix **G** and parity-check matrix **H**. Based on this pattern, use matrix manipulation to determine the following for a (4, 3) even SPC code

---

(a) The parity bit for the message 111
(b) The codeword for the message 001
(c) The decision (error or no error) on the received codeword 1011
(d) The decision on the received codeword 1100

---

1. c = mG:

$$\underbrace{[b_o m_o m_1 m_2 m_3 m_4 m_5 m_6]}_{\text{Codeword, } \mathbf{c}} = \underbrace{[m_o m_1 m_2 m_3 m_4 m_5 m_6]}_{\text{Message bits, } \mathbf{m}} \times \underbrace{\begin{bmatrix} 1&1&0&0&0&0&0&0 \\ 1&0&1&0&0&0&0&0 \\ 1&0&0&1&0&0&0&0 \\ 1&0&0&0&1&0&0&0 \\ 1&0&0&0&0&1&0&0 \\ 1&0&0&0&0&0&1&0 \\ 1&0&0&0&0&0&0&1 \end{bmatrix}}_{\text{Generator matrix, } \mathbf{G}}$$

2. $\mathbf{cH}^T = \mathbf{0}$:

$$\underbrace{[b_o m_o m_1 m_2 m_3 m_4 m_5 m_6]}_{\text{Codeword, } \mathbf{c}} \times \underbrace{[1\ 1\ 1\ 1\ 1\ 1\ 1\ 1]'}_{\text{Parity-check matrix, } \mathbf{H}} = \mathbf{0}$$

Fig. 11.21 Matrix specification of the (8, 7) even SPC

Based on the SPC matrix specification pattern in Fig. 11.21 and the relationships in Fig. 11.20b, the generator matrix **G**, coefficient matrix **P** and parity check matrix **H** of a (4, 3) even SPC are as follows:

$$\mathbf{G} = \begin{bmatrix} 1&1&0&0 \\ 1&0&1&0 \\ 1&0&0&1 \end{bmatrix} \equiv \left[P \vdots \mathbf{I}_k\right]; \quad \mathbf{P} = \begin{bmatrix} 1 \\ 1 \\ 1 \end{bmatrix}$$

$$\mathbf{H} = [1 \quad 1 \quad 1 \quad 1]$$

The required computations readily follow by matrix multiplication, recalling that cell (j, k) in the product of matrices **A** and **B** is the result of multiplying row j of matrix **A** by column k of matrix **B**, and that the summing of the products of corresponding elements in this row × column operation is modulo-2:

(a) Given $\boldsymbol{m} = [1\ 1\ 1]$, we obtain the parity vector **b** as

$$\mathbf{b} = \mathbf{m}P = [1 \quad 1 \quad 1] \times \begin{bmatrix} 1 \\ 1 \\ 1 \end{bmatrix} = 1 + 1 + 1 = 1$$

The operation therefore correctly computes the parity bit as 1.

(b) Given $\boldsymbol{m} = [0\ 0\ 1]$, we obtain the code vector **c** as

$$\mathbf{c} = \mathbf{m}G = [0 \quad 0 \quad 1] \times \begin{bmatrix} 1 & 1 & 0 & 0 \\ 1 & 0 & 1 & 0 \\ 1 & 0 & 0 & 1 \end{bmatrix}$$

$$= [0+0+1 \quad 0+0+0 \quad 0+0+0 \quad 0+0+1]$$

$$= [1 \quad 0 \quad 0 \quad 1]$$

The codeword is thus correctly generated as the parity bit followed by the message bits.

(c) Given the received vector $\mathbf{r} = [1\ 0\ 1\ 1]$, we base a decision on whether or not there is error in **r** on the value of the error syndrome vector **s**: If it is null, we conclude there is no error, otherwise we conclude there is error:

$$\mathbf{s} = \mathbf{r}\mathbf{H}^T = [1 \quad 0 \quad 1 \quad 1] \times [1 \quad 1 \quad 1 \quad 1]'$$

$$= [1 \quad 0 \quad 1 \quad 1] \times \begin{bmatrix} 1 \\ 1 \\ 1 \\ 1 \end{bmatrix} = 1 + 0 + 1 + 1 = 1$$

The syndrome computation therefore correctly flags an error.

(d) For the received vector $\mathbf{r} = [1\ 1\ 0\ 0]$, we proceed in a similar way to obtain

$$\mathbf{s} = [1 \quad 1 \quad 0 \quad 0] \times \begin{bmatrix} 1 \\ 1 \\ 1 \\ 1 \end{bmatrix} = 1 + 1 + 0 + 0 = 0$$

which indicates that received vector **r** is a valid codeword.

Worked Example 11.3: Hamming Codes

The *Hamming code* is a family of (n, k) linear block code with the following parameters:

Codeword length $n = 2^m - 1$
Number of parity bits $= m$
Number of message bits $k = n - m = 2^m - m - 1$, where $m \geq 3$
Number of errors that can be corrected $t = 1$

An appropriate coefficient matrix **P** for the (7, 4) Hamming code is given by

$$\mathbf{P} = \begin{bmatrix} 1 & 1 & 0 \\ 0 & 1 & 1 \\ 1 & 1 & 1 \\ 1 & 0 & 1 \end{bmatrix}$$

Determine:

(a) The generator matrix **G**
(b) The parity check matrix **H**
(c) The codeword for the message bits 1100
(d) The decoding table
(e) Whether an error in the last bit of the codeword in (c) would be detected and corrected
(f) What happens with two errors in the first two bits of this codeword

(a) Generator matrix is given by

$$\mathbf{G} = \left[\mathbf{P}\vdots\mathbf{I}_4\right] = \begin{bmatrix} 1 & 1 & 0 & 1 & 0 & 0 & 0 \\ 0 & 1 & 1 & 0 & 1 & 0 & 0 \\ 1 & 1 & 1 & 0 & 0 & 1 & 0 \\ 1 & 0 & 1 & 0 & 0 & 0 & 1 \end{bmatrix}$$

(b) Parity check matrix is

$$\mathbf{H} = \left[\mathbf{I}_3\vdots\mathbf{P}^T\right] = \begin{bmatrix} 1 & 0 & 0 & 1 & 0 & 1 & 1 \\ 0 & 1 & 0 & 1 & 1 & 1 & 0 \\ 0 & 0 & 1 & 0 & 1 & 1 & 1 \end{bmatrix}$$

(c) The codeword for message $\mathbf{m} = 1100$ is computed as

$$\mathbf{c} = \mathbf{m}G = [1100] \times \begin{bmatrix} 1 & 1 & 0 & 1 & 0 & 0 & 0 \\ 0 & 1 & 1 & 0 & 1 & 0 & 0 \\ 1 & 1 & 1 & 0 & 0 & 1 & 0 \\ 1 & 0 & 1 & 0 & 0 & 0 & 1 \end{bmatrix} = [1011100]$$

(d) The decoding table provides a list of syndrome values and their corresponding error patterns. Since the error-correction capability of this code is $t = 1$, the error patterns that can be handled will be seven-bit vectors with binary 1 in only one bit location and 0s at all other locations. To complete the table, we multiply each error pattern by $\mathbf{H}^T$ to obtain its corresponding error-syndrome vector. For example, for the error pattern 0100000 we compute

$$\mathbf{s} = \mathbf{e}\mathbf{H}^T = [0 \quad 1 \quad 0 \quad 0 \quad 0 \quad 0 \quad 0] \times \begin{bmatrix} 1 & 0 & 0 \\ 0 & 1 & 0 \\ 0 & 0 & 1 \\ 1 & 1 & 0 \\ 0 & 1 & 1 \\ 1 & 1 & 1 \\ 1 & 0 & 1 \end{bmatrix}$$

$$= [0 \quad 1 \quad 0]$$

Proceeding in this way yields the complete (7, 4) Hamming decoding table in Table 11.3.

Table 11.3 (7, 4) Hamming decoding table, in Worked Example 11.3*d*

Error Pattern, e	Syndrome, $\mathbf{s} = \mathbf{e}\mathbf{H}^T$
0000000	000
1000000	100
0100000	010
0010000	001
0001000	110
0000100	011
0000010	111
0000001	101

(e) The transmitted codeword is $\mathbf{c} = 1011100$. With error in the last bit, the received vector is $\mathbf{r} = 1011101$. The decoder computes the syndrome

$$\mathbf{s} = \mathbf{r}\mathbf{H}^T = [1 \quad 0 \quad 1 \quad 1 \quad 1 \quad 0 \quad 1] \times \begin{bmatrix} 1 & 0 & 0 \\ 0 & 1 & 0 \\ 0 & 0 & 1 \\ 1 & 1 & 0 \\ 0 & 1 & 1 \\ 1 & 1 & 1 \\ 1 & 0 & 1 \end{bmatrix}$$

$$= [1 \quad 0 \quad 1]$$

which (from the decoding table) corresponds to the error pattern $\mathbf{e} = 0000001$. To correct the error the decoder simply adds this error pattern to the received codeword. Thus the code actually sent is

$$\mathbf{c} = \mathbf{r} + \mathbf{e} = 1011101 + 0000001 = 1011100.$$

The decoder would then pass the message part $\mathbf{m} = 1100$ to the source decoder.

(f) The transmitted codeword is $\mathbf{c} = 1011100$. With error in the first two bits, the received vector is $\boldsymbol{r} = 0111100$. The decoder as usual computes the syndrome, which, proceeding as in (e), yields $\mathbf{s} = \mathbf{rH}^T = 110$. The decoding table gives the corresponding error pattern $\mathbf{e} = 0001000$, which the decoder adds to the received vector to obtain the corrected codeword $\mathbf{c} = \mathbf{r} + \mathbf{e} = 0110100$. Notice that this is not the transmitted codeword, so on this occasion the error has been flagged (via the non-zero syndrome computed) but is wrongly corrected. The decoder would pass the wrong message 0100 to the source decoder when 1100 is actually what was sent. This serves as a reminder that the (7, 4) Hamming code is able to detect up to two bit errors in a codeword but unable to correct more than one.

11.5.2 Cyclic codes

A binary block code is said to be cyclic if in addition to being *linear* it also satisfies the *cyclic* property, which states that any cyclic shift of a codeword in the code is also a codeword. For example, if 1011100 is a codeword from a cyclic code, then any seven-bit codeword obtained by reading the circular arrangement of bits shown in Fig. 11.22 going clockwise and starting at any bit is also a codeword. It is sometimes necessary to determine the codeword resulting from m cyclic shifts. To do this, you simply write the codeword into a circular ring as in Fig. 11.22, and go back (anti-clockwise) through m bit positions from the first bit of the given codeword and then read forward (clockwise) to obtain the shifted codeword. Thus a two-bit cyclic shift of 1011100 yields 0010111 as the new codeword.

A big attraction of cyclic codes is the ease with which their coding and decoding processes can be implemented using shift registers. To perform coding and decoding operations on paper, the use of polynomials to carry out the shifts involved is much more convenient than the matrix manipulations employed so far. In polynomial notation, an n-bit codeword $\mathbf{c} = [c_0, c_1, c_2, \ldots, c_{n-1}]$ is represented by a polynomial of order $n-1$ in X:

$$c(X) = c_0 + c_1X + c_2X^2 + \cdots + c_{n-1}X^{n-1}$$

where the coefficients $c_0, c_1, c_2, \ldots$ are binary so that the absence of any term in the polynomial is indicative of binary 0 for that bit position. For example, the

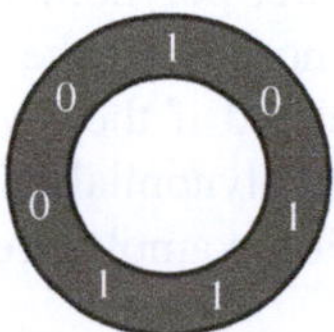

Fig. 11.22 If any seven-bit sequence read from this diagram going clockwise is the codeword of a cyclic code, then every possible seven-bit clockwise reading is also a codeword

codeword 1011100 is expressed as $c(X) = 1 + X^2 + X^3 + X^4$. Each code polynomial in an (n, k) cyclic code can be expressed as

$$c(X) = q(X)g(X)$$

where $q(X)$ is a polynomial of degree $k - 1$, and$g(X)$ is the *generator polynomial*. This generator polynomial has degree $n - k$ and is a factor of $X^n + 1$.

11.5.2.1 Encoding and decoding process

We have already learnt the method of matrix manipulation for coding and decoding. There is a different method which is applicable to cyclic codes and is based on the use of polynomials. The following steps are followed to encode an (n, k) cyclic code:

- Multiply the message polynomial $m(X)$ by X^{n-k}. This is to shift the message bits through $n-k$ places into the rightmost k bit positions of an n-bit codeword, leaving the leftmost $n-k$ bit positions for the parity bits generated next.
- Next, divide $X^{n-k}m(X)$ by $g(X)$. The remainder $b(X)$ produced by this division is the *parity polynomial* of degree $n-k-1$, which is one degree lower than the degree of the divisor $g(X)$. That is,

  $$X^{n-k}m(X) = q(X)g(X) + b(X)$$

  but we have no interest whatsoever in the quotient $q(X)$.
- The code polynomial is

  $$c(X) = b(X) + X^{n-k}m(X) \equiv \left[\text{Remainder} \vdots \text{Message}\right]$$

  Notice that the codeword is a concatenation of the parity bits (i.e. remainder) and the message bits because the polynomial factor X^{n-k} shifts the message out of the first $n - k$ bit positions. The corresponding code vector is $\mathbf{c} = [b_0, b_1, b_2, \ldots, b_{n-k-1}, m_0, m_1, m_2, \ldots, m_{k-1}]$.

Note that the codeword $c(X)$ as obtained above is divisible by $g(X)$ without a remainder, since

$$\begin{aligned}
c(X) &= b(X) + X^{n-k}m(X) \\
&= b(X) + q(X)g(X) + b(X) \\
&= q(X)g(X) + b(X)(1 + 1) \\
&= q(X)g(X)
\end{aligned}$$

Thus, systematic use of redundancy in cyclic codes involves sending only codewords that are multiples of a generator polynomial. At the receiver the decoder can therefore divide the received polynomial by the same generator polynomial used at the transmitter. An error is declared if the division leaves any remainder. The parity-check polynomial $h(X)$ is a polynomial of degree k that is also a factor of $X^n + 1$. Thus, $g(X)h(X) = X^n + 1$. For example, we know that

$$X^7 + 1 = \left(1 + X + X^3\right)\left(1 + X + X^2 + X^4\right) \equiv g(X)h(X)$$

It follows that $g(X) = 1 + X + X^3$ is a polynomial of order $3 \equiv n - k$ (as stated above), $n = 7$, and $k = 4$; so that this $g(X)$ can be used to generate $(n, k) \equiv (7, 4)$ cyclic codes. Alternatively, if we identify the other factor of $X^7 + 1$ as $g(X)$, we obtain a generator polynomial of order $4 \equiv n - k$, which yields $k = 3$, and hence $g(X) = 1 + X + X^2 + X^4$ can be used as a generator polynomial for a (7, 3) cyclic code.

The syndrome polynomial $s(X)$ is the remainder obtained by dividing the received code polynomial $r(X)$ by the generator polynomial $g(X)$. Given $r(X)$, the decoder estimates the transmitted code polynomial as

$$c(X) = r(X) + e(X)$$

where $e(X)$ is the error polynomial corresponding to $s(X)$ in the decoding table. An (n, k) cyclic encoder and its syndrome calculator are usually implemented using a linear feedback register with $n - k$ stages as discussed next.

11.5.2.2 Polynomial multiplication and division

Polynomial multiplication and polynomial division can be implemented using a circuit comprising a shift register and modulo-2 adders. Each stage of a shift register is a D-type flip-flop (FF) that works under the control of a common clock. A D-type FF performs a simple delay task, namely the signal at its input is transferred to its output at each instance (e.g. rising edge) of a clock signal. The circuit of Fig. 11.23 multiplies two polynomials

$$m(X) = m_0 + m_1X + m_2X^2 + \cdots + m_kX^k$$

and

$$g(X) = g_0 + g_1X + g_2X^2 + g_3X^3 + \cdots + g_pX^p$$

of orders k and p, respectively, shown in the diagram for $p = 3$. For polynomials with binary coefficients, the scale factors g_i, $i = 0, 1, \ldots, p$ in Fig. 11.23, and coefficients m_j, $j = 0, 1, 2, \ldots, k$ are either 1 or 0. This means that the ith adder is present in the circuit only if there is an X^i term in the polynomial $g(X)$. The operation of the circuit is as follows:

- With the shift register initially at zero state (i.e. FF1 = FF2 = FF3 = 0), the coefficients of $m(X)$ are clocked in starting from the most significant coefficient m_k.
- After $k + p$ clock cycles (completed by shifting in p trailing zeros beyond m_0 until the flip-flop returns to its zero initial state), the coefficients of $m(X)g(X)$ have been serially released at the output.

The above process is outlined in tabular form in Table 11.4 for $k = 2$. The arrows in the table indicate the input-to-output shift done by each flip-flop at each clock instance. Notice that $k + p + 1$ (= 6) rows of the final column of the

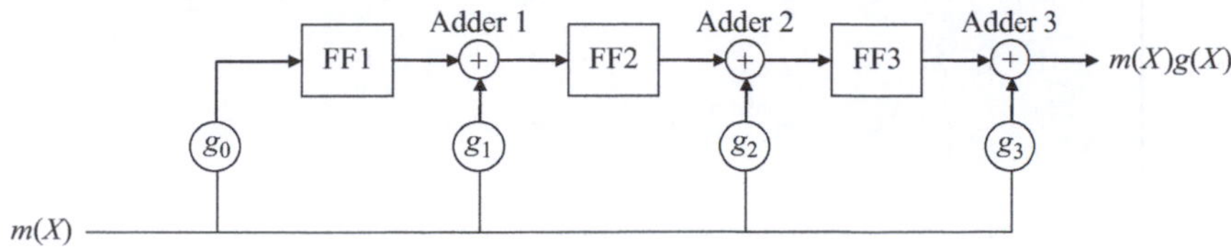

Fig. 11.23 Shift register circuit for polynomial multiplication

Table 11.4 Operation of the polynomial multiplication circuit in Fig. 11.20 on message $m(X) = m_0 + m_1X + m_2X^2$

CK	$m(X) = m_k$	$F1_{in} = m_k \times g_0$	$F1_{out}$	$A1_{out} = F2_{in} = m_kg_1 + F1_{out}$	$F2_{out}$	$A2_{out} = F3_{in} = m_kg_2 + F2_{out}$	$F3_{out}$	$m(X)g(X) = m_kg_3 + F3_{out}$
0	m_2	m_2g_0	0 (Initial)	m_2g_1	0 (Initial)	m_2g_2	0 (Initial)	m_2g_2
1	m_1	m_1g_0	m_2g_0	$m_1g_1 + m_2g_0$	m_2g_1	$m_1g_2 + m_2g_1$	m_2g_2	$m_1g_3 + m_2g_2$
2	m_0	m_0g_0	m_1g_0	$m_0g_1 + m_1g_0$	$m_1g_1 + m_2g_0$	$m_0g_2 + m_1g_1 + m_2g_0$	$m_1g_2 + m_2g_1$	$m_0g_3 + m_1g_2 + m_2g_1$
3	0	0	m_0g_0	m_0g_0	$m_0g_1 + m_1g_0$	$m_0g_1 + m_1g_0$	$m_0g_2 + m_1g_1 + m_2g_0$	$m_0g_2 + m_1g_1 + m_2g_0$
4	0	0	0	0	m_0g_0	m_0g_0	$m_0g_1 + m_1g_0$	$m_0g_1 + m_1g_0$
5	0	0	0	0	0	0	m_0g_0	m_0g_0
6	0	0	0	0	0	0	0	0

table (produced after $k+p$ clock cycles — see clock count in the first column) contain the coefficients of the product polynomial

$$\begin{aligned}(m_0 + m_1X + m_2X^2)(g_0 + g_1X + g_2X^2 + g_3X^3) \\ = m_0g_0 \\ + (m_0g_1 + m_1g_0)X \\ + (m_0g_2 + m_1g_1 + m_2g_0)X^2 \\ + (m_0g_3 + m_1g_2 + m_2g_1)X^3 \\ + (m_1g_3 + m_2g_2)X^4 \\ + m_2g_3X^5\end{aligned}$$

The circuit does indeed perform polynomial multiplication, with the additions involved in each of the above coefficients being done modulo-2. Table 11.4 is self-explanatory; however you may find the following steps useful in determining the output of any polynomial multiplication circuit such as Fig. 11.23. Use the example of Table 11.4 to make sure you understand each step of the instruction.

- Prepare a table with a column for $m(X)$ and for each input and each output of each stage of the shift register.
- Copy the coefficients of $m(X)$ into column 2 in the order from m_k to m_0 and append p zeros.
- Start by completing the row for clock cycle (CK) $= 0$, setting $\mathrm{F1}_{out} = \mathrm{F2}_{out} = \mathrm{F3}_{out} = 0$ (since the shift register is initially zero), and then carrying out the additions needed in the remaining cells of that row.
- Move on to the next row and set the output ($\mathrm{F1}_{out}$, etc.) of each stage of the register to its previous input ($\mathrm{F1}_{in}$, etc., as indicated by the arrows in Table 11.4) since the input of each stage is transferred to its output at each clock instance. Then complete the remaining cells by doing the additions stipulated by the circuit, which are also shown in the headers of the relevant columns.
- Continue in this way until the last row where the shift register is back to the all-zero state.
- At this point (after $k+p$ clock cycles) the last column is the required result: $m(X)g(X) = m_2g_3X^5 + (m_1g_3 + m_2g_2)X^4 + (m_0g_3 + m_1g_2 + m_2g_1)X^3 + \cdots + m_0g_0$

Whereas the cyclic encoder performs polynomial multiplication, the decoder carries out polynomial division using basically the same type of circuit as for multiplication, except that feedback is introduced. Fig. 11.24a shows a general circuit that divides $r(X)$ by $g(X) = g_0 + g_1X + g_2X^2 + g_3X^3 + \cdots + g_pX^p$, producing a quotient $q(X)$ and a remainder $s(X)$. All polynomial coefficients are binary and arithmetic is modulo-2, so addition and subtraction are identical operations. This is because $g_j + g_j = g_j(1 + 1) = 0$, which means that $g_j = -g_j$. Furthermore, the first and last coefficients of the divisor polynomial are required to be unity (i.e. $g_0 = g_p = 1$). Therefore the division circuit simplifies as shown in part (b) of the

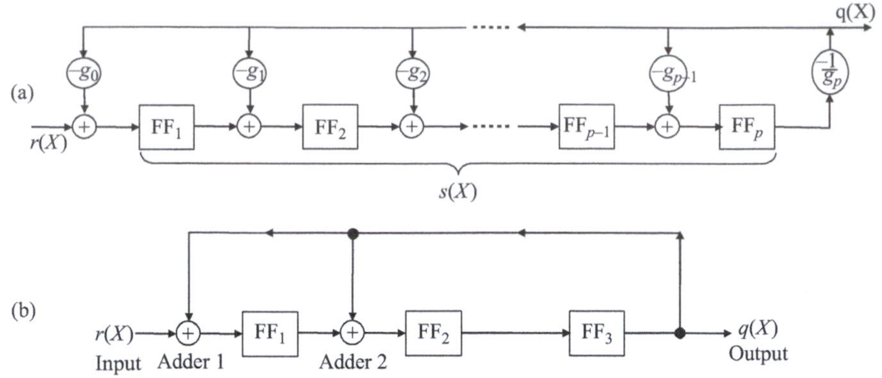

Fig. 11.24 Polynomial division: (a) general circuit; (b) circuit for dividing $r(X)$ by $g(X) = 1 + X + X^3$

figure for $g(X) = 1 + X + X^3$. The operation of the circuit of Fig. 11.24b is illustrated in Table 11.5 for $r(X) = 1 + X^2 + X^3 + X^4 + X^6$, which corresponds to the code vector 1011101, and is summarised below.

The steps involved in the circuit division of $r(X) = 1 + X^2 + X^3 + X^4 + X^6$ by $g(X) = 1 + X + X^3$ are as follows:

1. Copy the coefficients of $r(X)$ into column 2 in descending order from r_6 to r_0.
2. Initialise $F1_{out}$, $F2_{out}$ and $F3_{out}$ to zero in the first row below the header (i.e. row CK = 0).
3. Compute $F1_{in}$ and $F2_{in}$ in the current row using the summation stipulated by the circuit as also written in the table header.
4. Move to the next row and fill in cells $F1_{out}$, $F2_{out}$ and $F3_{out}$ by shifting the respective values of $F1_{in}$, $F2_{in}$ and $F3_{in}$ from the previous row, as indicated by the arrow in the table. Repeat step 3.
5. Repeat step 4 until row CK = 7, where only the register state [$F1_{out}$, $F2_{out}$, $F3_{out}$] needs to be filled in.
6. The register state in row CK = 7 gives the syndrome in ascending order, whereas the content of the last column from CK = 0 to CK = 6 gives the quotient in descending order.
7. The result in Table 11.5 is thus remainder $S(X) = 1 + X^2$ and quotient $q(X) = X^3$.

As a further example, the circuit operation steps involved in dividing $r(X) = X^3 + X^5 + X^6$ by $g(X) = 1 + X + X^3$ are tabulated in Table 11.6, from which we see that the remainder (last row) is $s(X) = 1$, and the quotient (reading the last column starting at row CK = 6 and going upwards) is $q(X) = 1 + X + X^2 + X^3$.

Based on our discussion of the coding and decoding operations of a cyclic code in the previous section, and the circuit implementations of polynomial multiplication and division presented above, we may design circuits for the (7, 4) Hamming code as shown in Fig. 11.25, where the circuit in part (a) generates the codeword, whereas part (b) computes the error syndrome.

Table 11.5 Operation of polynomial circuit division, illustrating the use of the circuit of Fig. 11.24*b to divide* $r(X) = 1 + X^2 + X^3 + X^4 + X^6$ *by* $g(X) = 1 + X + X^3$

Clock count (CK)	Input $r(X)$ $= r_n$	$F1_{in}$ $= r_n + q_n$	$F1_{out}$	$F2_{in}$ $= F1_{out} + q_n$	$F2_{out} =$ $F3_{in}$	Output $q(X)$ $= F3_{out} = q_n$
0	1	1	0 (initial)	0	0 (initial)	0 (initial)
1	0	0	1	1	0	0
2	1	1	0	0	1	0
3	1	0	1	0	0	1
4	1	1	0	0	0	0
5	0	0	1	1	0	0
6	1	1	0	0	1	0
7	–	–	1	–	**0**	**1**

Table 11.6 Using Fig. 11.24*b to divide* $r(X) = X^3 + X^5 + X^6$ *by* $g(X) = 1 + X + X^3$

CK	Input $r(X)$ $= r_n$	$F1_{in}$ $= r_n + q_n$	$F1_{out}$	$F2_{in}$ $= F1_{out} + q_n$	$F2_{out} = F3_{in}$	Output $r(X)$ $= F3_{out} = q_n$
0	1	1	0 (initial)	0	0 (initial)	0 (initial)
1	1	1	1	1	0	0
2	0	0	1	1	1	0
3	1	0	0	1	1	1
4	0	1	0	1	1	1
5	0	1	1	0	1	1
6	0	1	1	0	0	1
7	–	–	**1**	–	**0**	**0**

The code generator circuit (Fig. 11.25a) operates as follows:

- For the 1st k (= 4) clock cycles, switch S_1 is on and switch S_2 is down. The k message bits are therefore shifted into the code output, and are at the same time fed back into the shift register.
- The contents of the shift register at the end of this period are the parity bits.
- For the next $n - k$ (= 3) clock cycles, S_1 is off and S_2 is up. The parity bits are therefore shifted into the code output.
- In this way the n bits shifted out of the output terminal in n clock cycles represent the codeword.

(a)

S_1

FF_1 Adder 1 FF_2 FF_3 Adder 2

S_2

Code output

Message input

(b)

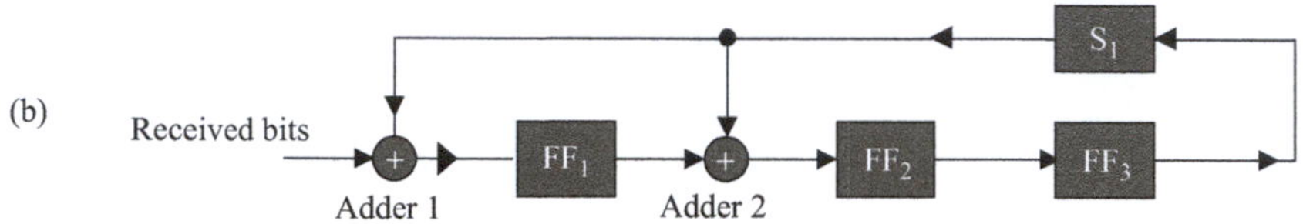

Fig. 11.25 (7, 4) Hamming code: (a) code generator using polynomial $g(X) = 1 + X + X^3$; (b) syndrome calculator

The syndrome calculator circuit operates as follows:

- The received bits are clocked into the shift register with feedback switch S_1 on.
- After n (= 7) clock cycles, when a codeword has been shifted in, the contents of the shift register represent the syndrome.

To complete our treatment of binary cyclic codes let us revisit the (7, 4) Hamming code problem which was earlier solved through a matrix approach, and this time solve it using a polynomial approach through both mathematical manipulation and circuit implementation.

Worked Example 11.4: (7, 4) Hamming Coding by Polynomial Method

(a) Using the generator polynomial $g(X) = 1 + X + X^3$ for the (7, 4) cyclic code
  (i) Encode the message sequence 1100.
  (ii) Decode the received codeword 1011101.

(b) Solve the above problems again, but this time use the circuits given in Fig. 11.25, presenting the circuit operation steps in tabular form.

(a) (i) The coding steps are straightforward. First, express the message $m = 1100$ in polynomial notation as $m(X) = 1 + X$. Second, form the product $X^{n-k}m(X) = X^3(1 + X) = X^3 + X^4$, since $n = 7$ and $k = 4$. Third, divide $X^{n-k}m(X)$ by $g(X) = 1 + X + X^3$ (see Fig. 11.26a for the layout of this long division) to obtain the remainder $b(X) = 1 + X^2$.

Notice in the long division that the divisor and dividend polynomials are written in descending order, and that addition replaces subtraction since they are identical in modulo-2 arithmetic. Fourth, the code polynomial is therefore $c(X) = b(X) + X^{n-k}m(X) = 1 + X^2 + X^3 + X^4$, which corresponds to the (seven-bit) codeword $\mathbf{c} = 1011100$. Note that if (as in this case) the highest power of $c(X)$ is not $n-1$, then trailing zeros must be used to complete the translation of the code from a polynomial to a bit sequence format.

(ii) The decoding steps are as follows. First, express the received code $\mathbf{r} = 1011101$ in polynomial form as $r(X) = 1 + X^2 + X^3 + X^4 + X^6$. Second, divide $r(X)$ by $g(X)$ as shown in Fig. 11.26b to obtain the remainder $s(X) = 1 + X^2$, which represents the syndrome vector $\mathbf{s} = 101$. Third, read the corresponding error pattern from the decoding table (Table 11.3). This is 0000001, which in polynomial form is $e(X) = X^6$. Fourth, add this error polynomial to the received code polynomial to obtain the transmitted code polynomial as

$$\begin{aligned} c(X) &= r(X) + e(X) = 1 + X^2 + X^3 + X^4 + X^6 + X^6 \\ &= 1 + X^2 + X^3 + X^4 + X^6(1+1) \\ &= 1 + X^2 + X^3 + X^4 \end{aligned}$$

which represents transmitted codeword $\mathbf{c} = 1011100$, and hence message $\mathbf{m} = 1100$.

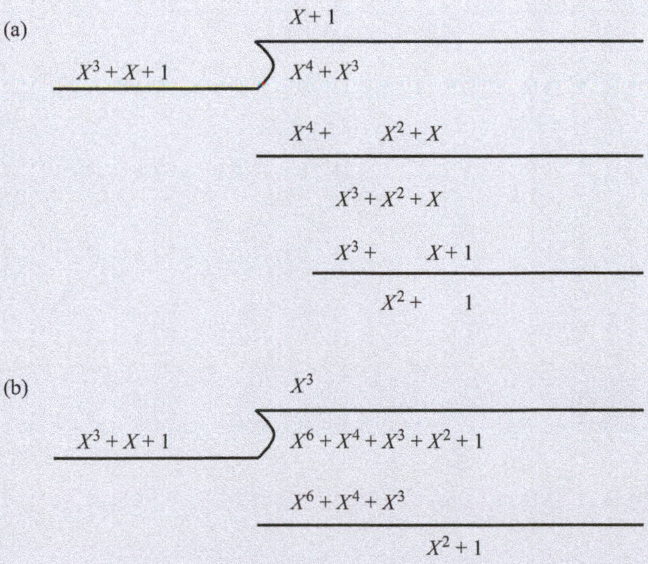

(a)

$$\begin{array}{r|l} & X+1 \\ \hline X^3+X+1 & X^4+X^3 \\ & X^4+\phantom{X^3+{}}X^2+X \\ \hline & \phantom{X^4+{}}X^3+X^2+X \\ & \phantom{X^4+{}}X^3+\phantom{X^2+{}}X+1 \\ \hline & \phantom{X^4+X^3+{}}X^2+\phantom{X+{}}1 \end{array}$$

(b)

$$\begin{array}{r|l} & X^3 \\ \hline X^3+X+1 & X^6+X^4+X^3+X^2+1 \\ & X^6+X^4+X^3 \\ \hline & \phantom{X^6+X^4+X^3+{}}X^2+1 \end{array}$$

Fig. 11.26 Long division of (a) $X^3 + X^4$, and (b) $1 + X^2 + X^3 + X^4 + X^6$ by $1 + X + X^3$ using modulo-2 arithmetic to obtain quotient $1 + X$ and remainder $1 + X^2$ in (a) and quotient X^3 and remainder $1 + X^2$ in (b)

Table 11.7 Worked Example 11.4b: (a) use of Fig. 11.25a to code message 1100; (b) use of Fig. 11.25b to decode received vector $\mathbf{r} = 1011101$

(a)

CK	Msg $m(X) = m_n$	$F1_{in} = m_n + F3_{out}$	$F1_{out}$	$F2_{in} = F1_{out} + F1_{in}$	$F2_{out} = F3_{in}$	$F3_{out}$
0	0	0	0 (initial)	0	0 (initial)	0 (initial)
1	0	0	0	0	0	0
2	1	1	0	1	0	0
3	1	1	1	0	1	0
$k = 4$	–	–	1	–	0	1

(b)

CK	$r(X) = r_n$	$F1_{in} = r_n + F3_{out}$	$F1_{out}$	$F2_{in} = F1_{out} + F3_{out}$	$F2_{out} = F3_{in}$	$F3_{out}$
0	1	1	0 (initial)	0	0 (initial)	0 (initial)
1	0	0	1	1	0	0
2	1	1	0	0	1	0
3	1	0	1	0	0	1
4	1	1	0	0	0	0
5	0	0	1	1	0	0
6	1	1	0	0	1	0
7	–	–	1	–	0	1

(b) (i) The encoding circuit operation is laid out in tabular form in Table 11.7a. Note in particular the reverse order in which the message bits 1100 are copied into column 2. Refer to previous discussions if in doubt. The parity bits are given by the state of the shift register after k (=4) shifts. Thus parity bits = 101, message bits = 1100 (given and shifted into the code output during the first k clock cycles), and so the generated codeword is 1011100.

(ii) The operation of the syndrome calculator circuit (Fig. 11.25b) is tabulated in Table 11.7b. The syndrome $s(X)$ is given by the state of the shift register after n (=7) shifts. Thus $s(X) \equiv 101 = 1 + X^2$, which, from the decoding table, corresponds to the error polynomial $e(X) = X^6$. The transmitted code polynomial is therefore estimated to be $c(X) = r(X) + e(X) = 1 + X^2 + X^3 + X^4 + X^6 + X^6 = 1 + X^2 + X^3 + X^4$, which corresponds to the seven-bit codeword = 1011100, and hence the message 1100.

11.5.2.3 Application and effectiveness

Binary (n, k) cyclic codes are widely used for error detection in ARQ schemes, where they are described as *cyclic redundancy check* (CRC), and the parity bits are called *check bits*. They can detect a wide combination of likely errors, including

- All error bursts of length less than or equal to the number of check bits $N = n - k$, where n is the codeword length, k is the number of message bits, and the number of check bits N is also the highest power of the generator polynomial $g(X)$, which if represented in bit sequence format contains $N + 1$ bits.
- All error patterns with an odd number of bit errors if the generator polynomial $g(X)$ for the code has an even number of non-zero coefficients.
- A large percentage of other error patterns.
- The probability of an error going undetected, which is the chance of a random error pattern adding to the transmitted codeword and changing it to a new value that just by coincidence is divisible by $g(X)$, is roughly $2^{-(N+1)}$.

A number of international standards have been adopted for CRC implementation, specifying the code polynomial $g(X)$ to use. Some of these include

- CRC-12 with $g(X) = 1 + X + X^2 + X^3 + X^{11} + X^{12}$
- CRC-16 with $g(X) = 1 + X^2 + X^{15} + X^{16}$, used in America
- CRC-ITU with $g(X) = 1 + X^5 + X^{12} + X^{16}$
- CRC-32 (with powers of X at 32, 26, 23, 22, 16, 12, 11, 10, 8, 7, 5, 4, 2, 1), used in IEEE 802 LAN standards

Note that in all cases $g(X)$ is an $(N + 1)$-bit prime binary number. For example, the ITU cyclic redundancy check standard has codeword size $n = 3860$ and number of check bits $N = 16$, and its generator polynomial $g(X)$ is a 17-bit prime number.

11.5.3 Reed-Solomon code

The Reed-Solomon (RS) code is a class of non-binary cyclic codes with a wide range of code rates. It makes efficient use of redundancy and operates on m-bit symbols, where $m > 2$, rather than on individual bits as for the codes discussed so far. Parameters of RS code include

Block length $n = 2^m - 1$ symbols (note that this is not bits)
Message size $= k$ symbols
Parity-check size $= n - k = 2t$ symbols
Minimum distance $d_{min} = 2t + 1$, where t is the number of correctable errors.

An RS (n, k) code takes a block of k message symbols at a time and adds $n - k$ parity symbols to produce a codeword containing n symbols. Each symbol is a group of m bits. For example, if $m = 8$ bits, then the coder forms an n-byte codeword from each block of k message bytes. The code has minimum distance d_{min} and is able to correct any pattern of t or fewer symbol errors, or any pattern of ρ or fewer symbol erasures, or to simultaneously correct α symbol errors and γ symbol erasures. These parameters are given by

$$\begin{aligned} d_{min} &= n - k + 1; \quad n = 2^m - 1 \\ t &= \left\lfloor \frac{d_{min} - 1}{2} \right\rfloor = \left\lfloor \frac{n - k}{2} \right\rfloor \\ \rho &= d_{min} - 1 = n - k \\ d_{min} &> 2\alpha + \gamma \end{aligned} \tag{11.31}$$

For a non-binary code such as this one, the distance between a pair of codewords is the number of symbol positions where they differ. A codeword consists of 2^m unique symbols in n positions, giving 2^{mn} possibilities, of which only 2^{mk} are legitimate codewords (assigned to the 2^{mk} unique k-symbol message blocks). Thus the fraction of legitimate codewords is $2^{mk}/2^{mn} = 1/2^{m(n-k)}$. We see that this fraction decreases exponentially with m and $n - k$ (the number of parity symbols). Note that the smaller this fraction, the larger the minimum distance d_{min} that can be achieved.

11.5.3.1 RS code effectiveness

A codeword with $t + 1$ or more symbol errors arriving at the decoder cannot be corrected. Thus the symbol error ratio (i.e. average fraction of errored symbols in n) at decoder output is given in terms of the channel's probability of symbol error, denoted by SER, as

$$\begin{aligned} \text{SER}_{\text{coded}} &= \frac{1}{n} \sum_{j=t+1}^{n} j \binom{n}{j} \text{SER}^j (1 - \text{SER})^{n-j} \\ &= \frac{1}{2^m - 1} \sum_{j=t+1}^{2^m - 1} j \binom{2^m - 1}{j} \text{SER}^j (1 - \text{SER})^{2^m - 1 - j} \end{aligned} \tag{11.32}$$

Since there are m bits per symbol, SER can be expressed in terms of the channel's probability of bit error BER as in Eq.(11.33) below, where the last term is the probability that all m bits in the symbol arrive without error:

$$\text{SER} = 1 - (1 - \text{BER})^m \tag{11.33}$$

To determine the bit error rate $\text{BER}_{\text{coded}}$ at the decoder output we must know the number of bit places at which the errored m-bit symbol differs from the correct symbol. Best case scenario is unit distance where the difference is in one bit position only; whereas worst case is when the correct symbol can be mistaken for any of the other $2^m - 1$ symbols with equal probability, yielding average distance (see also the comments leading to Eq. (10.76)):

$$\bar{d} = \frac{1}{2^m - 1}\sum_{j=1}^{m} j\binom{m}{j} = m\frac{2^{m-1}}{2^m - 1} \tag{11.34}$$

Since

$$\text{BER}_{\text{coded}} = \frac{\bar{d}}{m}\text{SER}_{\text{coded}}$$

it follows that

$$\text{BER}_{\text{coded}_{\max}} = \frac{2^{m-1}}{(2^m - 1)^2}\sum_{j=t+1}^{2^m-1} j\binom{2^m - 1}{j}\text{SER}^j(1 - \text{SER})^{2^m-1-j} \quad \text{where, } \text{SER} = 1 - (1 - \text{BER})^m \tag{11.35}$$

This result is plotted in Fig. 11.27 for various RS codes with codeword size $n = 31$ symbols. By comparing this graph with Fig. 11.16 or Eq. (11.25) for the (7, 4) Hamming code used for error correction, we see how vastly superior the RS code is in its error correction effectiveness. For example, the RS (31, 15) decoder reduces bit error rate from 4×10^{-3} at its input to 10^{-9} at its output, a reduction factor of 4×10^6, whereas the (7, 4) Hamming code (by Eq. (11.25)) would reduce the same BER at its input to 1.369×10^{-4}, a reduction factor of only 30.

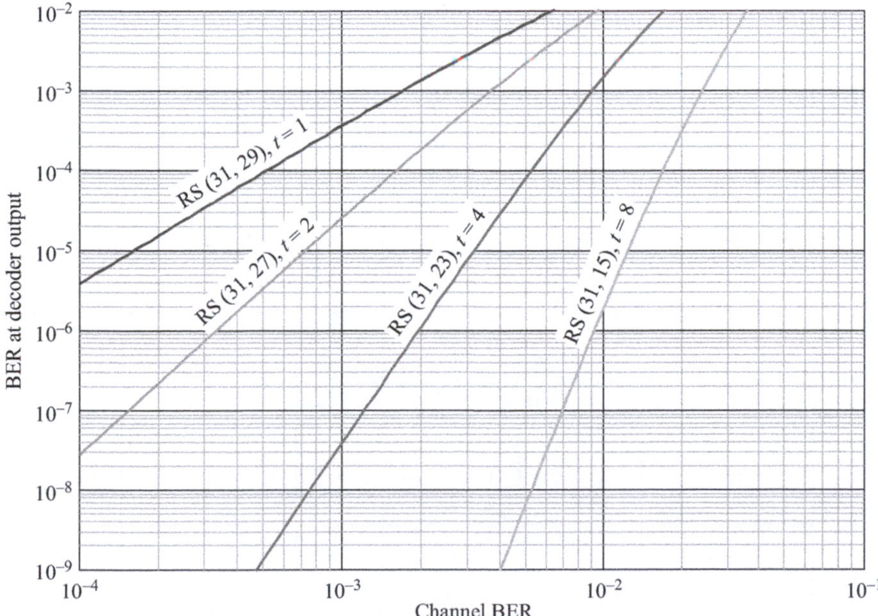

Fig. 11.27 BER at decoder output of various RS codes (with $n = 31$) versus input (channel) BER

Another important indicator of error correction effectiveness is coding gain, which we estimate as follows for RS codes based on Eq. (11.35). We know that for a given modulation scheme, channel BER is exclusively determined by the value of E_b/N_o, so we select the BPSK scheme (it could be any other scheme) and for each value of channel BER along the x-axis of Fig. 11.27 we determine the corresponding E_b/N_o in dB. We then plot the RS-decoded BER against E_b/N_o. Fig. 11.28 shows these results along with the BER of uncoded BPSK for comparison. The horizontal separation between the uncoded BPSK curve and any of the RS-coded curves at a given value of output bit error rate gives the coding gain of that coding scheme at that BER. For example, we see that the RS (255, 239) code has a coding gain of ~4 dB at BER = 10^{-7}, rising to ~5 dB at BER = 10^{-10}.

RS codes are particularly effective in correcting burst errors since it deals with symbol rather than bit errors, and all bits in a symbol being in error impacts on the error correction process in exactly the same way as if a single bit of the symbol were in error.

11.5.3.2 Finite fields

To understand the principle of operation of RS codes, we need a brief introduction to the theory of finite fields. A finite field is a field that has the following properties:

- It is closed under the operations of addition and multiplication. This means that the sum or product of any two elements of the field is another element.
- It contains the elements 0 (which is the additive identity element) and 1 (which is the multiplicative identity element).
- Every element has an additive inverse (i.e. another element which when the pair are added together yields zero); and every element (except 0) has a multiplicative inverse. An element multiplied by its multiplicative inverse yields 1.

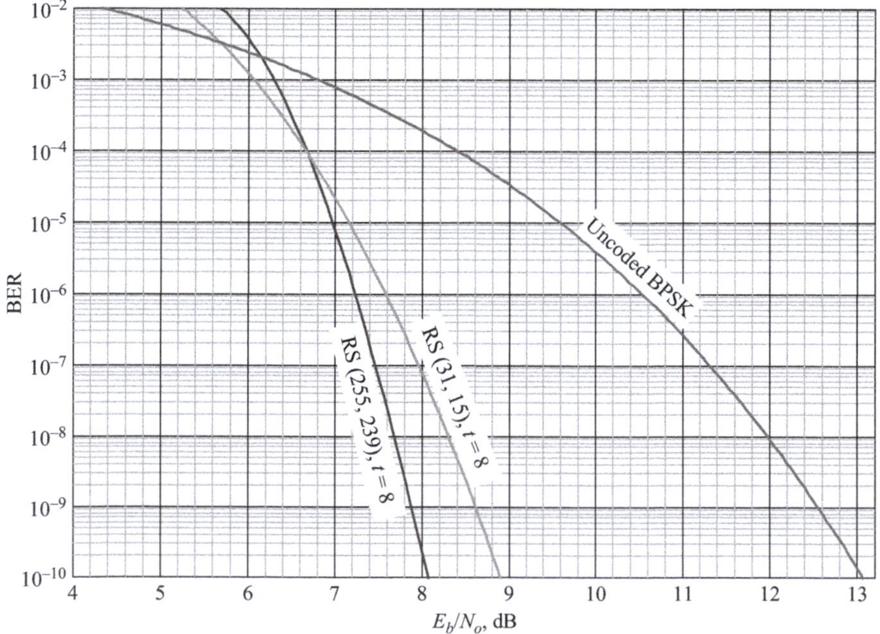

Fig. 11.28 BER of RS codes versus E_b/N_o in BPSK modulation

- The two operations of addition (+) and multiplication (×) that are defined on the field both obey the commutative, associative and distributive laws. That is, given that a, b and c are elements of the field, then

$$\begin{array}{lll} a+b=b+a & a+(b+c)=(a+b)+c & a\times(b+c)=a\times b+a\times c \\ a\times b=b\times a & a\times(b\times c)=(a\times b)\times c & \text{Distributive law} \\ \text{Commutative law} & \text{Associative law} & \end{array} \tag{11.36}$$

An example of such a field is one containing a prime number p of integer elements from 0 to $p-1$ in which addition and multiplication operations are defined modulo-p. This field is denoted by GF(p), which stands for *Galois field* of order p. Thus, 2 and 7 being prime numbers, GF(2) = {0, 1}, and GF(7) = {0, 1, 2, 3, 4, 5, 6}. You may verify that these fields obey all the properties stated above. For example, in GF(7) the additive inverse of 6 is 1 since $6+1=0$ (modulo-7); the multiplicative inverse of 6 is 6 (i.e. $6\times 6=1$, again modulo-7); the multiplicative inverse of 5 is 3; and so on. Each non-zero element of GF(p) can be expressed as an integer power of one of the elements, which is called a *primitive*. There is always at least one primitive in the field. For example, 3 and 5 are primitives of GF(7) since GF(7) = $\{0, 3^0, 3^2, 3^1, 3^4, 3^5, 3^3\} = \{0, 5^0, 5^4, 5^5, 5^2, 5^1, 5^3\}$. You should be able to confirm these results by evaluating each power modulo-7, e.g. $3^2=9\equiv 2$ (modulo-7), etc.

The Galois field GF(p^m) is an extension field of GF(p). In the special case where the prime number $p=2$, we have GF(2^m), which is an *extension field* of the binary field GF(2). Conversely, we say that GF(2) is a *subfield* of GF(2^m). To obtain the finite field

$$\text{GF}(2^m) = \{0,\ \alpha^0,\ \alpha^1,\ \alpha^2, \ldots,\ \alpha^{q-2}\}$$

comprising $q=2^m$ elements from the infinite extension field $\{0, \alpha^0, \alpha^1, \alpha^2, \ldots, \alpha^{q-2}, \alpha^{q-1}, \alpha^q, \ldots\}$, where α is a primitive, we must impose the condition, $\alpha^{q-1}=\alpha^0\ (=1)$ in order to force the powers of α to repeat after q elements. This condition is equivalent to $\alpha^{q-1}+1=0$, since $-1=+1$ in modulo-2 arithmetic. If we express $\alpha^{q-1}+1$ as a product of lower-order polynomials, then one or more of them must be zero (since $\alpha^{q-1}+1=0$). We choose as zero the one that is a *primitive polynomial* of degree m. An m-degree polynomial $f(\alpha)$ is said to be primitive if and only if

1. It is irreducible (i.e. it has no factors), and
2. It divides (without remainder) α^M+1 for $M=2^m-1$, but does not divide α^M+1 for $M<2^m-1$

The exclusion clause in the second condition is necessary otherwise the powers of α would repeat before all $q-1$ non-zero field elements have been generated. A polynomial $f(\alpha)$ that satisfies the above two conditions is then a field generator polynomial of GF(2^m). A total of m bits are needed to uniquely identify each of the elements of GF(2^m), which may then be expressed as a polynomial of degree $m-1$, whose m coefficients correspond to the values of the m bits for that element.

To clarify further, consider the Galois field GF(8), which is GF(2^m) with $m=3$ and $M=2^m-1=7$, so that

$$\alpha^M+1\equiv\alpha^7+1=(1+\alpha)(1+\alpha+\alpha^3)(1+\alpha^2+\alpha^3)$$

Table 11.8 Field elements for GF(8) with field generator $f(X) = 1 + X + X^3$

Element (exponential form)	Reduction modulo $f(x)$	Polynomial Representation	m-bit Binary ID $X^0\ X^1\ X^2$	Decimal Representation
0	–	0	0 0 0	0
α^0	–	1	1 0 0	1
α^1	–	X	0 1 0	2
α^2	–	X^2	0 0 1	4
α^3	$1+\alpha$	$1+X$	1 1 0	3
α^4	$\alpha+\alpha^2$	$X+X^2$	0 1 1	6
α^5	$1+\alpha+\alpha^2$	$1+X+X^2$	1 1 1	7
α^6	$1+\alpha^2$	$1+X^2$	1 0 1	5

Note that beyond α^6 the cycle of elements repeats, since $\alpha^7 = \alpha \times \alpha^6 = \alpha + \alpha^3 = 1 \equiv \alpha^0$; $\alpha^8 = \alpha^1$; etc.

The irreducible polynomial $f(\alpha) = 1 + \alpha + \alpha^3$, of degree $m = 3$, is primitive since it divides $\alpha^7 + 1$ (yielding $\alpha^4 + \alpha^2 + \alpha + 1$ without a remainder), but does not divide any of $\alpha^6 + 1$, $\alpha^5 + 1$, $\alpha^4 + 1$ or $\alpha^3 + 1$ without a remainder. The root α satisfies $f(\alpha) = 0$, which yields (modulo-2), $\alpha^3 = 1 + \alpha$, and hence (by repeated iteration) $\alpha^4 = \alpha + \alpha^2$; $\alpha^5 = 1 + \alpha + \alpha^2$; $\alpha^6 = 1 + \alpha^2$; $\alpha^7 = \alpha^0$. The elements of GF(8) and their representations are shown in Table 11.8. Note that since $f(X)$ divides $X^M + 1$, for $M = 2^m - 1$, it follows that α, being a root of $f(X)$, is also a root of $X^M + 1$, so that $\alpha^M + 1 = 0$, and hence $\alpha^M = 1 = \alpha^0$. It is essential then that $f(X)$ is not a factor of $X^M + 1$ when $M < 2^m - 1$. If it was, then α^K would equal α^0 for some $K < 2^m - 1$, and thus GF(2^m) would have less than 2^m unique elements, contrary to our starting premise. Since the non-zero elements of GF(8) are generated by repeated multiplication by α, they can be obtained using a linear feedback shift register (LFSR) connected as dictated by $f(X)$. An example for $f(X) = 1 + X + X^3$ is shown in Fig. 11.29. Starting at any non-zero initial state, all the elements of GF(8) will cyclically appear in the LFSR. We are now ready to turn our attention to the application of this theory to RS encoding and decoding.

11.5.3.3 RS encoding process

The generator polynomial $g(X) = g_0 + g_1X + g_2X^2 + \cdots + g_{2t-1}X^{2t-1} + X^{2t}$ is of degree $2t = n - k$, which is the number of parity bits, and may be expressed in terms of its $2t$ roots $\alpha, \alpha^2, \ldots, \alpha^{2t}$ as

$$g(X) = (X - \alpha)(X - \alpha^2)\cdots(X - \alpha^{2t}) = \prod_{j=1}^{2t}(X - \alpha^j)$$

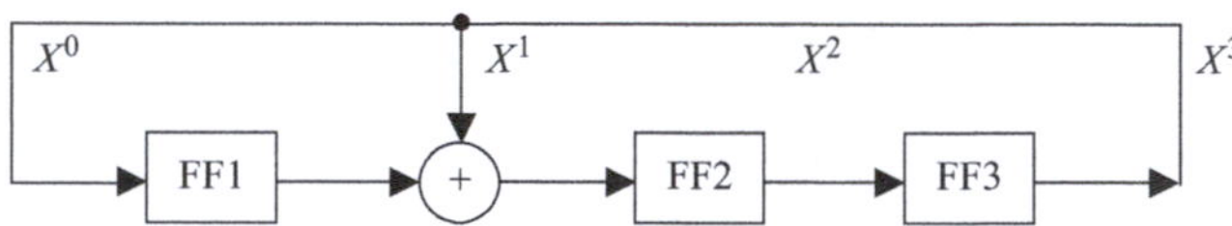

Fig. 11.29 LFSR for generating the elements of GF(8) having field generator polynomial $f(X) = 1 + X + X^3$

where α is a primitive in GF($n+1$) and $n = 2^m - 1$. For example, the RS (7, 3) code is defined over GF(2^3) and has $t = (n-k)/2 = 2$. Substituting into the above equation yields,

$$\begin{aligned} g(X) &= (X-\alpha)(X-\alpha^2)(X-\alpha^3)(X-\alpha^4) \\ &= X^4 \\ &\quad -X^3[\alpha+\alpha^2+\alpha^3+\alpha^4] \\ &\quad +X^2[\alpha^3+\alpha^4+\alpha^5+\alpha^5+\alpha^6+\alpha^7] \\ &\quad -X[\alpha^6+\alpha^7+\alpha^8+\alpha^9]+\alpha^{10} \\ &= \alpha^3+\alpha X+\alpha^0X^2+\alpha^3X^3+X^4 \end{aligned} \tag{11.37}$$

where we have used the relations in the GF(2^3) Table 11.8 as well as $\alpha^7 = \alpha^0$, and ignored all negative signs, since $-1 = +1$ in modulo-2 arithmetic. The RS codeword, like other cyclic codes, is obtained as

$$\mathbf{c}(X) = \mathbf{p}(X) + X^{n-k}\mathbf{m}(X)$$

where $\mathbf{m}(X)$ is the message polynomial, of degree $k-1$, with coefficients drawn from elements of the GF(2^m) field, $\boldsymbol{p}(X)$ is the remainder when $X^{n-k}\mathbf{m}(X)$ is divided by $g(X)$, and the code polynomial $\mathbf{c}(X)$ is of degree $n-1$ with coefficients drawn from elements of the GF(2^m) field. We follow the convention that bits are arranged in low to high polynomial order. For example, in the RS(7, 3) encoder the message bits $m = 001{,}110{,}100$ correspond to decimals 4, 3, 1 and GF(8) symbols α^2, α^3, α^0. Therefore, $m(X) = \alpha^2 + \alpha^3X + \alpha^0X^2$, where the first bit group (001) expressed using GF(8) symbol α^2 is the coefficient of X^0, the second bit group (110) is the coefficient of X^1 and the third bit group (100) is the coefficient of X^2. Note that the coding and decoding process will also work well with a reverse ordering of bits provided there is consistency of interpretation at both transmitter and receiver.

Worked Example 11.5: RS Encoding

We wish to determine the RS(7, 3) codeword for the message 100 000 101.

Given RS(7, 3) ≡ RS(n, k) ≡ RS($2^3 - 1$, 3). Thus $n = 7$, $m = 3$, $k = 3$, $t = 2$. So the coder uses 3 three-bit symbols as input from the GF(2^3) field to generate 7 three-bit symbols as the codeword output. The solution is based on the GF(2^3) Table 11.8 and the generator polynomial determined earlier. The input message $m = 100{,}000{,}101$ corresponds to 3 symbols α^0, 0, α^6 in GF(8). Thus,

$$\begin{aligned} \mathbf{m}(X) &= \alpha^0 + \alpha^6X^2 \\ X^{n-k}\mathbf{m}(X) &= X^4\mathbf{m}(X) \\ &= \alpha^0X^4 + \alpha^6X^6 \end{aligned}$$

Dividing $X^{n-k}\mathbf{m}(X)$ by $g(X)$ as shown in Fig. 11.30 yields the remainder

$$\mathbf{p}(X) = \alpha^2X^2 + \alpha^0X + \alpha^6$$

$$
\begin{array}{r|l}
 & \alpha^6X^2+\alpha^2X+\alpha^3 \\ \hline
X^4+\alpha^3X^3+\alpha^0X^2+\alpha X+\alpha^3 & \alpha^6X^6 \qquad\qquad +\alpha^0X^4 \\
 & \alpha^6X^6+\alpha^9X^5+\alpha^6X^4+\alpha^7X^3+\alpha^9X^2 \\ \cline{2-2}
 & \quad \alpha^2X^5+\alpha^2X^4+\alpha^0X^3+\alpha^2X^2 \\
 & \quad \alpha^2X^5+\alpha^5X^4+\alpha^2X^3+\alpha^3X^2+\alpha^5X \\ \cline{2-2}
 & \qquad \alpha^3X^4+\alpha^6X^3+\alpha^5X^2+\alpha^5X \\
 & \qquad \alpha^3X^4+\alpha^6X^3+\alpha^3X^2+\alpha^4X+\alpha^6 \\ \cline{2-2}
 & \qquad\qquad \alpha^2X^2+\alpha^0X+\alpha^6
\end{array}
$$

Fig. 11.30 Long division in GF(8) field. NOTE: In doing this long division (i) addition replaces subtraction; (ii) any power of $\alpha \geq 7$ is reduced by 7 since $\alpha^7 = \alpha^0$; (iii) To add coefficients of the polynomial, express them in binary using the GF(8) Table, XOR the binary representation, and convert the result back to exponential form. For example, $\alpha^2 + \alpha^5 = 001 + 111 = 110 = \alpha^3$

Hence

$$
\begin{aligned}
\mathbf{c}(X) &= \mathbf{p}(X) + X^{n-k}\mathbf{m}(X) \\
&= \alpha^6 + \quad \alpha^0X + \quad \alpha^2X^2 + \qquad\quad \alpha^0X^4 + \qquad \alpha^6X^6 \\
&\equiv 101 \quad 100 \quad 001 \quad 000 \quad 100 \quad 000 \quad 101
\end{aligned}
$$

where the last line is the encoder output bit stream obtained by translating the coefficients from exponential forms (column 1 of Table 11.8) to binary format (column 4 of the table). This completes the solution. Note that since $g(X)$ is a factor of $c(X)$ it follows that the roots of $g(X)$, namely α, α^2, α^3 and α^4 are also roots of $c(X)$. That is,

$$
\mathbf{c}(\alpha) = \mathbf{c}(\alpha^2) = \mathbf{c}(\alpha^3) = \mathbf{c}(\alpha^4) = 0 \tag{11.38}
$$

An LFSR circuit, similar to that of the (7, 4) cyclic code generator in Fig. 11.25a, except for non-binary weights will generate the RS(7,3) code discussed above. This circuit is shown in Fig. 11.31a, where the indicated multiplier weights are the coefficients of the encoder generator polynomial derived in Eq. (11.37). The operation of the circuit is straightforward as earlier discussed, but a tabulation of this operation leading to the output codeword is provided in part (b) of the figure for information. All data paths in the circuit are m bits wide. Furthermore, due to our adopted low to high order convention, the message polynomial is clocked into the circuit starting from the highest order coefficient.

11.5.3.4 RS decoding

The task of an RS decoder is first to compute the syndrome vector and then to determine an error pattern corresponding to a non-zero syndrome vector. We outline the essential elements of this task below, following after Sklar [1], and demonstrate the application of the results through a worked example. It would be okay to skip to the worked example at this point to first learn the steps involved in the decoding of an RS-encoded sequence.

(a)

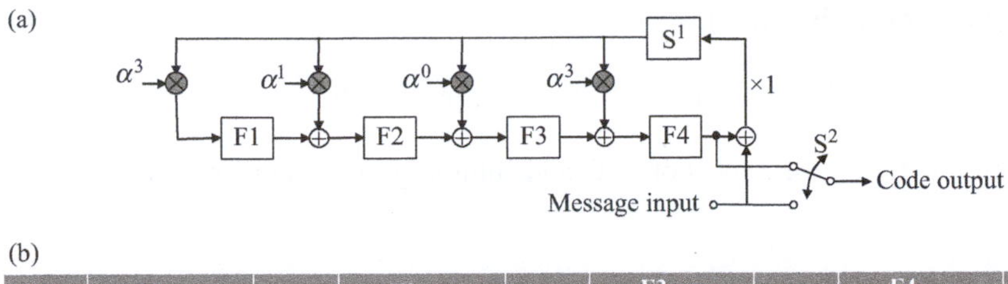

(b)

Msg (m)	$F1_{in} = \alpha^3(y+m)$	$F1_{out}$	$F2_{in} = F1_{out}+\alpha^1(y+m)$	$F2_{out}$	$F3_{in} = F2_{out}+\alpha^0(y+m)$	$F3_{out}$	$F4_{in} = F3_{out}+\alpha^3(y+m)$	$F4_{out}$ (y)
α^6	α^2	0	α^0	0	α^6	0	α^2	0
0	α^5	α^2	α^5	α^0	α^6	α^6	α^1	α^2
α^0	α^6	α^5	α^0	α^5	α^2	α^6	0	α^1
	p(X) →→	$\boldsymbol{\alpha^6}$		$\boldsymbol{\alpha^0}$		$\boldsymbol{\alpha^2}$		**0**

Code output:

	X^0	X^1	X^2	X^3	X^4	X^5	X^6
GF(8) symbolism	α^6	α^0	α^2	0	α^0	0	α^6
Binary	101	100	001	000	100	000	101

Fig. 11.31 (a) RS encoder circuit; (b) circuit operation

Since

$$c(X) = p(X) + X^{n-k}m(X) = p(X) + q(X)g(X) + p(X)$$
$$= q(X)g(X)$$

it means that each transmitted codeword is divisible by $g(X)$ and therefore all the $2t$ roots of $g(X)$, namely $\alpha, \alpha^2, \ldots, \alpha^{2t}$ are also roots of $c(X)$. The received vector is

$$r(X) = c(X) + e(X)$$

where $e(X)$ is an error polynomial of degree $n - 1$ with non-zero coefficients at no more than t positions. Evaluating $r(X)$ at each root yields the syndrome S, which if non-zero indicates error in $r(X)$ of value and location to be determined. Let there be errors $\varepsilon_1, \varepsilon_2, \varepsilon_3, \ldots, \varepsilon_t$ at respective positions $z_1, z_2, z_3, \ldots, z_t$; and let S_i denote the syndrome obtained at root α^i. We obtain the following $2t$ syndrome equations which may be solved for up to t errors and their corresponding t locations.

$$S_i = \sum_{j=1}^{t} \varepsilon_j (\alpha^{z_j})^i = \sum_{j=1}^{t} \varepsilon_j \left(\beta_j\right)^i \qquad (11.39)$$

where $i = 1, 2, 3, \ldots, 2t$; and $\beta_j = \alpha^{z_j} \equiv$ error locator

Let us define the error-locator polynomial $\sigma(X)$ with roots at $1/\beta_1, 1/\beta_2, \ldots, 1/\beta_t$

$$\sigma(X) = (1 + \beta_1 X)(1 + \beta_2 X) \cdots (1 + \beta_t X)$$
$$= 1 + \sigma_1 X + \sigma_2 X^2 + \cdots + \sigma_t X^t$$

Since $1/\beta_1$ is a root, it means that

$$1 + \sigma_1(1/\beta_1) + \sigma_2(1/\beta_1)^2 + \cdots + \sigma_t(1/\beta_1)^t = 0$$

Multiplying through by $\varepsilon_1(\beta_1)^{i+t}$:

$$\varepsilon_1\beta_1^{i+t} + \sigma_1\varepsilon_1\beta_1^{i+t-1} + \sigma_2\varepsilon_1\beta_1^{i+t-2} + \cdots + \sigma_t\varepsilon_1\beta_1^{i} = 0$$

Repeating this for all the t roots of $\sigma(X)$ and summing the equations (which equate to zero):

$$\sum_{j=1}^{t}\varepsilon_j\left(\beta_j\right)^{i+t} + \sigma_1\sum_{j=1}^{t}\varepsilon_j\left(\beta_j\right)^{i+t-1} + \sigma_2\sum_{j=1}^{t}\varepsilon_j\left(\beta_j\right)^{i+t-2} + \cdots + \sigma_t\sum_{j=1}^{t}\varepsilon_j\left(\beta_j\right)^{i+t} = 0$$

$$\Rightarrow S_{i+t} \quad + \quad \sigma_1 S_{i+t-1} \quad + \sigma_2 S_{i+t-2} \quad + \cdots + \sigma_t S_i \quad = 0$$

Letting the index i in the equation on the last line above take on values from 1 to t, we see that we have t linear simultaneous equations in t unknowns $\sigma_1, \sigma_2, \ldots, \sigma_t$, which is written in matrix form as

$$\begin{bmatrix} S_t & S_{t-1} & \cdots & S_2 & S_1 \\ S_{t+1} & S_t & \cdots & S_3 & S_2 \\ & & \vdots & & \\ S_{2t-2} & S_{2t-3} & \cdots & S_t & S_{t-1} \\ S_{2t-1} & S_{2t-2} & \cdots & S_{t+1} & S_t \end{bmatrix} \begin{bmatrix} \sigma_1 \\ \sigma_2 \\ \vdots \\ \sigma_{t-1} \\ \sigma_t \end{bmatrix} = \begin{bmatrix} S_{t+1} \\ S_{t+2} \\ \vdots \\ S_{2t-1} \\ S_{2t} \end{bmatrix} \tag{11.40}$$

These equations involve the $2t$ syndromes $S_1, S_2, \ldots, S_{2t}$ whose values are known from the received vector, since

$$S_i = r(\alpha^i), \quad i = 1, 2, 3, \ldots, 2t$$

Therefore we can solve for the coefficients $\sigma_1, \sigma_2, \ldots, \sigma_t$ using the largest dimensioned matrix that has a non-zero determinant. Doing this gives us the error-locator polynomial

$$\sigma(X) = 1 + \sigma_1 X + \sigma_2 X^2 + \sigma_3 X^3 + \cdots + \sigma_t X^t \tag{11.41}$$

By trial and error (i.e. by substituting each element of GF$(n+1)$ into $\sigma(X)$ to see which one yields zero) we obtain the roots $1/\beta_1, 1/\beta_2, \ldots,$ of this polynomial and hence the error locators $\beta_1, \beta_2, \ldots,$ and positions $z_1, z_2, \ldots$ Assuming that we have determined v error locators $\beta_1, \beta_2, \ldots, \beta_t$, where $v \le t$, then we return to the original $2t$ syndrome equations and use any v of these as simultaneous equations from which we obtain the error values $\varepsilon_1, \varepsilon_2, \ldots, \varepsilon_v$. Using the first v equations yields

$$\begin{bmatrix} \beta_1 & \beta_2 & \beta_3 & \cdots & \beta_v \\ \beta_1^2 & \beta_2^2 & \beta_3^2 & \cdots & \beta_v^2 \\ & & \vdots & & \\ \beta_1^v & \beta_2^v & \beta_3^v & \cdots & \beta_v^v \end{bmatrix} \begin{bmatrix} \varepsilon_1 \\ \varepsilon_2 \\ \vdots \\ \varepsilon_v \end{bmatrix} = \begin{bmatrix} S_1 \\ S_2 \\ \vdots \\ S_v \end{bmatrix} \tag{11.42}$$

So we now have all the error value and error locator pairs $(\varepsilon_1, \beta_1), (\varepsilon_2, \beta_2), \ldots,$ (ε_v, β_v) and hence the error value and error position pairs $(\varepsilon_1, z_1), (\varepsilon_2, z_2), \ldots,$

(ε_v, z_v), since $\beta = \alpha^z$ (as earlier stated). The decoder's estimate of the error polynomial is therefore given by

$$\hat{\mathbf{e}}(X) = \varepsilon_1 X^{z_1} + \varepsilon_2 X^{z_2} + \varepsilon_3 X^{z_3} + \cdots + \varepsilon_v X^{z_v} \tag{11.43}$$

Finally, the decoder delivers its estimate of the transmitted codeword as

$$\begin{aligned}\hat{\mathbf{c}}(X) &= \mathbf{c}(X) + \mathbf{e}(X) + \hat{\mathbf{e}}(X) \\ &= \mathbf{r}(X) + \hat{\mathbf{e}}(X) \\ &= r_0 + r_1 X + r_2 X^2 + r_3 X^3 + \cdots + r_{n-1} X^{n-1} + \varepsilon_1 X^{z_1} + \varepsilon_2 X^{z_2} + \cdots + \varepsilon_v X^{z_v}\end{aligned} \tag{11.44}$$

This estimate will be identical to the transmitted codeword if the error polynomial computed by the decoder correctly represents the error $\mathbf{e}(X)$ that occurred in transit, which will always be the case if there were t or fewer errored symbols in the received vector $\boldsymbol{r}(X)$. The next worked example makes use of the following matrix inversion relation which is useful when solving RS decoding problems involving 2-symbol error correcting codes such as RS(7, 3), RS(15, 11), etc.

$$\begin{bmatrix} a & b \\ c & d \end{bmatrix}^{-1} = \frac{1}{(ad - bc)} \begin{bmatrix} d & -b \\ -c & a \end{bmatrix} \tag{11.45}$$

Worked Example 11.6: RS(7, 3) Decoding

Imagine that the RS(7, 3) codeword of the previous worked example is corrupted by error in the 4th and 6th symbols so that the decoder receives the bit stream

101 100 001 111 100 111 101

Determine the codeword estimated by the decoder and hence its output message bits.

For this code, $n = 7$ and $k = 3$ so that $2t = n - k = 4$. Thus the decoder can correct up to $t = 2$ symbol errors. The generator polynomial $g(X)$ has $2t$ roots α, α^2, α^3 and α^4 which are also the roots of each transmitted codeword. If the received vector $r(X)$ arrived without an error, it would evaluate to zero at each of the four roots. More generally evaluating $r(X)$ at each root yields four syndromes. We follow the following steps to find a solution.

1. Use the GF(8) Table 11.8 to read the exponential form of each three-bit element of the received vector, which is then written in polynomial form as follows

$$r(X) = \alpha^6 + \alpha^0 X + \alpha^2 X^2 + \alpha^5 X^3 + \alpha^0 X^4 + \alpha^5 X^5 + \alpha^6 X^6$$

2. Compute the 4 syndromes, using the GF(8) table to guide addition of elements:

$$S_1 = r(\alpha^1) = \alpha^6 + \alpha^0\alpha^1 + \alpha^2\alpha^2 + \alpha^5\alpha^3 + \alpha^0\alpha^4 + \alpha^5\alpha^5 + \alpha^6\alpha^6$$
$$= \alpha^6 + \alpha^3 + \alpha^5 = \alpha^0$$

$$S_2 = r(\alpha^2)$$
$$= \alpha^6 + \alpha^0\alpha^2 + \alpha^2\alpha^4 + \alpha^5\alpha^6 + \alpha^0\alpha^8 + \alpha^5\alpha^{10} + \alpha^6\alpha^{12} = \alpha^2$$

$$S_3 = r(\alpha^3) = \alpha^6 + \alpha^0\alpha^3 + \alpha^2\alpha^6 + \alpha^5\alpha^9 + \alpha^0\alpha^{12} + \alpha^5\alpha^{15} + \alpha^6\alpha^{18}$$
$$= \alpha^1 + \alpha^0 + \alpha^5 = \alpha^2$$

$$S_4 = r(\alpha^4)$$
$$= \alpha^6 + \alpha^0\alpha^4 + \alpha^2\alpha^8 + \alpha^5\alpha^{12} + \alpha^0\alpha^{16} + \alpha^5\alpha^{20} + \alpha^6\alpha^{24} = \alpha^6$$

3. Armed with the syndrome values, we may now solve Eq. (11.40) for the coefficients σ_1 and σ_2 of the error locator polynomial $\sigma(X) = 1 + \sigma_1 X + \sigma_2 X^2$, employing Eq. (11.45) to invert the 2-by-2 matrix involved. Thus,

$$\begin{bmatrix} \sigma_1 \\ \sigma_2 \end{bmatrix} = \begin{bmatrix} S_2 & S_1 \\ S_3 & S_2 \end{bmatrix}^{-1} \begin{bmatrix} S_3 \\ S_4 \end{bmatrix}$$
$$= \frac{1}{S_2^2 + S_3 S_1} \begin{bmatrix} S_2 & S_1 \\ S_3 & S_2 \end{bmatrix} \begin{bmatrix} S_3 \\ S_4 \end{bmatrix}$$
$$= \frac{1}{\alpha} \begin{bmatrix} S_2 S_3 + S_1 S_4 \\ S_3^2 + S_2 S_4 \end{bmatrix} = \frac{1}{\alpha} \begin{bmatrix} \alpha^3 \\ \alpha^2 \end{bmatrix} = \begin{bmatrix} \alpha^2 \\ \alpha^1 \end{bmatrix}$$

4. Now obtain the roots of the error locator polynomial

$$\sigma(X) = 1 + \sigma_1 X + \sigma_2 X^2$$

by trying all the elements of GF(8). We find that $\sigma(\alpha^2) = 0$ and $\sigma(\alpha^4) = 0$. Thus $1/\beta_1 = \alpha^2$, $1/\beta_2 = \alpha^4$, which means $\beta_1 = \alpha^{-2} = \alpha^5$ and $\beta_2 = \alpha^{-4} = \alpha^3$. Since, from Eq. (11.39), $\beta = \alpha^z$, where z is the error location, it follows that there is error at symbol numbers $z_1 = 5$ and $z_2 = 3$
5. Next, substitute the values of β_1 and β_2 obtained above into Eq. (11.42) and solve for the corresponding values ε_1 and ε_2 of the above errors:

$$\begin{bmatrix} \varepsilon_1 \\ \varepsilon_2 \end{bmatrix} = \begin{bmatrix} \beta_1 & \beta_2 \\ \beta_1^2 & \beta_2^2 \end{bmatrix}^{-1} \begin{bmatrix} S_1 \\ S_2 \end{bmatrix} = \frac{1}{\beta_1 \beta_2^2 + \beta_1^2 \beta_2} \begin{bmatrix} \beta_2^2 & \beta_2 \\ \beta_1^2 & \beta_1 \end{bmatrix} \begin{bmatrix} S_1 \\ S_2 \end{bmatrix}$$
$$= \frac{1}{\alpha^{11} + \alpha^{13}} \begin{bmatrix} \alpha^6 & \alpha^3 \\ \alpha^{10} & \alpha^5 \end{bmatrix} \begin{bmatrix} \alpha^0 \\ \alpha^2 \end{bmatrix}$$
$$= \frac{1}{\alpha^3} \begin{bmatrix} \alpha^6 & \alpha^3 \\ \alpha^{10} & \alpha^5 \end{bmatrix} \begin{bmatrix} \alpha^0 \\ \alpha^2 \end{bmatrix} = \begin{bmatrix} \alpha^3 & \alpha^0 \\ \alpha^0 & \alpha^2 \end{bmatrix} \begin{bmatrix} \alpha^0 \\ \alpha^2 \end{bmatrix}$$
$$= \begin{bmatrix} \alpha^3 + \alpha^2 \\ \alpha^0 + \alpha^4 \end{bmatrix} = \begin{bmatrix} \alpha^5 \\ \alpha^5 \end{bmatrix}$$

What this says is that there is error $\varepsilon_1 = \alpha^5$ at location $z_1 = 5$; and there is error $\varepsilon_2 = \alpha^5$ at location $z_2 = 3$.

6. Finally, the estimates of error polynomial and hence of transmitted codeword are

$$\hat{\mathbf{e}}(X) = \varepsilon_1 X^{z_1} + \varepsilon_2 X^{z_2} = \alpha^5 X^5 + \alpha^5 X^3$$

$$\begin{aligned}\hat{\mathbf{c}}(X) &= r(X) + \hat{\mathbf{e}}(X) = \alpha^6 + \alpha^0 X + \alpha^2 X^2 + \alpha^5 X^3 + \alpha^0 X^4 \\ &\quad + \alpha^5 X^5 + \alpha^6 X^6 + \alpha^5 X^5 + \alpha^5 X^3 \\ &= \alpha^6 + \alpha^0 X + \alpha^2 X^2 + (\alpha^5 + \alpha^5) X^3 + \alpha^0 X^4 + (\alpha^5 + \alpha^5) X^5 + \alpha^6 X^6 \\ &= \alpha^6 + \alpha^0 X + \alpha^2 X^2 + 0 \cdot X^3 + \alpha^0 X^4 + 0 \cdot X^5 + \alpha^6 X^6 \\ &= 101\ \ 100\ \ 001\ \ \ \ 000\ \ \ \ 100\ \ \ \ 000\ \ 101\end{aligned}$$

We see that this codeword is indeed what was sent by the transmitter. So the decoder has correctly determined the error locations as well as the value of each error. The output message bit stream of this RS(7, 3) decoder is provided by the last 3 symbols of the above estimated codeword as $m = 100\ 000\ 101$.

For brevity, we based most of our discussion on the GF(8) field that supports RS codes with symbol size $m = 3$. RS codes exist with symbol size of one byte, supported by GF(256), and larger. Field elements for GF(16) with field generator $f(X) = 1 + X + X^4$ are presented in Table 11.9.

Table 11.9 Field elements for GF(16) with field generator $f(X) = 1 + X + X^4$

Element (exponential form)	**Reduction modulo $f(X)$**	**Polynomial Representation**	***m*-bit Binary ID X^0 X^1 X^2 X^3**	**Decimal representation**
0	–	0	0 0 0 0	0
α^0	–	1	1 0 0 0	1
α^1	–	X	0 1 0 0	2
α^2	–	X^2	0 0 1 0	4
α^3	–	X^3	0 0 0 1	8
α^4	$1+\alpha$	$1+X$	1 1 0 0	3
α^5	$\alpha+\alpha^2$	$X+X^2$	0 1 1 0	6
α^6	$\alpha^2+\alpha^3$	X^2+X^3	0 0 1 1	12
α^7	$1+\alpha+\alpha^3$	$1+X+X^3$	1 1 0 1	11
α^8	$1+\alpha^2$	$1+X^2$	1 0 1 0	5
α^9	$\alpha+\alpha^3$	$X+X^3$	0 1 0 1	10
α^{10}	$1+\alpha+\alpha^2$	$1+X+X^2$	1 1 1 0	7
α^{11}	$\alpha+\alpha^2+\alpha^3$	$X+X^2+X^3$	0 1 1 1	14
α^{12}	$1+\alpha+\alpha^2+\alpha^3$	$1+X+X^2+X^3$	1 1 1 1	15
α^{13}	$1+\alpha^2+\alpha^3$	$1+X^2+X^3$	1 0 1 1	13
α^{14}	$1+\alpha^3$	$1+X^3$	1 0 0 1	9

Beyond α^{14} the cycle of elements repeats, since $\alpha^{15} = \alpha(\alpha^{14}) = \alpha + \alpha^4 = 1 \equiv \alpha^0$; $\alpha^{16} = \alpha^1$; etc.

11.6 Summary

Error control coding is an indispensable tool in the design of digital communication systems that operate close to the Shannon-Hartley benchmark for signal power and bandwidth utilisation. Rather than attempt to study the huge variety of error control methods in any detail, an endeavour that would require several volumes of an advanced textbook, our emphasis in this chapter was on gaining a complete understanding of the general principles, applications and limitations of error detection and correction, including the manner in which they facilitate a more efficient trade-off between signal power and bandwidth in the design of reliable communication systems. We did, however, learn the technique of linear block codes in detail, using several different approaches for interpretation and analysis, including graphical, matrix, polynomial, and shift register circuits. We also examined the use of interleaving to deal with bursty error channels and the serial concatenation of two codes to improve overall coding effectiveness and increase coding gain.

The use of automatic repeat request (ARQ) techniques is widespread. We discussed four ARQ protocols in detail and derived expressions for link utilisation in terms of various system, link and protocol parameters. Some of the graphical results of that analysis have been presented in this chapter and may be consulted to help with the selection of the correct protocol as well as optimum combination of parameter values for an intended application.

One of the most useful quantities in error control coding for a system designer is the coding gain of a given forward error correction method. How coding gain is used in link design is covered in the concluding chapter.

11.7 Reference

[1] Sklar, B. (2001). *Digital communications fundamentals and applications*. Upper Saddle River, NJ: Prentice Hall PTR

Chapter 12

Digital transmission link analysis and design

In Chapter Twelve

✓ Bringing it all together for link analysis and design.

12.1 Introduction

The purpose of this brief concluding chapter is to introduce a useful tool called the link power budget, and to summarise the steps and considerations involved in communications link design and analysis.

12.2 Link analysis

Figs. 12.1 and 12.2 show two types of communication links, one an optical communication link and the other a satellite communication link. Analysis of such communication links typically involves the following computations and steps:

- Calculate the signal power at a reference point in the receiver (usually low noise amplifier (LNA) input). This was covered in Section 10.3.
- Calculate the noise power at the same reference point of the receiver. Again this was covered in Section 10.3.
- Obtain signal to noise ratio C/N and hence power efficiency E_b/N_o. The relationship between C/N and E_b/N_o depends on modulation scheme and is given by Eq. (10.29).
- Using the value of E_b/N_o obtained above, determine the bit error rate (BER) of the transmission link and assess whether this is acceptable for the communication service. BER versus E_b/N_o curves were provided in Chapter 10 from which readings may be made, or the BER may be calculated using the equations derived for the different M-ary transmission systems in Chapter 10 and the table of complementary error function in Appendix D.
- Alternatively, compare the value of C/N calculated above to the threshold requirement $(C/N)_{th}$ of the transmission scheme and determine the link margin as $C/N-(C/N)_{th}$. If this is positive it means that the link can support good quality communication, but if it is negative it means that the link as it stands cannot support communication at a quality demanded by the service.

The satellite link shown in Fig. 12.2 comprises two links, an uplink carrying transmission from the transmit earth station to the satellite, and a downlink from the satellite to the receive earth station. Assuming a transparent transponder in which the RF signal from the transmit earth station is not demodulated at the satellite, the

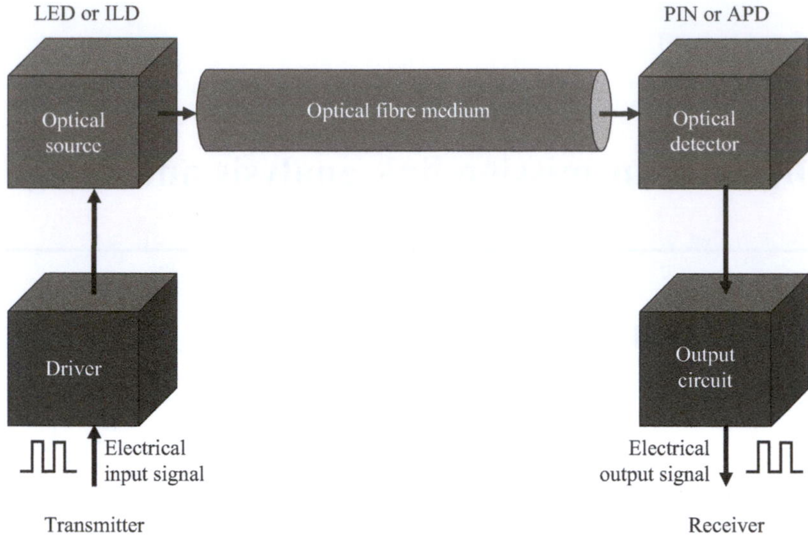

Fig. 12.1 Optical communication link

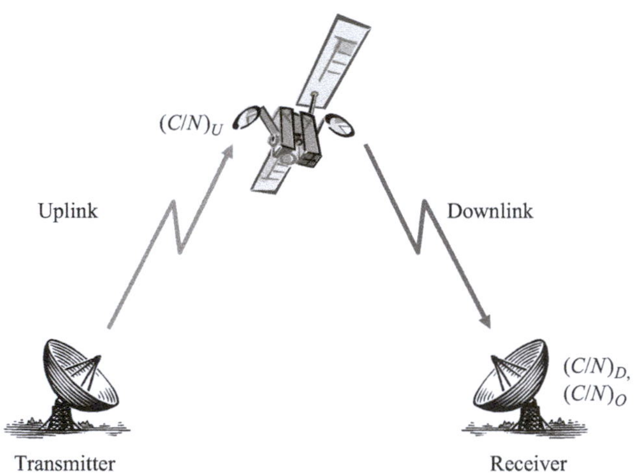

Fig. 12.2 Satellite communication link

C/N needed in the above steps is the overall value, denoted by $(C/N)_o$. This is given in terms of the *C/N* observed at the satellite receiver, denoted by $(C/N)_U$ for uplink carrier-to-noise ratio, the downlink *C/N*, denoted by $(C/N)_D$, and the system carrier-to-interference ratio, denoted by *C/I*, as follows:

$$(C/N)_o = -10\log_{10}\left\{10^{-[(C/N)_U/10]} + 10^{-[(C/N)_D/10]} + 10^{-[(C/I)/10]}\right\}\ \text{dB} \quad (12.1)$$

where all quantities are expressed in dB. A small value of *C/I* on a satellite is typically due to intermodulation distortions arising from transponder nonlinearity as evaluated in Chapter 3. If *C/I* is not known, it may be assumed that there is sufficient transponder power back-off so that $C/I \gg (C/N)_U$, $(C/N)_D$ and may therefore be dropped from the above equation which then simplifies to

$$(C/N)_o = -10\log_{10}\left\{10^{-[(C/N)_U/10]} + 10^{-[(C/N)_D/10]}\right\} \text{ dB} \tag{12.2}$$

It is convenient to lay out calculations leading to signal power and noise power on a transmission link in a tabular format referred to as the link power budget. Typical templates for the uplink and downlink power budgets of a satellite communication system are presented in Tables 12.1 and 12.2. All quantities must be specified in

Table 12.1 Satellite communication uplink power budget template

Parameter	Comment	Value
Uplink signal power		
P_t	Earth station output power	
G_t	Earth station antenna gain	
G_r	Satellite antenna gain	
L_s	Free space path loss at uplink frequency	
L_{ant}	Edge of beam loss	
L_a	Clear air atmospheric loss	
L_{rain}	Rain attenuation at specified availability	
L_m	Miscellaneous losses	________
P_r	Signal power received by satellite	
Uplink noise power		
k	Boltzmann's constant	
T_{sys}	Transponder system noise temperature	
B	Noise bandwidth	________
N	Noise power at satellite	
$\mathbf{(C/N)_U}$	$P_r - N$	

Table 12.2 Satellite communication downlink power budget template

Parameter	Comment	Value
Downlink signal power		
P_t	Transponder saturated output power	
B_o	Transponder output backoff	
G_t	Satellite antenna gain	
G_r	Earth station antenna gain	
L_s	Free space path loss at downlink frequency	
L_{ant}	Edge of beam loss	
L_a	Clear air atmospheric loss	
L_{rain}	Rain attenuation at specified availability	
L_m	Miscellaneous losses	________
P_r	Signal power at receiver reference point	
Downlink noise power		
k	Boltzmann's constant	
T_{sys}	Clear air system noise temperature	
B	Noise bandwidth	
ΔN_{rain}	Increase in noise power due to rain	________
N	Noise power at receiver reference point	
$\mathbf{(C/N)_D}$	$P_r - N$	________
$\mathbf{(C/N)_O}$		

logarithmic units, taking care to ensure consistency; for example signal power and noise power must both be expressed in the same units of either dBW or dBm. If they are in dBW then Boltzmann's constant k must be consistently expressed as −228.6 dBW/K/Hz, but if they are in dBm, then k is −198 dBm/K/Hz. Furthermore, all losses must be entered as negative values and all gains as positive. For example, a free space path loss of 200 dB must be entered as −200, whereas an antenna gain of 40 dB is entered as 40. Signal power is then given by the algebraic sum of all the entries above it in the link budget table. Note that this is simply a tabular evaluation of Eq. (10.31). Similarly, noise power is given by the algebraic sum of the entries above it in its section of the link budget table. Again, this is a tabular evaluation of Eq. (10.30). The bandwidth used must be the noise equivalent bandwidth given in terms of transmission symbol rate by

$$B = \begin{cases} \left(1 - \dfrac{\alpha}{4}\right) R_s & M\text{-ary ASK, PSK, QAM} \\ \left(M + 1 - \dfrac{\alpha}{2}\right) \dfrac{R_s}{2} & M\text{-ary FSK} \end{cases} \tag{12.3}$$

where β is the roll-off factor of the raised cosine filter, a standard processing component in all digital transmission systems as discussed in Chapter 8. If the value of β is not known, $\beta = 0$ should be used.

The amount of clear air atmospheric loss to use in the link power budget may be determined from Fig. 12.3 which is based on the ITU-R model for gaseous attenuation [1]. If rain attenuation is not included, then the analysis is applicable to clear air performance only and the link margin computed is called the clear air link margin. There will be further comments on rain attenuation in the next Section. The satellite parameters in the link budget, such as satellite receive antenna gain and satellite

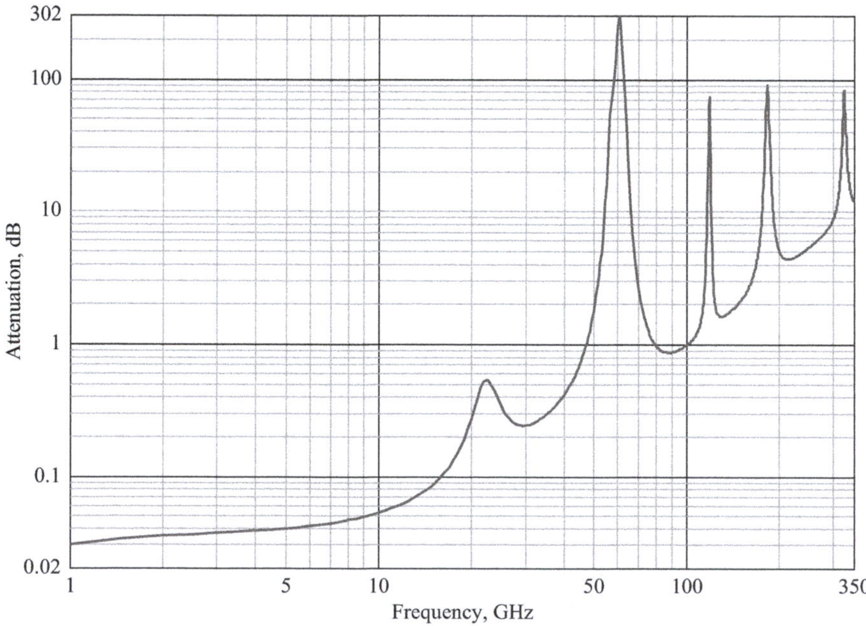

Fig. 12.3 Zenith atmospheric gaseous attenuation based on the ITU-R model [1] at 15°C and 58.5% relative humidity. For other path elevation angles $\theta \geq 5°$, divide reading by sin θ

receiver noise temperature, should be read from the receive coverage maps published by the satellite owners, usually as contours of figure of merit G/T_{sys} for the service area. Similarly, satellite EIRP ($=P_t+G_t$) may be read from the transmit coverage maps.

Notice in the downlink power budget template that there is an entry ΔN_{rain} for increase in noise power due to rain. This is because when there is rain attenuation (say A dB) on the downlink, the downlink receive antenna is located in or near the rain volume and will therefore pick up noise radiated by the absorbing rain volume. There is no such effect on the uplink since the uplink receive antenna (i.e. satellite antenna) is located in space very far removed from the rain medium. The satellite antenna is in any case pointed at a warm earth and therefore sees a roughly constant warm earth noise source of ~300 K as discussed in Chapter 10. This keeps the system noise temperature of the satellite on-board receiver at roughly 500 K. The following steps should be followed in calculating ΔN_{rain} when link analysis is being carried out for a specified level of rain attenuation A_{rain}.

1. Note clear air system noise temperature T_{sysca}.
2. Calculate clear air antenna noise temperature T_{aca} from clear air attenuation A_{ca} (dB):

   $$T_{aca} = 275(1 - 10^{-A_{ca}/10}) \text{ K}$$

   Recall that A_{ca} may be determined from Fig. 12.3.
3. Calculate antenna noise temperature T_{arain} during rain from total attenuation during rain, $A_{ca}+A_{rain}$:

   $$T_{arain} = 275[1 - 10^{-(A_{ca}+A_{rain})/10}] \text{ K}$$

4. Calculate increase in earth station system noise temperature during rain:

   $$\Delta T_{sys} = (T_{arain} - T_{aca})/L_{fr}$$

   where L_{fr} is the loss of the cable feed from antenna to LNA input of receiver.
5. Earth station system noise temperature during rain is

   $$T_{sysrain} = T_{sysca} + \Delta T_{sys}$$

6. Therefore the increase in noise power during rain is

   $$\Delta N_{rain} = 10 \log_{10}\left(T_{sysrain}/T_{sysca}\right) \text{ dB}$$

The link power budget templates in Tables 12.1 and 12.2 and most of the discussions so far are specific to a satellite communication link. Other communication links may also be analysed with the aid of a link power budget, using a template that is customised to the particular link and contains components of losses and gains that are significant in such links. See two worked examples at the end of this chapter, one on a mobile radio link and the other on an optical fibre link.

12.3 Link design

A digital satellite communication link is usually designed for a specified availability, which stipulates the percentage P of time in an average year that the BER must be below a given threshold BER_{th}. Other links (e.g. optical fibre) not subject to significant variation in losses may be designed to achieve a given fixed link margin, which is the amount in dB by which the link's overall carrier-to-noise ratio exceeds

Table 12.3 Annual link availability figures

Availability (%)	Outage (%)	Problem duration per year on average
90	10	36 days and 13 hours
95	5	18 days and 6 hours
99	1	3 days and 16 hours
99.5	0.5	1 day and 20 hours
99.9	0.1	9 hours
99.99	0.01	1 hour

the minimum required to achieve an acceptable link BER. A satellite link may occasionally be designed to achieve a given clear air margin, although this practice should be discouraged and the link should always be designed to achieve a specified availability. Table 12.3 lists typical annual availability values and the corresponding cumulative total link down time or outage in an average year. In most climates, the effect of rain attenuation may be ignored altogether on occasional-use links (having availability of 95% or lower) since one is prepared to accept the link not being available for use over a cumulative period of up to 18 days in a year.

Link outage is said to occur whenever BER exceeds BER_{th}. By specifying an availability of P%, it means that it is acceptable for link outage to happen for $p = (100 - P)$% of the year. The link must be able to cope with all losses, noise and atmospheric attenuation occurring during P% of the year. In other words, the link's overall E_b/N_o must exceed a threshold $(E_b/N_o)_{\text{th}}$, which is the value required by the adopted modulation scheme in order to operate at $\text{BER} = \text{BER}_{\text{th}}$. Like the analysis discussed in the previous section, this link design is usually presented in the form of a tabular layout of the link power budget featuring:

- All gains deployed (coding gain, diversity gains, antenna gains, etc.).
- All fixed losses (modem implementation loss, spreading loss, edge of beam loss, clear air atmospheric loss and miscellaneous losses).
- The value of variable losses exceeded for p% of the year, including rain attenuation (or cloud attenuation if it exceeds rain attenuation) and scintillation. Reliable estimates of rain attenuation may be obtained using the ITU-R prediction model in recommendation P.618 [2]. This is where the specification of availability impacts on link design. You may find that it is difficult to balance the link power budget, i.e. to achieve a non-negative link margin, if availability is set so high that the losses included in this section of the link power budget becomes excessive. Rain attenuation varies with climate and depends on a number of link parameters such as path elevation angle and link frequency but it will in general be the largest component of additional loss if link frequency is above 10 GHz and link availability is ≥ 99.9%.
- All noise contributions (represented as system noise temperature), including receiver noise, atmospheric noise exceeded at p% of the year, extra-terrestrial noise, noise from the Earth's surface (if downlink), cross-polar signal power (if the system employs orthogonal polarisation frequency reuse) and interference signal if significant.

The use of error control coding delivers a coding gain G_c as discussed in Chapter 11. This may be incorporated in link design by revising the value of E_b/N_o needed to achieve the specified BER, say BER_{th}. Suppose that the E_b/N_o value read from the theoretical BER curves (see Chapter 10) is denoted by $(E_b/N_o)_o$, and the modem has an implementation loss L_{mil} (due to small departures from perfect phase synchronisation in the coherent detection process, timing jitter in the sampling instant at the detection point, etc.) and that forward error correction (FEC) delivers a coding gain of G_c, then the threshold E_b/N_o value that must be achieved on the link in order to operate at a BER no higher than BER_{th} is

$$\left(\frac{E_b}{N_o}\right)_{\text{th}} = \left(\frac{E_b}{N_o}\right)_o - G_c + L_{mil} \text{ dB} \tag{12.4}$$

In summary the link design steps are as follows:

- Determine the value of E_b/N_o required for a desired BER, say BER_{th}, taking account of modulation technique and coding gain. In other words, read the value of the BER versus E_b/N_o curve at the point $\text{BER} = \text{BER}_{\text{th}}$ for the modulation scheme being used. Call this reading $(E_b/N_o)_o$ and use it in Eq. (12.4) to obtain $(E_b/N_o)_{\text{th}}$, the threshold E_b/N_o that must be satisfied on the link. The value of modem implementation loss to use depends on modulation scheme, but 1 dB is a typical and safe value to assume.
- Complete all entries of the tabulated link power budget and check whether the resulting E_b/N_o exceeds threshold. It is unlikely that the first set of values (i.e. preliminary design) will be satisfactory, so you may have to adjust entries or link parameters that are under your control (e.g. antenna size — to increase gain, or transmit power, or availability requirement, etc.) until a satisfactory outcome is achieved.
- If the previous step fails to lead to a positive link margin, consider changing to (i) a different modulation technique that yields an acceptable BER at a more affordable E_b/N_o, and/or (ii) a more powerful FEC technique (with concatenation and interleaving if necessary) that delivers a higher coding gain and/or (iii) a lower link availability requirement.

It is always good practice to allow a positive link margin to ensure continued reliable operation of the link in the face of small estimation errors in the losses and gains included in the link power budget, as well as departures of the system from the average conditions assumed. The size of this margin depends on the type of link. On terrestrial mobile radio links it may need to be in the region of 15 dB but on fixed satellite links a more modest value of ~2 dB is sufficient.

Worked Example 12.1: Link Power Budget of a Mobile Radio System

A mobile radio transmission system has the following specifications:

Coverage range = 4 km; free space propagation break point = 100 m; path loss exponent in non-free-space propagation region is $n = 4$; carrier frequency = 950 MHz; bandwidth = 25 kHz; transmit antenna gain = 12 dB; feed losses at transmitter = 3 dB; receive antenna gain = –4 dB; noise figure of receiver = 5 dB.

If the required minimum carrier-to-noise ratio is 10 dB, determine the minimum transmit power needed to provide a fade margin of 16 dB.

The solution is laid out in tabular form in Table 12.4. This is a link design problem in which we need to specify transmit power P_t. This is therefore

Table 12.4 Link power budget of mobile radio system and solution to Worked Example 12.1

Quantity	Computation	Value
Signal power budget		
P_t, Transmit output power	To be determined	P_t, dBm
G_t, Transmit antenna gain	Given	+12 dB
L_{ft}, Transmit feed losses	Given	3 dB
L_{pm}, Mean path loss	$92.44 + 20 \log f_{c,\mathrm{GHz}} + 10\, n \log d - 10(n-2)\log d_1$ where $d_1 = 0.1$ km, $d = 4$ km, $f_c = 0.95$ GHz, $n = 4$	136.1 dB
M, Fade margin	Given	16 dB
G_r, Receive antenna gain	Given	+(4) dB
P_r, Received signal power	Algebraic sum of above six quantities	$P_t - 147.1$ dBm
Noise power budget		
k, Boltzmann's constant	1.3810^{23} W/K/Hz $= 1.3810^{20}$ mW/K/Hz	198.6 dBm/K/Hz
T_{sys}, System noise temperature	$T_a + T_e = 300 + T_o(F_e - 1) = 300 + 290(10^{5/10} - 1) = 927.1$ K	29.7 dBK
bandwidth	25 kHz (Given)	44 dBHz
N, Noise power	Algebraic sum of above three quantities	125 dBm
C/N, Carrier-to-noise ratio	$P_r - N = P_t - 147.1 + 125 = P_t - 22.1$	10 dB (Given)
	Thus $P_t - 22.1 = 10$, which means $P_t = 32.1$ dBm $= 1.622$ W	

entered in the link power budget as an unknown — to be determined (TBD) — quantity. The table includes comments on how each entry was calculated. At the end of the link power budget we are able to equate the link *C/N* (with P_t still unknown) to the desired *C/N* = 10 dB and hence solve for P_t. Notice also how the required fade margin was entered as a loss since it is a potential loss that the link is to be designed to be able to cope with. Notice also that the receive antenna gain is negative. Omnidirectional hand-held antenna units in the mobile radio environment have very small gains, which may even be negative as in this case.

Worked Example 12.2: Link Power Budget of an Optical Fibre Link

An optical fibre communication system has the following specifications:

Transmit power = 1 mW, receiver sensitivity = −42 dBm, maximum length of continuous fibre section $s = 2$ km, splicing loss $\beta = 0.5$ dB per splice, coupling loss = 1 dB at each end, fibre loss = 0.2 dB/km.

Determine the maximum length of the fibre link that can be operated without a repeater.

The task in this case is also a design issue to calculate an unknown maximum link length (denoted by *d*) necessary to ensure that received signal power is no lower than the receiver sensitivity level of −42 dBm. The solution is laid out in tabular form in Table 12.5 with the unknown link length carried through to the end of the link power budget where it features in the expression for received power P_r. Equating this to −42 dBm yields $d = 90$ km. Notice that by specifying receiver sensitivity, there is no noise power computation in the link budget. The specified receiver sensitivity is the amount of received signal power which when reduced by the noise power at the receiver meets the *C/N* threshold needed for reliable signal detection.

Table 12.5 Link power budget of optical fibre link and solution to Worked Example 12.2

Quantity	Computation	Value
P_t, Transmit output power	Given (1 mW) = 10 log10(1) dBm	0 dBm
Total fibre loss	Fibre loss per kilometer multiplied by total link length	0.2d dB
Total coupling loss	Sum of coupling loss at either end (1 + 1 dB)	2 dB
Total splicing loss	(Number of sections −1)α = (d/s −1)α	0.5(d/2 − 1) dB
M, Fade margin	This is zero at maximum link length	0 dB
P_r, Received signal power	Algebraic sum of above five quantities	(0.45d + 1.5) dBm
	P_r = Receiver sensitivity = −42 dBm (Given). Thus, −(0.45d + 1.5) = −42, which yields d = 90 km	

12.4 References

[1] ITU-R Recommendation P.676-9. (2012). *Attenuation by atmospheric gases*. Geneva: International Telecommunications Union

[2] ITU-R Recommendation P.618-10. (2009). *Propagation data and prediction methods required for the design of Earth-space telecommunication systems*. Geneva: International Telecommunications Union

Appendix A
Character codes

Table A.1 International Morse code

Character	Morse code	Character	Morse code
A	• —	2	• • — — —
B	— • • •	3	• • • — —
C	— • — •	4	• • • • —
D	— • •	5	• • • • •
E	•	6	— • • • •
F	• • — •	7	— — • • •
G	— — •	8	— — — • •
H	• • • •	9	— — — — •
I	• •	: (colon)	— — — • • •
J	• — — —	, (comma)	— — • • — —
K	— • —	; (semicolon)	— • — • — •
L	• — • •	?	• • — — • •
M	— —	. (period)	• — • — • —
N	— •	’ (apostrophe)	• — — — — •
O	— — —	“	• — • • — •
P	• — • —	/	— • • — •
Q	— — • —	- (hyphen)	— • • • • —
R	• — •	=	— • • • —
S	• • •	) or (	— • — — • —
T	—	Attention	— • — • —
U	• • —	Break	— • • • — • —
V	• • • —	End of message	• — • — •
W	• — —	Error	• • • • • • • •
X	— • • —	Go ahead	— • —
Y	— • — —	OK	• — •
Z	— — • •	SOS	• • • — — — • • •
0	— — — — —	End of contact	• • • — • —
1	• — — — —	Wait	• — • • •

Table A.2 International Telegraph Alphabet No. 2 (ITA-2): A modification of Baudot-Murray code

Character		ITA-2 code	Character		ITA-2 code
Letters	**Figures**		**Letters**	**Figures**	
A	-	11000	Q	1	11101
B	?	10011	R	4	01010
C	:	01110	S	' (US[1] = Bell)	10100
D	WRU[2] (US[1] = $)	10010	T	5	00001
E	3	10000	U	7	11100
F	! (UD[3])	10110	V	= (US[1] = ;)	01111
G	& (UD[3])	01011	W	2	11001
H	£ (US[1] = #) (UD[3])	00101	X	/	10111
I	8	01100	Y	6	10101
J	Bell (US[1] = ')	11010	Z	+ (US[1] = ")	10001
K	(	11110	Letter shift		11111
L	)	01001	Figure shift		11011
M	.	00111	Space		00100
N	,	00110	Carriage return		00010
O	9	00011	Line feed		01000
P	0	01101	Blank (Null)		00000

Notes:
[1]Where US standard differs from ITA-2, this is indicated in brackets as (US = . . .).
[2]WRU ≡ Who are you?
[3]Three codes were left undefined (UD) to allow for national variants.

Table A.3 EBCDIC code

First hexadecimal digit ($0 \rightarrow F \equiv b_4b_3b_2b_1$)	Second hexadecimal digit ($0 \rightarrow F \equiv b_8b_7b_6b_5 = 0000 \rightarrow 1111$)															
	0 0000	**1 0001**	**2 0010**	**3 0011**	**4 0100**	**5 0101**	**6 0110**	**7 0111**	**8 1000**	**9 1001**	**A 1010**	**B 1011**	**C 1100**	**D 1101**	**E 1110**	**F 1111**
0	NUL	DLE	DS		SP	&	-						(	)	\	0
1	SOH	DC1	SOS		RSP		/		a	j	-		A	J	NSP	1
2	STX	DC2	FS	SYN					b	k	s		B	K	S	2
3	ETX	DC3	WUS	IR					c	l	t		C	L	T	3
4	SEL	RES	BYP	PP					d	m	u		D	M	U	4
5	HT	NL	LF	TRN					e	n	v		E	N	V	5
6	RNL	BS	ETB	NBS					f	o	w		F	O	W	6
7	DEL	POC	ESC	BOT					g	p	x		G	P	X	7
8	GE	CAN	SA	SBS					h	q	y		H	Q	Y	8
9	SPS	EM	SPE	IT					i	r	z		I	R	Z	9
A	RPT	UBS	SM	RFF	¢	!	^	:								
B	VT	CU1	CSP	CU3	.	$	,	#								
C	FF	IFS	MFA	DC4	<	*	%	@								
D	CR	IGS	ENQ	NAK	(	)	-	'								
E	SO	IRS	ACK		+	;	>	=								
F	SI	IUS	BEL	SUB	|	¬	?	"								

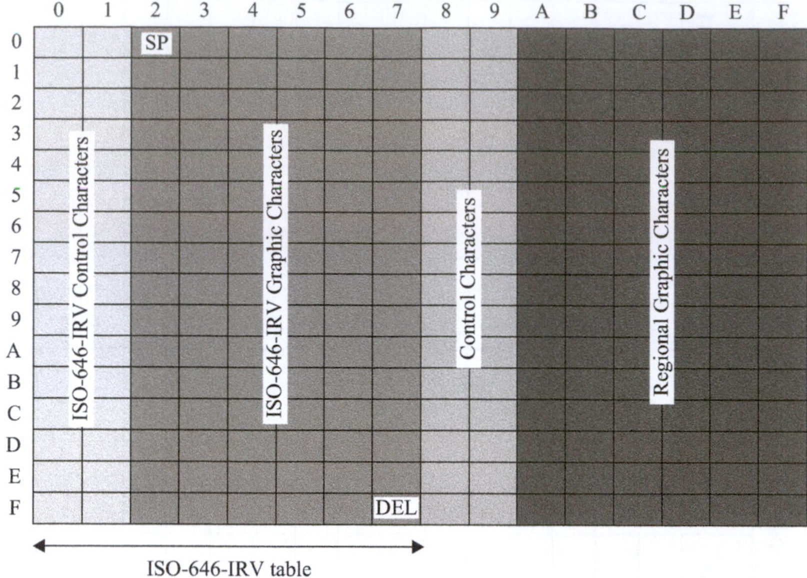

Fig. A.1 Layout of eight-bit character code table

Table A.4 ASCII code

$b_7b_6b_5$ →		hex→ **0**	**1**	**2**	**3**	**4**	**5**	**6**	**7**
$b_4b_3b_2b_1$ ↓**hex**	↓**bin**	bin→ **000**	**001**	**010**	**011**	**100**	**101**	**110**	**111**
0	0000	^{0}NUL	^{16}DLE	^{32}SP	^{48}0	^{64}@	^{80}P	^{96}`	^{112}p
1	0001	^{1}SOH	^{17}DC1	^{33}!	^{49}1	^{65}A	^{81}Q	^{97}a	^{113}q
2	0010	^{2}STX	^{18}DC2	^{34} "	^{50}2	^{66}B	^{82}R	^{98}b	^{114}r
3	0011	^{3}ETX	^{19}DC3	^{35}#	^{51}3	^{67}C	^{83}S	^{99}c	^{115}s
4	0100	^{4}EOT	^{20}DC4	^{36}$	^{52}4	^{68}D	^{84}T	^{100}d	^{116}t
5	0101	^{5}ENQ	^{21}NAK	^{37}%	^{53}5	^{69}E	^{85}U	^{101}e	^{117}u
6	0110	^{6}ACK	^{22}SYN	^{38}&	^{54}6	^{70}F	^{86}V	^{102}f	^{118}v
7	0111	^{7}BEL	^{23}ETB	^{39} '	^{55}7	^{71}G	^{87}W	^{103}g	^{119}w
8	1000	^{8}BS	^{24}CAN	^{40}(	^{56}8	^{72}H	^{88}X	^{104}h	^{120}x
9	1001	^{9}HT	^{25}EM	^{41})	^{57}9	^{73}I	^{89}Y	^{105}i	^{121}y
A	1010	^{10}LF	^{26}SUB	^{42}*	^{58}:	^{74}J	^{90}Z	^{106}j	^{122}z
B	1011	^{11}VT	^{27}ESC	^{43} +	^{59};	^{75}K	^{91}[	^{107}k	^{123}{
C	1100	^{12}FF	^{28}FS	^{44},	^{60} <	^{76}L	^{92}\	^{108}l	$^{124}\vert$
D	1101	^{13}CR	^{29}GS	^{45} –	^{61} =	^{77}M	^{93}]	^{109}m	^{125}}
E	1110	^{14}SO	^{30}RS	^{46}.	^{62}>	^{78}N	^{94}^	^{110}n	^{126}~
F	1111	^{15}SI	^{31}US	^{47}/	^{63}?	^{79}O	^{95}_	^{111}o	^{127}DEL

Note: The number at the top left corner of each cell is the decimal value of the ASCII code for the character or control signal.
ACK, acknowledge; BEL, bell or alarm; BS, backspace; CAN, cancel; CR, carriage return; DC1 . . . 4, device control 1 . . . 4; DEL, delete; DLE, data link escape; EM, end of medium; ENQ, enquiry; EOT, end of transmission; ESC, escape; ETB, end of transmission block; ETX, end of text; FF, form feed; FS, file separator; GS, group separator; HT, horizontal tab; LF, line feed; NAK, negative acknowledge; NUL, null or all zeros; RS, record separator; SI, shift in; SO, shift out; SOH, start of heading; SP, space; STX, start of text; SUB, substitute; SYN, synchronous idle; US, unit separator; VT, vertical tab.

Table A.5 ISO-8859-1 or Latin-1 alphabet for Western Europe

hex →		**A**	**B**	**C**	**D**	**E**	**F**
bin →		**1010**	**1011**	**1100**	**1101**	**1110**	**1111**
↓hex	**↓bin**						
0	0000	A0 NBSP	B0 °	C0 À	D0 Đ	E0 à	F0 ð
1	0001	A1 ¡	B1 ±	C1 Á	D1 Ñ	E1 á	F1 ñ
2	0010	A2 ¢	B2 2	C2 Â	D2 Ò	E2 â	F2 ò
3	0011	A3 £	B3 3	C3 Ã	D3 Ó	E3 ã	F3 ó
4	0100	A4 ¤	B4 /	C4 Ä	D4 Ô	E4 ä	F4 ô
5	0101	A5 ¥	B5 µ	C5 Å	D5 Õ	E5 å	F5 õ
6	0110	A6 ¦	B6 ¶	C6 Æ	D6 Ö	E6 æ	F6 ö
7	0111	A7 §	B7 ·	C7 Ç	D7 ×	E7 ç	F7 ÷
8	1000	A8 ¨	B8 è	C8 È	D8 Ø	E8 è	F8 ø
9	1001	A9 ©	B9 1	C9 É	D9 Ù	E9 é	F9 ù
A	1010	AA a	BA o	CA Ê	DA Ú	EA ê	FA ú
B	1011	AB «	BB »	CB Ë	DB Û	EB ë	FB û
C	1100	AC ¬	BC ¼	CC Ì	DC Ü	EC ì	FC ü
D	1101	AD -	BD ½	CD Í	DD Ý	ED í	FD ý
E	1110	AE ®	BE ¾	CE Î	DE Þ	EE î	FE þ
F	1111	AF -	BF ¿	CF Ï	DF ß	EF ï	FF ÿ

Note: Only columns A → F (hex) are shown. Columns 0 → 7 are identical to Table A.4, whereas columns 8 and 9 define so-called C1 control characters. (i) NBSP ≡ Non-breaking space; (ii) ¨ is the diaeresis character used for example in naïve.

Appendix B
Trigonometric identities

Let us start from the following compound angle relations, which can be readily obtained from the solution of triangles:

$$\sin(A+B) = \sin A\cos B + \cos A\sin B \quad \text{(B.1)}$$

$$\sin(A-B) = \sin A\cos B - \cos A\sin B \quad \text{(B.2)}$$

$$\cos(A+B) = \cos A\cos B - \sin A\sin B \quad \text{(B.3)}$$

$$\cos(A-B) = \cos A\cos B + \sin A\sin B \quad \text{(B.4)}$$

We obtain identities for the multiplication of sinusoids as follows:

(i) subtract Eq. (B.3) from Eq. (B.4) to obtain Eq. (B.5).
(ii) add Eqs. (B.3) and (B.4) to obtain Eq. (B.6).
(iii) add Eqs. (B.1) and (B.2) to obtain Eq. (B.7).

Thus

$$\sin A\sin B = \frac{1}{2}[\cos(A-B) - \cos(A+B)] \quad \text{(B.5)}$$

$$\cos A\cos B = \frac{1}{2}[\cos(A+B) + \cos(A-B)] \quad \text{(B.6)}$$

$$\sin A\cos B = \frac{1}{2}[\sin(A+B) + \sin(A-B)] \quad \text{(B.7)}$$

Noting that Eqs. (B.1) to (B.7) hold for all values of A and B, we can obtain other very useful identities as follows:

(i) replace B with A in Eqs. (B.1), (B.3), (B.5), (B.6) and (B.4) to obtain Eqs. (B.8) to (B.12), respectively.
(ii) substitute $B = 180°$ and $B = 90°$ in Eqs. (B.1) to (B.4) to obtain Eqs. (B.13) to (B.16), respectively.
(iii) finally, substitute $A = 0°$ in Eqs. (B.2) and (B.4) to obtain Eqs. (B.17) and (B.18), respectively.

Thus

$$\sin 2A = 2\sin A\cos A \quad \text{(B.8)}$$

$$\cos 2A = \cos^2 A - \sin^2 A \quad \text{(B.9)}$$

$$\sin^2 A = \frac{1}{2}(1 - \cos 2A) \quad \text{(B.10)}$$

$$\cos^2 A = \frac{1}{2}(1 + \cos 2A) \quad \text{(B.11)}$$

$$\sin^2 A + \cos^2 A = 1 \tag{B.12}$$

$$\sin(A \pm 180°) = -\sin A \tag{B.13}$$

$$\cos(A \pm 180°) = -\cos A \tag{B.14}$$

$$\sin(A \pm 90°) = \pm\cos A \tag{B.15}$$

$$\cos(A \pm 90°) = \mp\sin A \tag{B.16}$$

$$\sin(-B) = -\sin B \tag{B.17}$$

$$\cos(-B) = \cos B \tag{B.18}$$

Making repeated use of Eqs. (B.11) and (B.6) yields the following useful identities for integer powers of a cosine function:

$$\begin{aligned}
\cos^3 A &= \frac{1}{4}(\cos 3A + 3\cos A) \\
\cos^4 A &= \frac{1}{8}(\cos 4A + 4\cos 2A + 3) \\
\cos^5 A &= \frac{1}{16}(\cos 5A + 5\cos 3A + 10\cos A) \\
\cos^6 A &= \frac{1}{32}(\cos 6A + 6\cos 4A + 15\cos 2A + 10) \\
\cos^7 A &= \frac{1}{64}(\cos 7A + 7\cos 5A + 21\cos 3A + 35\cos A) \\
&\text{etc.}
\end{aligned} \tag{B.19}$$

Similar identities can be obtained for a sine function. For example, for odd powers

$$\begin{aligned}
\sin^3 A &= \frac{1}{4}(-\sin 3A + 3\sin A) \\
\sin^5 A &= \frac{1}{16}(\sin 5A - 5\sin 3A + 10\sin A) \\
\sin^7 A &= \frac{1}{64}(-\sin 7A + 7\sin 5A - 21\sin 3A + 35\sin A)
\end{aligned} \tag{B.20}$$

Appendix C
Fourier transform

A statement of the Fourier theorem is that any periodic signal $g_T(t)$ of period T, such as shown in Fig. C.1(a), can be expressed as the sum of sinusoids with frequencies at integer multiples (called harmonics) of the fundamental frequency, and with appropriate amplitudes and phases. Thus we have the Fourier series expression for $g_T(t)$:

$$\begin{aligned} g_T(t) &= A_o + \sum_{n=1}^{\infty} A_n \cos(2\pi n f_o t + \phi_n) \\ &\equiv A_o + \sum_{n=1}^{\infty} a_n \cos(2\pi n f_o t) + \sum_{n=1}^{\infty} b_n \sin(2\pi n f_o t) \end{aligned} \qquad f_o = \frac{1}{T} \tag{C.1}$$

The coefficients in Eq. (C.1) are obtained as follows: Integrating both sides of Eq. (C.1) over an interval of one period T yields the average value (or DC component) A_o of the periodic signal. First multiplying both sides by $\cos(2\pi m f_o t)$ before integrating over one period yields the mth cosine coefficient a_m; and multiplying first by $\sin(2\pi m f_o t)$ before integrating over one period gives the mth sine coefficient b_m:

$$\begin{aligned} A_o &= \frac{1}{T} \int_{-T/2}^{T/2} g(t)\, dt \\ a_n &= \frac{2}{T} \int_{-T/2}^{T/2} g(t) \cos(2\pi n f_o t)\, dt \\ b_n &= \frac{2}{T} \int_{-T/2}^{T/2} g(t) \sin(2\pi n f_o t)\, dt \\ A_n &= \sqrt{a_n^2 + b_n^2}, \qquad \text{Amplitude of } n^{\text{th}} \text{ harmonic component} \\ \phi_n &= -\arctan(b_n/a_n), \qquad \text{Phase of } n^{\text{th}} \text{ harmonic component} \end{aligned} \tag{C.2}$$

Employing Euler's formula $\exp(j2\pi n f_o t) = \cos(2\pi n f_o t) + j\sin(2\pi n f_o t)$ in Eq. (C.1) leads to the exponential form of Fourier series:

$$g(t) = \sum_{n=-\infty}^{\infty} C_n e^{j2\pi n f_o t}$$

where

$$C_n = \frac{1}{2}(a_n - jb_n) = \frac{1}{T} \int_{-T/2}^{T/2} g(t) e^{-j2\pi n f_o t} dt \tag{C.3}$$

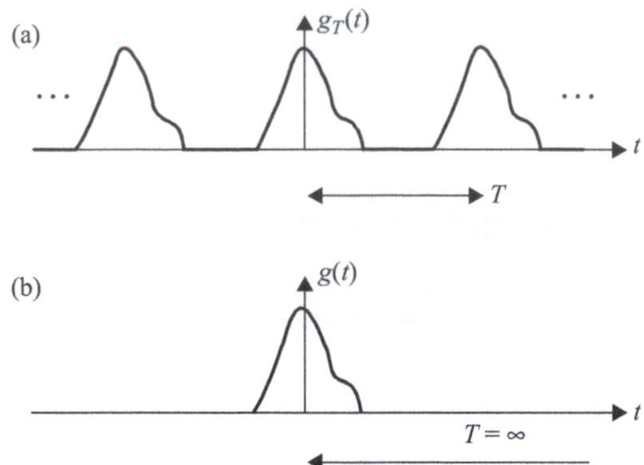

Fig. C.1 (a) Periodic signal $g_T(t)$; (b) nonperiodic signal $g(t)$ as a limiting case of $g_T(t)$ when $T \to \infty$

As illustrated in Fig. C.1(b), a nonperiodic signal $g(t)$ can be treated as a limiting case of a periodic signal $g_T(t)$ when $T \to \infty$. That is, $g(t)$ repeats after an infinitely long time and therefore has period $T \to \infty$. In this limit, spectral spacing f_o ($=1/T$) becomes infinitesimally small (denoted by df) and the discrete harmonic frequencies nf_o become a continuum of frequencies f. Therefore a nonperiodic waveform $g(t)$ contains a continuum of sinusoids, and we now need to define a function $G(f)$ that gives the sinusoidal amplitude contained in each unit frequency interval of $g(t)$ at frequency location f. This function is called the *Fourier transform* of $g(t)$ and it follows from this definition of $G(f)$ and the expression for C_n in the 2nd line of Eq. (C.3) that

$$\begin{aligned} G(f) &= \lim_{T\to\infty} \frac{C_n}{f_o} = \lim_{T\to\infty} C_n T \\ &= \int_{-\infty}^{\infty} g(t) e^{-j2\pi ft} dt \quad \text{V/Hz} \end{aligned} \tag{C.4}$$

where it is assumed that $g(t)$ is a voltage signal in volts (V), which leads to the unit of its Fourier transform $G(f)$ being V/Hz. Furthermore, in this limit the summation on the RHS of the 1st line of Eq. (C.3) becomes integration and, upon replacing C_n by $G(f)f_o \equiv G(f)df$ and nf_o by f, we see that the Fourier series for a periodic signal takes the following form for a nonperiodic signal:

$$g(t) = \int_{-\infty}^{\infty} G(f) e^{j2\pi ft} df \tag{C.5}$$

$G(f)$ is in general a complex number which may be written in polar form as

$$G(f) = |G(f)| \exp(j\phi_g(f))$$

The magnitude $|G(f)|$ of $G(f)$ gives the *amplitude spectrum* of $g(t)$, whereas the angle $\angle G(f)$ of $G(f)$, denoted $\phi_g(f)$, gives the *phase spectrum* of $g(t)$.

C.1 Fourier transform properties

The following properties or relationships of the Fourier transform can be readily proved from the definition of the Fourier transform. Given the transform pair

$$\begin{aligned} G(f) &= \int_{-\infty}^{\infty} g(t)e^{-j2\pi ft}dt, \quad \text{Fourier transform (FT)} \\ g(t) &= \int_{-\infty}^{\infty} G(f)e^{j2\pi ft}df, \quad \text{Inverse Fourier transform (IFT)} \end{aligned} \tag{C.6}$$

where $G(f)$ is the FT of $g(t)$, and $g(t)$ is the IFT of $G(f)$, a relationship denoted by $g(t) \rightleftharpoons G(f)$, and also given that

$$g_1(t) \rightleftharpoons G_1(f)$$
$$g_2(t) \rightleftharpoons G_2(f)$$

and that a, a_1 and a_2 are constants, then the following transform pairs and relationships also hold by virtue of the property identified.

C.1.1 Linearity

$$a_1g_1(t) + a_2g_2(t) \rightleftharpoons a_1G_1(f) + a_2G_2(f) \tag{C.7}$$

Eq. (C.7) is the *superposition principle*: The Fourier transform is a linear operator whose output is the result of always doing the same thing to every input and then adding them together.

C.1.2 Time shifting

$$g(t - t_o) \rightleftharpoons G(f)\exp(-j2\pi ft_o) \tag{C.8}$$

The implication of Eq. (C.8) is that delaying a signal by time t_o does not affect the signal's amplitude spectrum but shifts the phase of each frequency component by $-2\pi ft_o$.

C.1.3 Frequency shifting

$$g(t)\exp(j2\pi f_ct) \rightleftharpoons G(f - f_c) \tag{C.9}$$

That is, multiplying a signal by $\exp(j2\pi f_ct)$ has the effect of translating its entire spectrum by amount f_c in the positive frequency direction. And since

$$\cos(2\pi f_ct) = \frac{1}{2}[\exp(j2\pi f_ct) + \exp(-j2\pi f_ct)]$$

it follows from Eq. (C.9) and the linearity property that

$$g(t)\cos(2\pi f_ct) \rightleftharpoons \frac{1}{2}G(f - f_c) + \frac{1}{2}G(f + f_c) \tag{C.10}$$

Thus the effect of multiplication by $\cos(2\pi f_c t)$ — an operation carried out for example in a *mixer* circuit — is to scale the signal spectrum by half and translate it by $\pm f_c$ to two new locations along the frequency axis.

C.1.4 Time scaling

$$g(at) \rightleftharpoons \frac{1}{|a|} G(f/a) \qquad \text{(C.11)}$$

The signal $g(at)$ is scaled in time by the factor a, which means that if the separation between two events in the signal $g(t)$ is τ, then the separation between the same events in the signal $g(at)$ is τ/a. For example, the period of the signal $\cos(4\pi ft)$ — i.e. $\cos(2\pi f \times 2t)$ — is half that of $\cos(2\pi ft)$, and the duration of the pulse $\text{rect}\left(\frac{1}{3}t/\tau\right)$ is three times that of $\text{rect}(t/\tau)$.

Eq. (C.11) states that compressing a signal in the time domain (e.g. by shortening its duration) causes its spectrum to expand by the same factor, and vice versa. This is a statement of the inverse relationship that exists between the description of a signal in the time and frequency domains.

C.1.5 Time reversal

$$g(-t) \rightleftharpoons G(-f) \qquad \text{(C.12)}$$

Note that the waveform of the time-reversed signal $g(-t)$ is obtained by flipping the waveform of $g(t)$ horizontally about $t=0$. Eq. (C.12) states that reversal in the time domain corresponds to reversal in the frequency domain. We know that if $g(t)$ is real, then its amplitude spectrum is even — and hence unchanged by a reversal operation — whereas its phase spectrum is odd and therefore changed in sign by a reversal operation. Thus time reversal does not alter the amplitude spectrum of a real signal, but multiplies its phase spectrum by the factor -1.

C.1.6 Complex conjugation

$$g^*(t) \rightleftharpoons G^*(-f) \qquad \text{(C.13)}$$

Thus complex conjugation in the time domain is equivalent to complex conjugation plus reversal (i.e. flipping about $f=0$) in the frequency domain. If $g(t)$ is a real signal then $g(t)=g^*(t)$ and it follows that

$$\begin{aligned} G^*(-f) &= G(f) \\ g(-t) &\rightleftharpoons G^*(f) \end{aligned} \qquad \text{(C.14)}$$

C.1.7 Duality

$$G(t) \rightleftharpoons g(-f) \qquad \text{(C.15)}$$

Eq. (C.15) states that if we consider a new signal $g_2(t)=G(t)$, which has a waveform that is identical in shape to the spectrum of $g(t)$, then the spectrum of this new signal is $G_2(f)=g(-f)$, and it has a shape that is identical to a time-reversed version of the signal $g(t)$. For example, since a rectangular pulse has a sinc spectrum it follows from this that a sinc pulse will have a rectangular spectrum.

C.1.8 Differentiation

$$\frac{d}{dt}g(t) \rightleftharpoons j2\pi f G(f)$$
$$\frac{d^n}{dt^n}g(t) \rightleftharpoons (j2\pi f)^n G(f) \tag{C.16}$$

Thus differentiation in the time domain is a kind of high pass-filtering operation. It boosts the high-frequency components by multiplying the amplitude spectrum by a factor $2\pi f$, which increases with frequency. This relatively reduces the amplitudes of the low-frequency components. Furthermore, the factor j indicates that the phase of each frequency component is advanced by 90°.

C.1.9 Integration

$$\int_{-\infty}^{t} g(t)dt \rightleftharpoons \frac{G(f)}{j2\pi f} \quad \text{provided } G(0) = 0 \tag{C.17}$$

Integration in the time domain is therefore a low pass-filtering operation. The amplitude of each frequency component is multiplied by a factor $1/2\pi f$, which is inversely proportional to frequency. This attenuates the high-frequency components relative to the low-frequency components. The phase of each component is also reduced by 90°.

C.1.10 Multiplication in the time domain

$$g_1(t)g_2(t) \rightleftharpoons \int_{-\infty}^{\infty} G_1(\gamma)G_2(f-\gamma)d\gamma = G_1(f) * G_2(f) \tag{C.18}$$

Thus multiplication of two signals in the time domain is equivalent to a convolution of their Fourier transforms in the frequency domain.

C.1.11 Convolution in the time domain

$$g_1(t) * g_2(t) \rightleftharpoons G_1(f)G_2(f) \tag{C.19}$$

Thus convolution of two signals or functions in the time domain is equivalent to a multiplication of their Fourier transforms in the frequency domain.

C.1.12 Areas

$$g(0) = \int_{-\infty}^{\infty} G(f)df \quad \text{(a)}$$
$$G(0) = \int_{-\infty}^{\infty} g(t)dt \quad \text{(b)} \tag{C.20}$$

Thus the total area under the spectrum of a signal gives the value of the signal at time $t = 0$, whereas the total area under the waveform of the signal gives the value of the spectrum at $f = 0$.

C.2 Fourier transform pairs

Signal	Fourier transform
Unit impulse $\delta(t)$, defined by $\delta(t) = 0, \quad t \neq 0$ $\int_{-\infty}^{\infty} \delta(t)dt = 1$	1
A constant (DC) signal of value K	$K\delta(f)$
$\delta(t - t_o)$	$\exp(-j2\pi f t_o)$
$\exp(j2\pi f_c t)$	$\delta(f - f_c)$
Signum function sgn(t), defined by $\text{sgn}(t) = \begin{cases} 1, & t > 0 \\ 0, & t = 0 \\ -1, & t < 0 \end{cases}$	$\dfrac{1}{j\pi f}$
$1/\pi t$	$-j\text{sgn}(f)$
Unit step function $u(t)$, defined by $u(t) = \begin{cases} 1, & t > 0 \\ 0, & t < 0 \end{cases}$	$\dfrac{1}{2}\delta(f) + \dfrac{1}{j2\pi f}$
Rectangular pulse $\text{rect}(\frac{t}{\tau})$, defined by $\text{rect}\left(\dfrac{t}{\tau}\right) = \begin{cases} 1, & -\frac{\tau}{2} \leq t \leq \frac{\tau}{2} \\ 0, & \|t\| > \frac{\tau}{2} \end{cases}$	$\tau\text{sinc}(f\tau)$
Sinc function sinc($2t/\tau$) defined by $\text{sinc}\left(\dfrac{2t}{\tau}\right) = \dfrac{\sin(2\pi t/\tau)}{2\pi t/\tau}$	$\dfrac{\tau}{2}\text{rect}\left(f\dfrac{\tau}{2}\right)$
Triangular pulse trian(t/τ), defined by $\text{trian}\left(\dfrac{t}{\tau}\right) = \begin{cases} 1 - 2\|t\|/\tau, & -\tau/2 \leq t \leq \tau/2 \\ 0, & \|t\| > \tau/2 \end{cases}$	$\dfrac{\tau}{2}\text{sinc}^2\left(f\dfrac{\tau}{2}\right)$
$\cos(2\pi f_c t)$	$\frac{1}{2}\delta(f - f_c) + \frac{1}{2}\delta(f + f_c)$
$\sin(2\pi f_c t)$	$\frac{1}{2j}\delta(f - f_c) - \frac{1}{2j}\delta(f + f_c)$
$\exp(-\pi t^2)$	$\exp(-\pi f^2)$
$\exp(-at)u(t), \quad a > 0$	$\dfrac{1}{a + j2\pi f}$
$\exp(-a\|t\|), \quad a > 0$	$\dfrac{2a}{a^2 + (2\pi f)^2}$

Appendix D
Tables and constants

D.1 Complementary error function erfc(x)

$$\mathrm{erfc}(x) = \frac{2}{\sqrt{\pi}} \int_x^\infty \exp(-y^2)\,dy$$

For $x > 5.59$ (not covered in these tables), use the approximation

$$\mathrm{erfc}(x) \simeq \frac{\exp(-x^2)}{\sqrt{\pi}x}$$

x	**erfc(x)**	x	**erfc(x)**	x	**erfc(x)**	x	**erfc(x)**
0	1.0000	0.20	0.7773	0.40	0.5716	0.60	0.3961
0.01	0.9887	0.21	0.7665	0.41	0.5620	0.61	0.3883
0.02	0.9774	0.22	0.7557	0.42	0.5525	0.62	0.3806
0.03	0.9662	0.23	0.7450	0.43	0.5431	0.63	0.3730
0.04	0.9549	0.24	0.7343	0.44	0.5338	0.64	0.3654
0.05	0.9436	0.25	0.7237	0.45	0.5245	0.65	0.3580
0.06	0.9324	0.26	0.7131	0.46	0.5153	0.66	0.3506
0.07	0.9211	0.27	0.7026	0.47	0.5063	0.67	0.3434
0.08	0.9099	0.28	0.6921	0.48	0.4973	0.68	0.3362
0.09	0.8987	0.29	0.6817	0.49	0.4883	0.69	0.3292
0.10	0.8875	0.30	0.6714	0.50	0.4795	0.70	0.3222
0.11	0.8764	0.31	0.6611	0.51	0.4708	0.71	0.3153
0.12	0.8652	0.32	0.6509	0.52	0.4621	0.72	0.3086
0.13	0.8541	0.33	0.6407	0.53	0.4535	0.73	0.3019
0.14	0.8431	0.34	0.6306	0.54	0.4451	0.74	0.2953
0.15	0.8320	0.35	0.6206	0.55	0.4367	0.75	0.2888
0.16	0.8210	0.36	0.6107	0.56	0.4284	0.76	0.2825
0.17	0.8100	0.37	0.6008	0.57	0.4202	0.77	0.2762
0.18	0.7991	0.38	0.5910	0.58	0.4121	0.78	0.2700
0.19	0.7882	0.39	0.5813	0.59	0.4041	0.79	0.2639

x	erfc(x)	x	erfc(x)	x	erfc(x)	x	erfc(x)
0.80	0.2579	1.20	8.9686e-02	1.60	2.3652e-02	2.00	4.6777e-03
0.81	0.2520	1.21	8.7044e-02	1.61	2.2793e-02	2.01	4.4752e-03
0.82	0.2462	1.22	8.4466e-02	1.62	2.1962e-02	2.02	4.2805e-03
0.83	0.2405	1.23	8.1950e-02	1.63	2.1157e-02	2.03	4.0937e-03
0.84	0.2349	1.24	7.9495e-02	1.64	2.0378e-02	2.04	3.9142e-03
0.85	0.2293	1.25	7.7100e-02	1.65	1.9624e-02	2.05	3.7419e-03
0.86	0.2239	1.26	7.4764e-02	1.66	1.8895e-02	2.06	3.5765e-03
0.87	0.2186	1.27	7.2486e-02	1.67	1.8190e-02	2.07	3.4178e-03
0.88	0.2133	1.28	7.0266e-02	1.68	1.7507e-02	2.08	3.2656e-03
0.89	0.2082	1.29	6.8101e-02	1.69	1.6847e-02	2.09	3.1195e-03
0.90	0.2031	1.30	6.5992e-02	1.70	1.6210e-02	2.10	2.9795e-03
0.91	0.1981	1.31	6.3937e-02	1.71	1.5593e-02	2.11	2.8452e-03
0.92	0.1932	1.32	6.1935e-02	1.72	1.4997e-02	2.12	2.7164e-03
0.93	0.1884	1.33	5.9985e-02	1.73	1.4422e-02	2.13	2.5930e-03
0.94	0.1837	1.34	5.8086e-02	1.74	1.3865e-02	2.14	2.4747e-03
0.95	0.1791	1.35	5.6238e-02	1.75	1.3328e-02	2.15	2.3614e-03
0.96	0.1746	1.36	5.4439e-02	1.76	1.2810e-02	2.16	2.2528e-03
0.97	0.1701	1.37	5.2688e-02	1.77	1.2309e-02	2.17	2.1489e-03
0.98	0.1658	1.38	5.0984e-02	1.78	1.1826e-02	2.18	2.0494e-03
0.99	0.1615	1.39	4.9327e-02	1.79	1.1359e-02	2.19	1.9541e-03
1.00	0.1573	1.40	4.7715e-02	1.80	1.0909e-02	2.20	1.8628e-03
1.01	0.1532	1.41	4.6148e-02	1.81	1.0475e-02	2.21	1.7756e-03
1.02	0.1492	1.42	4.4624e-02	1.82	1.0057e-02	2.22	1.6921e-03
1.03	0.1452	1.43	4.3143e-02	1.83	9.6532e-03	2.23	1.6122e-03
1.04	0.1414	1.44	4.1703e-02	1.84	9.2641e-03	2.24	1.5358e-03
1.05	0.1376	1.45	4.0305e-02	1.85	8.8890e-03	2.25	1.4627e-03
1.06	0.1339	1.46	3.8946e-02	1.86	8.5275e-03	2.26	1.3929e-03
1.07	0.1302	1.47	3.7627e-02	1.87	8.1793e-03	2.27	1.3261e-03
1.08	0.1267	1.48	3.6346e-02	1.88	7.8438e-03	2.28	1.2623e-03
1.09	0.1232	1.49	3.5102e-02	1.89	7.5207e-03	2.29	1.2014e-03
1.10	0.1198	1.50	3.3895e-02	1.90	7.2096e-03	2.30	1.1432e-03
1.11	0.1165	1.51	3.2723e-02	1.91	6.9101e-03	2.31	1.0876e-03
1.12	0.1132	1.52	3.1587e-02	1.92	6.6218e-03	2.32	1.0345e-03
1.13	0.1100	1.53	3.0484e-02	1.93	6.3443e-03	2.33	9.8380e-04
1.14	0.1069	1.54	2.9414e-02	1.94	6.0774e-03	2.34	9.3543e-04
1.15	0.1039	1.55	2.8377e-02	1.95	5.8207e-03	2.35	8.8927e-04
1.16	0.1009	1.56	2.7372e-02	1.96	5.5737e-03	2.36	8.4522e-04
1.17	9.8000e-02	1.57	2.6397e-02	1.97	5.3363e-03	2.37	8.0321e-04
1.18	9.5163e-02	1.58	2.5453e-02	1.98	5.1080e-03	2.38	7.6314e-04
1.19	9.2392e-02	1.59	2.4538e-02	1.99	4.8886e-03	2.39	7.2494e-04

x	erfc(x)	x	erfc(x)	x	erfc(x)	x	erfc(x)
2.40	6.8851e-04	2.80	7.5013e-05	3.20	6.0258e-06	3.60	3.5586e-07
2.41	6.5380e-04	2.81	7.0693e-05	3.21	5.6354e-06	3.61	3.3025e-07
2.42	6.2072e-04	2.82	6.6610e-05	3.22	5.2694e-06	3.62	3.0642e-07
2.43	5.8920e-04	2.83	6.2750e-05	3.23	4.9261e-06	3.63	2.8426e-07
2.44	5.5917e-04	2.84	5.9102e-05	3.24	4.6044e-06	3.64	2.6365e-07
2.45	5.3058e-04	2.85	5.5656e-05	3.25	4.3028e-06	3.65	2.4448e-07
2.46	5.0335e-04	2.86	5.2401e-05	3.26	4.0202e-06	3.66	2.2667e-07
2.47	4.7743e-04	2.87	4.9327e-05	3.27	3.7554e-06	3.67	2.1011e-07
2.48	4.5276e-04	2.88	4.6424e-05	3.28	3.5074e-06	3.68	1.9472e-07
2.49	4.2929e-04	2.89	4.3684e-05	3.29	3.2752e-06	3.69	1.8043e-07
2.50	4.0695e-04	2.90	4.1098e-05	3.30	3.0577e-06	3.70	1.6715e-07
2.51	3.8571e-04	2.91	3.8657e-05	3.31	2.8541e-06	3.71	1.5482e-07
2.52	3.6550e-04	2.92	3.6355e-05	3.32	2.6636e-06	3.72	1.4337e-07
2.53	3.4629e-04	2.93	3.4183e-05	3.33	2.4853e-06	3.73	1.3274e-07
2.54	3.2802e-04	2.94	3.2134e-05	3.34	2.3185e-06	3.74	1.2288e-07
2.55	3.1066e-04	2.95	3.0203e-05	3.35	2.1625e-06	3.75	1.1373e-07
2.56	2.9416e-04	2.96	2.8382e-05	3.36	2.0166e-06	3.76	1.0524e-07
2.57	2.7849e-04	2.97	2.6666e-05	3.37	1.8801e-06	3.77	9.7359e-08
2.58	2.6360e-04	2.98	2.5049e-05	3.38	1.7526e-06	3.78	9.0055e-08
2.59	2.4946e-04	2.99	2.3526e-05	3.39	1.6334e-06	3.79	8.3282e-08
2.60	2.3603e-04	3.00	2.2090e-05	3.40	1.5220e-06	3.80	7.7004e-08
2.61	2.2329e-04	3.01	2.0739e-05	3.41	1.4179e-06	3.81	7.1185e-08
2.62	2.1119e-04	3.02	1.9466e-05	3.42	1.3207e-06	3.82	6.5793e-08
2.63	1.9971e-04	3.03	1.8268e-05	3.43	1.2299e-06	3.83	6.0798e-08
2.64	1.8882e-04	3.04	1.7141e-05	3.44	1.1452e-06	3.84	5.6171e-08
2.65	1.7849e-04	3.05	1.6080e-05	3.45	1.0661e-06	3.85	5.1886e-08
2.66	1.6869e-04	3.06	1.5082e-05	3.46	9.9220e-07	3.86	4.7919e-08
2.67	1.5940e-04	3.07	1.4143e-05	3.47	9.2329e-07	3.87	4.4246e-08
2.68	1.5059e-04	3.08	1.3260e-05	3.48	8.5900e-07	3.88	4.0847e-08
2.69	1.4224e-04	3.09	1.2429e-05	3.49	7.9903e-07	3.89	3.7702e-08
2.70	1.3433e-04	3.10	1.1649e-05	3.50	7.4310e-07	3.90	3.4792e-08
2.71	1.2684e-04	3.11	1.0915e-05	3.51	6.9095e-07	3.91	3.2101e-08
2.72	1.1974e-04	3.12	1.0226e-05	3.52	6.4234e-07	3.92	2.9612e-08
2.73	1.1301e-04	3.13	9.5780e-06	3.53	5.9703e-07	3.93	2.7310e-08
2.74	1.0665e-04	3.14	8.9696e-06	3.54	5.5482e-07	3.94	2.5183e-08
2.75	1.0062e-04	3.15	8.3982e-06	3.55	5.1548e-07	3.95	2.3217e-08
2.76	9.4918e-05	3.16	7.8617e-06	3.56	4.7885e-07	3.96	2.1400e-08
2.77	8.9520e-05	3.17	7.3581e-06	3.57	4.4473e-07	3.97	1.9721e-08
2.78	8.4413e-05	3.18	6.8854e-06	3.58	4.1296e-07	3.98	1.8171e-08
2.79	7.9582e-05	3.19	6.4419e-06	3.59	3.8339e-07	3.99	1.6739e-08

x	erfc(x)	x	erfc(x)	x	erfc(x)	x	erfc(x)
4.00	1.5417e-08	4.40	4.8917e-10	4.80	1.1352e-11	5.20	1.9249e-13
4.01	1.4197e-08	4.41	4.4695e-10	4.81	1.0291e-11	5.21	1.7314e-13
4.02	1.3071e-08	4.42	4.0829e-10	4.82	9.3279e-12	5.22	1.5570e-13
4.03	1.2031e-08	4.43	3.7291e-10	4.83	8.4530e-12	5.23	1.3999e-13
4.04	1.1073e-08	4.44	3.4052e-10	4.84	7.6586e-12	5.24	1.2584e-13
4.05	1.0188e-08	4.45	3.1089e-10	4.85	6.9375e-12	5.25	1.1310e-13
4.06	9.3727e-09	4.46	2.8378e-10	4.86	6.2831e-12	5.26	1.0163e-13
4.07	8.6207e-09	4.47	2.5898e-10	4.87	5.6893e-12	5.27	9.1307e-14
4.08	7.9276e-09	4.48	2.3630e-10	4.88	5.1506e-12	5.28	8.2014e-14
4.09	7.2887e-09	4.49	2.1557e-10	4.89	4.6620e-12	5.29	7.3653e-14
4.10	6.7000e-09	4.50	1.9662e-10	4.90	4.2189e-12	5.30	6.6131e-14
4.11	6.1577e-09	4.51	1.7929e-10	4.91	3.8172e-12	5.31	5.9365e-14
4.12	5.6582e-09	4.52	1.6347e-10	4.92	3.4531e-12	5.32	5.3282e-14
4.13	5.1981e-09	4.53	1.4901e-10	4.93	3.1230e-12	5.33	4.7812e-14
4.14	4.7746e-09	4.54	1.3580e-10	4.94	2.8240e-12	5.34	4.2895e-14
4.15	4.3847e-09	4.55	1.2374e-10	4.95	2.5531e-12	5.35	3.8477e-14
4.16	4.0258e-09	4.56	1.1273e-10	4.96	2.3077e-12	5.36	3.4506e-14
4.17	3.6956e-09	4.57	1.0268e-10	4.97	2.0855e-12	5.37	3.0940e-14
4.18	3.3919e-09	4.58	9.3503e-11	4.98	1.8844e-12	5.38	2.7736e-14
4.19	3.1124e-09	4.59	8.5133e-11	4.99	1.7023e-12	5.39	2.4859e-14
4.20	2.8555e-09	4.60	7.7496e-11	5.00	1.5375e-12	5.40	2.2277e-14
4.21	2.6192e-09	4.61	7.0531e-11	5.01	1.3883e-12	5.41	1.9958e-14
4.22	2.4021e-09	4.62	6.4179e-11	5.02	1.2534e-12	5.42	1.7878e-14
4.23	2.2025e-09	4.63	5.8387e-11	5.03	1.1314e-12	5.43	1.6011e-14
4.24	2.0191e-09	4.64	5.3108e-11	5.04	1.0211e-12	5.44	1.4336e-14
4.25	1.8506e-09	4.65	4.8297e-11	5.05	9.2131e-13	5.45	1.2834e-14
4.26	1.6958e-09	4.66	4.3913e-11	5.06	8.3113e-13	5.46	1.1487e-14
4.27	1.5537e-09	4.67	3.9919e-11	5.07	7.4963e-13	5.47	1.0280e-14
4.28	1.4232e-09	4.68	3.6281e-11	5.08	6.7599e-13	5.48	9.1972e-15
4.29	1.3034e-09	4.69	3.2969e-11	5.09	6.0947e-13	5.49	8.2271e-15
4.30	1.1935e-09	4.70	2.9953e-11	5.10	5.4938e-13	5.50	7.3578e-15
4.31	1.0926e-09	4.71	2.7207e-11	5.11	4.9512e-13	5.51	6.5792e-15
4.32	1.0000e-09	4.72	2.4708e-11	5.12	4.4613e-13	5.52	5.8817e-15
4.33	9.1516e-10	4.73	2.2435e-11	5.13	4.0191e-13	5.53	5.2572e-15
4.34	8.3732e-10	4.74	2.0366e-11	5.14	3.6200e-13	5.54	4.6980e-15
4.35	7.6594e-10	4.75	1.8485e-11	5.15	3.2599e-13	5.55	4.1975e-15
4.36	7.0052e-10	4.76	1.6774e-11	5.16	2.9351e-13	5.56	3.7496e-15
4.37	6.4056e-10	4.77	1.5219e-11	5.17	2.6421e-13	5.57	3.3488e-15
4.38	5.8561e-10	4.78	1.3805e-11	5.18	2.3779e-13	5.58	2.9903e-15
4.39	5.3528e-10	4.79	1.2520e-11	5.19	2.1396e-13	5.59	2.6696e-15

D.2 Useful constants

Acceleration due to gravity g	9.80665 m/s^2
Avogadro's number N	6.022×10^{26}/(kg mol)
Base of natural logarithm e	2.718281828459046
Bohr magneton β	$9.2740154 \times 10^{-24}$ A m^2
Boltzmann's constant k	$1.3806488 \times 10^{-23}$ J/K
Characteristic impedance of free space Z_o	120π ohm
Earth's mean equatorial radius R_e	6378.137 km
Electron charge q_e	$1.60217733 \times 10^{-19}$ C
Electron rest mass m_e	$9.1093897 \times 10^{-31}$ kg
Faraday constant F	9.64853399×10^{4} C mol^{-1}
Permeability of free space μ_o	$4\pi \times 10^{-7}$ H/m
Permittivity of free space ε_o	$8.854187817 \times 10^{-12}$ F/m
Pi π	3.141592653589793
Planck's constant h	$6.6260755 \times 10^{-34}$ J s
Proton rest mass m_p	$1.6726231 \times 10^{-27}$ kg
Speed of light in vacuum c	2.99792458×10^{8} m/s
Stefan Boltzmann's constant σ	5.670373×10^{-8} W m^{-2} K^{-4}
Universal constant of gravitation G	6.67384×10^{-11} N m^2/kg^2
Universal gas constant R_o	8.3144621 J K^{-1} mol^{-1}

D.3 SI units

	Quantity	Unit	Symbol	Equivalent
Basic Units	Area	square metre	m^2	—
	Electric current	ampere	A	—
	Length	metre	m	—
	Luminous intensity	candela	cd	—
	Mass	kilogram	kg	—
	Plane angle	radian	rad	—
	Temperature	kelvin	K	—
	Time	second	s	—
	Volume	cubic metre	m^3	—
Derived Units	Capacitance	farad	F	A s/V
	Charge	coulomb	C	A s
	Electric field strength	—	—	V/m
	Electric flux density	—	—	C/m^2
	Electric potential	volt	V	W/A
	Energy, work or heat	joule	J	N m
	Force	newton	N	Kg m/s^2
	Frequency	hertz	Hz	s^{-1}
	Inductance	henry	H	V s/A = N m/A^2
	Magnetic field strength	—	—	A/m
	Magnetic flux	weber	Wb	V s = N m/A
	Magnetic flux density	tesla	T	Wb/m^2
	Power	watt	W	J/s
	Pressure or stress	pascal	Pa	N/m^2
	Resistance	ohm	Ω	V/A

Index

RECEIVED
JUN 17 2014